TAURINE IN HEALTH AND DISEASE

ADVANCES IN EXPERIMENTAL MEDICINE AND BIOLOGY

Recent Volumes in this Series

Volume 357
LACTOFERRIN: Structure and Function
Edited by T. William Hutchens, Sylvia V. Rumball, and Bo Lönnerdal

Volume 358
ACTIN: Biophysics, Biochemistry, and Cell Biology
Edited by James E. Estes and Paul J. Higgins

Volume 359
TAURINE IN HEALTH AND DISEASE
Edited by Ryan J. Huxtable and Dietrich Michalk

Volume 360
ARTERIAL CHEMORECEPTORS: Cell to System
Edited by Ronan G. O'Regan, Philip Nolan, Daniel S. McQueen, and David J. Paterson

Volume 361
OXYGEN TRANSPORT TO TISSUE XVI
Edited by Michael C. Hogan, Odile Mathieu-Costello, David C. Poole, and Peter D. Wagner

Volume 362
ASPARTIC PROTEINASES: Structure, Function, Biology, and Biomedical Implications
Edited by Kanji Takahashi

Volume 363
NEUROCHEMISTRY IN CLINICAL APPLICATION
Edited by Lily C. Tang and Steven J. Tang

Volume 364
DIET AND BREAST CANCER
Edited under the auspices of the American Institute for Cancer Research; Scientific Editor: Elizabeth K. Weisburger

Volume 365
MECHANISMS OF LYMPHOCYTE ACTIVATION AND IMMUNE REGULATION V: Molecular Basis of Signal Transduction
Edited by Sudhir Gupta, William E. Paul, Anthony DeFranco, and Roger Perlmutter

---

TAURINE IN HEALTH AND DISEASE

Edited by

Ryan J. Huxtable
University of Arizona
College of Medicine
Tucson, Arizona

and

Dietrich Michalk
University of Cologne
Cologne, Germany

PLENUM PRESS • NEW YORK AND LONDON

Library of Congress Cataloging-in-Publication Data

On file

Proceedings of an International Symposium on Taurine in Health and Disease,
held August 29–September 1, 1993, in Cologne, Germany

ISBN 0-306-44812-2

A Division of Plenum Publishing Corporation
233 Spring Street, New York, N. Y. 10013

Printed in the United States of America

PREFACE

Taurine, or 2-aminoethane sulfonic acid, has long been known to be the major organic product formed from the breakdown of the sulfur-containing amino acids, methionine and cysteine. It is excreted as such in the urine or as bile acid conjugates in the bile. Research over the last 25 years, however, has uncovered an amazing variety of phenomena involving taurine. Despite the fact that the mechanism of action of taurine in some of these phenomena remains to be elucidated, it is clear that taurine is of far more physiological relevance than being the mere metabolic debris of sulfur amino acid catabolism. It also has numerous pharmacological actions that are beginning to be developed, as exemplified by some of the studies reported in this volume.

This volume reports the proceedings of the latest of a series of international symposia, continuing on from the initial meeting held in Tucson, Arizona, in 1975. The latest symposium was the first to be held in Germany, in Cologne from August 29 through September 1, 1993. It continued the multidisciplinary approach that has been the norm for taurine symposia. We hope the publication of the proceedings will stimulate further investigations on this simple but fascinating compound.

The editors thank the members of the local organizing committee: Peter Wingenfeld, Marianne Blomeier, Stefan Strübind and Ute Gehrmann. We also thank Taisho Pharmaceutical Co., of Japan, and Fresenius, Milupa, Humana, Aponti and Kabi-Pfrimmer of Germany for their help and generosity. Finally, we thank Sue Brandes for her editorial assistance.

Ryan J. Huxtable

Dietrich Michalk

CONTENTS

Part I: Chemistry, Biochemistry and Metabolism of Taurine

Part II: Taurine and the Liver

Part III: Taurine and the Kidney

Part IV: Taurine in Muscle and the Cardiovascular System

Part V: Taurine and the Nervous System

Part VI: Taurine Deficiency

Part VII: Taurine - Clinical Studies

POSSIBLE RELATIONSHIPS BETWEEN TAURINE DERIVATIVES AND PRODUCTS OF THE METABOLISM OF KETIMINES

Silvestro Duprè, Francesco Pinnen[1], Mario Fontana, Raffaella Coccia, Laura Pecci and Doriano Cavallini

Dipartimento di Scienze Biochimiche "A. Rossi Fanelli"
Università di Roma "La Sapienza"
Roma, Italy
Centro di Biologia Molecolare, C.N.R. Roma
[1]Istituto di Chimica Farmaceutica
Università di Catania
Catania, Italy

INTRODUCTION

Within the class of six- and seven-membered sulfur-containing cyclic ketimines and their derivatives, some had been synthesized and chemically characterized long before a biochemical role was evidenced (10,11). Biochemically, cyclothionine has been first demonstrated to be present in the urine of a cystathioninuric patient (12). Lanthionine ketimine and cystathionine ketimine have been detected in normal human urine together with 1,4-thiomorpholine-3,5-dicarboxylic acid (TMDA), 1,4-hexahydro-thiazepine-3,5-dicarboxylic acid (cyclothionine) and 1,4-thiomorpholine-3-carboxylic acid (TMA), the reduced forms respectively of lanthionine ketimine, cystathionine ketimine and S-aminoethylcysteine ketimine (15,16,20). The presence of these compounds in normal and pathological human urine indicates their natural origin but gives no informations about their physiological role. The successive identification and recovery of lanthionine ketimine, cystathionine ketimine, S-aminoethylcysteine ketimine and of the corresponding reduced forms of lanthionine ketimine and cystathionine ketimine in bovine brain, where they are present in the μM range, show that these compounds are normal metabolites in mammals (3,4,19,24,25). Biosynthetic and degradative reactions have also been identified which justify the in vivo presence of these compounds. A transamination activity leads to the formation of ketimines from their parent amino acids, which could be considered as the putative precursors in vivo (6,7,23). A NAD(P)H-dependent specific reductase has been isolated and characterized (17,18). A recent review collects many of these data (5).

Ketimines are stable in the solid form. In diluted aqueous solution they are less stable and undergo various spontaneous changes, depending on the pH. Autooxidation becomes

Taurine in Health and Disease, Edited by R. Huxtable
and D.V. Michalk, Plenum Press, New York, 1994

important at pH > 7 and the rate increases with the increase in pH. Ketimines reduce many reagents, such as the Folin-Marenzi reagent, the Fehling and the Benedict reagents, ammoniacal silver nitrate and 2,6-dichloroindophenol solutions. Spontaneous reactions of S-aminoethylcysteine ketimine in neutral and slight alkaline conditions have been studied in more detail, with auto-oxidation and dimerization being the main reactions (21). S-aminoethylcysteine ketimine, as well as the other ketimines, displays reducing properties towards dioxygen (21), cytochrome c and other hemoproteins (28), permanganate, hydrogen peroxide and performic acid (unpublished results). As the result of the reducing activity, ketimines are converted into oxidation products of not yet fully understood nature. The main oxidation product obtained from S-aminoethylcysteine ketimine at pH 8.5 under O_2 bubbling has been identified as S-aminoethylcysteine ketimine sulfoxide (21). Other products deserve identification.

After reduction of the double bond of ketimines with $NaBH_4$ the corresponding reduced compounds are obtained, which are stable at alkaline pH and do not autooxidize. TMA is further oxidized by stronger oxidants, such as performic acid. One of the oxidation products has been identified as TMA sulfoxide This compound, which has been synthesized (2,8), has been found in algae (13,29) and is known in the literature under the name of "chondrine" (it was first called "yunaine"). Other very acidic oxidation products are observed on amino acid analysis: among them apparently taurine and cysteic acid are present.

In order to allow the identification of unknown compounds obtained by oxidation of ketimines or of their reduced counterparts, the synthesis of some derivatives of cysteic acid and taurine has been carried out. In this paper the synthesis of the derivative with cysteic acid (carboxylated tauropine) will be described in detail and some properties will be reported. The same procedure performed with taurine gives a new taurine derivative which has not been described until now.

MATERIALS AND METHODS

Products

L-cysteic acid and taurine are Sigma products; R(+)-2-bromopropionic acid is a Fluka product. DL-TMA is synthesized as reported (2).

Analyses

^{1}H NMR spectra were obtained with a Varian EM-390. ^{13}C NMR spectra were determined with dioxane as internal standard using a Varian XL-300 instrument. The amino acid analyzer was a Carlo Erba 3A30. Thin layer chromatography (TLC) was performed on cellulose alluminium sheets (0.1 mm) Merck F with methanol:butanol:acetic acid:water (40:15:5:40). Detection was with ninhydrin or with UV lamp (Spectroline).

Syntheses

N-[(R)-1-Carboxyethyl]-L-Cysteic Acid. A stirred solution of L-cysteic acid (1.5 g, 8.9 mmol) and R(+)-2-bromopropionic acid (1.64 ml, 18 mmol) in water (2 ml) is titrated to pH 8.5 by the addition of 2 M NaOH. The temperature is maintained at 40° C and the pH at 8.5 by contually adding 2 M NaOH. After 48 h, additional R(+)-2-bromopropionic acid (1.64 ml, 18 mmol) is added and the disappearance of cysteic acid is monitored by TLC examination. After 96 h at 40° C, the reaction mixture is acidified to pH 4 with 4 M

HCl and evaporated to dryness under vacuum. The resulting white solid is dissolved in water (10 ml) and repeatedly washed with ether. Evaporation under vacuum of the aqueous solution followed by crystallization from water-methanol of the resulting residue affords the title compound (2.0 g) which is dried in a vacuum desiccator over P_2O_5.

N-Bis[(R)-1-Carboxyethyl]-Taurine Barium Salt. The reaction of taurine (1.13 g, 9 mmol) and R(+)-2-bromopropionic acid (1.64 ml, 18 mmol), performed according to the above procedure, gives an oily product which is dissolved in water and chromatographed through a Dowex 50-X8 H^+ column. After neutralization of the eluate with $Ba(OH)_2$, white crystals precipitate in the cold which are collected and dried (0.8 g).

Performic Acid Oxidation. DL-TMA (4.5 mg) are added with 2 ml performic acid solution (10 volumes 30% hydrogen peroxide and 90 volumes 99% formic acid left at room temperature for 120 min) and left in ice for 150 min. After dilution with 40 ml water the solution is lyophilized. The residue is dissolved in the running buffer and submitted to amino acid analyzer separation.

RESULTS

N-[(R)-1-Carboxyethyl]-L-Cysteic Acid

Chemical and spectroscopic properties of the product of the above synthesis are in accordance with the assigned structure. In the ^1H NMR spectrum (D_2O) the methyl group appears at $\delta = 1.5$ as doublet (J = 7.0 Hz) coupled with the adjacent CH proton resonating at $\delta = 3.75$. The αCH proton of the cysteiyl residue is present at $\delta = 4.0$ coupled with the adjacent CH_2 group (at $\delta = 3.4$). In the ^{13}C NMR spectrum (D_2O) two sp^2 carbon atoms are present resonating at $\delta = 177.4$ and 173.8 ppm.

The yield of the reaction is more than 90%. The compound has a melting point of 230-235° C (uncorrected). On TLC the spot (located by UV light) has a R_f value of 0.56 (cysteic acid $R_f = 0.24$). On amino acid analysis, the compound can not be detected, the color yield being very low (less than 1% of the color yield of cysteic acid).

N-Bis[(R)-1-Carboxyethyl]-Taurine Barium Salt

The synthesis of rhodoic acid (N-[(R)-1-carboxyethyl]-taurine), as described by Kuriyama (14), is reported with insufficient detail to be reproduced. Sato reports later the synthesis in a Japanese journal of difficult access (26) and the synthesis is not described in Chemical Abstracts. Also the synthesis reported by Gäde (9) is not detailed at all. Rhodoic acid is reported to give a faint reaction with ninhydrin (14,26), and it can not be seen on TLC as a coloured spot. We tried to perform the synthesis of rhodoic acid as described above for cysteic acid starting from R(+)-2-bromopropionic acid. NMR examination of the crude product obtained by us before ion-exchange chromatography reveals the presence of large amounts of a compound different from rhodoic acid. On cellulose F one spot may be seen with UV light, located at $R_f = 0.62$ (taurine $R_f = 0.40$). It has also a low color yield on amino acid analysis and a run of the reaction products before crystallization shows the presence of a symmetric peak centered at 4.6 min (taurine = 7.5 min).

Spectra of the crystalline product obtained as the barium salt, as reported above, are in accordance with the structure of the dialkyl derivative of taurine. In the ^1H NMR spectrum (D_2O) of N-bis[(R)-1-carboxyethyl]-taurine as the Ba^{2+} salt the two methyl groups appear at δ = 1.25 and 1.4 as doublets (J = 7.5 Hz) coupled with the respective adjacent CH proton resonating at δ 4.15 and 3.7, respectively. A multiplet attributable to the α- and β-CH_2 protons of the tauryl residue is present in the range 3.2-3.5 δ. The structure is confirmed by the ^{13}C NMR spectrum (data not shown).

Performic Acid Oxidation of TMA

Fig. 1 shows a representative amino acid analyzer elution profile of DL-TMA after performic acid oxidation. The relative amount of the various ninhydrin-positive peaks varies with the oxidation time and with the ratio between performic acid and TMA. The two main peaks observed correspond to the elution time of taurine and, approximately, of carboxymethylcysteamine sulfone; the position of TMA sulfone, which has not been identified, is tentatively assigned; the first small peak, which becomes very large with excess performic acid, is close to the position of cysteic acid. One of the small peaks has been identified certainly as TMA sulfoxide (chondrine).

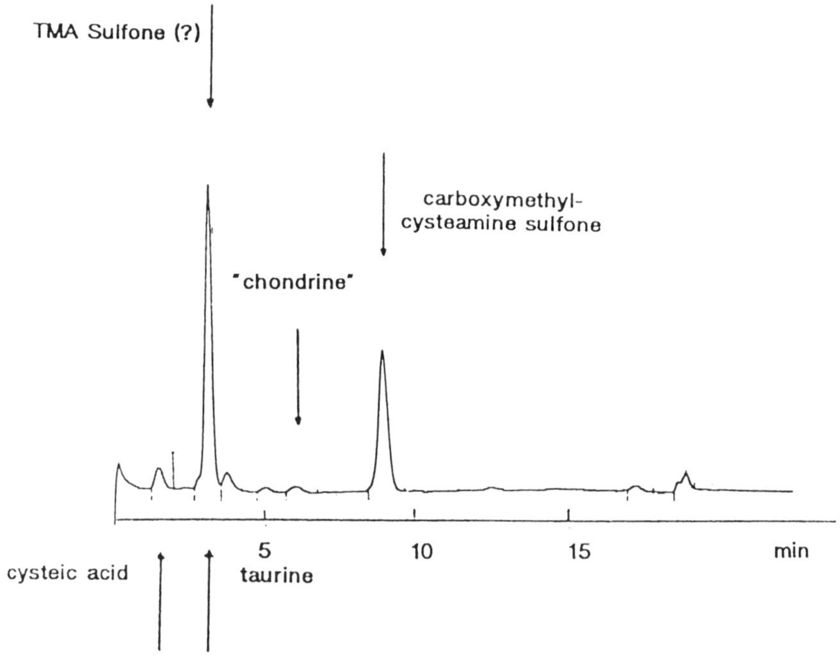

Figure 1. Amino acid elution profile of the oxidation products of DL-TMA after performic acid oxidation. Arrows below the trace indicate the elution positions of cysteic acid and taurine; the other peaks have been identified as TMA sulfoxide (chondrine) and carboxymethylcysteamine sulfone. The position of TMA sulfone is unknown

DISCUSSION

The compounds discussed in this note are shown in Fig. 2, together with some other taurine derivatives which could be putative oxidation end products of saturated sulfur, nitrogen-containing natural compounds.

The report of various syntheses of rhodoic acid, as can be found in the literature, are in our opinion, not detailed enough to be reproduced. As reported for the synthesis of histopine (1), the hydrolysis of 2-bromopropionic acid may strongly affect the yield of the

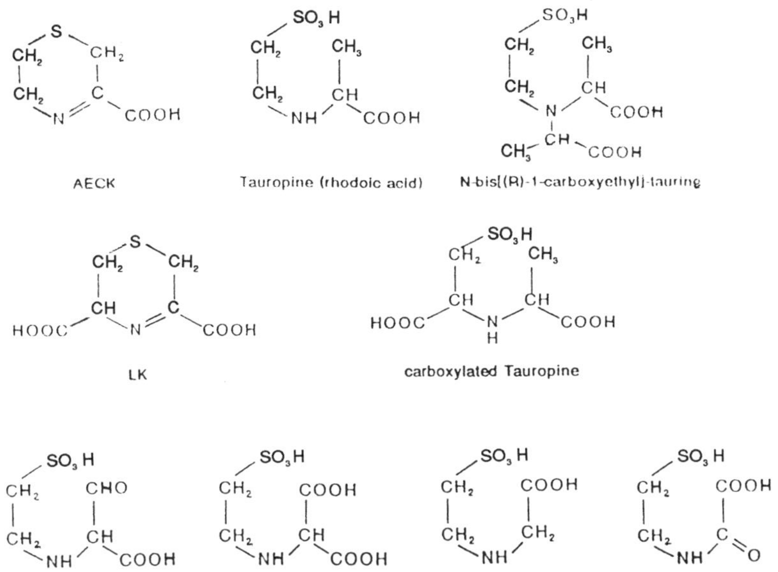

Figure 2. Structures of compounds discussed in the text and of some taurine derivatives which could be oxidation products of S-aminoethylcysteine ketimine

reaction, the extent of which also depends from the amino reagent. For the synthesis of the derivative of cysteic acid, it is essential to use the optical active form of both reagents in order not to obtain a mixture of diastereoisomers with different chromatographic and spectroscopic properties. The derivative of taurine with DL-bromo reagent forms a racemic mixture, and D-rhodoic acid should be obtained with the R(+) form of the bromo reagent. The dialkyl derivative of taurine has been purified and characterized. The complete purification of the monoalkyl derivative is in progress and a detailed synthesis will be reported later. The monoalkyl derivatives have a low color yield and are not easily detected on amino acid analysis. Derivatization with phenylisothiocyanate prior to HPLC separation seems to be a better method for quantitative determination, as has been recently reported for acidic opines extracted from mollusks (27).

The oxidation of S-aminoethylcysteine ketimine at pH 8.5 in the presence of oxygen produces at least five compounds absorbing at 250 nm. Two of them have been undoubtedly identified as the sulfoxide of S-aminoethylcysteine ketimine and as thiomorpholine-3-one (21). The cyclic structure seems to be conserved and the sulfur atom is one of the targets of this mild oxidation process. Hydrolysis with 6N HCl and some derivatization reactions of the oxidation products gave more insights into the reaction. The presence after hydrolysis of carboxymethylcysteamine indicates that oxidation may also occur at the unsaturated carbon atom adjacent to nitrogen, through mechanisms which have not been investigated. We suppose that an oxidative decarboxylation takes place. The presence, after HCl hydrolysis, of glyoxylic acid and of mesoxalic acid semialdehyde, identified as dinitrophenylhydrazine derivatives, the presence of cysteamine and cystamine, identified as monobromobimane derivatives, and of taurine after oxidation with H_2O_2 and ammonium molybdate either before or after acid hydrolysis indicate that the cyclic molecule of the ketimine can also be opened at the level of the thioether bond. It has been demonstrated (22) that ketimines in solution are predominantly in the eneiminic form. Oxidation of the eneimine form of S-aminoethylcysteine ketimine may explain the oxidation at carbon 2, with the formation of a labile thioemiacetal or thioester bond.

After borohydride reduction of the product of mild oxidation of S-aminoethylcysteine ketimine, a compound was obtained which was identical to the one isolated from *Chondria crassicaulis* and from *Undaria pinnatifida* (14,29) and identified as TMA sulfoxide. The formation of taurine, cysteic acid, cystine and TMA from chondrine (TMA sulfoxide) after exhaustive acid hydrolysis has been explained (8) through a disproportionation reaction.

The amino acid profile of the products obtained from TMA after performic acid oxidation shows the presence of acidic compounds, apparently running as taurine and cysteic acid, together with chondrine and larger amounts of an unknown derivative, which could tentatively be identified as carboxymethylcysteamine sulfone. Performic acid oxidation can not easily explain the breakage of the bond between the secondary amine and the carboxyl-bearing carbon atom, which in TMA is not a double bond as in S-aminoethylcysteine ketimine. Therefore, some other very acidic compounds, running similarly to taurine and cysteic acid, are probably present. As possible oxidation products, some taurine derivatives can be envisaged. We tried to synthesize one of these compounds, known as tauropine or rhodoic acid, by a reported procedure: a mixture of rhodoic acid and of the dialkyl derivative was obtained. The related compound derived from cysteic acid has been purified and characterized. This latter compound could come from the oxidation of lanthionine ketimine, which also shows high lability at alkaline pH and is easily oxidized by many reagents (unpublished results). Apparently the reactivity of the amino group of these compounds is quite different: cysteic acid is easily monoalkylated with a very good yield without the formation of the dialkyl derivative. On the contrary, taurine gives apparently both derivatives. The synthesis and characterization of the dialkyl derivative of taurine has not been reported until now. The presence of various amounts of the monoalkyl (rhodoic acid) and of the dialkyl-derivative after the synthetic procedures described in the literature has also not been described in the past. Studies are in progress in our laboratory in order to optimize the reaction conditions for the synthesis of rhodoic acid and related compounds.

REFERENCES

1. Bates, H.A., Kaushal, A., Deng, P.-N. and Sciaky, D., 1984, Structure and synthesis of histopine, a histidine derivative produced by crown gall tumors, *Biochemistry* 23:3287-3290.
2. Carson, J.F. and Wong, F.F., 1964, The synthesis of L-thiazane-3-carboxylic acid 1-oxide, *J. Org. Chem.* 29:2203-2205.
3. Cavallini, D., Matarese, R.M., Pecci, L. and Ricci, G., 1985, 1,4-Thiomorpholine-3,5-dicarboxylic acid, a novel cyclic imino acid detected in bovine brain, *FEBS Lett.* 192:247-250.
4. Cavallini, D., Pecci, L., Matarese, R.M., Ricci, G. and Achilli, M., 1985, Gas-chromatographic mass-spectrometric detection of 1,4-hexahydrothiazepine-3,5-dicar-boxylic acid (cyclothionine) in bovine brain, *J. Biol. Chem.* 260:15577-15579.
5. Cavallini, D., Ricci, G., Duprè, S., Pecci, L., Costa, M., Matarese, R.M., Pensa, B., Antonucci, A., Solinas, S.P. and Fontana, M., 1991, Sulfur-containing cyclic ketimines and iminoacids, *Eur. J. Biochem.* 202:217-223.
6. Costa, M., Pensa, B., Fontana, M., Foppoli, C. and Cavallini, D., 1986, Transa-mination of L-cystathionine and related compounds by a bovine liver enzyme. Possible identification with glutamine transaminase, *Biochim. Biophys. Acta* 881:314-320.
7. Costa, M., Pensa, B., Di Costanzo, B., Coccia, R. and Cavallini, D., 1987, Transamination of L-cystathionine and related compounds by bovine brain glutamine transaminase, *Neurochem. Int.* 10:377-382.
8. Däbritz, E. and Virtanen, A.I., 1965, S-vinyl-cysteine-S-oxide, ein Homologes zur Vorstufe der tränentreibenden Substanz der Zwiebel, *Chem. Ber.* 98:781-788.
9. Gäde, G., 1986, Purification and properties of tauropine dehydrogenase from the shell adductor muscle of the ormer, *Haliotis lamellosa*, *Eur. J. Biochem.* 160:311-318.
10. Hermann, P., 1961, Zur Reaktion von Halogenbrenztraubensäure mit thiolaminen, *Chem. Ber.* 94:442-445.

11. Hermann, P., 1981, Chemistry and biochemistry of thia-analogues of amino acids, *in:* "IUPAC, Organic Sulfur Chemistry", Freidlina, R.K., and Skorova, A.E., eds., Pergamon Press, Oxford, p. 51.
12. Kodama, H., Ishimoto, J., Shinomura, M. Hirota, T. and Ohmori, S., 1975, Isolation of two new sulfur-containing amino acids from the urine of a cystathioninuric patient, *Physiol. Chem. Phys.* 7:147-152.
13. Kuriyama, M., Takagi, M. and Murata, K., 1960, Ninhydrin reactive substances in marine algae. *Bull. Fac. Fisheries, Hokkaido Univ.* 11:58-63
14. Kuriyama, M., 1961, New ninhydrin-reactive substances from red algae, *Nature,* 192:969.
15. Matarese, R.M., Pecci, L., Ricci, G., Nardini, M., Antonucci, A. and Cavallini, D., 1987, Hexahydro-1,4-thiazepine-3,5-dicarboxylic acid and thiomorpholine-3,5-dicar-boxylic acid are present in normal human urine, *Proc. Natl. Acad. Sci. USA* 84:5111-5114.
16. Matarese, R.M., Solinas, S.P., Montefoschi, G., Ricci, G. and Cavallini, D., 1989, Identification of 1,4-thiomorpholine-3-carboxylic acid (TMA) in normal human urine, *FEBS Lett.* 250:75-77.
17. Nardini, M., Ricci, G., Caccuri, A.M., Solinas, S.P., Vesci, L. and Cavallini, D., 1988, Purification and characterization of a ketimine reducing enzyme, *Eur. J. Biochem.* 173:689-694.
18. Nardini, M., Ricci, G., Vesci, L., Pecci, L. and Cavallini, D., 1988, Bovine brain ketimine reductase, *Biochim. Biophys. Acta* 957:286-292.
19. Nardini, M., Matarese, R.M., Pecci, L., Antonucci, A., Ricci, G. and Cavallini, D., 1990, Detection of 2H-1,4-thiazine-5,6-dihydro-3-carboxylic acid (aminoethylcysteine ketimine) in the bovine brain, *Biochem. Biophys. Res. Commun.* 166:1251-1256.
20. Pecci, L., Antonucci, A., Nardini, M. and Cavallini, D., 1988, Detection of cystathionine and lanthionine ketimines in human urine, *Biochem. Internat.* 17:877-883.
21. Pecci, L., Solinas, S.P., Antonucci, A., Montefoschi, G., Blarzino. C. and Cavallini, D., 1993, The oxidation of sulfur containing cyclic ketimines. The sulfoxide is the main product of S-aminoethyl-cysteine ketimine autoxidation, *Amino Acids* 5:23-32.
22. Ricci, G., Federici, G. Lucente, G., Achilli, M. and Cavallini, D., 1982, L-Lanthioni-ne oxidation by snake venom L-amino acid oxidase, *Physiol. Chem. Phys.* 14:193-199.
23. Ricci, G., Nardini, M. Federici, G. and Cavallini, D., 1986, The transamination of L-cystathionine, L-cystine and related compounds by a bovine kidney transaminase, *Eur. J. Biochem.* 157:57-63.
24. Ricci, G., Vesci, L., Nardini, M., Arduini, A., Storto, S., Rosato, N. and Cavallini, D., Detection of 2H-1,4-thiazine-5,6-dihydro-3,5-dicarboxylic acid (lanthionine keti-mine) in the bovine brain by a fluorometric assay, 1989, *Biochim. Biophys. Acta* 990:211-215.
25. Ricci, G., Vesci, L., Matarese, R.M., Antonucci, A., Maggio, A., Pecci, L. and Cavallini, D., 1990, Detection of cystathionine ketimine in bovine cerebellum, *J. Neurochem.* 55:1599-1602.
26. Sato, M., Kanno, N. and Sato, Y.,1985, Isolation of D-rhodoic acid from the abalone muscle, *Nippon Suisan Gakkaishi* 51:1681-1683; *C.A.* 104:17840 (1986).
27. Sato, M., Suzuki, S., Yasuda, Y., Kawauchi, H., Kanno, N. and Sato, Y., 1988, Quantitative HPLC analysis of acidic opines by phenylthiocarbamyl derivatization, *Anal. Biochem.* 174:623-627.
28. Solinas, S.P., Pecci, L., Montefoschi, G. and Cavallini, D., 1992, The reducing activity of S-aminoethyl-cysteine ketimine and similar sulfur-containing ketimines, *Biochem. Biophys. Res. Commun.* 183:481-486.
29. Tominaga, F. and Oka, K.,1963, On the isolation and identification of 1,4-thiazane-3-carboxylic acid S-oxide from the brown alga *Undaria pinnatifida, J. Biochem. (Tokyo)* 54:222-224.

THE INHIBITORY EFFECTS OF TAURINE ON PROTEIN PHOSPHORYLATION: COMPARISON OF VARIOUS CHARACTERISTICS OF THE TAURINE-AFFECTED PHOSPHOPROTEINS PRESENT IN RAT RETINA, BRAIN AND HEART

John B. Lombardini

Departments of Pharmacology and
Ophthalmology & Visual Sciences
Texas Tech University
Health Sciences Center
Lubbock, Texas 79430

INTRODUCTION

The role of taurine in the animal kingdom has been questioned for many years since its discovery in ox bile in 1827 by Tiedemann and Gmelin (20). Eventually, the seminal review by Jacobsen and Smith published in 1968 (1) provided the impetus to a small but determined segment of the scientific community to study this ß-amino sulfonic acid systematically for the past 25 years. Since 1975 some 10 international symposia have been devoted to taurine (18). Nevertheless, the basic question as to the function of taurine remains unanswered. In the early years of the study of taurine phenomenology was the norm, while in recent years many of the questions posed are directed towards the mechanisms of action of taurine at the molecular level. Consequently, my laboratory has pursued the understanding, on a mechanistic basis, of the effects of taurine on Ca^{2+} uptake and protein phosphorylation in excitable tissues. This chapter summarizes the recent and on-going studies of the effects of taurine on protein phosphorylation.

METHODS

Phosphorylation Assay

A mitochondrial fraction of the rat retina and a synaptosomal fraction of the rat cortex (P_2) were prepared as previously described by Lombardini and colleague (3,11). The rat heart mitochondrial fraction was prepared by homogenizing cardiac tissue in 10 vol of Krebs-bicarbonate buffer (below) with a Polytron homogenizer (6 s). The homogenate was centrifuged at 500 x g for 10 min. The pellet was discarded and the supernatant containing mitochondria was centrifuged at 3000 x g for 10 min. The pellet containing mitochondria was washed in Krebs-bicarbonate buffer and recentrifuged at 3500 x g for 15 min. The mitochondrial pellet was then suspended in buffer. The subcellular tissue fractions, i.e., retina, brain or heart (~0.1-1.0 mg), were preincubated for 2 min at 37°C in a test tube containing either bicarbonate buffer (retina: $NaHCO_3$, 50 mM; KCl, 50 mM; NaCl, 50 mM; KH_2PO_4, 1.2 mM; $MgCl_2$, 1.17 mM; and $CaCl_2$, 10 μM; pH 7.4) or Krebs-bicarbonate buffer (brain: NaCl, 118 mM; KCl, 4.7 mM; KH_2PO_4, 1.2 mM; $MgSO_4$, 1.17; $CaCl_2$, 2.5 mM; $NaHCO_3$, 25 mM; pH 7.4; heart: same buffer as for brain but without $CaCl_2$). The phosphorylation reaction was initiated with 20 μCi of [γ-^{32}P]ATP (10 μM) and the tubes were then incubated for a specified time period: retina and heart, 6 min; brain, 1 min. Final volume was 0.25 ml. The reaction was stopped by the addition of 0.5 ml of sodium dodecyl sulfate (SDS) solubilization buffer [0.05 M Tris-HCl (pH 6.8), 2% SDS, 10% glycerol, 2 mM mercaptoethanol, 0.00125% bromophenol blue]. The incubation tubes were agitated on a Vortex mixer and then immediately boiled for 5 min. Aliquots of the boiled incubation mixtures were subjected to 1-dimensional polyacrylamide gel electrophoresis (PAGE) according to the method of Laemmli (2). The gels were dried and exposed to X-ray film to visualize the incorporation of radioactive phosphate into the various proteins. Quantitation of the radioactivity in specific proteins was determined by densitometry.

RESULTS

Studies in my laboratory over the last few years have been directed towards understanding the inhibitory effect of taurine on protein phosphorylation (3-14). Subcellular fractions from three rat tissues have been used in these studies (Table 1): a mitochondrial fraction prepared from the retina, a P_2 synaptosomal fraction from the brain cortex, and a mitochondrial fraction from the heart. In each of these preparations, taurine primarily inhibits the phosphorylation of one specific protein. Taurine also affects the phosphorylation of other proteins to varying degrees - both as an inhibitor and, in a few instances such as in the mitochondrial fraction of the retina, as a stimulator (11). However, because of technical problems and, in certain cases, due to the episodic nature of the taurine effect, we have opted to study the inhibitory effect of taurine on one protein from each tissue. The individual proteins had the following molecular weights (Fig. 1): retina, ~20 K; brain, ~20 K (migrates slightly faster on PAGE than the brain ~10 K phosphoprotein); heart, ~44 K. [Note: the inhibitory effect of 10 mM taurine on the ~20 K molecular weight protein in the rat cortex preparation is more pronounced in 2-dimensional PAGE (3)].

The taurine effect has also been quantitated by densitometric measurements of the autoradiograms. This manipulation of the data allowed us to calculate the concentration of taurine necessary to inhibit the phosphorylation of each protein by 50% (IC_{50}). The IC_{50} values were: retina, 34 mM; brain, 4 mM; and heart, 9 mM. These IC_{50} values for taurine are in a physiological range well within the in vivo concentrations of taurine for each specific tissue. Fig. 2 shows the concentration-response relationship of the effect of taurine on the phosphorylation of the ~20 K molecular weight protein in the mitochondrial preparation obtained from the rat retina (13).

Analogues of taurine with an altered amino group, such as guanidinoethane sulfonic acid (amino moiety replaced with a guanidino group or isethionic acid (amino moiety replaced with a hydroxyl group), had no effect on the phosphorylation of the three proteins in question in any of the tissues. A sulfone analogue of taurine [3-aminotetrahydrothiophene-1,1-dioxide (ATS)] was also inhibitory in the rat retina and heart preparations, with an IC_{50}

Table 1. Comparisons between retina, brain, and heart

Parameter	Tissue		
	Retina	Brain	Heart
Taurine concentration in tissue:	~50 mM	~ 5-7 mM	~25-35 mM
Subcellular location:	mitochondria	synaptosomes	mitochondria
Apparent molecular weight:	~20 K M_r[b]	~20 K M_r[b]	~44 K M_r
Taurine/analogues (IC_{50}[a]):			
Taurine	inhibit (34 mM)	inhibit (4 mM)	inhibit (9 mM)
Guanidinoethane sulfonate	no effect	no effect	no effect
Isethionate	no effect	no effect	no effect
ATS	inhibit	NT	inhibit (16 mM)
TAPS	inhibit	NT	no effect
Calcium:	no effect	stimulate	inhibit
Potassium chloride:	no effect	stimulate	no effect
Kinase activators:			
cyclic AMP	no effect	no effect	no effect
cyclic GMP	no effect	no effect	no effect
calmodulin	no effect	no effect	stimulate
phorbol ester	no effect	stimulate	no effect
Kinase inhibitors:			
Staurosporine	inhibit	NT[d]	no effect
Chelerythrine	stimulate	NT	inhibit
W-7[c]	inhibit	NT	stimulate
Phosphatase inhibitor:			
Okadaic acid	no effect	NT	stimulate
Nature of phosphate bond:	phosphoester	NT	phosphoester
Phosphorylated amino acid:	serine>>threonine	NT	serine
Isoelectric point:	> pH 6.0	pH 5.6	> pH 7.5
Family of proteins:	3 proteins	1 protein	~7 proteins
Phase separation:			
Aqueous phase	present (100%)	present (>95%)	present (>100%)
Detergent phase	absent	present (< 5%)	absent

Comparisons are with respect to: a) endogenous taurine concentrations; b) the subcellular location of the specific proteins whose phosphorylation is inhibited by taurine; and c) the effects of various treatments on the phosphorylation of the specific proteins whose phosphorylation is inhibited by taurine.The retinal ~20 K phosphoprotein migrates slightly faster on PAGE than the brain ~20 K phosphoprotein. [a] IC_{50} = concentration required to inhibit phosphorylation by 50%; [b]M_r = apparent molecular weight; [c]W-7 = N-(6-aminohexyl)-5-chloro-1-naphthalenesulfonamide. [d]NT = not tested. ATS = 3-aminotetrahydrothiophene-1,1-dioxide; TAPS: (±)-*trans*-2-aminocyclopentane sulfonic acid of 16 mM in the latter. This compound is of interest primarily for two reasons: 1) ATS is a sulfone rather than a sulfonic acid and thus lacks a negative charge, and 2) the taurine part of the ATS structure (S-C-C-N) is partially rigidified in a 5-membered ring. A second rigid structural analogue of taurine, (±)*trans*-2-aminocyclopentanesulfonic acid (TAPS), was an inhibitor of the phosphorylation of the retinal ~20 K molecular weight protein (6) but had no effect in the heart mitochondrial preparation.

Ions such as Ca^{2+} and K^{+} were found to stimulate the phosphorylation of the brain ~20 K molecular weight protein (4) but had no effect in the retinal preparation. In the heart preparation Ca^{2+} inhibited the phosphorylation of the ~44 K molecular weight protein by more than 50% (12). Addition of K^{+} to the heart phosphorylation system had no effect.

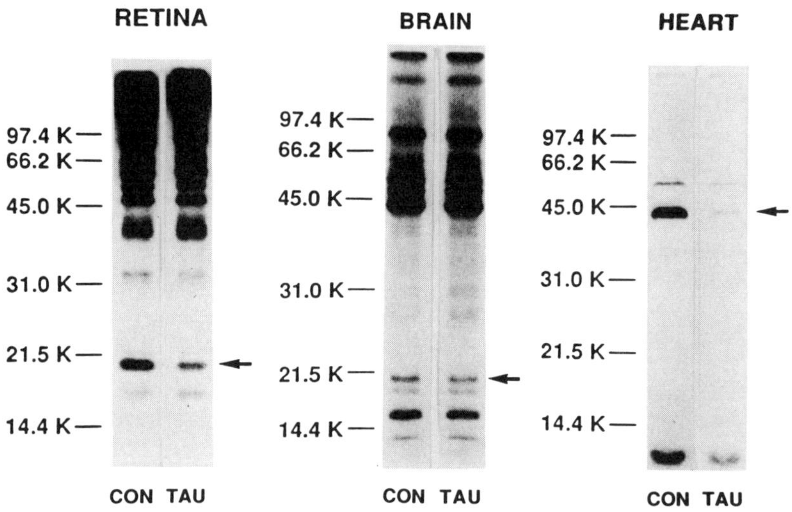

Figure 1. Autoradiographs from 1-D PAGE of the effects of taurine on the phosphorylation of specific proteins in various tissue preparations. Conditions of the phosphorylation assay and PAGE are described in the Methods section. Marker proteins with molecular weights ranging from 14.4 to 97.4 K are indicated. Arrow designates location of specific phosphoproteins discussed in text. Retina (mitochondrial fraction): CON = control, TAU = 20 mM taurine (adapted figure reproduced from ref. 13 with permission). Brain (P_2 fraction): TAU = 10 mM taurine (figure reproduced from ref. 12 with permission). Heart (mitochondrial fraction): TAU = 20 mM

Several known kinase activators and inhibitors were also tested for their effects on the phosphorylation of the specific proteins in the three tissues. Cyclic AMP and cyclic GMP had no effect. Phorbol ester stimulated the phosphorylation of the ~20 K molecular weight protein in the cortex synaptosomal preparation indicating that protein kinase C was involved (4). In the heart preparation, calmodulin appeared to stimulate the phosphorylation of the ~44 K molecular weight protein, an effect possibly due to a calmodulin-Ca^{2+} interaction, since it was observed that Ca^{2+} inhibits phosphorylation of the ~44 K molecular weight protein (12).

Staurosporine, a broad spectrum kinase inhibitor, inhibited the phosphorylation of the retinal ~20 K molecular weight protein but had no effect on the heart ~44 K molecular weight protein. Chelerythrine, a potent inhibitor of protein kinase C, stimulated the phosphorylation of the retinal protein and inhibited the phosphorylation of the heart protein. Interestingly, W-7, a calmodulin antagonist, had the opposite effect; that is, it inhibited the retinal ~20 K molecular weight protein while stimulating the phosphorylation of the heart ~44 K molecular weight protein. The stimulatory effects of chelerythrine in the retina and of W-7 in the heart so far have not been explained. While it appears that there may be a message in these data, it is not understood.

The incorporation of radioactive phosphate from [γ-^{32}P]ATP was determined to take place through a phosphoester linkage in both the retinal ~20 K molecular weight and cardiac ~44 K molecular weight phosphoproteins. In the retinal ~20 K molecular weight

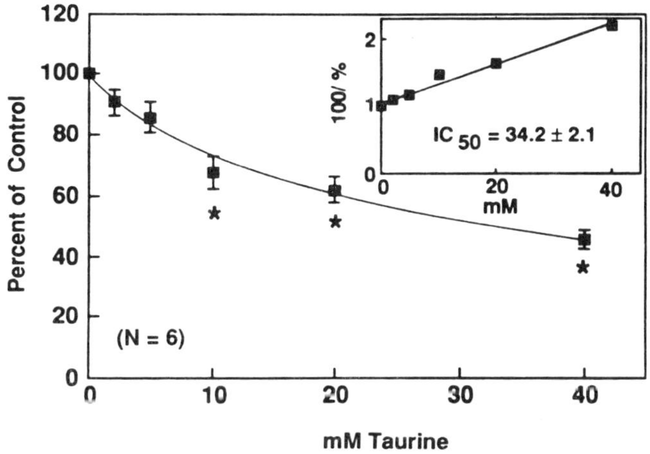

Figure 2. Concentration-response relationship of the effect of taurine on the phosphorylation of a ~20 K molecular weight protein in a mitochondrial fraction of the rat retina. Data generated by densitometry measurements of the autoradiographs are presented as means ± standard errors and represent 6 experiments. The inset contains the data plotted as the reciprocal of the percent of control (x100) vs. the concentration of taurine (mM). Statistical differences (asterisks) between the control (0 taurine) and various concentrations of taurine were determined by non-parametric Kruskal-Walis one-way analysis of variance ($P < 0.05$). (Reprinted with permission from ref. 13)

phosphoprotein, serine and threonine residues were phosphorylated with serine being the predominant acceptor (Fig. 3). In the heart ~44 K molecular weight phosphoprotein only serine residues were phosphorylated.

Two-dimensional PAGE was also used to separate potential isoforms of the phosphoproteins and to determine their isoelectric points. This technique revealed three phosphoproteins with molecular weights of ~20 K and isoelectric points greater than pH 6.0 for the retinal mitochondrial preparation (12), one ~20 K molecular weight phosphoprotein with an isoelectric point of pH 5.6 for the brain cortex preparation (3), and a family of at least 7 phosphoproteins with molecular weights of ~44 K for the heart mitochondrial fraction. The positions of the multiple ~44 K molecular weight heart phosphoproteins on the 2-D PAGE gel indicate that their isoelectric points were greater than pH 7.5.

Phase separation experiments were also conducted to determine if the phosphoproteins from the three tissues were integral membrane proteins. All three phosphoproteins, i.e., the retina ~20 K, the brain ~20 K (Fig. 4) and the heart ~44 K molecular weight proteins, partitioned primarily in the aqueous phase indicative of soluble proteins. A slight (less than 5%) amount of the brain ~20 K molecular weight phosphoprotein partitioned into the detergent phase.

DISCUSSION

The physiological function of the inhibitory effect of taurine on protein phosphorylation is only speculation at this time. However, two laboratories other than mine have also reported that taurine inhibits the in vitro incorporation of radioactive phosphate into specific proteins.

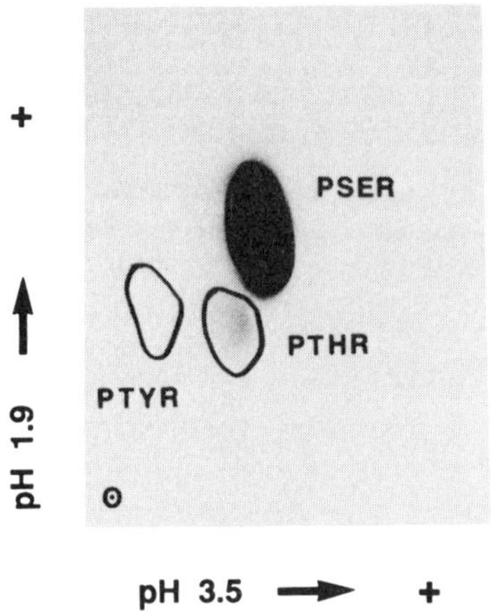

Figure 3. Autoradiograph of the phosphoamino acid analysis of the ~20 K molecular weight phosphoprotein obtained from the mitochondrial fraction of the rat retina. Locations of the standard phosphoamino acids [phosphoserine (PSER), phosphothreonine (PTHR), and phosphotyrosine (PTYR)] were determined by ninhydrin stain and are indicated by circles. Radioactivity in the labelled phosphoserine and phosphothreonine positions is indicated by autoradiography. (Figure reproduced from ref. 13 with permission)

In a synaptosomal membrane preparation of the cat cortex, Sturman and Gargano (19) reported that 50 mM taurine significantly inhibited the incorporation of radioactive phosphate into specific proteins in 8-week-old kittens and adult cats, as shown by 1-D PAGE. However, total phosphate incorporation into trichloroacetic acid precipitates, while reduced in the presence of taurine, was not statistically different from controls. In younger animals such as one-week-old newborns, exogenous taurine had no effect on radioactive phosphate incorporation into trichloroacetic acid precipitates. In these studies, Sturman and Gargano (19) demonstrated that in vitro total radioactive phosphate incorporation into trichloroacetic acid precipitates was reduced (though not significantly) in adult taurine-deficient cats and also in 8-week old kittens from taurine deficient mothers compared to their controls. It was also observed that young cats (4-week-old, one-week-old and newborn) had a greater incorporation of radioactive phosphate into trichloroacetic acid precipitates than older cats (8-week-old and adult).

These data (19) suggest that endogenous taurine may be controlling phosphorylation by allowing in vivo phosphorylation with endogenous ATP to proceed to a greater extent in taurine-deficient animals than in control taurine-normal animals and also to a greater extent in adult animals, known to have lower cortical levels of taurine, than in young animals. If in vivo turnover of these specific phosphoproteins is slow, then the amount of

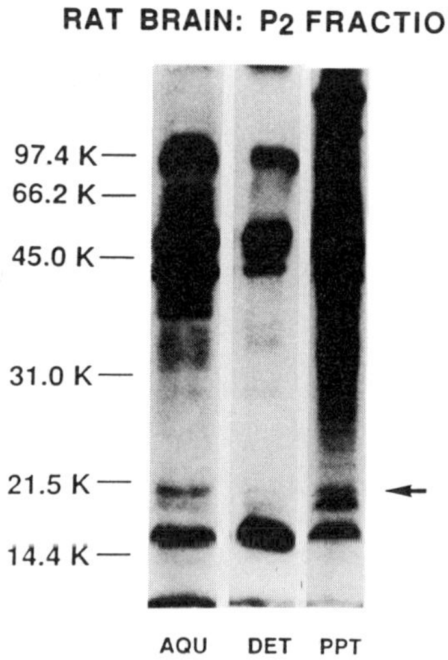

Figure 4. Autoradiographs of a PAGE analysis of proteins from a P_2 synaptosomal fraction of rat brain cortex after phase separation with Triton X-114. The location of the ~20 K molecular weight phosphoprotein is designated by the arrow. Marker proteins with molecular weights ranging from 14.4 to 97.4 K are indicated. AQU = aqueous phase, DET = detergent phase, PPT = detergent-insoluble precipitate

radioactive phosphate incorporated in the in vitro assay from [γ-^{32}P]ATP will be reduced because of the acceptor amino acid moieties being saturated with nonradioactive phosphate in the taurine-deficient animal and in the older animal which has less endogenous taurine to inhibit in vivo phosphorylation. Sturman and Gargano (19) also speculated that synaptosomal membranes may have a different protein composition in the young animal and thus have an increased capacity for phosphorylation compared to membranes in the adult animal.

Finally, Schaffer and colleagues (16) reported in 1990 that 20 mM taurine inhibits the calmodulin-dependent phosphorylation of three proteins (molecular weights: ~44 K, ~54 K, and ~190 K) in rat myocardial cell membranes (sarcolemmal preparation). The inhibitory effect of taurine appears to be specific for the calmodulin-dependent protein kinase since taurine had no effects on cyclic AMP-dependent or cyclic GMP-dependent protein kinases in the sarcolemmal rat heart preparation. Thus, Schaffer and colleagues (16,17) proposed a taurine-calmodulin hypothesis which states that taurine acts in excitable tissues by inhibiting calmodulin-linked proteins including calmodulin-dependent protein kinases. If this hypothesis is proven correct, it may answer the questions as to the function of taurine in myocardial calcium regulation; i.e., calcium channels and the sarcolemmal calcium pump.

CONCLUSIONS

In the studies presented herein involving the inhibitory effects of taurine on phosphorylation of specific proteins in the rat retina, brain and heart, the obvious but unanswered questions remain as to the identity of each phosphoprotein. In rat heart mitochondria there are at least four known enzymes with a molecular weight of ~44 K: creatine kinase, aspartate aminotransferase, serine-pyruvate aminotransferase and acetyl-CoA acetyltransferase. The enzyme creatine kinase is also reported to be phosphorylated, thereby altering the K_m value for creatine phosphate, a known physiologically important metabolite in ATP energetics (15). It is, therefore, tempting to speculate that taurine may be inhibiting the phosphorylation of a well-known enzyme. Studies are currently in progress to purify and isolate the cardiac ~44 K molecular weight phosphoprotein.

ACKNOWLEDGEMENTS

This work was supported in part by NEI Grant EY04780, the Tarbox Institute of Texas Tech University Health Sciences Center, and the South Plains Foundation of Lubbock, Texas. I wish to thank Mrs. Josie Aleman for typing the manuscript, and Mr. Darren Elliott and Mr. Nick Moutos for the photography. Mrs. Ingrid Greene is thanked for her critical comments of the manuscript.

REFERENCES

1. Jacobsen, J.G., and Smith, L.L.H., 1968, Biochemistry and physiology of taurine and taurine derivatives, *Physiol. Rev.* 48:424-511.
2. Laemmli, U.K., 1970, Cleavage of structural proteins during the assembly of the head of bacteriophage T_4, *Nature* 227:680-685.
3. Li, Y.-P., and Lombardini, J.B., 1990, Taurine inhibits the phosphorylation of two endogenous proteins (M_r ~140,000 and ~20,000) in subcellular preparations of rat cortex, *Neurochem. Int.* 17:389-399.
4. Li, Y.-P., and Lombardini, J.B., 1991, Taurine inhibits protein kinase C-catalyzed phosphorylation of specific proteins in a rat cortical P_2 fraction, *J. Neurochem.* 56:1747-1753.
5. Li, Y.-P., and Lombardini, J.B., 1991, Inhibition by taurine of the phosphorylation of specific synaptosomal proteins in the rat cortex: effects of taurine on the stimulation of calcium uptake in mitochondria and inhibition of phosphoinositide turnover, *Brain Res.* 553:89-96.
6. Liebowitz, S.M., Lombardini, J.B., and Allen, C.I., 1988, Effects of aminocycloalkanesulfonic acid analogs of taurine on ATP-dependent calcium ion uptake and protein phosphorylation, *Biochem. Pharmacol.* 37:1303-1309.
7. Liebowitz, S.M., Lombardini, J.B., and Allen, C.I., 1989, Sulfone analogues of taurine as modifiers of calcium uptake and protein phosphorylation in rat retina, *Biochem. Pharmacol.* 38:399-406.
8. Lombardini, J.B., 1985, Effects of taurine on calcium ion uptake and protein phosphorylation in rat retinal membrane preparations, *J. Neurochem.* 45:268-275.

9. Lombardini, J.B., 1985, Inhibition by taurine of the phosphorylation of rat retinal membranes, *in:* "Taurine: Biological Actions and Clinical Perspectives", Oja, S.S., Ahtee, L., Kontro, P., and Paasonen, M.K., eds., Alan R. Liss Press, New York, *Prog. Clin. Biol. Res.* 179:383-396.
10. Lombardini, J.B., 1991, Taurine: retinal function, Brain Research Reviews 16:157-169.
11. Lombardini, J.B., 1992, Effects of taurine on the phosphorylation of specific proteins in subcellular fractions of the rat retina, *Neurochem. Res.* 17:821-824.
12. Lombardini, J.B., 1992, Effect of taurine on protein phosphorylation in mammalian tissues, *in:* "Taurine: Nutritional Value and Mechanisms of Action", Lombardini, J.B., Schaffer, S.W., and Azuma, J., eds., Plenum Press, New York, *Adv. Exp. Med. Biol.* 315:309-318.
13. Lombardini, J.B., 1993, Partial characterization of a ~20 K retinal protein whose phosphorylation is inhibited by taurine, Biochem. Pharmacol. (in press).
14. Lombardini, J.B., and Liebowitz, S.M., 1989, Taurine modifies calcium ion uptake and protein phosphorylation in rat heart, *in:* "Taurine and the Heart", Iwata, H., Lombardini, J.B., and Segawa, T., eds., Kluwer Academic Publishers, Boston, pp. 117-137.
15. Quest, A.F.G., Soldati, T., Hemmer, W., Perriard, J.-C., Eppenberger, H.M., and Wallimann, T., 1990, Phosphorylation of chicken brain-type creatine kinase affects a physiologically important kinetic parameter and gives rise to protein microheterogeneity in vivo, *FEBS Letters* 269:457-464.
16. Schaffer, S.W., Allo, S., Harada, H., and Azuma, J., 1990, J. Regulation of calcium homeostasis by taurine: role of calmodulin, *in:* "Taurine: Functional Neurochemistry, Physiology, and Cardiology," Pasantes-Morales, H., Martin, D.L., Shain, W., and Martín del Río, R., eds., Wiley-Liss, New York, pp. 217-225.
17. Schaffer, S.W., Kramer, J.H., Lampson, W.G., Kulakowski, E., and Sakane, Y., 1983, Effect of taurine on myocardial metabolism: role of calmodulin, *in:* "Sulfur Amino Acids: Biochemical and Clinical Aspects," Kuriyama, K., Huxtable, R.J., and Iwata, H., eds., New York, Alan R. Liss, pp. 39-50.
18. Sturman, J., 1992, Review: Taurine deficiency and the cat, *in:* "Taurine: Nutritional Value and Mechanisms of Action", Lombardini, J.B., Schaffer, S.W., and Azuma, J., eds., Plenum Press, New York, *Adv. Exp. Med. Biol.* 315:1-5.
19. Sturman, J.A., and Gargano, A.D., 1990, Phosphorylation of cat cortical membranes in vitro during development: effect of taurine, *Pharmacol. (Life Sci. Adv.)* 9:595-602.
20. Tiedemann, F., and Gmelin, L., 1827, Einige neue Bestandteile der Galle des Ochsen, *Ann. Physik. Chem.* 9:326-337.

EFFECT OF PHOTOLYTICALLY GENERATED RIBOFLAVIN RADICALS AND OXYGEN ON HYPOTAURINE ANTIOXIDANT FREE RADICAL SCAVENGING ACTIVITY

Terrence R. Green[1,2] and Jack H. Fellman[2]

[1]Department of Pathology
[2]Department of Biochemistry and Molecular Biology
Oregon Health Sciences University
Portland, Oregon 97201

INTRODUCTION

Hypotaurine is a unique amino sulfinate found in narrowly circumscribed sites in mammalian tissue. These tissues are burdened with high oxidant exposure and have helped focus the investigation of the role of hypotaurine as an antioxidant and free radical trapping agent. The human neutrophil in its resting state contains upwards of 0.8 mM hypotaurine, and even greater concentrations of its metabolic product, taurine (7). When this cell is stimulated, a marked increase in oxygen consumption occurs initiated by a NADPH oxidase system (3). The products of this process include superoxide and hydrogen peroxide which can generate by Fenton chemistry hydroxyl radicals (1,12). Evidence for their appearance by this system and the ability of hypotaurine to quench oxidants released by the neutrophil has been presented (7,12). Evidence of free radical-mediated oxidation of hypotaurine has also been reported using liver microsomal NADPH oxidase as a generator of oxygen-centered free radicals (5). Additionally, brain, lung and sexual tissue contain high levels of hypotaurine. In the seminal fluid of the mammalian male and the uterine washings from the female, especially at esterus, high concentrations of this sulfinate have been observed (9,11). The precise function of hypotaurine has not been delineated but there is evidence to support the contention that hypotaurine can protect these cells from oxidative damage.

We present here data that strongly support the view that the sulfinyl radical generated during hypotaurine quenching of carbon- or oxygen-based free radicals can, itself, react with oxygen, removing dissolved oxygen from the microenvironment undergoing oxidative stress. This observation extends the role of hypotaurine as a biological antioxidant.

Taurine in Health and Disease, Edited by R. Huxtable and D.V. Michalk, Plenum Press, New York, 1994

MATERIALS AND METHODS

Hypotaurine, taurine, riboflavin, nitroblue tetrazolium, superoxide dismutase, xanthine oxidase, and xanthine were all purchased from Sigma Chemical Co. All other chemicals were of analytical reagent grade commonly available in the laboratory.

The generation of organic and oxygen centered radicals and metabolites was accomplished by continuous photolysis of stock 5 mM riboflavin made up in 50 mM potassium phosphate buffer, pH 7.8, containing 0.1 mM EDTA (8). Radical production was initiated by illuminating test solutions with a 500 Watt halogen lamp positioned approximately 24 inches from test solutions. Total free radical production was monitored kinetically by following the conversion of nitroblue tetrazolium (NBT, 0.8 mM) to nitroblue formazan (NBF) caused by collision of freshly generated radicals with NBT. Superoxide-independent free radical production (i.e., from organic and organoperoxy) was determined by measuring NBF formation in the presence of excess superoxide dismutase (150 U/ml) in test reaction mixtures to ensure dismutation of any generated superoxide before it could react with NBT. The difference in the rate of reduction of NBT to NBF measured in the presence and absence of superoxide dimutase was taken as an indication of superoxide production. NBF formation was quantitated spectrally at 540 nm on a microtiter plate reader (BioTech-Fisher, Model BT2000). Free radical production was also measured indirectly using the same reaction mixtures and techniques as in the riboflavin experiments, but with omission of NBT, by continuous monitoring of oxygen uptake triggered with photolysis of riboflavin using a water-jacketed Clark cell and oxygen electrode connected to a Gilson KI-IC oxygen analyzer (6).

Superoxide and hydrogen peroxide were also generated independently without production of riboflavin radicals by substituting xanthine (200 μM) in place of riboflavin and starting free radical production with addition of xanthine oxidase to final test solutions (see text). In the latter instance, reactions were monitored by oxygen uptake analysis with exclusion of NBT from test reaction mixtures. The modulating activities of hypotaurine and taurine on radical products were evaluated by examining the effect of each on free radical reactions initiated by either photolysis of riboflavin or enzymatic generation of superoxide and hydrogen peroxide as assessed by alterations in the normal rates of NBT reduction and oxygen consumption as described below.

RESULTS

Under atmospheric oxygen tensions, and with illumination of riboflavin reaction mixtures, hypotaurine markedly stimulated the rate of NBF formation over that of control assays in which hypotaurine was omitted. NBF formation was not detected in control assays excluding riboflavin, nor in complete reaction mixtures containing all reactants, including hypotaurine, but kept in the dark. Furthermore, taurine substituted in place of hypotaurine had no stimulatory nor inhibitory effect on NBF formation. Enhanced rates of NBT reduction caused by inclusion of hypotaurine in reaction mixtures subjected to photo-illumination were traceable to enhanced superoxide-independent free radical formation. This was evident upon comparing NBT reduction rates with superoxide dimutase present to those obtained with exclusion of superoxide dimutase from the final reaction mixtures (Figs. 1 and 2).

Fig. 3 shows the dependence of superoxide-independent NBF formation caused by photolysis of riboflavin in the presence of varying concentrations of hypotaurine and fixed amounts of excess superoxide dimutase sufficient to scavenge any superoxide formed during the photolytic experiments. Sufficient superoxide dimutase to scavenge all superoxide formed in reaction mixtures was verified by titrating reaction mixtures subjected to photo-

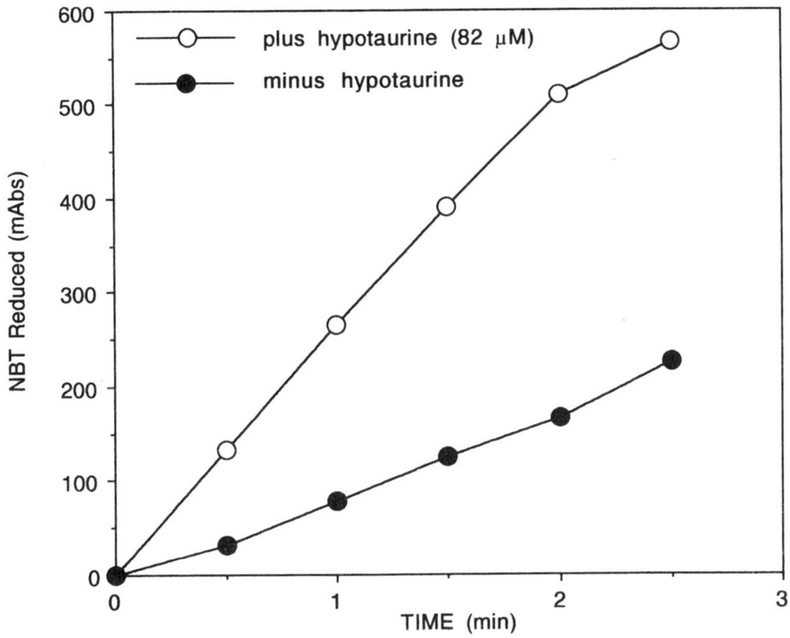

Figure 1. Organoradical reduction of NBT with photolysis of riboflavin.

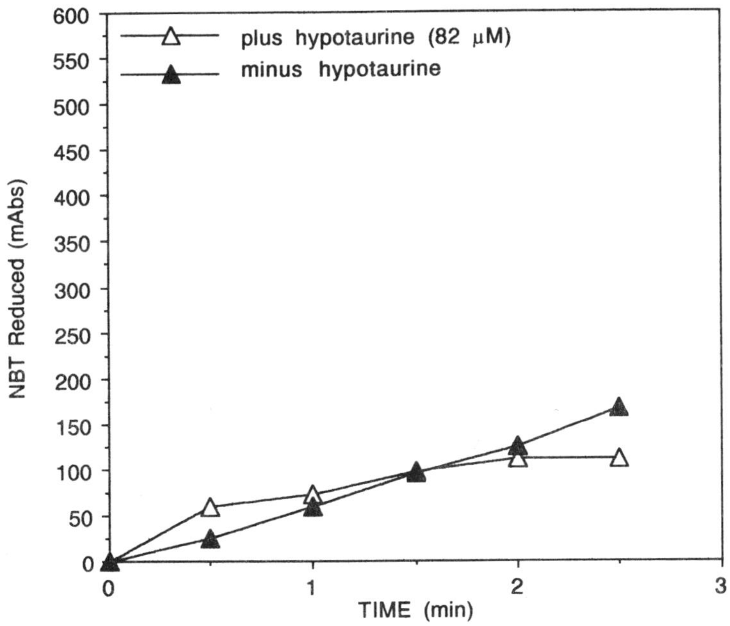

Figure 2. Superoxide-dependent reduction of NBT with photolysis of riboflavin.

lysis with varying levels of superoxide dimutase to the point where NBT reduction was observed to depend no longer upon increasing levels of superoxide dimutase included in the reaction mixtures. The basal rate of NBT reduction induced with photolysis under the latter conditions was taken as the rate of superoxide- independent radical generating activity, reflecting the relative rates of organic and peroxy radical species generated in the photolytic experiments. From these experiments we observed the rate of NBF formation in the photolytic experiments increased linearly until hypotaurine reached a level of ~300 μM at which point further increases in NBF formation abruptly ceased (Fig. 3). The abrupt cessation in enchanced NBF formation seen near 300 μM hypotaurine was not attributable to insufficient NBT in the reaction mixtures. Test solutions were exposed for only 60 sec to light and then immediately read in the microtiter plate reader against control reaction mixtures unexposed to light yielding readings in the range of 350 milliabsorbance units. This contrasts with earlier kinetic readings made on test samples wherein we observed linearity in the readings up to ~500 milliabsorbance units (Fig. 1).

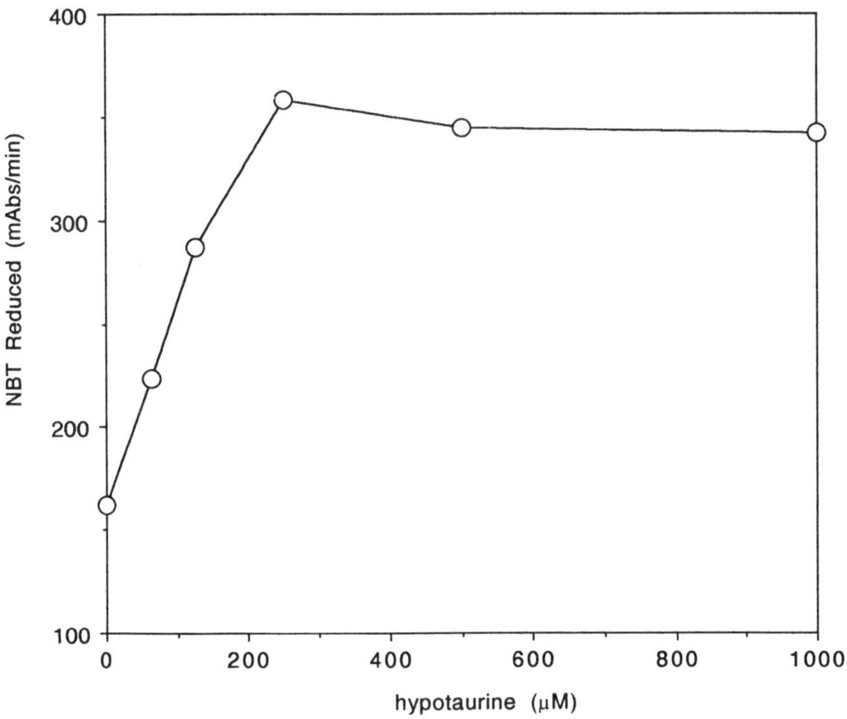

Figure 3. Aerobic rate of radical detection v. hypotaurine concentration.

The cessation of enhanced rates of NBF formation, in excess of the basal rate of NBF formation associated with photolysis of riboflavin, caused by inclusion of hypotaurine in reaction mixtures subjected to photolysis, and truncating at a hypotaurine concentration of ~300 μM, is very close to the dissolved oxygen content of the reaction mixtures, which at room temperature (~20° C) is ~270 μM (6). This observation thus led us to suspect that enhanced superoxide-independent rates of NBF formation seen with inclusion of hypotaurine in reaction mixtures must be linked with molecular oxygen, and thus to hypothesize that molecular oxygen was a reactant in hypotaurine-mediated reduction of NBT catalyzed initially by photolysis of riboflavin solutions. To test this hypothesis, we examined oxygen consumption rates of reaction mixtures subjected to photolysis in the presence and absence of hypotaurine. We anticipated that if oxygen was participating in hypotaurine-mediated free radical reactions of the type observed in the NBT experiments, then a similar pattern of hypotaurine-dependent enhanced oxygen consumption should be seen above residual baseline rates of oxygen consumption concomitant with photolysis of riboflavin solutions. We thus placed test solutions in a Clark cell so that the solutions could be monitored for oxygen consumption rates upon initiating photolysis of the riboflavin, including and excluding hypotaurine in test solutions, but excluding NBT. The participation of oxygen in the hypotaurine-mediated radical reactions was easily confirmed, as evident from the data shown in Table I. Basal rates of oxygen consumption associated with photo-illumination of

Table 1. Oygen-dependent free radical-mediated oxidation of hypotaurine

	O_2 Consumption[1] (nmoles/min/ml)	
Free Radical Catalyst[2]	-hypotaurine	+hypotaurine
riboflavin + light	(6) 24 ± 1	(5) 113 ± 2
xanthine + xanthine oxidase	(3) 4 ± 0	(3) 4 ± 0

[1]Oxygen consumption measured at 21°C in a Clark cell using a Gilson KI-IC oxygen analyzer and oxygen electrode. [2]Stock solutions of 50 mM potassium phosphate, 0.1 mM EDTA ± 100 μM hypotaurine, pH 7.8, supplemented with free radical generator, as indicated. For generation of organic and oxygen centered radicals, 5 mM riboflavin, added to reaction mixtures, was illuminated with a 500 Watt lamp and oxygen consumption associated with photolysis of the riboflavin measured. For oxygen-centered radical production independent of organic radical generation (vis., superoxide and hydrogen peroxide), 200 μM xanthine and xanthine oxidase were substituted in place of riboflavin and light. No oxygen consumption was observed with omission of either riboflavin or xanthine and xanthine oxidase free radical generators from the assay mixtures. Values shown represent the average ± SD with the number of repeat measurements in parenthesis.

riboflavin were approximately one-fourth of the rates observed with hypotaurine (100 μM) included in test solutions. The enhanced rate of oxygen consumption caused by hypotaurine was thus observed to agree closely with the results of NBF experiments (Fig. 1).

When hypotaurine was exposed to superoxide and hydrogen peroxide through generation of these oxygen metabolites by the xanthine oxidase reaction, no differences in oxygen uptake rates were observed (Table 1). Further proof that hydrogen peroxide was not involved in free radical reactions associated with hypotaurine was obtained by examining the kinetics of oxygen consumption linked with photo-illumination of riboflavin solutions spiked with hypotaurine and subjected to alternate cycles of light and dark. Fig. 4 shows a continuous tracing of oxygen consumption measured in the Clark cell for a 100 μM solution of hypotaurine subjected to alternating light and dark cycles of photo-illumination wherein it can be seen that oxygen uptake was alternately robust and ceased as the test sample was alternately exposed and blocked from light. Since hydrogen peroxide is known to be a byproduct in the photolysis of riboflavin, the abrupt cessation of oxygen uptake seen with blockage of light from test reaction mixtures (Fig. 4), even after an extended interval of photo-illumination sufficient to build a pool of hydrogen peroxide in the reaction mixtures, was interpreted by us as evidence against hydrogen peroxide as a participant in the hypotaurine-mediated free radical reactions.

These results thus indicated that hypotaurine, subjected to photolysis of riboflavin under atmospheric conditions, reacts with oxygen in an apparent 1:1 ratio as evident by the linear rise in NBF formation linked with additions of hypotaurine and enhanced oxygen consumption to a threshhold limit of ~300 μM, after which further reaction with oxygen ceases with depletion of dissolved oxygen from the reaction mixtures. Concomitant with the depletion of oxygen, the rate of radical production caused with photolysis of riboflavin appears to reach a steady state level as reflected by a near constant rate of NBF formation independent of further increases in hypotaurine levels in reaction mixtures (Fig. 3).

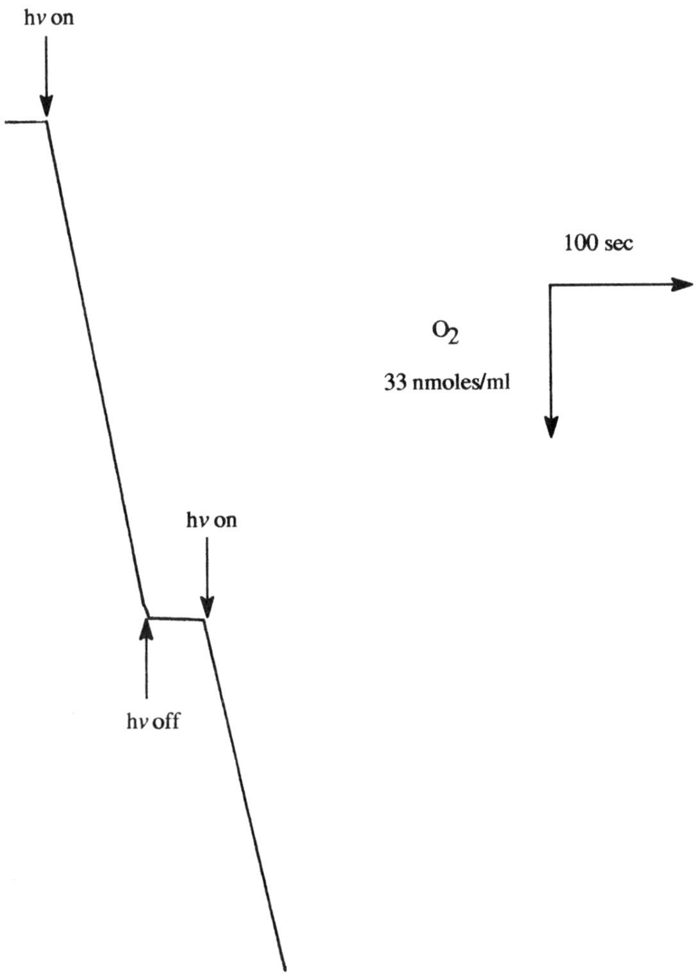

Figure 4. Effect of light on oxygen consumption of riboflavin solutions supplemented with hypotaurine (100 μM).

Since oxygen thus appeared to be an important factor in the reactivity of hypotaurine as a modulator of radical species generated in the riboflavin experiments, we chose to re-examine the radical scavenging activity of hypotaurine under anaerobic conditions through repetition of the initial hypotaurine titration experiments. Riboflavin solutions, spiked with varying concentrations of hypotaurine, were therefore subjected to photo-illumination while simultaneously purging solutions of oxygen under a stream of nitrogen gas. Under these conditions hypotaurine was easily seen to curtail the rate of NBF formation (Fig. 5). In the absence of oxygen hypotaurine thus appeared to markedly quench organoradicals formed with photolysis of riboflavin in a dose-dependent manner.

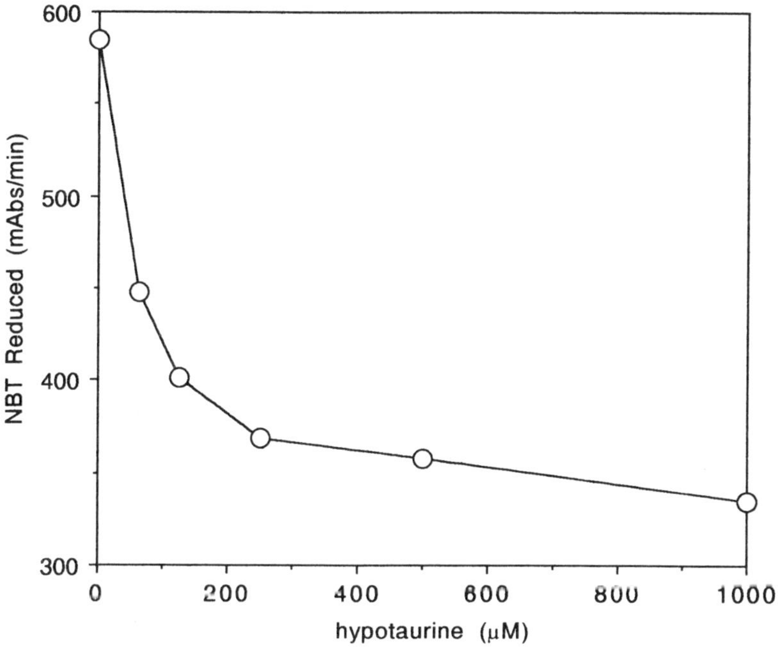

Figure 5. Anaerobic rate of radical detection versus hypotaurine concentration.

DISCUSSION

On photolysis of riboflavin (RH), superoxide ($O_2^\bullet$) and riboflavin organoradicals ($R^\bullet$) are formed (8). In addition, diatomic oxygen can fix into photolytically generated riboflavin radicals leading to formation of organoperoxy radicals ($ROO^\bullet$). The uptake of oxygen associated with photolysis of riboflavin reflects superoxide formation from diatomic oxygen in addition to the fixation of oxygen into freshly formed organoradicals. These reactions can be summarized as in equations 1 through 3.

(1) $hv + RH + O_2 \rightarrow R^\bullet + O_2^\bullet$

(2) $RH + O_2^\bullet \rightarrow R^\bullet + H_2O_2$

(3) $R^\bullet + O_2 \rightarrow ROO^\bullet$, etc.

Radical production can thus be monitored by oxygen uptake analysis using a Clark cell and oxygen analyzer, or through the use of a radical trapping agent such as NBT. In the latter instance, NBT becomes reduced to NBF, and thus can be used to detect free radical products through tracking the rate at which it is reduced. Although NBT is capable of scavenging organic, organoperoxy and superoxide radicals, it is possible to discriminate between superoxide and other organoradical species by including superoxide dimutase in reaction mixtures. In the presence of superoxide dismutase, superoxide is rapidly dismutated to hydrogen peroxide at a rate much faster than the reaction of superoxide with NBT, and thus, in the presence of excess superoxide dimutase, NBF formation accompanying photolysis of riboflavin reflects radical species other than superoxide which are colliding with NBT in converting it to NBF. Hence, in examining the antioxidant capabilities of hypotaurine as a scavenger of free radicals, we chose to examine its ability to modulate radical species through measurements reflecting its capacity to modulate the rate of NBF formation and oxygen consumption induced with photolysis of riboflavin.

Although superoxide can act as either a reductant or oxidant, and thus its role in propagating organoradical production is less certain, peroxy radicals are known to be particularly reactive in propagating free radical reactions (2,10). In order to effectively blunt radical propagation, it thus follows that effective antioxidants must exhibit the capability of intercepting organoradical species before they can combine with molecular oxygen in further propagating the series of chain reactions leading to peroxy radical products (equation 3). Hence, in examining the radical scavenging activity of hypotaurine, we initially anticipated that it might quench oxygen uptake and NBF formation through its capacity to donate an electron equivalent to organoradicals formed in the initial photolysis of riboflavin. We thus expected to observe a decreased, dose-dependent, diminution in NBF formation and oxygen consumption with progressively increasing concentrations of hypotaurine. This anticipated result was predicated on earlier observations made in our laboratory on the hydroxyl radical scavenging activity of hypotaurine in which hypotaurine was shown to be converted to an α-disulfone upon exposure to hydroxyl radicals generated by Fenton chemistry (4).

Surprisingly, we observed instead that with inclusion of hypotaurine in reaction mixtures subjected to photolysis that the apparent rate of radical generation increased in a dose-dependent manner as measured by NBT and by oxygen uptake analysis, and that the enhanced rates of radical generation reflected organoradical production rather than enhanced superoxide production (Figs. 1 and 2 and Table 1). Moreover, enhanced radical production caused by inclusion of hypotaurine in riboflavin solutions subjected to photolysis was clearly linked with enhanced oxygen consumption as evident by titration studies demonstrating a near 1:1 relationship between available oxygen and hypotaurine in reaction mixtures (see Fig. 3 and proportional rates of oxygen consumption shown in Table I versus NBF formation with hypotaurine additions to reaction mixture summarized in Fig. 1).

The apparent increased rate of radical production seen with hypotaurine included in reaction mixtures is not likely to be caused by a secondary series of reactions propagated by metabolites of oxygen such as hydrogen peroxide (via dismutation of superoxide) or hydroxyl radicals (via Fenton chemistry) since the increased rates seen were strictly dependent upon organoradical production as evident by the tight coupling of reaction rates with continuous illumination of reaction mixtures (Fig. 4). Peroxides would be expected to continuously accumulate under the conditions of these experiments, and thus after allowing for their formation as in the experiment illustrated in Fig. 4, cessation of light would not have been expected to shut down oxygen consumption in hypotaurine-supplemented reaction mixtures had peroxides been involved in the enhanced rates of apparent radical production.

In the absence of oxygen, hypotaurine was clearly seen to act as predicted as a radical scavenger of organoradicals, competing with NBT for organoradicals formed, and thus decreasing the levels of organoradicals in solution before they could react with NBT in a dose-dependent manner (Fig. 5). These experiments taken together with the earlier observations under atmospheric conditions indicate that the radical scavenging activity of hypotaurine is profoundly influenced by the oxygen tension in which radicals are formed.

Figure 6 outlines a metabolic pathway for the organoradical scavenging activity of hypotaurine consistent with the experimental observations presented in this study. We propose that formation of the riboflavin radical with photolysis results in the formation of a radical species which is capable of oxidizing the sulfinyl group of hypotaurine to a sulfinyl radical, and that under anaerobic conditions this sulfinyl radical must therefore degenerate, probably to an α-disulfone as indicated in Fig. 6. This would explain the effectiveness of hypotaurine in quenching NBF formation under anaerobic conditions (Fig. 5). On the other hand, under atmospheric oxygen tensions, the sulfinyl radical could react with oxygen to form a peroxy sulfinyl radical which itself reacts to reduce NBT to NBF. Depending upon the reactivity and stability of these radical species, it is possible that the apparent enhanced NBF formation seen with hypotaurine included in the reaction mixtures is actually a reflection of increased shunting of radicals via the peroxy sulfinyl pathway and accelerated rates of NBF formation arising through collision of this radical with NBT. The reactivity of the sulfinyl radical with oxygen accounts for the apparent 1:1 stoichiometry between hypotaurine and oxygen consumption. Hence, in the titration experiments under aerobic conditions, cessation of apparent enhanced NBF formation seen to occur near the upper limit of dissolved oxygen content in reaction mixtures, and matching hypotaurine levels where the cutoff was seen to occur, reflect conversion of the reaction to a semi-anaerobic condition where further enhanced NBF formation is not possible because of diversion of the sulfinyl radical metabolite toward the a-disulfone pathway. While other metabolites of hypotaurine might be formed, we have noted in Fig. 6 the possible formation of a persulfonate arising through condensation of the primary sulfinyl radical with the peroxy sulfinyl radical intermediate. Hence, it might be predicted that under anaerobic conditions the α-disulfone would be a significant end product whereas under higher oxygen tensions the persulfonyl adduct would be formed. At intermediate oxygen tensions a mixture of the two products might be observed.

The actual fate of the metabolic intermediates of hypotaurine formed by free radical oxidants in biological tissues is largely unknown. However, in cell-free assays of free radical reactants, efforts at detecting radical scavenging activities as estimated by the ability

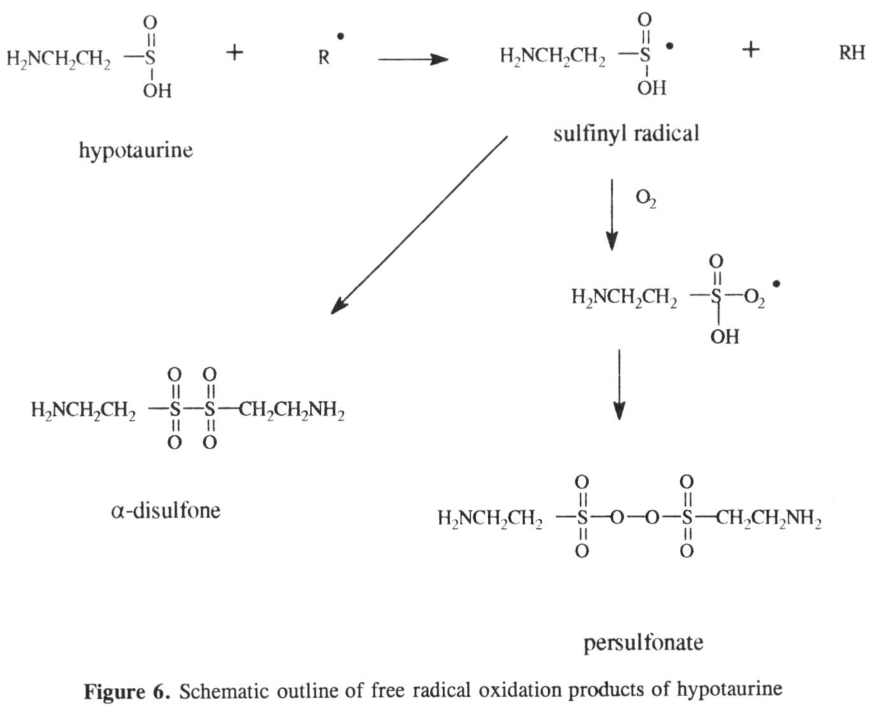

Figure 6. Schematic outline of free radical oxidation products of hypotaurine

of a presumed antioxidant to quench radical production by measuring NBF formation, or quenching of oxygen consumption, can yield unexpected results as in the case of the effect of hypotaurine in the photolytic experiments with riboflavin. Whereas we had initially anticipated hypotaurine might quench NBF formation, and oxygen consumption, accompanying photolysis of riboflavin, the opposite effect was seen. However, in retrospect it is apparent that whether NBF formation is quenched or stimulated, and whether the free radical intermediate formed in the initial oxidative reaction reacts with molecular oxygen, will largely depend upon its chemical reactivity toward diatomic oxygen and NBT. For example, Fig. 7 illustrates a few radical species which might compete in the reduction of NBT of which only superoxide can be differentiated through the use of superoxide dimutase as was done in the experiments of this study. If the reactivity of the peroxysulfinyl radical with NBT is very high in comparison to riboflavin radicals, one might anticipate that formation of the sulfinyl radical would indeed result in an apparent higher rate of NBF formation than in control assays, contrary to what might be expected of an efficient antioxidant. Hence, whereas we observed a higher rate of NBF formation with hypotaurine included in the cell-free riboflavin radical generating system, we cannot ascertain from these measurements alone whether or not the sulfinyl radical species ordinarily dissipate through nondestructive pathways, or, alternatively, act as prooxidants in biological tissues. It is conceivable that hypotaurine, like β-carotene (2), exhibits both antioxidant and prooxidant activities, depending upon the oxygen tension in which its radical metabolites are formed within respiring tissues.

Whereas it remains to be determined through structural analysis of the radical products of these reactions whether the peroxysulfinyl intermediate and persulfonate end product depicted in Fig. 6 are formed, we have previously demonstrated by mass spectral analysis the formation of a stable α-disulfone (4). These earlier observations, and the present observations on the reactivity of hypotaurine with oxygen upon its exposure to organoradicals, suggest that hypotaurine may have a the unique capability of ameliorating oxidative stress in response to free radicals formed in biological tissues. First, it appears capable of reacting with water-soluble organoradicals and thus exhibits the potential of truncating chain propagating reactions leading to oxidative damage. Secondly, it is interesting that, in reacting with organoradicals, hypotaurine becomes converted to an intermediate species capable of scrubbing dissolved oxygen from the immediate reaction environment. This secondary reaction may have the effect of eliminating further oxygen centered radical species from forming through dropping the oxygen tension to a point where the latter reactions proceed too slowly to be of any significance. These two antioxidant functions of hypo-

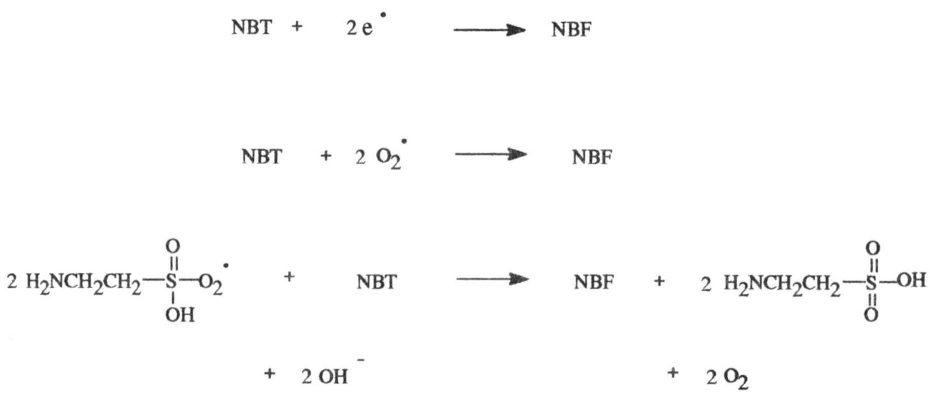

Figure 7. Examples of varying radical species capable of reducing NBT to NBF

taurine may account for its rather peculiar distribution in tissues typically subjected to high oxidant fluxes where its ability to drop the oxygen tension, and quench organoradical production, would logically confer some advantage to cell viability. Hence, hypotaurine is typically found in sperm, neutrophils, liver and neural tissues, including retinal tissue subjected to photolytic oxidative stress, all of which likewise contain very high levels of taurine (9).

It could be that the high levels of taurine in these tissues actually reflect the turnover of hypotaurine via free radical reactions of the type presented in this study. Hence, from this view, the relative abundance of taurine in these tissues, as an oxidation end product of hypotaurine metabolism, might be more logically viewed as an indirect measure of the oxidative stress associated with such tissues. Clearly, the role of oxygen tension in such tissues would be expected to effect the metabolic fate and turnover of hypotaurine radicals, and requires further study in gaining a better insight into the antioxidant activity of hypotaurine in tissues susceptible to oxidant stress. Moreover, the detection of characteristic free radical metabolites of hypotaurine such as the α-disulfone, or persulfonate, in tissues, and the development of reliable techniques in quantitating these radical metabolites, may prove of some value in assessing to what degree specific tissues have undergone oxidative free radical attack. The detection of these latter products could prove to be useful as diagnostic indicators of radical damage in specific tissues indicative of certain disease processes or of past exposure to toxic oxidants.

ACKNOWLEDGEMENTS

This work was supported in part by the Medical Research Foundation of Oregon.

REFERENCES

1. Britigan, B.E., Cohen, M.S., and Rosen, G.M. 1987, Detection of the production of oxygen-centered free radicals by human neutrophils using spin trapping techniques: A critical perspective, *J.Leukoc.Biol.*, 41:349-362.
2. Burton, G.W. and Ingold, K.V. 1984, β-carotene: an unusual type of lipid antioxidant, *Science*, 224:569-573.
3. Curnutte, J.T. 1990, Recent advances in chronic granulomatous disease, *Curr.Opin.Pediatr.*, 2:907-915.
4. Fellman, J.H., Green, T.R., and Eicher, A.L. 1987, The oxidation of hypotaurine to taurine: *Bis*-aminoethyl-α-disulfone, a metabolic intermediate in mammalian tissue, *in:* "The Biology of Taurine: Methods and Mechanisms", Huxtable, R.J., Franconi, F. and Giotti, A., eds., Plenum Press, New York, pp. 39-48.
5. Fellman, J.H. and Roth, E.S. 1985, The biological oxidation of hypotaurine to taurine: hypotaurine as an antioxidant, *in:* "Taurine: Biological Actions and Clinical Perspectives", Oja, S.S., Ahtee, L., Kontro, P. and Paasonen, M.K., eds., Alan R. Liss, New York, pp. 71-82.
6. Green, M.J. and Hill, H.A.O. 1984, Chemistry of dioxygen, *Methods Enzymol.*, 105:3-35.
7. Green, T.R., Fellman, J.H., Eicher, A.L., and Pratt, K.L. 1991, Antioxidant role and subcellular location of hypotaurine and taurine in human neutrophils, *Biochim.Biophys.Acta Gen.Subj.*, 1073:91-97.
8. Hassan, H.M. 1984, Determination of microbial damage caused by oxygen free radicals, and the protective role of superoxide dismutase, *Methods Enzymol.*, 105:404-412.
9. Huxtable, R.J. 1987, From heart to hypothesis: a mechanism for the calcium modulatory actions of taurine, *in:* "The Biology of Taurine: Methods and Mechanisms", Huxtable, R.J., Franconi, F. and Giotti, A., eds., Plenum Press, New York, pp. 371-388.
10. Palozza, P. and Krinsky, N.I. 1992, Antioxidant effects of carotenoids in vivo and in vitro: An overview, *Methods Enzymol.*, 213:403-420.
11. van der Hoerst, C.J.G. and Brand, A. 1969, Occurrence of hypotaurine and inositiol in the reproductive tract of the ewe and its regulation by pregnenolane and progesterone, *Nature*, 223:67-68.
12. Weiss, S.J. 1989, Tissue destruction by neutrophils, *N.Engl.J.Med.*, 320:365-376.

TAURINE PROTECTS AGAINST OXIDANT-INDUCED LUNG INJURY: POSSIBLE MECHANISM(S) OF ACTION

Georgia Schuller-Levis,[1] Michael R. Quinn,[2]
Charles Wright,[2] and Eunkyue Park[1]

[1]Department of Immunology
[2]Developmental Biochemistry
New York State Institute for Basic Research
in Developmental Disabilities
Staten Island, New York

INTRODUCTION

Ozone (O_3) is one of the six major air pollutants for which National Ambient Air Quality Standards have been designated under the Clean Air Act (8). Ozone is the most prevalent photochemical oxidant and an important component of "smog". Ozone has been demonstrated to elicit the release of bronchoactive chemical mediators and to induce inflammation in the lungs of various species, including humans. Although the precise mechanism(s) of lung injury induced by O_3 is unknown, this oxidant gas may interact directly with tissues, can generate free radicals, and clearly produces a lung inflammatory response.

The immunologically mediated events that result in pathological changes in O_3 exposed lung include increased production of inflammatory mediators such as tumor necrosis factor (TNF-α), nitric oxide (NO), interleukin-6 and prostaglandin E_2. Exposure of humans and experimental animals to oxidant gases such as NO_2 or O_3 results in pulmonary injury similar to that observed during hyperoxia (3,4,12). Even though taurine is a free amino acid present in high concentrations within most mammalian tissues, preincubation of rat alveolar macrophages with taurine enhanced the viability of O_3-exposed cells, decreased lipid peroxidation, and decreased the leakage of glutathione (1,2). Taurine has also been demonstrated to be effective in protecting against lung damage in other models as well, such as amiodarone-induced lung fibrosis in hamsters (17). Of particular interest, Gordon et al. (7) showed that taurine prevents lung damage, assessed histopathologically, in hamsters that inhaled NO_2 for short periods. Lung damage induced in hamsters by intratracheal administration of bleomycin is also attenuated by taurine (16).

Alveolar macrophages and polymorphonuclear leukocytes are activated by O_3 exposure and subsequently produce large amounts of nitric oxide, a major mediator of lung damage (14). Some inflammatory cells, i.e. polymorphonuclear leukocytes, eosinophils, and monocytes, possess a halide-dependent peroxidase that oxidizes halides (9,21). Since

Taurine in Health and Disease, Edited by R. Huxtable
and D.V. Michalk, Plenum Press, New York, 1994

leukocytes contain a high concentration of taurine (6,8,11,20), the primary chloramine produced by polymorphonuclear leukocytes is N-chlorotaurine (taurine chloramine) (9,19), which is relatively nontoxic compared to other chloramines.

In this study we demonstrate that pretreatment of rats with taurine prior to acute O_3 exposure results in decreased oxidant-induced lung damage. In addition, we determined the effects of N-chlorotaurine and its precursor, taurine, on the production of nitric oxide and TNF-α by RAW 264.7 cells, a macrophage cell-line, activated with lipopolysaccharide and recombinant interferon. Our data demonstrate that N-chlorotaurine inhibits production of nitric oxide and TNF-α by RAW 264.7 cells. It possible that this mechanism(s) may account, at least in part, for the prophylactic effects of taurine in vivo.

MATERIALS AND METHODS

Ozone Exposure

Female Sprague Dawley rats (150-250 g) were purchased from Taconic Farms, Germantown, NY. Each animal was randomly assigned to one of the following two groups: 1) taurine plus O_3, or 2) water only plus O_3. The group supplemented with taurine received 0.5% taurine (v/v) in the drinking water for five days, followed by 1% taurine for five additional days prior to O_3 exposure. There were no differences in either the body weight or the amount of fluid consumed between groups, indicating that rats consumed taurine in the drinking water without difficulty. Rats were exposed to O_3 in 5.5 cubic foot Plexiglas exposure chambers. The O_3 was formed from pure oxygen passing through an ozonator (OREC Model 03VI-0 Ozonator, O_3 Research and Equipment Corp., AZ). The flow rate through the chamber was 500 ml per min. The concentration of O_3 inside the chamber was monitored continuously by UV absorption with an O_3 analyzer (Model LC-1). Rats were exposed to 2 ppm O_3 for 3 h at 23°C and ambient relative humidity.

Twenty-four, 48, and 72 h following O_3 exposure, animals were lightly anesthetized with sodium pentobarbital and then exsanguinated via the aorta. The chest cavity was opened, the lungs and trachea were removed en bloc and inflation-fixed for 24 h with 3% glutaraldehyde containing a 0.2M Na^+ Cacodylate buffer, pH 7.4. After overnight fixation, the lungs were dissected, sectioned (4-5 μm), and stained with methylene blue and azure II for histologic examination.

Cell Culture Preparation

RAW 264.7 cells were obtained from ATCC (Rockville, MD) and maintained in RPMI 1640 containing 10% fetal calf serum, penicillin (100 U/ml) and streptomycin (100 μg/ml). Cells were washed with complete medium and $2x10^5$ cells/well were used in 96 well plates (Corning Glass Works, Corning, NY). Nonadherent cells (< 5%) were removed after overnight adherence in 5% CO_2. Lipopolysaccharide (10 μg/ml) and recombinant interferon (50 U/ml) were added either alone or with various concentrations of taurine or N-chlorotaurine in a final volume of 200 μl. After 24 h, 100 or 50 μl of supernatant from each well was used for NO_2^- or TNF-α determinations, respectively. Synthesis of NO was determined by assaying for nitrite (NO_2^-), a stable reaction product of NO with molecular oxygen. Nitrite and TNF-α assays were conducted with complete medium which contained Dulbecco's minimal essential medium, 10% fetal calf serum, 2 mM glutamine, penicillin and streptomycin without phenol red. Nitrite concentration and TNF-α production was measured as previously described (10,13).

RESULTS AND DISCUSSION

In Vivo Studies: Ozone Exposure

No changes could be identified in the group pretreated with taurine and then exposed to O_3 at any of the times examined (Fig. 1b). In fact, the morphological characteristics of lung tissue from O_3-exposed rats pretreated with taurine were indistinguishable from control rats that were exposed to room air with or without taurine (not shown). In rats exposed to O_3 without taurine pretreatment the following sequelae were observed. Rats sacrificed 24 h after O_3 exposure exhibited a mild inflammatory infiltrate only in the areas of the terminal and respiratory bronchioles and alveolar ducts accompanied by a few mitotic bronchiolar cells. By 48 h there was a significant inflammatory infiltrate composed almost exclusively of macrophages (Fig. 1a). At 72 h there was a focal hyperplasia in the respiratory bronchioles.

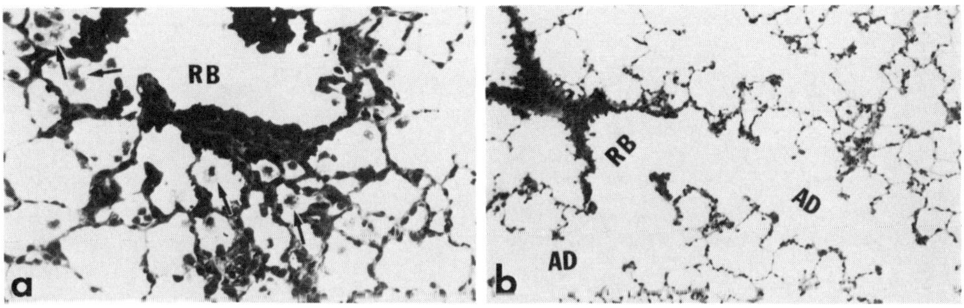

Figure 1. a) Light micrograph of a one micron epon section from the lung of a rat exposed to O_3 and allowed to recover for 48 h. Arrows show many vacuolated macrophages in the alveolar spaces in the respiratory bronchiole (RB), alveolar duct (AD) and surrounding alveoli. X400 (original magnification). **b)** Light micrograph of a one micron epon section from the lung of a rat treated for 10 days with taurine and then exposed to O_3 and allowed to recover for 48 h. There is no evidence of an inflammatory infiltrate within the interstitial spaces of the respiratory bronchiole (RB). Sections were stained with methylene blue and azure II. X100 (original magnification).

Taurine provided in the drinking water 10 days prior to O_3 exposure significantly protected rats against O_3-induced oxidant damage as demonstrated by the absence of infiltrating inflammatory cells in their lungs. Although the mechanism through which taurine affords this protection is unclear, neutrophils containing halide-dependent myeloperoxidase are recruited to the lungs of ozone-exposed rats and could produce significant localized concentrations of N-chlorotaurine. Lung lavage cells, especially from O_3-exposed rats, consist of a complex mixture of cell populations in various stages of differentiation and activation and in situ conditions are difficult to monitor and manipulate at the level necessary for mechanistic studies. However, these obstacles are easily overcome by using the clonal macrophage cell line RAW 264.7. Since these cells do not have a halide-dependent myeloperoxidase system, the concentrations of N-chlorotaurine can be manipulated directly.

In Vitro Studies: Nitrite Production by RAW 264.7 Cells

RAW 264.7 cells were stimulated with lipopolysaccharide (10 μg/ml) and/or recombinant interferon (50 U/ml) for various times before NO_2^- accumulation was measured (Table 1). At least 6 h was required before a significant amount of NO_2^- was detected when cells were stimulated with lipopolysaccharide, rIFN-γ, or both combined. Maximal amounts of NO_2^- were produced after 48 h of stimulation and no further accumulation occurred thereafter. Less than 0.7 nmol/well of NO_2^- were detected in the absence of lipopolysaccharide and recombinant interferon. lipopolysaccharide and recombinant interferon synergistically stimulated NO_2^- production as previously reported (17,18). In the following experiments all NO_2^- measurements were made 24 h post activation unless noted otherwise.

Table 1. Lipopolysaccharide and/or recombinant interferon induce nitrite production in RAW 264.7 cells

Incubation Times (h)	Activators: Lipopolysaccharide	rIFN-γ	Lipopolysaccharide + rIFN-γ
6	0	0.7 ± 0.2	1.3 ± 0.0
18	3.4 ± 0.7	2.5 ± 0.2	11.8 ± 0.5
24	4.8 ± 0.2	4.7 ± 0.3	12.2 ± 0.3
48	5.6 ± 0.1	7.3 ± 0.5	20.1 ± 0.6
72	3.9 ± 0.1	6.8 ± 1.2	17.5 ± 1.3

Time course of NO_2^- production by RAW 264.7 cells activated with lipopolysaccharide (10 μg/ml), recombinant interferon (rIFN-γ) (50 U/ml), or lipopolysaccharide and rIFN-γ combined. Supernatants were removed and assayed for nitrite at times indicated. Values are expressed as nmol/well and represent the mean ± SD for triplicate samples.

Inhibition of Nitrite Production by N-Chlorotaurine

Concentrations ranging from 0.1 mM to 1 mM of either N-chlorotaurine or taurine were used to examine the effects on NO_2^- production by RAW 264.7 cells activated with lipopolysaccharide and recombinant interferon (Fig. 2). N-chlorotaurine or taurine were added simultaneously with the activators and NO_2^- accumulation was measured 24 h later. N-chlorotaurine inhibited NO_2^- production in a dose-dependent manner whereas taurine had no effect (Fig. 2). It is not surprising that taurine was without effect because RAW 264.7 cells do not have a halide-dependent myeloperoxidase system. Viability, monitored over the range of 0.1 mM to 1.0 mM N-chlorotaurine, was measured by trypan blue dye exclusion and was unaffected.

We suspect that N-chlorotaurine acts intracellularly because activated RAW 264.7 cells have an uptake system for N-chlorotaurine (Km = 50 ± 8 μM; Vmax = 85 ± 7 pmol/min-/10^6 cells) (13). This uptake system for N-chlorotaurine is temperature, energy, and sodium-dependent, and is separate and distinct from that for taurine (13).

Inhibition of Nitric Oxide Synthase by N-Chlorotaurine

The direct effects of N-chlorotaurine on nitric oxide synthase activity was determined using a supernatant fraction prepared from 4-5 x 10^7 RAW 264.7 cells after 12 h activation with recombinant interferon and lipopolysaccharide (Fig. 3). Full activity of nitric oxide synthase relied upon the addition of several cofactors (NADPH, (6R)-5,6,7,8-tetrahydrobiopterin and FAD) (Fig. 3). Addition of 300 μM monomethylarginine a competitive substrate inhibitor, or 1 mM N-chlorotaurine in the assay also inhibited nitric oxide synthase activity. Monomethylarginine, similar to N-chlorotaurine, also inhibited the production of nitric oxide by RAW 264.7 cells activated in culture (data not shown). The cytosolic preparation was pretreated with 1 mM N-chlorotaurine to determine whether the latter inhibited nitric oxide synthase activity by interacting with the enzyme or by inactivating cofactors. After a preincubation with N-chlorotaurine, unreacted N-chlorotaurine was removed from the preparation, and nitric oxide synthase activity was subsequently measured. Nitric oxide synthase was irreversibly inactivated during the preincubation period which demonstrates a direct effect of N-chlorotaurine on the enzyme rather than on cofactors in the assay mixture. In control samples similarly preincubated at room temperature for 10 min either alone or with 1 mM taurine, or with 1 mM N-chloroserine before being assayed, nitric oxide synthase activity was unaffected (data not shown).

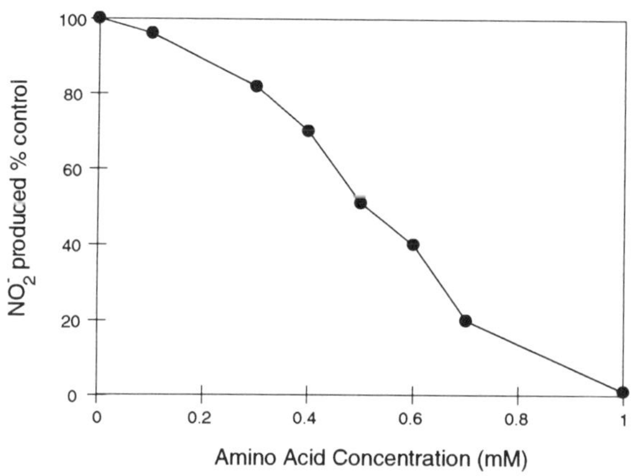

Figure 2. N-chlorotaurine inhibits NO_2^- production. RAW 264.7 cells were cultured with lipopolysaccharide (10 μg/ml) and recombinant interferon (rIFN-γ) (50 U/ml) in the presence of different concentrations of N-chlorotaurine (●). Supernatants were removed after 24 h and assayed for nitrite. Values are expressed as means for triplicate samples. Similar results were obtained in 10 independent experiments.

Inhibition of TNF-α Release from Activated RAW 264.7 Cells by N-Chlorotaurine

RAW 264.7 cells activated with lipopolysaccharide and recombinant interferon also secrete TNF-α into the medium. Release of TNF-α, a potent mediator of tissue damage, was inhibited by N-chlorotaurine in a dose-dependent manner (Fig. 4) while taurine had no effect at any dose tested. Since the release of TNF-α and the production of nitric oxide by activated RAW 264.7 cells are mediated by separate intracellular events, N-chlorotaurine may be inhibiting the activation of these cells in general. This possibility is under further investigation in our laboratories.

Kinetics of N-Chlorotaurine Inhibition of Nitric Oxide Production

To determine whether the effects of N-chlorotaurine were mediated by inhibition of nitric oxide synthase activity directly or by inhibition of the induction of the enzyme, N-chlorotaurine was added to RAW 264.7 cells at various times after activation with lipopolysaccharide and recombinant interferon and NO_2^- was measured 24 h later (Table 2). When N-chlorotaurine was added simultaneously with lipopolysaccharide and recombinant interferon the greatest inhibition of NO_2^- accumulation was observed.
There was some inhibition of NO_2^- production when N-chlorotaurine was added 1-3 h after activation. However, N-chlorotaurine added 6 h after activation did not inhibit nitric oxide production (Table 2) which is of particular interest since nitric oxide synthase activity is induced 6 h after activation with lipopolysaccharide and recombinant interferon (Fig. 2). These data suggest that N-chlorotaurine may be inhibiting activation of the cells, perhaps by interfering with the lipopolysaccharide and/or IFN-γ signal transduction pathway.

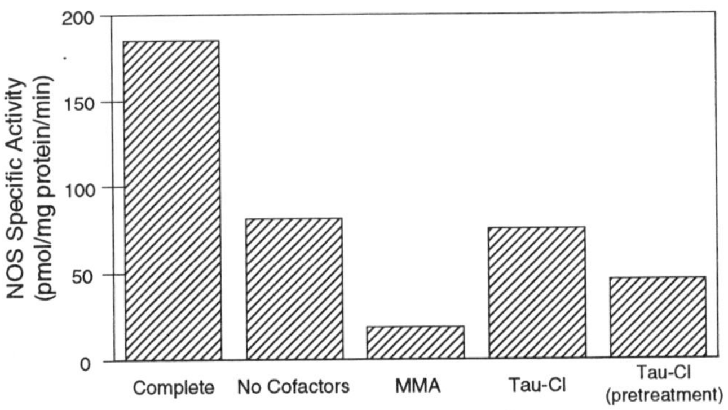

Figure 3. Inhibition of nitric oxide synthase by N-chlorotaurine. "Complete" indicates that the assay contained all constituents necessary for full activity (13). "No cofactors" indicates the absence of added NADPH, 5,6,7,8-tetrahydrobiopterin and FAD. Monomethylarginine (MMA) was present in the assay at 300 μM. N-Chlorotaurine (1 mM) was either added to the assay or used to pretreat the preparation before assaying ("pretreatment"). All values represent the mean of triplicate samples. Similar results were obtained in 3 independent experiments.

In vitro macrophage cell culture studies from our laboratories demonstrate that T-Cl inhibits production of nitrite and TNF-α. Although N-chlorotaurine can irreversibly inhibit nitric activity directly (Fig. 3), its action on intact activated RAW 264.7 cells is more complex. N-Chlorotaurine appears to inhibit the activation of RAW 264.7 cells because it has little effect on nitrite production when added 6 h after activation with recombinant interferon and lipopolysaccharide (Table 2). This notion is supported by the concomitant inhibition by N-chlorotaurine of TNF-α secretion into the medium. These observations cannot be accounted for by interaction of N-chlorotaurine with lipopolysaccharide and/or recombinant interferon directly because preincubation of activators with N-chlorotaurine alone, for 6 h at 37°C, only slightly diminishes the activity of the activators (not shown). It seems likely that more than one mechanism of N-chlorotaurine inhibition is involved. Whether N-chlorotaurine inhibits the activation of these cells in general or more specifically inhibits the induction of nitric oxide synthase at signal transduction, transcription, or translation remains to be determined. Studies in progress suggest that N-chlorotaurine does not inhibit macrophage activation in general but does inhibit nitric oxide synthase protein exprcssion.

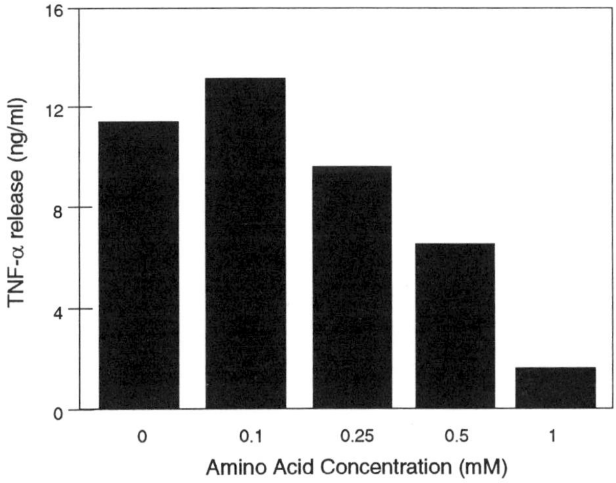

Figure 4. N-chlorotaurine inhibits TNF-α release. RAW 264.7 cells were cultured with lipopolysaccharide (10 μg/ml) and recombinant interferon (50 U/ml) in the presence of different concentrations of N-chloro taurine. Supernatants were collected and TNF-α content measured by ELISA. Values are expressed as means for triplicate samples. Similar results were obtained in three independent experiments.

Table 2. Inhibition of nitrite production by N-chlorotaurine added at various times post-activation

Time of N-chlorotaurine Addition (h)	Nitrite produced (nmole/well)
0	2.9 ± 0.54*
1	7.5 ± 0.43*
3	10.0 ± 0.01*
6	13.0 ± 0.3*
24	12.2 ± 0.3*

Time course of inhibition by N-chlorotaurine of nitrite production. N-chlorotaurine (1 mM) was added at the times indicated after RAW 264.7 cells were activated with lipopolysaccharide and recombinant interferon. Controls were not exposed to N-chlorotaurine. Supernatants were collected and assayed for nitrite. Values represent the mean ± SD for triplicate samples. Asterisk denotes a significant difference ($p < .05$) from the control value of 14.6 ± 0.12.

SUMMARY

It is thought that oxidant-induced tissue damage is not a direct effect of the oxidant per se, but rather results from the inflammatory response that occurs thereafter. As a result of inflammation following oxidant exposure, there are neutrophils, monocytes, and macrophages with myeloperoxidase-H_2O_2-halide activity in the lung. Leukocytes and especially neutrophils contain high intracellular concentrations (22-50mM) of taurine (6,8,11,20). Taurine acts as a trap for toxic hypochlorous acid (HOCl) and forms the less reactive metabolite, N-chlorotaurine (5-6). Thus, the biological activity of halide-dependent myeloperoxidase may be regulated by endogenous taurine. Although taurine had no effect in the present study, polymorphonuclear leukocytes have an active myeloperoxidase system capable of producing N-chlorotaurine (9,19) and would be present at the site of inflammation in oxidant-exposed lungs. Our data suggest that taurine via N-chlorotaurine formation may protect the lung from oxidant injury, at least in part, by inhibiting production of nitrite and TNF-α. Moreover, lavage cells isolated from rats pretreated with taurine and exposed to O_3 have a significant decrease in the production of nitrite and TNF-α, compared with lavage cells from rats exposed to O_3 without taurine supplementation (preliminary studies). Both the concentration of taurine and the effects of N-chlorotaurine strengthen the potential impact of this chlorinated amine in vivo. N-Chlorotaurine may protect against oxidant-induced lung injury by inhibiting production of nitrite and the release of TNF-α which are both known to be directly linked to tissue injury.

ACKNOWLEDGMENT

This work was supported by the Office of Mental Retardation and Developmental Disabilities of the State of New York and the New York Lung Association. We thank Ron Gordon, Ph.D., for evaluation of histological sections of the lung and Debra Laskin, Ph.D., for use of the ozone chamber. We thank Joanne Lopez for secretarial assistance.

REFERENCES

1. Banks, M.A., Porter, D.W., Martin, W.G., and Castranova, V., 1991, Ozone-induced lipid peroxidation and membrane leakage in isolated rat alveolar macrophages: protective effects of taurine, *J. Nutr. Biochem.* 2:308-313.
2. Banks, M.A., Porter, D.W., Martin, W.G., and Castranova, V., 1990, Effects of in vitro ozone exposure on peroxidative damage. Membrane damage, leakage, and taurine contents of rat alveolar macrophages, *Tox. Appl. Pharmacol.* 105:55-65.
3. Chow, C.K., Hussain, M.Z., Cross, C.E., Dungworth, D.L., and Mustafa, M.G., 1976, Effect of low levels of ozone on rat lungs, *I. Biochemical responses during recovery and reexposure, Exp. Mol. Pathol.* 25:182-188.
4. Crapo, J.D., Barry, B.E., Foscue, H.A., and Shelburne, J., 1980, Structural and biochemical changes in rat lungs occurring during exposures to lethal and adaptive doses of oxygen, *Amer. Rev. Rspir. Dis.* 122:123-143.
5. Ding, A., Nathan, C.F., and Stuehr, D.J., 1988, Release of reactive nitrogen intermediates and reactive oxygen intermediates from mouse peritoneal macrophages, *J. Immunol.* 141:2407-2412.
6. Fukuda, K., Hirai, Y., Yoshida, H., Hakajima, T., and Usii, T., 1982, Free amino acid content of lymphocytes and granulocytes compared, *Clin. Chem.* 28:1758-1761.
7. Gordon, R.E., Shaked, A.A., and Solano, D.F., 1986, Taurine protects hamster bronchiols from acute NO_2-induced alteration, *Am. J. Path.* 125:585-600.
8. Green, T., Fellman, J.H., Eicher, A.L., and Pratt, K.J., 1991, Anti-oxidant role and subcellular location of hypotaurine and taurine in human neutrophils, *Biochem. Biophys. Acta,* 1073:91-97.
9. Grisham, M.B., Jefferson, M.M., Melton, D.F., and Thomas, E.L., 1984, Chlorination of endogenous amines by isolated neutrophils, *J. Biol. Chem.* 259:10404-10413.
10. Kroencke, K.D., Kolb-Bachofen, V., Berschick, B., Burkhart, V., and Kolb, H., 1991, Activated macrophages kill pancreatic syngeneic isolet cells via arginine-dependent nitric oxide generation, *Biochem. Biophy. Res. Comm.* 175:752-758.
11. Learn, D.B., Fried, V.A., and Thomas, E.L., 1990, Taurine and hypotaurine contents of human leukocytes, *J. Leuko. Biol.* 48:174-182.
12. Mustafa, M.G., and Tierney, D.F., 1978, Biochemical and metabolic changes in the lung with oxygen, ozone, and nitrogen dioxide toxicity, *Amer. Rev. Respir. Dis.* 118:1061–1090.
13. Park, E., Quinn, M., Wright, C., and Schuller-Levis, G., 1993, Taurine chloramine inhibits the synthesis of nitric oxide and the release of tumor necrosis factor in activated RAW 264.7 cells, *J. Leuk. Biol.* 54:119-224.

14. Pendino, K., Punjabi, C., Gardner, C., Laskin, J., and Laskin, D., 1992 Ozone-induced alteration in reactive nitrogen and oxygen intermediate production by rat alveolar macrophages, *Toxicologist* 12:291 (abstract).
15. National Research Council, 1991, Rethinking the ozone problem in urban and regional air pollution. National Academy Press, Washington D.C.
16. Wang, Q.J., Giri, S.N., Hyde, D.M., and Li, C., 1991, Amelioration of bleomycin-induced pulmonary fibrosis in hamsters by combined treatment with taurine and niacin, *Biochem. Pharmacol.* 42:1115-1122.
17. Wang, Q.J., Hollinger, M.A., and Giri, S.N., 1992, Attenuation of amiodarone-induced lung fibrosis and phopholipidosis in hamsters by taurine and/or niacin treatment, *J. Pharmacol. Exp. Ther.* 262:127-132.
18. Wang, Q.J., Giri, S.N., Hyde, D.M., and Nakashima, J.M., 1989, Effects of taurine on bleomycin-induced lung fibrosis in hamsters, *Proc. Soc. Exp. Biol. Med.* 190:330-338.
19. Witko, V., Nguyen, A.T., and Descamp-Latscha, B., 1992, Microtiter plate assay for phagocyte-derived taurine chloramines, *J. Clin. Lab. Anal.* 6:47-53.
20. Wright, C.E., Lin, T.T., Lin, Y.Y., Sturman, J., and Gaull, G., 1985, Taurine scavenges oxidized chlorine in biological systems, *Prog. Clin. Biol. Res.* 179:113-123.
21. Zglicznski, J.M., Selvaraj, R.J., Paul, B.B., Stelmaszynska, T., Poskitt, P.K.F., and Sbarra, A.J., 1977, Chlorination by the myeloperoxidase-H202-C1 antimicrobial system at acid and neutral pH, *Proc. Soc. Exp. Biol. Med.* 154:418-422.

SULPHUR AMINO ACIDS AND THE METABOLIC RESPONSE TO CYTOKINES

R.F. Grimble

Institute of Human Nutrition
University of Southampton
Southampton SO9 3TU. U.K.

INTRODUCTION

The metabolic response to trauma and infection is characterised by major increases in urinary excretion of nitrogen and sulphur containing compounds. The increase arises as a consequence of qualitative and quantitative changes in protein and amino acid metabolism. The changes are mediated by the production of pro-inflammatory cytokines from macrophages, endothelial cells and fibroblasts. An increasing number of cytokines are being identified which activate and modulate the activity of the immune system. However only three, interleukins 1 and 6, and tumour necrosis factor, have been shown to have major metabolic effects. These cytokines play a pivotal role in the response to trauma and infection as they initiate processes which provide substrate for the activated immune system, and for the synthesis of substances that are important in the control of the inflammatory process (6). Such control is of major importance, as many of the materials and cellular events associated with the process have the potential to damage the host. Inappropriate or excessive production of cytokines and free radicals and other oxidant molecules have been closely associated with morbidity and mortality in a wide range of conditions in which the immune system has become activate. These include rheumatoid arthritis, inflammatory bowel disease, sepsis, adult respiratory disease, cerebral malaria and meningitis (6).

Cytokines and oxidant molecules have the potential to enhance each others production. interleukin 1 and tumour necrosis factor stimulate oxygen free radicals, hydrogen peroxide and nitric oxide production from a range of phagocytic and endothelial cells (4,10). In turn oxygen radicals and hydrogen peroxide enhance interleukin 1 and tumour necrosis factor production by activation of the nuclear transcription factor NFkB. The latter effect can be inhibited by the endogenous antioxidant glutathione and by a range of exogenous antioxidants (16).

The metabolic changes initiated by cytokines provide a mechanism for controlling and preventing overexpression of cytokines. Interleukin 1 and tumour necrosis factor enhance the production of a number of antioxidant enzymes and glutathione (7). A number of acute

Taurine in Health and Disease, Edited by R. Huxtable and D.V. Michalk, Plenum Press, New York, 1994

phase proteins may exert an inhibitory control on cytokine production since α-1-acid glycoprotein and α-1-antitrypsin at physiological concentrations reduce interleukin 1 and tumour necrosis factor production by human monocytes (15). Concentrations of interleukin 1 receptor antagonist and soluble tumour necrosis factor receptor antagonist increase substantially in the circulation in response to interleukin 1 and tumour necrosis factor production thereby limiting the biological potency of the cytokines (11).

Aminoacids released from peripheral tissues provide substrate for the synthesis of the molecules which exert a regulatory influence on cytokine production and potency following an inflammatory challenge . It is therefore important that the supply of amino acids from the periphery, should match the requirements of visceral tissue for synthesis of these substances, particularly as anorexia is a common feature of the response to trauma and infection.

AMINO ACID REQUIREMENTS DURING INFLAMMATION

There are a number of pieces of circumstantial evidence which suggest that the demands for the sulphur amino acids, cysteine and methionine, and glycine and serine which are metabolically interrelated with them, may increase during inflammation. The proportions of the four amino acids which occur in glutathione, acute phase proteins and cytokine receptor antagonists are shown in Table 1. It should be noted that while all of these substances contain a substantial proportion of the four related amino acids, that glutathione, serum amyloid A, metallothionein and tumour necrosis factor receptor antagonist are particularly rich in them. In quantitative terms glutathione synthesis may play an important part in the increased demand for cysteine. In the rat, if the assumption is made that liver contains 180 mg/g and 5 μmol/g of protein and glutathione respectively, and that protein contains 0.13 mmol of cysteine per g , the distribution of cysteine between glutathione and protein will be in the ratio of 1:5. In rats not exposed to inflammatory stimuli the replacement rate of liver protein is 60% per day compared with values of 300% for glutathione (18). The relative incorporation rates of the amino acid into protein and glutathione in liver may thus be close to parity. Inflammatory agents will bring about a doubling in the rate of protein turnover, however. While no data are available for the turnover of hepatic glutathione in animals undergoing similar treatment, it is likely that demand for cysteine, to synthesise glutathione, will increase.

In patients with severe trauma and sepsis, decreases in plasma serine, glycine and taurine concentrations of 11, 35 and 45% have been observed (1). Glycine is an important precursor of serine. As serine forms the carbon skeleton of cysteine, and taurine is a major end product of cysteine, the fall in plasma amino acids may represent increased flux of substrate through the biosynthetic pathway for cysteine, and diversion of cysteine into cysteine-rich proteins and glutathione, rather than into taurine synthesis. Thus changes in plasma taurine concentration may be indicative of the adequacy of sulphur amino acid intake to support the increased demands for the synthesis of hepatic proteins and glutathione during inflammation. An alternative explanation for the fall in plasma serine, glycine and taurine may reside in the amino aciduria which occurs in infected or traumatised patients. However Jeeevanandam et al. calculated that the generalised fourfold increase in excretion represented only 0.54% of amino acids arising from tissue protein breakdown (9). Furthermore, taurine excretion is decreased if tissue concentrations fall due to a reduction in intake. Huxtable et al. showed that in rats fed a taurine- deficient diet, excretion was reduced, rather than synthesis enhanced, in an attempt to maintain tissue concentrations (8). Enhanced taurine excretion thus seems unlikely in patients in whom plasma concentrations are declining as a consequence of infection and trauma.

Preferential retention of sulphur amino acids may occur in response to inflammation. Following fractures and burns, urinary sulphur excretion is enhanced to a lesser extent than that of nitrogen (3,12). We examined the excretion of urinary inorganic sulphate and nitrogen in patients that had undergone an abdominal aorta aneurism repair operation. Significant decreases in the ratio of sulphur to nitrogen occured during the three days following the operation as shown in Figure 1.

Table 1. Amino acid composition of proteins and peptides that show increased rates of synthesis during inflammation

	% of amino acids			
	Glycine	Serine	Methionine and Cysteine	Total
Caeruloplasmin	7	6	3	13
C reactive protein	8	10	3	21
Serum amyloid A	12	9	4	25
Metallothionein	11	13	32	56
Glutathione	33	0	33	66
IL1 receptor agonist	6	7	7	20
TNF receptor antagonist	10	8	9	27

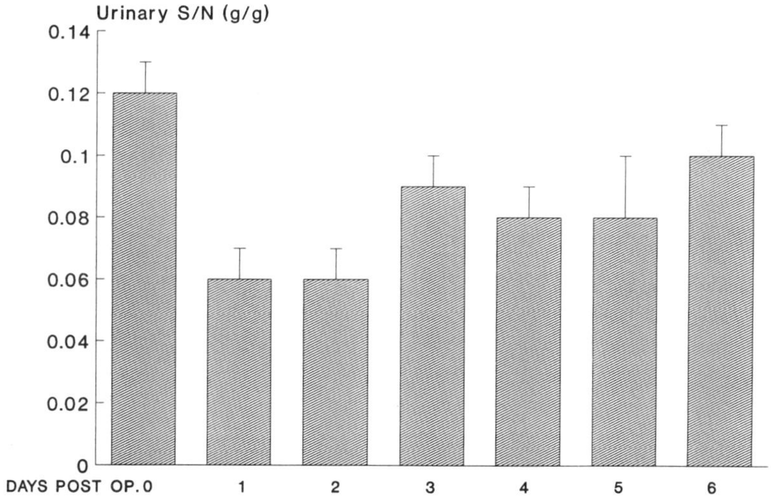

Figure 1. Changes in the ratio of urinary nitrogen and inorganic sulphate in patients after aortic aneurism repair operation (Hird and Grimble unpublished data).

The ability to release sulphur amino acids, glycine and serine from peripheral tissues, and to utilise them for the synthetic processes in visceral tissues, may have a bearing on the prognosis of patients with trauma and sepsis. Rosenblatt et al. (14) measured the output of amino acids from peripheral tissue and uptake by visceral tissues in patients with sepsis. When rates of output and clearance in patients who died are compared with rates in survivors, it is apparent that in non-survivors, serine and methionine output was less than 50% of that in survivors, and glycine, cysteine and methionine uptake was less than 60% The phenomenon did not represent a generalised difficulty in utilising amino acids. Indeed visceral clearance rates of branched chain amino acids, threonine, glutamine arginine and threonine were normal in patients who died.

PARTITIONING OF SULPHUR AMINO ACIDS IN RESPONSE TO DIETARY ADEQUACY AND INFLAMMATION

Protein, glutathione, taurine and sulphate are the main sulphur-containing products of cysteine and methionine. The synthesis of these products is influenced by the dietary intake of sulphur-containing amino acids. As outlined earlier, the requirements for these amino acids, to facilitate the synthesis of acute phase proteins and glutathione by liver during inflammation, are likely to be increased as a consequence of enhanced cytokine production. Dietary intake is likely to be reduced due to the anorexia which often accompanies cytokine production. However, increased release of amino acids from peripheral tissues is also likely to occur, thereby reducing the extent of the shortfall in sulphur amino acid supplies to the liver. To what extent therefore do cytokines influence the partitioning of sulphur amino acids into end products?

In vitro studies indicate that partitioning of sulphur amino acids between glutathione, sulphate and taurine, in the absence of cytokines, is influenced by cysteine concentration and availability (17). Production of all three materials increased in concert with cysteine concentration in the medium, however low concentrations favour glutathione production, while high concentrations favour inorganic sulphate and taurine production. At concentrations of 50 μM, the percentage of metabolised cysteine sulphur terminating in glutathione, sulphate and taurine is 43%, 51% and 6% respectively. At cysteine concentrations of 1 mM, values of 31%, 48% and 21% respectively were observed. The rates of production of the three metabolites increased by 7.5, 9.9 and 36 fold for GSH, sulphate and taurine respectively over the range of substrate concentrations from 50 μM to 1 mM.

In addition to the overall availability of sulphur amino acids determining the proportions of sulphur being incorporated into glutathione, sulphate and taurine by hepatocytes, the location of the cells may influence the proportions of end products formed (13). Metabolised sulphur, from cysteine, is incorporated in periportal hepatocytes as protein, sulphate and taurine, in the proportions of 1:2.2:0.04 respectively. In perivenous hepatocytes the respective proportions are 1:1.9:0.28. While incorporation rates into protein were identical in cells from both locations, incorporation into glutathione was 20% higher in periportal than perivenous hepatocytes.

During infection and following trauma, cytokines will modulate the rate of delivery of sulphur amino acids to the liver, and the proportions entering the organ via the hepatic portal vein and hepatic artery, by causing anorexia and mobilisation of amino acids from peripheral tissues. Anorexia will reduce cysteine and methionine delivery via the portal vein and peripheral protein mobilisation will increase delivery via the hepatic artery. If it is necessary to instigate total parenteral nutrition, then amino acid will be supplied predominantly by the latter route. Theoretically therefore, such changes, in amount and rate of substrate delivery, will relatively favour taurine synthesis against that of glutathione. Unfortunately there are no direct data on how cytokines modify the metabolic fate of the sulphur moiety of methionine and cysteine. To elucidate the situation, we examined the characteristics of the partitioning of sulphur amino acids into protein, GSH and inorganic sulphate in intact rats, in response to tumour necrosis factor, and variations in the intake of cysteine and metabolically related amino acids. The amino acids studied were serine , the source of the carbon skeleton of cysteine; methionine, the source of the sulphur moiety of cysteine; and taurine the main nitrogenous metabolite of cysteine. In the studies, the amino acids were incorporated in isonitrogenous amounts into low protein diets, containing 8% casein. Alanine was also included (6 g/kg) in an 8% casein diet to act as an isonitrogenous, free amino acid control, that was devoid of sulphur. Animals receiving a diet containing 8% casein containing alanine, showed a reduced ability to respond to tumour necrosis factor when compared to the responses of animals receiving a high protein diet

Table 2. Influence of supplementary serine, methionine, cysteine and taurine on the response of protein-depleted rats to tumour necrosis factor

	Liver protein (g/liver)		Liver GSH (μmol/liver)		Lung (GSH) (μmol/lung)	
	S	T	S	T	S	T
Diet						
20% Casein + cysteine	1.20	1.46	26	49	0.81	0.72
8% Casein + alanine	0.73	0.76	7	11	0.38	0.21
8% Casein + cysteine	0.79	1.19	14	38	0.74	1.26
8% Casein + methionine	0.76	1.06	4	27	0.42	1.21
8% Casein + serine	0.71	0.76	10	10	0.61	0.38
8% Casein + taurine	0.88	0.88	12	10	0.51	0.38

After receiving the diets for one week, rats received an i.p. injection of either 50 μg/kg of recombinant human tumour necrosis factor (T), or saline (S). Measurements were made 24h after injection. Saline animals were pair fed intakes of T groups (Hunter and Grimble unpublished data).

Table 3. Influence of supplementary sulphur amino acids on changes in liver protein synthesis and glutathione content and sulphate excretion in protein-depleted rats after receiving tumour necrosis factor

	Protein synthesis (g/day/liver)		GSH (μmol/liver)		Urinary sulphate (μmol/24 h)	
	S	T	S	T	S	T
Diet						
20% Casein + Cys	0.91	1.46	10	41	456	379
8% Casein + ala	0.36	1.03	5	9	37	67
8% Casein + cys	0.70	1.40	8	38	309	339
8% Casein + met	0.80	1.24	6	27	118	345

After receiving the diets for one week rats received an i.p. injection of either recombinant human tumour necrosis factor (T) 50 μg/kg, or saline (S) i.p. Measurements were made 24 h after injection. S animals were pair fed the intakes of the corresponding T group post injection. Protein synthesis was measured by the "massive dose" technique which gives a composite value for rates of synthesis of hepatic structural and export proteins (Hunter and Grimble unpublished data.

containing 20% casein. Three studies were conducted in which liver protein and glutathione content, liver protein synthesis and glutathione content, and urinary sulphate excretion respectively, were compared in young animals receiving the supplemented low protein diets, or high protein diet, for a week prior to an injection of saline, as a control, or tumour necrosis factor. The results are shown in Tables 2 and 3.

Feeding an 8% casein diet, containing alanine, impaired growth and resulted in livers with a lower GSH and protein content associated with a smaller rate of protein synthesis, than in rats receiving the 20% casein diet. The low protein diet containing alanine resulted in a reduced lung GSH content. Addition of cysteine and related amino acids to low protein diets, in place of alanine, influenced the various parameters to differing extents. Cysteine increased liver and lung glutathione content, rates of liver protein synthesis and sulphate excretion towards the values found in the high protein group. Methionine supplementation produced a smaller effect on liver and lung GSH content and sulphate excretion, but was as equally potent as cysteine in restoring liver protein synthesis rates towards normal values. Serine and taurine enhanced liver and lung glutathione content more effectively than methionine, but less so than cysteine.

Exposure to the inflammatory effect of tumour necrosis factor changed and enhanced some of the modulatory effects of the sulphur amino acids, and related compounds, on the metabolic parameters studied. The increase in liver protein and glutathione content and protein synthetic rate was reduced in magnitude by the 8% casein diet containing alanine. Lung GSH fell by 45 %. With supplemental cysteine a normal increase in liver protein synthetic rate, and protein and glutathione content occured. Furthermore lung GSH content was enhanced. Methionine had a similar influence on lung GSH content, but was less effective than cysteine in increasing liver GSH and protein content towards those seen in animals consuming the 20% casein diets. Serine and taurine supplementation did not restore liver GSH and protein content to normal. However they ameliorated the fall in lung GSH.

Sulphate excretion responded to tumour necrosis factor injection in a complex manner. While a fall occcured in animals receiving the 20% casein diet, increases occurred in rats receiving 8% casein diets supplemented with alanine, cysteine or methionine. The change was greatest in animals receiving the methionine supplement. The data indicate that the ability to enhance hepatic protein synthesis, and enhance and maintain GSH content in liver and lung, is dependent upon sufficient dietary cysteine. However, methionine, the provider of the sulphur in cysteine, is unable to replace cysteine completely in this respect. In the present studies, this phenomenon may be due to the low protein intake of the animals, as under these circumstances, homocysteine, arising from methionine, is more likely to be remethylated to methionine than to transfer its sulphur moiety to serine in the formation of cystathionine (the immediate precursor of cysteine) (5). Indeed, in the saline injected animals, the levels of urinary inorganic sulphate and hepatic GSH, which are products of cysteine, provides indirect evidence of this effect. Under the influence of tumour necrosis factor, however, an increased amount of methionine-derived sulphur is diverted via the transsulphuration pathway to cysteine, as evidenced by enhanced hepatic GSH concentrations and urinary inorganic sulphate excretion. Supplementation of the low protein diets with serine, which provides the carbon skeleton of cysteine, has only a small influence on the impaired response to tumour necrosis factor. Finkelstein and Martin showed that trebling the serine content of the medium during in vitro incubation of rat hepatocytes caused a 69% increase in the conversion of cystathionine to cysteine (5). The influence of serine supplementation on liver and lung GSH concentration indicates that the same phenomenon may occur, to a minor extent, in vivo. In the saline injected control animals, serine raised liver and lung GSH levels towards values seen in animals fed the 20% casein diet. However, when tumour necrosis factor injection resulted in increased demands for cysteine, due to the stimulatory actions of the cytokine on GSH synthesis together with increased consumption of GSH in response to the increased levels of superoxide induced by tumour necrosis factor, hepatic GSH was unable to increase further to levels seen in animals fed a 20% casein diet. Nonetheless lung GSH was maintained at a higher concentration than in animals fed the 8% casein diet supplemented with alanine. A similar effect was produced by taurine supplementation. The modulatory effect may occur in a more indirect manner than that exerted by cysteine or serine. Taurine has antioxidant properties (2). Tissue concentrations are influenced by taurine intake (8). Dietary supplementation may therefore enhance antioxidant defences in lung, thereby reducing the oxidative impact of free radicals on GSH in the organ. The ability to sustain GSH synthesis is dependent upon the availablity of cysteine and in particular of its sulphur moiety (19). In liver, the synthesis of GSH, protein and inorganic sulphur are the main destinations of sulphur derived from cysteine. In vitro studies suggest that while taurine is an important product of cysteine in physiological terms, that it is likely to remain a minor product quantitatively (13,17).

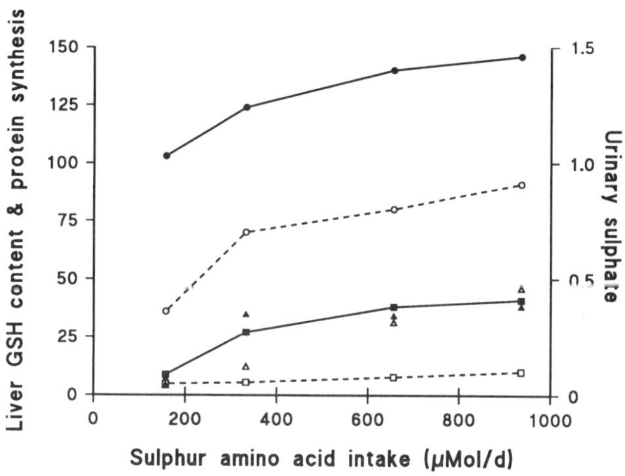

Figure 2. Relationship of dietary sulphur amino acid intake with hepatic GSH content and protein synthesis, and inorganic sulphate excretion in rats. Measurements made 24 h after saline or tumour necrosis factor injections. Values for animals receiving tumour necrosis factor and saline are indicated by closed and open symbols respectively. GSH (μmol/liver) ■,□; Protein synthesis (g/day) ●,○; Sulphate excretion (mmol/day) ▲,△ (Hunter and Grimble unpublished data).

In Fig. 2, the total hepatic protein synthesis, GSH content and inorganic sulphate excretion, described in Table 3, are shown in relation to sulphur amino acid intake. In control animals, receiving saline, inorganic sulphate excretion and hepatic GSH increase in a linear fashion as intake increases. Hepatic protein synthesis increases in a curvilinear fashion in relation to intake. Although inorganic sulphate excretion and hepatic GSH do not necessarily relate to hepatic synthetic rates of the two molecules, it can be seen that sulphate excretion increases to a greater extent than GSH content or liver protein synthetic rate. Increases of 12.3, 2.0 and 2.5 fold respectively occur over the range of intakes of sulphur amino acids in the study. The in vivo data thus indicate a similar propensity to synthesise sulphate, rather than GSH, under conditions of high sulphur amino acid availability, as occurs in vitro (17). Thus, under conditions of high availability of sulphur amino acids hepatic sulphate production appears to be favoured over that of GSH and protein. Under the influence of cytokines, however, the relationships between dietary sulphur amino acid intake and product formation changes. Increases of 4.6 and 1.4 fold occur in hepatic GSH and protein synthesis, while sulphate excretion fails to show major rises in any dietary group, with the exception of the methionine-supplemented animals. Thus, increased diversion of cysteine sulphur from sulphate production into substances of crucial importance in the inflammatory response occur under the influence of cytokines. Furthermore, GSH synthesis may be favoured more than protein synthesis in liver, highlighting the importance of cysteine in maintaining antioxidant synthesis under conditions of inflammation (7). Indeed, the relative fall in inorganic sulphate in patients, reported in Figure 1, may indicate that such a metabolic manoeuver occurs in response to cytokines elicited by surgery.

ACKNOWLEDGMENTS

The author is grateful to the Meat and Livestock Commission for financial support for the studies reported in Tables 2 and 3 and Figure 2.

REFERENCES

1. Askanazi, J., Carpentier, Y.A., Michelsen, C.B., Elwyn, D.H., Furst, P., Kantrowitz, L.R., Gimp, F.E., and Kinney, J.M. 1980, Muscle and plasma amino acids following injury: influence of intercurrent infection, *Ann.Surg.*, 192:78-95.
2. Canas, P., Guerra, R., and Valenzuela, A. 1989, Antioxidant properties of hypotaurine: comparison with taurine, glutathione and β alanine, *Nutr.Rep.Int.*, 39:433-438.
3. Cuthbertson, D.P. 1931, The distribution of N and S in urine during conditions of increased catabolism, *Biochem.J.*, 25:236.
4. Farante, A., Nandoskar, M., Walz, A., Goh, D.H.B., and Kowanko, I.C. 1988, Effects of tumor necrosis factor alpha and interleukin 1 alpha and beta on human neutrophil migration, respiratory burst and degranulation, *Int.Arch.Allerg.Appl.Immunol.*, 86:82-91.
5. Finkelstein, J.D. and Martin, J.J. 1984, Methionine metabolism in mammals. Distribution of homocysteine between competing pathways, *J.Biol.Chem.*, 259:9508-9513.
6. Grimble, R.F. 1992, Dietary manipulation of the inflammatory response, *Proc.Nutr.Soc.*, 51:285-294.
7. Grimble, R.F. 1993, The maintenance of antioxidant defences during inflammation, in: "Update in Intensive Care and Emergency Medicine", Vincent, J.L., ed., Springer Verlag, Berlin, pp. in press.
8. Huxtable, R.J. and Lippincott, S.E. 1982, Sources and turnover rates of taurine in newborn, weanling, and mature rats, in: "Taurine in Nutrition and Neurology", Huxtable, R.J. and Pasantes-Morales, H., eds., Plenum Press, New York, pp. 23-45.
9. Jeevanandam, M., Young, D.H., Ramais, L., and Schiller, W.R. 1990, Effect of major trauma on plasma free amino acid concentrations in geriatric patients, *Am.J.Clin.Nutr.*, 51:1040-1050.
10. Lamas, S., Michel, T., Brenner, B.M., and Marsden, P.A. 1991, Nitric oxide synthesis in endothelial cells: evidence for a pathway inducible by TNF-α, *Am.J.Physiol.*, 261:C674-C641.
11. Larrick, J.W. 1989, Native interleukin 1 inhibitors, *Immunol.Today*, 10:61-66.
12. Larsson, J., Liljedahl, S., Martensson, J., Nordstrom, H., Schildt, B., and Sarbo, B. 1983, Urinary excretion of sulfur amino acids and sulfur metabolites in burned patients receiving parenteral nutrition, *J.Trauma*, 2:656-663.
13. Penttila, K.E. 1990, Role of cysteine and taurine in regulating glutathione synthesis by periportal and perivenous hepatocytes, *Biochem.J.*, 269:659-664.
14. Rosenblatt, S., Clowes, G.H.A., and George, B.C. 1983, Exchange of amino acids by muscle and liver in sepsis, *Arch.Surg.*, 118:167-175.

15. Scuderi, P., Dorr, R.T., Liddil, J.D., Finley, P.R., Meltzer, P., Raitano, A.B., and Rybski, A.B. 1989, Alpha globulins suppress human leukocyte tumour necrosis factor secretion, *Eur.J.Immunol.*, 19:939-942.
16. Staal, F.J.T., Roederer, M., and Herzenberg, L.A. 1990, Intracellular thiols regulate activation of nuclear factor kB and transcription of human immunodeficiency virus, *Proc.Natl.Acad.Sci.USA*, 87:9943-9947.
17. Stipanuk, M.H., Coloso, R.M., Garcia, R.A.G., and Banks, M.F. 1992, Cysteine concentration regulates cysteine metabolism to glutathione, sulfate and taurine in rat hepatocytes, *J.Nutr.*, 122:420-427.
18. Taniguchi, M., Hirayama, K., Yamaguchi, K., Tateishi, N., and Suzuki, M. 1993, Nutritional aspects of glutathione metabolism and function, in: "Glutathione: Chemical,Biochemical and Medical Aspects, Part B", Dolphin, D., Poulson, R. and Avramovic, O., eds., Wiley Interscience, pp. 645-727.
19. Tateishi, N., Higashi, T., Shinya, S., Naruse, A., and Sakamoto, Y. 1974, Studies on the regulation of glutathione level in rat liver, *J.Biochem.*, 75:93-103.

EXPRESSION AND REGULATION OF THE TAURINE TRANSPORTER IN CULTURED CELL LINES OF HUMAN ORIGIN

Vadivel Ganapathy, and Frederick H. Leibach

Department of Biochemistry and Molecular Biology
Medical College of Georgia
Augusta, GA 30912-2100, U.S.A.

INTRODUCTION

Taurine is the most abundant free amino acid in several tissues (5,9,28). The ability of tissues to accumulate taurine is primarily determined by the activity of a transporter that is specific for taurine and other β-amino acids. The findings that certain tissues can concentrate taurine to levels as high as 40-50 mM with the plasma levels of taurine being in the range of only 50-80 μM point to the uniqueness of the transporter in terms of its concentrative ability. Three different driving forces, namely a Na^+ gradient, a Cl^- gradient, and membrane potential, energize the transport system. The Na^+: Cl^-: taurine stoichiometry is 2:1:1. The transporter has been recently cloned from MDCK (Madin-Darby Canine Kidney) cells (32), rat brain (27), mouse brain (15), human FRTL-5 thyroid cells (10) and human placenta (21,25). A comparison of nucleotide sequences of the taurine transporter cDNAs with those of the other cloned transporters indicates that the taurine transporter belongs to a gene family which encodes Na^+- and Cl^--coupled transporters (2,3).

The nutritional requirements for taurine in man are met partly by dietary sources and partly by biosynthesis from methionine and cysteine. The biosynthetic capacity, however, exhibits an interesting developmental pattern. It is nonexistent or very low at fetal and neonatal stages, and progressively increases with age to reach adult levels (1,8,30). In marked contrast, the tissue levels of taurine decrease with age, being highest in fetal life and lowest in adult life (1,26,29). Therefore, it seems very likely that taurine plays an important role in fetal and neonatal development in man (28). Paradoxically however, the human fetus and the newborn have the least ability to produce this amino acid endogenously. Milk contains high concentrations of taurine, and intestinal absorption from this source is the primary route by which the newborn obtains this amino acid. The brush border membrane of the intestinal mucosal cells expresses the Na^+- and Cl^-- coupled taurine transporter that is responsible for absorption of taurine from dietary sources (18). The only mechanism available for meeting the nutritional requirements for taurine in fetal life is the placental transfer from mother. The maternal-facing brush border membrane of the human placenta

Taurine in Health and Disease, Edited by R. Huxtable and D.V. Michalk, Plenum Press, New York, 1994

possesses a highly active taurine transporter (16) which participates in the transfer of taurine from mother to fetus across the placenta. There is unquestionable evidence for a role of taurine in retinal function (20) and the retinal pigment epithelium which is present between the choroid and the neural retina is obligatorily involved in the transfer of taurine in the choroid-to-neural retina direction as well as in the neural retina-to- choroid direction. We have demonstrated that the apical membrane of the retinal pigment epithelium which lies apposed to the neural retina expresses the taurine transporter (17).

Our laboratory has been interested in the transport of taurine in three tissues: placenta, small intestine, and retinal pigment epithelium. To investigate the regulatory aspects of the taurine transporter, we have identified several cell lines of human origin which were originally derived from these three tissues and characterized the taurine transporter in these cell lines. Subsequent regulatory studies with these cell lines have explored the role of protein kinase C, cyclic AMP, cyclic GMP, and calmodulin in the modulation of the transporter function.

EXPRESSION OF THE TAURINE TRANSPORTER IN HUMAN CELL LINES DERIVED FROM PLACENTA, INTESTINE, AND RETINAL PIGMENT EPITHELIUM

Functional expression of the taurine transporter was studied in confluent monolayer cultures of JAR and BeWo cells (placenta), HT-29 and Caco-2 cells (intestine) and HRPE cells (retinal pigment epithelium) (4,13,14,31). All these cell lines constitutively express taurine transport activity. Competition experiments reveal that the transporter responsible for taurine uptake interacts selectively with β-amino acids and exhibits no or little affinity for α-amino acids. Detailed kinetic analysis on HT-29 cells provide strong evidence for the participation of a single common transporter in the uptake of taurine and β-alanine (31). The catalytic function of the taurine transporter is dependent on the presence of Na^+ as well as Cl^-. Substitution of Cl^- with anions such as I^-, F^-, NO_3^- and gluconate markedly reduces the activity. SCN^- can substitute for Cl^- to some extent. Replacement of Na^+ with choline almost totally abolishes the activity. In all cell lines studied, the activity measured in the presence of Cl^- shows a sigmoidal relationship with Na^- concentration, suggesting that multiple Na^+ ions are involved per transport cycle. The Na^+/taurine coupling ratio appears to be 2. In contrast, the activity measured in the presence of Na^+ shows a hyperbolic relationship with Cl^- concentration and the value for the Cl^-/taurine coupling ratio is 1. This Na^+: Cl^-: taurine stoichiometry of 2:1:1 renders the transport system electrogenic. There is a significant difference in the affinity of the transporter for taurine in different cell lines. The value for the Michaelis-Menten constant varies from 0.9 ± 0.1 μM is Caco-2 cells to 10.6 ± 0.3 μM in HT-29 cells.

Recently, we have cloned a cDNA from human placenta which encodes a taurine transporter (21,25). This cDNA when functionally expressed in HeLa cells induces Na^+- and Cl^--dependent taurine transport activity whose characteristics are similar to those of the taurine transporter in cultured cell lines. We have used this cDNA as a probe to detect the presence of taurine transporter-specific mRNA transcripts in the human cell lines. The cDNA probe identifies several species of mRNA transcripts in the human placenta under high stringency conditions. The size of the major transcript is about 6.9 Kb. Other hybridizing transcripts are 11.5, 8.5, 4.5, 2.3, and 1.1 Kb in size, but the levels of these transcripts are many-fold lower than that of the 6.9 Kb size transcript. Analysis of poly(A)$^+$ RNA isolated from cultured human cell lines with this cDNA probe indicates the presence of the 6.9 Kb primary transcript in JAR, BeWo, HT-29, Caco-2, and HRPE cells. The presence of the other minor transcripts is also evident in all of these cell lines.

REGULATION OF THE TAURINE TRANSPORTER IN CULTURED HUMAN CELL LINES

Protein Kinase C

The first indication that the function of the taurine transporter in human cell lines is regulated by protein kinase C came from studies with JAR placental choriocarcinoma cells (13). Treatment of these cells with phorbol 12-myristate 13-acetate, an activator of protein kinase C, results in the inhibition of taurine transport in a dose-dependent manner. Maximal inhibition is observed within 60 min of incubation and the concentration of phorbol ester necessary for half-maximal inhibition is approximately 10 nM. However, the inhibition does not exceed 35-40%. Protein kinase C is implicated in the action of phorbol ester because 4α-phorbol 12, 13-didecanoate, which is unable to activate protein kinase C, has no effect on taurine transport. Staurosporine, is a potent inhibitor of protein kinase C, blocks the inhibitory effect of phorbol ester.

These studies have been extended in greater detail to the intestinal cell lines, HT-29 and Caco-2. In both cell lines, phorbol ester causes a decrease in taurine transport activity (4). As observed in JAR cells, the onset of phorbol ester-induced inhibition in HT-29 cells is rapid with maximal inhibition occurring within 2 h of incubation. The extent of inhibition, however, is much greater in HT-29 cells (75-80%) than in JAR cells (35-40%). The inhibition caused by phorbol ester treatment on taurine transport in HT-29 cells is specific, as similar treatment does not inhibit leucine and lysine transport and stimulates aspartate transport in these cells. Similarly, phorbol ester, which inhibits taurine transport in JAR cells, causes a significant stimulation of leucine transport under similar conditions (24). The influence of phorbol ester on the kinetic parameters of taurine transport has been investigated in HT-29 cells. Phorbol ester decreases the maximal velocity of the transport system as well as the affinity of the system for taurine. In control cells, the Michaelis-Menten constant for the transport process is 10.6 ± 0.3 μM and the maximal velocity is 954 ± 26 pmol/mg of protein/10 min. The corresponding values in phorbol ester-treated cells are 13.3 ± 1.0 μM and 676 ± 28 pmol/mg of protein/10 min.

We have investigated the possible mechanisms of phorbol ester- induced inhibition of the taurine transport activity in HT-29 cells (4). To determine whether the phorbol ester effect is due to inhibition of the synthesis of the transporter protein at the level of transcription and/or translation, we studied the effect of phorbol ester in the presence and absence of inhibitors of transcription (actinomycin D) and translation (cycloheximide). While these compounds effectively block the incorporation of uridine and leucine respectively into trichloroacetic acid-insoluble fraction, they do not affect the inhibitory action of phorbol ester on taurine transport. Similarly, colchicine, an inhibitor of the function of microtubules, and cytochalasin D, which interferes with the function of microfilaments, also have no influence on the phorbol ester effect. These results show that the inhibitory effect of phorbol ester on taurine transport is neither due to inhibition of synthesis of the transporter protein nor due to inhibition of recruitment of preformed transporter protein into the plasma membrane. In view of these findings. we hypothesize that the catalytic function of the taurine transporter is regulated by phosphorylation/dephosphorylation and that protein kinase C leads to the phosphorylation of the transporter causing inhibition of the transporter function. Directly relevant to this hypothesis is the observation that the primary amino acid sequence of the human taurine transporter predicted by the coding region of the transporter cDNA (25) indicates the presence of six recognition sites for protein kinase C-dependent phosphorylation in the transporter protein.

Involvement of protein kinase C in the regulation of taurine transport has also been studied by Jones et al. (11) in LLC-PK_1 cells, a cell line of porcine kidney origin. These investigators have shown that treatment of LLC-PK$_1$ cells with phorbol ester results in a decrease in taurine uptake. Interestingly, this decrease appears to be primarily a result of phorbol ester-induced increase in taurine efflux. However, whether the taurine transporter is responsible for the efflux of taurine from the cells as it is for the influx of taurine into the cells is not known at this time. Available evidence indicates that efflux of taurine that is observed, at least under certain conditions, is not mediated by the taurine transporter (12).

CYCLIC AMP

The role of cyclic AMP in the regulation of taurine transport has been investigated in JAR cells (7,22). Treatment of the cells with cholera toxin for either a short time or a prolonged time does not affect taurine transport activity, even though cyclic AMP content of the cells increases several hundredfold (7). Under similar conditions, the activity of the serotonin transporter is stimulated markedly by cholera toxin. In agreement with the absence of any change in taurine transport activity, the steady state levels of the taurine transporter mRNA transcripts also remain unaffected by cholera toxin treatment. In contrast, the increase in serotonin transport caused by cholera toxin is accompanied by a marked increase in the transporter-specific mRNA levels (22). Interestingly, treatment with cholera toxin or cyclic AMP for 3 days leads to an increase in taurine transport in FRTL-5 cells, a human thyroid cell line (10). In contrast, the taurine transport activity in HT-29 and Caco-2 cells is inhibited significantly following treatment of the cells with cholera toxin for 24 h. Apparently, there are cell line-specific differences in the response of taurine transport activity to an increase in intracellular cyclic AMP levels.

CYCLIC GMP

Studies currently underway in our laboratory show that the activity of the taurine transporter in the Caco-2 cell line is under the regulation of cyclic GMP-dependent cellular processes. Treatment of the cells with *E. coli* heat-stable enterotoxin, which interacts with the membrane-bound guanylate cyclase to increase the intracellular levels of cyclic GMP, leads to a decrease in taurine transport. This inhibitory effect is evident within 30 min of treatment. The effect is specific because the transport of two other amino acids, leucine and phenylalanine, and that of the dipeptide glycylsarcosine are not affected under identical conditions. Kinetic analysis has revealed that the toxin-induced inhibition is due to a decrease both in the maximal velocity of the transport process and in the affinity of the transporter for taurine. This effect is reproducible with guanylin, an endogenous peptide activator of the membrane-bound guanylate cyclase. Again, we have observed cell line-specific differences in the process. Treatment of JAR cells with *E. coli* heat-stable enterotoxin does not influence taurine transport.

CALMODULIN

The role of calmodulin in the regulation of the taurine transporter has been studied in JAR, HT-29, and HRPE cells. In all of these cell lines, blockade of calmodulin action with calmodulin antagonists results in a marked decrease in the activity of the taurine transporter. The involvement of calmodulin in the modulation of taurine transport activity was recognized first in JAR cells where it was found that treatment of the cells with cyclosporin A impaired taurine uptake (23). Cyclosporin A is known to be a calmodulin antagonist (6).

The inhibitory effect of cyclosporin A on taurine transport in JAR cells is reproducible with the widely used calmodulin antagonists W-7 and calmidazolium. The effect is specific because the uptake of leucine is stimulated and the uptake of alanine remains unaffected under similar conditions.

Subsequent studies on the role of calmodulin in the regulation of the taurine transporter were carried out with CGS 9343 B, a very selective calmodulin antagonist (19). Most of the compounds currently used as calmodulin antagonists, such as W-7, chlorpromazine, trifluoperazine, and pimozide, also interact with protein kinase C. Since protein kinase C is known to modulate taurine transport activity, interpretation of results obtained with these rather nonselective compounds is difficult with respect to the differential involvement of protein kinase C and calmodulin in the observed effects of these compounds. The highly selective nature of CGS 9343 B as a calmodulin antagonist makes it a desirable tool to investigate the modulation of the taurine transporter function by calmodulin-dependent cellular processes. Treatment of JAR, HT-29, and HRPE cells with CGS 9343 B leads to a marked decrease in taurine transport. In all three cell lines, the inhibitory effect is selective and is seen with treatment time as short as 1 h. The effect of CGS 9343 B is not associated with any change in the steady state levels of the taurine transporter-specific mRNA transcripts in these cells. The levels of the primary transcript (6.9 Kb in size) remain the same in control cells and in cells treated with CGS 9343 B.

CONCLUSION

The various cellular mechanisms involved in the regulation of the function of the taurine transporter constitute an important aspect in the area of biology of taurine. Availability of cultured human cell lines which constitutively express the taurine transporter is no doubt a great help for progress in this field. Available evidence indicates that multiple cellular mechanisms involving different second messenger pathways participate in the modulation of the taurine transporter function. The role of protein kinase C and calmodulin in the regulation of the transporter is particularly interesting for a number of reasons. Unlike the effects of cyclic GMP and cyclic AMP, which appear to vary depending on the cell line used, the effects of calmodulin and protein kinase C are similar in all cell lines studied. Since protein kinase C and calmodulin are dependent on Ca^{2+} for their activities, this implies that the catalytic function of the taurine transporter is intimately related to intracellular Ca^{2+} levels. Activation of protein kinase C leads to inhibition of the taurine transporter. Activation of calmodulin, on the other hand, appears to stimulate the transporter function because it is the antagonism of calmodulin that inhibits the transporter. Thus, these functions of protein kinase C and calmodulin in the regulation of the taurine transporter are reciprocal. Since the Ca^{2+}-sensitivity of protein kinase C is different from that of calmodulin, it can be envisaged that the activities of these two regulatory pathways are differentially modulated by intracellular Ca^{2+} depending on its concentration. Such a role for Ca^{2+} in determining the activity of the taurine transporter, and hence the intracellular levels of taurine, is interesting because one of the most widely accepted functions of taurine is its role in the maintenance of cellular Ca^{2+} homeostasis. This would suggest a functional cross-talk between taurine and Ca^{2+} inside the cell.

ACKNOWLEDGEMENTS

This work was supported in part by the National Institutes of Health Grant HD 24451 and by a grant from Taisho Pharmaceutical Company, Japan. The authors thank Mary Ann Roupp for expert secretarial assistance.

REFERENCES

1. Agrawal, H.C., Davison, A.N., and Kaczmarek, L.K., 1971, Subcellular distribution of taurine and cysteinesulphinate decarboxylase in developing rat brain, *Biochem. J.*, 122: 759-763.
2. Amara, S.G., and Pacholczyk, T., 1991, Sodium-dependent neurotransmitter reuptake systems, *Curr. Opin. Neurobiol.*, 1: 84-90.
3. Blakely, R.D., 1992, Advances in molecular biology of neurotransmitter transporters, *Curr. Opin. Psychiatry* 5:69-73.
4. Brandsch, M., Miyamoto, Y., Ganapathy, V., and Leibach, F.H., 1993, Regulation of taurine transport in human colon carcinoma cell lines (HT-29 and Caco-2) by protein kinase C, *Am. J. Physiol.*, 264: G939-G946.
5. Chesney, R.W., 1985, Taurine: its biological role and clinical implications, *Adv. Pediatrics*, 32: 1-42.
6. Colombani, P.J., and Hess, A.D., 1987, Fluorometric and flow cytometric analysis of the cyclosporine-calmodulin interaction in T lymphocytes, *Methods Enzymol.*, 139: 677-690.
7. Cool, D.R., Leibach, F.H., Bhalla, V.K., Mahesh, V.B., and Ganapathy, V., 1991, Expression and cyclic AMP-dependent regulation of a high affinity serotonin transporter in the human placental choriocarcinoma cell line (JAR), *J. Biol. Chem.*, 266: 15750-15757.
8. Gaull, G., Sturman, J.A., and Raiha, N.C.R., 1972, Development of mammalian sulfur metabolism: absence of cystathionase in human fetal tissues, *Pediatr. Res.*, 6: 538-647.
9. Huxtable, R.J., 1989, Taurine in the central nervous system and the mammalian actions of taurine, *Prog. Neurobiol.*, 32: 471-533.
10. Jhiang, S.M., Fithian, L., Smanik, P., McGill, J., Tong, Q., and Mazzaferri, E.L., 1993, Cloning of the human taurine transporter and characterization of taurine uptake in thyroid cells, *FEBS Lett.*, 318: 139-144.
11. Jones, D.P., Miller, L.A., Dowling, C., and Chesney, R.W., 1991, Regulation of taurine transporter activity in LLC-PK_1 cells: role of protein synthesis and protein kinase C activation, *J. Am. Soc. Nephrol.*, 2: 1021-1029.
12. Kirk, K., and Kirk, J., 1993, Volume-regulatory taurine release from a human lung cancer line. Evidence for amino acid transport via a volume-activated chloride channel, *FEBS Lett.*, 336: 153-158.
13. Kulanthaivel, P., Cool, D.R., Ramamoorthy, S., Mahesh, V.B., Leibach, F.H., and Ganapathy, V., 1991, Transport of taurine and its regulation by protein kinase C in the JAR human placental choriocarcinoma cell line, *Biochem. J.*, 277: 53-58.
14. Leibach, J.W., Cool, D.R., Del Monte, M.A., Ganapathy, V., Leibach, F.H., and Miyamoto, Y., 1993, Properties of taurine transport in a human retinal pigment epithelial cell line, *Curr. Eye Res.*, 12: 29-36.
15. Liu, Q.R., Lopez-Corcuera, B., Nelson, H., Mandiyan, S., and Nelson, N., 1992, Cloning and expression of a cDNA encoding the transporter of taurine and β-alanine in mouse brain, *Proc. Natl. Acad. Sci. (USA)* 89: 12145-12149.
16. Miyamoto, Y., Balkovetz, D.F., Leibach, F.H., Mahesh, V.B., and Ganapathy, V., 1988, Na^+ + Cl^--gradient-driven, high-affinity, uphill transport of taurine in human placental brush-border membrane vesicles, *FEBS Lett.*, 231: 263-267.
17. Miyamoto, Y., Kulanthaivel, P., Leibach, F.H., and Ganapathy, V., 1991, Taurine uptake in apical membrane vesicles from the bovine retinal pigment epithelium, *Invest. Ophthalmol. Vis. Sci.*, 32: 2542-2551.
18. Miyamoto, Y., Tiruppathi, C., Ganapathy, V., and Leibach, F.H., 1989, Active transport of taurine in rabbit jejunal brush-border membrane vesicles, *Am. J. Physiol.*, 257: G65-G72.
19. Norman, J.A., Ansell, J., Stone, G.A., Wennogle, L.P., and Wasley, J.W.F., 1987, CGS 9343 B, a novel, potent, and selective inhibitor of calmodulin activity, *Mol. Pharmacol.*, 31: 535-540.
20. Pasantes-Morales, H., 1986, Current concepts on the role of taurine in the retina, *in*: *Progress in Retinal Research*, N. Osborne and J. Chader, eds., Pergamon Press, New York, pp. 207-229.
21. Ramamoorthy, S., Ainol, L.S., Leibach, F.H., Blakely, R.D., and Ganapathy, V., 1993, Molecular Cloning, expression, chromosomal localization, and regulation of the human placental taurine transporter, *Placenta*, 14: A.60.
22. Ramamoorthy, S., Cool, D.R., Mahesh, V.B., Leibach, F.H., Melikian, H.E., Blakely, R.D., and Ganapathy, V., 1993, Regulation of the human serotonin transporter. Cholera toxin-induced stimulation of serotonin uptake in human placental choriocarcinoma cells is accompanied by increased serotonin transporter mRNA levels and serotonin transporter-specific ligand binding, *J. Biol. Chem.* 268: 21626-21631.
23. Ramamoorthy, S., Leibach, F.H., Mahesh, V.B., and Ganapathy, V., 1992, Selective impairment of taurine transport by cyclosporin A in a human placental cell line, *Pediatr. Res.*, 32: 125-127.

24. Ramamoorthy, S., Leibach, F.H., Mahesh, V.B., and Ganapathy, V., 1992, Modulation of the activity of amino acid transport system L by phorbol esters and calmodulin antagonists in a human placental choriocarcinoma cell line, *Biochim. Biophys. Acta*, 1136: 181-188.
25. Ramamoorthy, S., Leibach, F.H., Mahesh, V.B., Han, H., Yang-Feng, T., Blakely, R.D., and Ganapathy, V., 1994, Functional characterization and chromosomal localization of a cloned taurine transporter from human placenta, *Biochem. J.*, in press.
26. Ryan, W.L., and Carver, J.J., 1966, Free amino acids of human fetal and adult liver, *Nature*, 212: 292-293.
27. Smith, K.E., Borden, L.A., Wang, H.D., Hartig, P.R., Branchek, T.A., and Weinshank, R.L., 1992, Cloning and expression of a high affinity taurine transporter from rat brain, *Mol. Pharmacol.* 42: 563-569.
28. Sturman, J.A., 1993, Taurine in development, *Physiol. Rev.*, 73: 119-147.
29. Sturman, J.A., and Gaull, G.E., 1975, Taurine in the brain and liver of the developing human and monkey, *J. Neurochem.*, 25: 831-835.
30. Sturman, J.A., Gaull, G., and Raiha, N.C.R., 1970, Absence of cystathionase in human fetal liver: is cystine essential?, *Science*, 169: 74-76.
31. Tiruppathi, C., Brandsch, M., Miyamoto, Y., Ganapathy, V., and Leibach, F.H., 1992, Constitutive expression of the taurine transporter in a human colon carcinoma cell line, *Am. J. Physiol.*, 263: G625-G631.
32. Uchida, S., Kwon, H.M., Yamauchi, A., Preston, A.S., Marumo, F., and Handler, J.S., 1992, Molecular cloning of the cDNA for an MDCK cell Na^+- and Cl^--dependent taurine transporter that is regulated by hypertonicity, *Proc. Natl. Acad. Sci.*, (USA), 89: 8230-8234.

DIETARY TAURINE REQUIREMENT OF CATS IS DETERMINED BY MICROBIAL DEGRADATION OF TAURINE IN THE GUT

James G. Morris, Quinton R. Rogers,
Seungwook W. Kim, and Robert C. Backus

Department of Molecular Biosciences
School of Veterinary Medicine
University of California, Davis CA 95616

INTRODUCTION

Taurine is an essential dietary constituent for cats because in vivo synthesis is limited, and cats have an obligatory requirement for taurine for the conjugation of bile salts. Dogs and cats use only taurine to conjugate bile acids, but dogs unlike cats, appear to have a rate of synthesis adequate to meet their needs. The basis for the lower synthetic capacity in cats than in dogs and other animals has not been fully elucidated. The activity of cysteine dioxygenase is low in the liver of cats which results in limited production of cysteine sulfinate and synthesis of taurine. Other contributing factors are the low activity of cysteine sulfinate decarboxylase and the high activity of aspartate aminotransferase which transaminates cysteine sulfinate to pyruvate, rather than allowing it to be decarboxylated to hypotaurine and hence to taurine (29).

Taurine is essential for the maintenance of normal function of the visual, auditory, cardiac and reproductive systems of cats, as well as for development (32,37,39). The concentration of taurine in the diet that will maintain a eutaurine status is dependent on the type of diet. For expanded dry diets, a concentration of 1000 mg taurine/kg diet is adequate, whereas for canned diets concentrations as high as 2500 mg taurine/kg dry matter are required (5,30). Taurine in both the canned and expanded diets is in the free state and can be extracted by water, so the difference between diets is not one of indigestibility of taurine rendering it unavailable for absorption. This paper will focus on the role of the gastrointestinal tract and dietary components, in particular protein, which have a dominant role in determining the utilization of taurine in a diet, and hence the dietary requirement of cats for taurine.

Taurine in Health and Disease, Edited by R. Huxtable
and D.V. Michalk, Plenum Press, New York, 1994

TAURINE BALANCE

As taurine is not degraded by mammalian enzymes, taurine balance of cats at equilibrium might be expected to be zero; that is, the taurine in feces and urine should be equal to taurine ingested and synthesized. When taurine balances measurements are taken on cats given diets containing fixed amounts of taurine for long periods, less than half the dietary taurine is recovered in the feces and urine. Examples of such balance studies are shown in Tables 1a and b. The proportion of ingested taurine recovered in the urine and feces varied from about half (0.55) for a purified diet based on casein, to only 0.07 to 0.08 for the commercial diets. The partition of the excreted taurine between urine and feces indicated that diets with the highest recoveries had the greatest urinary taurine excretions, whereas those with the lowest recoveries had the lowest urine taurine excretion.

Table 1a. Taurine balance of cats given two commercial canned diets.

	Food A	Food B
	Taurine (mg/day)	
Intake	85.8 ± 8.7	56.9 ± 2.4
Excretion		
Urine	30.9 ± 8.0	9.8 ± 4.0
Feces	11.1 ± 7.8	7.3 ± 1.7
Recovery (feces + urine)	40.2 ± 8.0	17.2 ± 3.0
Proportion recovered	0.47	0.30

From reference (3). Data are means ± SEM.

Table 1b. Taurine balance of four groups of cats given either one of two purified diets or a commercial diet before and after thermal processing

	Casein purified	Soy purified	Cooked commercial	Frozen commercial
	Taurine (mg/day)			
Intake	60.7 ± 12.0	52.0 ± 13.0	101 ± 8.4	104 ± 14.3
Excretion				
Urine	33.1 ± 9.1	3.2 ± 1.0	1.8 ± 0.4	4.0 ± 0.9
Feces	1.5 ± 0.9	2.7 ± 1.0	6.2 ± 2.1	2.7 ± 0.9
Recovery (feces + urine)	34.7 ± 9.5	5.9 ± 1.7	8.0 ± 2.3	6.7 ± 0.9
Proportion recovered	0.55 ± 0.08	0.13 ± 0.04	0.08 ± 0.02	0.07 ± 0.00

All diets contained 1.5 g taurine/kg dry matter. Data are means ± SEM.

In contrast to animals, some microbes are able to degrade taurine. Kondo and associates (23) reported that extracts from a bacterium able to grow on taurine as the sole source of carbon, nitrogen and sulfur converted taurine to sulfoacetaldehyde ($HO_3S\text{-}CH_2\text{-}CHO$). Two organisms with taurine aminotransferases have been studied. A taurine:α-ketoglutarate aminotransferase (EC 2.6.1.55) has been isolated and characterized from *Achromobacter superficialis* (40). This enzyme catalyzes the transamination of taurine with α-ketoglutarate to yield sulfoacetaldehyde and L-glutamate. A strain of *Pseudomonas aeruginosa* TAU-5 has a taurine:pyruvate aminotransferase that catalyses the production of alanine and sulfoacetaldehyde. The latter then undergoes a lyase reaction to produce sulfite and acetate (33). The sulfite is presumably oxidized and excreted as sulfate (38). Sulfoacetaldehyde may also be metabolized to isethionate (7).

When a pulse dose of ^{14}C labeled taurine was given to cats and the expired CO_2 collected and analyzed, it was found that 100 times more label was recovered in the CO_2 from cats given a thermally processed canned diet (that caused taurine depletion) than from cats given the same diet in the unprocessed state (16). These observation indicated that processing a diet had a marked effect on the extent of taurine degradation in the gut. It was not possible from these observations to quantify the taurine degraded. $^{14}CO_2$ could have come from CO_2 produced by the microbes directly or from oxidation of products of taurine degradation such as acetate which enters the body pool and may only be partially oxidized in the period of observation.

Similar overall recoveries of taurine from purified and cooked and frozen diets are presented in Table 1b. However, these diets supported very different blood concentrations of taurine. The whole blood taurine concentration of cats given the four diets are presented in Table 2. While three of the diets (two purified and frozen commercial) produced only a slight fall in taurine concentration in the blood, the cooked diet resulted in marked depletion in whole blood taurine concentration. These results indicate that the cooked diet, relative to uncooked or purified diets, was associated with a greater degree of degradation of taurine in the gut.

Table 2. Concentration of taurine (μmol/liter) in the whole blood from groups of cats given purified diets containing different protein sources or a commercial diet before and after thermal processing

Sampling time	Casein purified	Soy purified	Cooked commercial	Frozen commercial
Initial	554 ± 47	521 ± 23	448 ± 50	446 ± 67
After 42 days	518 ± 62	493 ± 1	186 ± 34	387 ± 38
Difference	28 ± 36	28 ± 31	261 ± 28*	58 ± 44

All diets contained 1.5 g taurine/kg dry matter. Data are means ± SEM. *Significantly different from other 3 groups ($p < 0.05$).

Binding of bile salts to indigestible dietary constituents was considered as a possible mechanism whereby some diets deplete cats of taurine. Anion exchange resins such as Cholestyramine® or Colestipol® bind bile acids (24) in the intestinal lumen and have been used as hypocholestrolemic agents in humans (13). When these resins are included in the diet of cats, rapid depletion of taurine occurs, as the resins bind taurocholate in the lumen of the intestine, prevent it participating in the enterohepatic recycling, and resulting in its excretion in the feces.

The effect of the inclusion of 20 g of Cholestyramine®/kg dry matter in a purified diet containing 1500 mg taurine/kg dry matter is shown in Figure 1. Both plasma and whole blood concentrations of taurine declined rapidly in the group receiving the resin. Supplementation with additional taurine to bring the total concentration to 3500 mg taurine/kg diet overcame the effect of the resin on whole blood and plasma taurine concentrations. Bile salt turn over time in humans is shortened when Cholestyramine® is included in the diet.

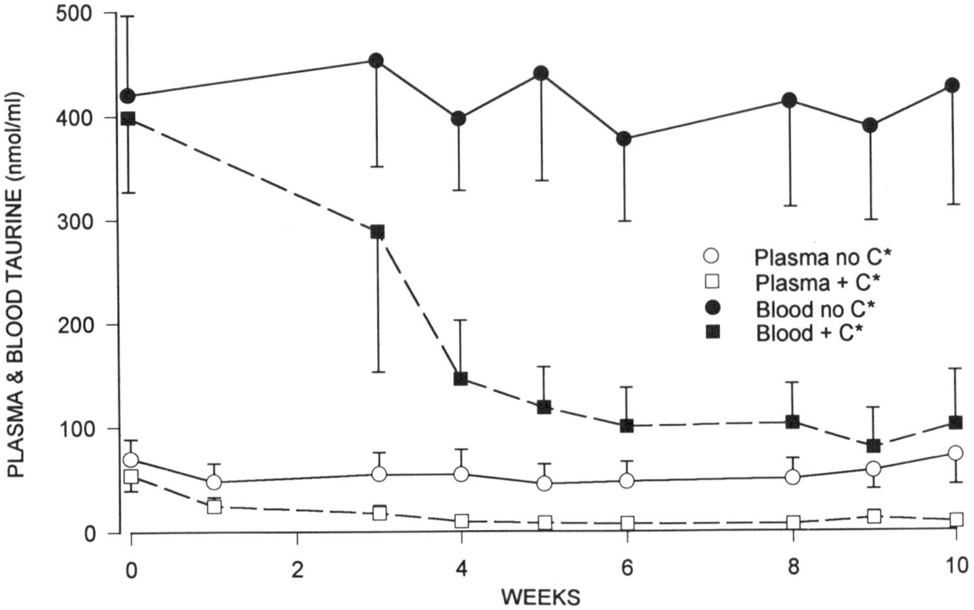

Figure 1 Changes in plasma and whole blood concentrations of taurine in cats given a purified diet containing 1500 mg of taurine/kg dry matter with and without 20 g Cholestyramine®/kg diet. C* in legend = Cholestyramine®. (Morris, M.L., Rogers, Q.R and Morris, J.G. unpublished data).

EFFECT OF LEVEL AND SOURCE OF PROTEIN IN THE DIET ON TAURINE STATUS OF CATS

When cats are given expanded (dry) diets and canned diets containing similar concentrations of taurine on a dry matter basis, those cats receiving the expanded diets consistently maintain higher concentrations of taurine in plasma and whole blood than cats given the canned diets. As canned diets generally contain higher levels of protein and fat than expanded diets, and the protein is of different origin, the effect of level and source of protein in the diet on the taurine status of cats was examined. We had previously reported that the level and source of fat in the diet had only a minor effect on the concentration of taurine in plasma (4). Two experiments were conducted using purified diets containing 1000 mg taurine/kg diet dry matter and different sources and levels of protein. In the first experiment, the proteins examined were isolated soy protein versus casein. Two groups each of 4 mature cats were given purified diets containing 100 mg taurine and either 250 or 500 g casein/kg of dry matter. It can be seen from Table 3 that after 6 weeks there were no significant differences in plasma concentration of taurine from initial values or between the two groups, although the group of cats given the 500 g casein/kg diet had a slightly lower plasma concentration. The protein source of the diet was then changed from casein to isolated soy protein maintaining the same levels of protein in the diet. There was a marked decline in the concentration of taurine in the plasma of the cats given the 500 g soy protein/kg diet whereas the concentration in the plasma of the group given the 250 g soy protein /kg diet declined only slightly.

In a similar experiment, five groups each of 6 mature cats (3 males and 3 females) were given purified diets containing either a low level of protein (300 g casein/kg diet dry

Table 3. Plasma concentrations of taurine (μmol/liter) in cats given diets containing casein for 0 to 6 weeks then isolated soy protein from 6 to 12 weeks

Protein in diet (g/kg)	Initial	6 weeks	12 weeks
	<-----Casein------>	<---Isolated soy protein--->	
	Plasma taurine (μmol/liter)		
250	91.4 ± 15.2	84.1 ± 10.3	71.0 ± 11.8
500	101.5 ± 14.3	73.7 ± 12.3	22.3 ± 3.8*

Data are means ± SEM. *Significantly different from 12 week value for 250 g/kg diet.

matter with either 0, 20 or 40 g of soybean cotyledon fiber) or a high level of protein (600 g of casein or isolated soybean protein/kg diet dry matter) for 14 weeks. Food intake was controlled to 26 g dry matter/kg body weight/day to equalize taurine intakes. Cats given the low casein diet had the highest concentration of taurine in plasma (86.8 ± 7.1 μmol/liter), which was significantly greater than cats receiving the high casein diet (63.9 ± 7.2 μmol/liter) which in turn was significantly greater than cats given the high soy group (34.2 ± 2.6 μmol/liter). Whole blood concentrations of taurine were not significantly different.

The addition of soy cotyledon fiber (Fibrim®), which had been shown to have a hypocholestrolemic effect and to increase fecal bulk in humans (28,34), had no significant effect on plasma taurine concentration. These levels of fiber would include those present in normal cat foods, other than low energy cat foods. This finding is in agreement with another report where soluble fiber was not found to have an effect on plasma taurine concentration of cats (1).

When the results were analyzed for the effect of sex on plasma taurine no sex effect was present at zero time. However, by 14 weeks there was a significant ($P < 0.05$) sex effect, males having a higher plasma concentration of taurine than females. These results indicate the need for care in the design of experiments involving cats of different sexes.

The results from the above experiments demonstrated a marked effect of the source and level of protein in the diet on the taurine status of cats. Three hypotheses were advanced to explain the protein effect. Certain proteins (a) modify the microflora in the gut resulting in greater degradation of taurine, or (b) bind taurine or taurocholic acid and render it unavailable for absorption, or (c) cause greater production of bile acids through stimulation of cholecystokinin release. In order to test the first hypothesis the following experiment was undertaken. Two groups of mature cats were given a heat-processed canned diet containing 1500 mg taurine/kg diet dry matter. This diet was known to cause depletion of taurine in cats. The food intake of both groups was restricted to 18 g dry matter/kg body weight /day to equalize the intake of taurine. One group of cats had 50 mg of tetracycline and 25 mg of procaine penicillin G added daily to the diet, the other group received no antibiotics. After 5 weeks the treatments were crossed over. Plasma and whole blood concentrations of taurine at the initial, 5 week (cross-over) and final 10 week samplings are shown in Table 4. In the first phase of the crossover, plasma taurine concentration declined more rapidly in the cats not receiving antibiotics than those receiving antibiotics. Within one week after the cross-over, the antibiotic treatment resulted in an increase in plasma taurine concentration which continued until 10 weeks. In contrast, the taurine concentration in the plasma of cats that had been receiving the antibiotic treatment declined more rapidly after the antibiotics were withdrawn.

The concentration of taurine in whole blood followed a similar response to plasma showing that there was a depletion of taurine in the formed elements of blood as well as in plasma. As whole blood and muscle taurine are correlated (6, 26) this decrease in whole blood taurine would indicate depletion of taurine in muscle, the largest pool of body

Table 4. Concentrations of taurine in plasma and whole blood of cats and urinary taurine excretion in a cross-over experiment with and without antibiotics added to the diet

Initial	Treatment	Cross-over (5 weeks)	Treatment	End (10 weeks)
		Plasma taurine (μmol/liter)		
116 ± 6	No antibiotics	26 ± 6	Antibiotics	67 ± 10
109 ± 6	Antibiotics	77 ± 7*	No antibiotics	16 ± 2*
		Whole blood taurine (μmol/liter)		
546 ± 8	No antibiotics	292 ± 30	Antibiotics	368 ± 21
559 ± 11	Antibiotics	431 ± 20*	No antibiotics	226 ± 11*
		Urinary taurine (μmol/day)		
365 ± 34	No antibiotics	54 ± 8	Antibiotics	95 ± 10
396 ± 23	Antibiotics	135± 19*	No antibiotics	35 ± 5*
		Fecal taurine (μmol/day)		
92 ± 10	No antibiotics	175 ± 15	Antibiotics	93 ± 6
88 ± 8	Antibiotics	53 ± 4*	No antibiotics	205 ± 8*

Data are means ± SEM. *Significantly different from equivalent value for no antibiotic group.

taurine. Urinary excretion of taurine by cats is an index of taurine adequacy (31). In both treatments, urinary excretion declined in the first 5 weeks, but the rate of decline was less in the group of cats receiving antibiotics. After the cross-over, the urinary excretion of cats receiving antibiotics increased, whereas the group now without antibiotics continued to decline. In contrast to urinary taurine excretion, the fecal excretion of taurine was greatest in cats not receiving antibiotics. Fecal excretion of bile acids by cats was also decreased by the antibiotic treatment (Table 5). The reason for the reduction in fecal taurine and bile acid excretion is not known. Possibly the lower activity of cholyltaurine hydrolase (Table 5) resulted in less taurocholate being cleaved and more absorbed.

This experiment highlights the importance of the microflora in determining the taurine status of cats. When microbial activity was inhibited by antibiotics there was a decreased cholyltaurine hydrolase activity in the feces, and presumably in the intestine, which decreased the rate of taurocholic acid hydrolysis and taurine degradation, and allowed more taurine to be recycled by the enterohepatic circulation. Greater recovery of taurine was reflected in the higher plasma and whole blood concentrations of taurine and in the greater urinary excretion of taurine in the antibiotic-treated cats.

Measurements of the half life of taurocholate and taurine in anaerobic cultures from the colon of cats indicates that hydrolysis proceeds at a faster rate than taurine degradation. The $t_{1/2}$ for taurocholate was 2.7 h compared to 15.5 h for taurine. Antibiotics have a profound effect on the activity of the cholyltaurine hydrolase which in turn probably reduces the quantity of taurine degraded in the intestine. Batt and co-workers (21) have reported that clinically healthy cats fed a commercial diet have high bacterial counts in the proximal small intestine at endoscopy. The numbers of bacteria in these cats fulfilled the established criteria for small intestinal bacterial overgrowth in man and dogs.

In another experiment 12 adult cats were given a purified diet containing 435 g casein and 800 mg taurine/kg. The cats were paired into two groups, one group of cats was given the diet ad libitum, while the other group had the food intake controlled to half that of the ad libitum pair mate. Food intake had no effect on plasma or whole blood taurine status of the cats. The concentration of taurine in the diet was then reduced to 500

Table 5. The effect of antibiotics in the diet on fecal excretion of bile acids and cholyltaurine hydrolase activity

Initial	Treatment	Cross-over (5 weeks)	Treatment	End (10 weeks)
		Fecal bile acids (μmol/day)		
122 ± 15	No antibiotics	235 ± 18	Antibiotics	184 ± 5
131 ± 9	Antibiotics	106 ± 11*	No antibiotics	358 ± 17*
		Cholyltaurine hydrolase activity[a]		
220 ± 38	No antibiotics	279 ± 54	Antibiotics	94 ± 19
191 ± 41	Antibiotics	42 ± 10*	No antibiotics	331 ± 29*

[a] nmol of cholic acid released/min/g dry feces. * Significantly different ($p < 0.05$) from equivalent value for no antibiotics group.

mg/kg diet which reduced plasma taurine concentration but there was no difference between treatments. These results support the thesis that the amount of taurine degraded in the gut is proportional to intake of food.

PROTEIN SOURCE AND EFFECT OF PROCESSING

The above studies indicated that the difference between diets in their ability to maintain the taurine status of cats was related to the source of protein in the diet and the type of processing it had undergone. Four groups of mature cats were given different purified nutritionally complete diets each containing 400 g casein and 800 mg taurine/kg. The diets were prepared by four processing methods:

(1) Unprocessed diet (control) containing starch as the carbohydrate source.
(2) Same diet as (1) + increased vitamins, moistened and autoclaved for 30 minutes.
(3) Same diet as (1) + increased vitamins except that 50 g glucose was substituted for starch in each kg of diet, moistened and autoclaved for 30 minutes.
(4) Casein component of the diet was separately autoclaved with glucose and the resulting brown material containing Maillard product was ground and added to the diet as the source of protein.

Each cat was individually housed and given 18 g dry diet/kg body weight to maintain a constant intake of taurine. Food refusals were negligible. After 5 weeks, antibiotics (50 mg of tetracycline and 25 mg of procaine penicillin G) were added daily to the diets for a further 5 weeks.

The effect of the dietary treatments on plasma, whole blood, fecal and urinary taurine is shown in Table 6. The results for weeks 0 to 5 show that processing the protein in a purified diet has a profound effect on the plasma and whole blood taurine concentrations of cats. The concentrations of taurine increased in the groups of cats given the unprocessed (control) and the autoclaved complete diet in which the carbohydrate was starch. This indicates that heat processing a purified diet without glucose (reducing sugar) has only a minor effect on taurine status. In contrast, the concentrations of taurine declined in plasma and in whole blood of cats given the diets autoclaved with glucose. The effect of autoclaving with glucose was independent of whether the complete diet was autoclaved, or only the casein and glucose were autoclaved separately to produce Maillard products. The difference in the response of plasma and whole blood taurine concentration of cats given the autoclaved starch and the autoclaved glucose-containing diets is presumed to be a consequence of the Maillard products formed in both the glucose-containing diets. Reducing sugars in the presence of heat and moisture readily form Maillard products with proteins, especially those high in lysine such as casein. Maillard products are known to be of lower digestibility than untreated protein.

Table 6. Concentrations of taurine in plasma and whole blood from cats given purified diets containing 400 g casein and 800 mg taurine per kg

Method of processing	Initial <--no antibiotics-->	Week 5 <--antibiotics--> added	Week 10
	Plasma taurine concentration (μmol/liter)		
Control	78 ± 13	90 ± 6^{a}	107 ± 5^{a}
Starch autoclaved	80 ± 11	70 ± 11^{a}	89 ± 4^{b}
Glucose autoclaved	79 ± 6	36 ± 2^{b}	84 ± 4^{b}
Maillard product	78 ± 8	46 ± 5^{b}	$94 \pm 4^{a,b}$
	Whole blood taurine concentration (μmol/liter)		
Control	509 ± 24	541 ± 25^{a}	579 ± 14
Starch autoclaved	543 ± 34	524 ± 35^{a}	580 ± 9
Glucose autoclaved	535 ± 41	357 ± 11^{b}	556 ± 15
Maillard product	507 ± 12	368 ± 7^{b}	577 ± 13

Means in the same column with different superscripts are significantly different ($P<0.05$). Each diet was processed by one of four methods. From week 0 to week 5 the diet did not contain antibiotics, whereas from week 5 to week 10 antibiotics were included.

The addition of antibiotics to these diets (weeks 5 to 10) restored the plasma and whole blood concentrations of taurine in the cats receiving the two diets containing casein autoclaved with glucose, indicating that the taurine-depleting mode of action of these casein-glucose diets had a large microbial component. Antibiotics also slightly increased plasma, but not whole blood, taurine of the control and autoclaved diet without added glucose.

When the diets did not contain antibiotics (weeks 0 to 5), fecal excretion of taurine by cats given the two glucose-containing diets was over twice that of cats given the control and autoclaved starch diet. The effect of the treatments on urinary taurine excretion was a mirror image of the fecal losses. As less taurine was lost in the feces from cats fed the control and starch autoclaved diets than the glucose-containing autoclaved diets, more was available for urinary excretion in these cats. When antibiotics were added to the diets, fecal excretion of taurine was greatly reduced to a level that was similar for cats in all treatments. Similarly, urinary output of taurine was enhanced by antibiotics, presumably reflecting the decreased degradation of taurine in the gut.

The cholyltaurine hydrolase activity of the feces of the cats (Table 7) show that activity was higher in the feces from cats receiving the glucose-containing diets than the control and starch autoclaved diet. The Maillard products supported either a greater microbial mass or changed the population to one with a greater cholyltaurine hydrolase activity. Antibiotics added to the diets caused a marked reduction in cholyltaurine hydrolase activity of the feces such that the activity was similar across all diets. The red-

Table 7. The effect of dietary processing on the cholyltaurine hydrolase activity in the feces and fecal bile acid output of cats given four casein containing diets with and without antibiotics

Method of processing	No antibiotics	Antibiotics
	Cholytaurine hydrolase (nmol cholate/min/g dry feces)	
Control	260 ± 34[a]	108 ± 33
Starch autoclaved	308 ± 23[a]	142 ± 37
Glucose autoclaved	520 ± 41[b]	125 ± 22
Maillard	568 ± 36[b]	149 ± 44
	Fecal bile acid excretion (μmol/day)	
Control	46 ± 4[a]	20 ± 4
Starc autoclaved	88 ± 2[b]	32 ± 7
Glucose autoclaved	120 ± 4[c]	44 ± 5
Maillard	124 ± 8[c]	38 ± 7

Means with different superscripts are significantly different ($P < 0.05$).

duction in activity was greatest for cats with the highest initial activity; that is, cats receiving the glucose-containing diets.

Fecal bile acid excretion of the cats not receiving antibiotics virtually doubled for those given the starch autoclaved versus control diet, and again increased when Maillard products were produced by autoclaving the complete diet or only the protein portion with glucose. It is not possible to distinguish which of the rate variables (bile secretion, absorption, or degradation) changed, but the Maillard products probably promoted a greater rate of bile secretion through greater stimulation of cholecystokinin release. Intact proteins are required for stimulation of cholecystokinin release in rats (25), but release in dogs is stimulated by a simple solution of two amino acids (phenylalanine and tryptophan) and fat. In rats, cholecystokinin release is inhibited by both the presence of pancreatic enzymes (12, 27) and presence of bile acids (11) in the intestine. As Maillard products are poorly digested, they would probably bind proteolytic enzymes and free cholecystokinin-releasing peptide (monitor peptide) (9) found in pancreatic juice to bind to sites in the duodenum and jejunum (41) and stimulate cholecystokinin release by interacting directly with the cholecystokinin-containing cells (26). If cholecystokinin had a major role, then maintenance of a greater difference between the control and heat processed diets would be expected when antibiotics were given.

The decreased total bile acid output in feces of cats given antibiotics is in agreement with the observed lower output in germ-free versus conventional rats (14). The half life of bile acids reported for germ-free rats is considerably longer (11 days) than for conventional rats (2 days) fed the same diet (14). Diet has a marked effect on bile acid excretion in germ-free as well a conventional rats. Fecal output of bile acids is increased two to three fold when the diet is changed from a purified to a rat chow diet (15).

Most values for urinary taurine excretion in the literature are for free taurine excretion. However, taurine in urine, as well as feces, is present both as free and as bound taurine. Total taurine (free + bound) may be measured after the sample has been digested for 12 h in 6N HCl. We have found variable proportions of free to bound taurine in the urine of cats given both commercial and purified diets. Across diets, the proportion of bound taurine ranged from 34 to 95% (5,16,17,18,19). The proportion bound did not appear to be related to total taurine excretion as suggested by Glass et al. (10). These authors reported that, for cats given purified diets based on casein, increments of crystalline taurine added to the diet increased free taurine excretion while bound taurine remained relatively constant.

We have not investigated the nature of the taurine conjugates in cat urine or feces. Taurine is known to be a constituent of a number of small peptides such as gamma-glutamyl-taurine which acts as a hormone in the parathyroid gland (8). Quinaldylglycyltaurine, a metabolite of quinaldic acid and kynurenic acid has been identified in cat urine (22). When physiological doses of retinoic acid were administered to rats, 60% of the dose appeared in the bile as retinotaurine and 10% and 2% of the dose was excreted in urine and feces, respectively (35). It appears that retinotaurine is an excretory form of retinoic acid, as it has no biological activity in rats (36). However, if all the retinol ingested by cats was excreted as retinotaurine, it would not make a significant contribution to taurine loss.

Many animals use taurine to conjugate xenobiotics before excretion. As cats have a reduced ability to undertake the glucuronide conjugation of small phenols and aromatic acids, conjugation with taurine may be more extensively used by cats than other species to eliminate xenobiotics. Cats given phenylacetic acid excrete it in urine as a taurine conjugate (20) and taurine is used by rats, mice and ferrets to conjugate 2-naphthylacetic acid (2) before urinary excretion. Taurine is a major conjugator for phenylacetic acid in ferrets and birds, but strangely glycine, and not taurine, was reported to be its main conjugator found in cat urine.

CONCLUSIONS

Over half the dietary taurine requirements of adult cats is required to replace the taurine degraded by the intestinal microbes. The quantity of taurine degraded in the gut is primarily dependent on the source and amount of protein in the diet and the method by which the protein is processed. In purified diets, casein maintains higher concentrations of taurine in plasma than isonitrogenous levels of isolated soy protein. Addition of oral antibiotics to a taurine-depleting diet decreased taurine loss in feces, increased plasma and whole blood concentrations of taurine, and increased urinary taurine excretion compared to the diet without antibiotics. Current knowledge suggests that thermal processing in the presence of reducing sugars produces Maillard products which increases microbial degradation of taurine in the intestine. Maillard products may either provide an environment that favors higher numbers of taurine-degrading bacteria, or increase taurine exposure to bacteria. Maillard products may bind proteolytic enzymes in the intestine which permits monitor peptide to stimulate cholecystokinin-containing cells to release cholecystokinin, which in turn increases bile entry into the small intestine, and exposure of taurine to microbial degradation.

REFERENCES

1. Anantharaman, G., Vuichoud, J., Gicquello, P., and Rabot, R., 1992, The effect of dietary gelifying agents on bile acid excretion and taurine status in cats, *Veterinary International* 1:45-46.
2. Caldwell J., 1981, The current status of attempts to predict species differences in drug metabolism, *Drug Metabolism Reviews* 12:221-237.
3. Cooke, J.A., Rogers, Q.R., and Morris, J.G., 1988, Urinary and fecal excretion of taurine by cats fed commercial canned diets, *FASEB J.* 2: A1617 (abs).
4. O'Donnell J.A. Rogers, Q.R., and Morris, J.G., 1981, Effect of diet on plasma taurine in the cat, *J. Nutr.* 111:1111-1116.
5. Earl, K.E., and Smith, P.M., 1991, The effect of dietary taurine content on the plasma taurine concentration of the cat, *Brit. J. Nutr.* 66:227-235.

6. Edgar, S.E., Hickman, M.A., Marsden, M.M., Morris, J.G., and Rogers, Q.R., 1993, Dietary cysteic acid serves as a precursor of taurine in cats, *J. Nutr.*, (in press).
7. Fellman, J.H., Roth, E.S., and Fujita, T.S., 1978, Taurine is not metabolized to isethionate in mammalian tissue, *in:* "Taurine and Neurological Disorders", Barbeau, A., and Huxtable, R.J., eds., Raven Press, New York, pp. 19-24.
8. Feuer, L., Torok, L.J., Kapa, E., and Csaba, G., 1978, The effect of gamma-glutamyl-taurine (litoralon) on the amphibian metamorphosis, *Comp. Biochem. Physiol.* 61C:67-71.
9. Fushiki, T., Kajiura, H., Fukuoka, S-I., Kido, K., Semba, T., and Iwai, K., 1989, Evidence for an intraluminal monitor in rat pancreatic enzyme secretion: reconstitution of the pancreatic response with dietary protein, trypsin and the monitor peptide, *J. Nutr.* 119:622-627.
10. Glass, E.N., Odle, J., and Baker, D.H., 1992, Urinary taurine excretion as a function of taurine intake in adult cats, *J. Nutr.* 122: 1135-1142.
11. Gomez, G., Upp, J.R., Lluis, F., Alexander, R.W., Poston, G.J., Greeley, G.H., and Thompson, J.C., 1988, Regulation of the release of cholecystokinin by bile salts in dogs and humans, *Gastroenterology* 94:1036-1046.
12. Green, G.M., and Lyman, R.L., 1972, Feedback regulation of pancreatic enzyme secretion as a mechanism for trypsin inhibitor-induced hypersecretion in rats, *Proc. Soc. Exp. Biol. Med.,* 140:6-12.
13. Grundy S.M., 1972, Treatment of hyercholestrolemia by interference with bile acid metabolism, *Arch. Internal Medicine* 130:638-648.
14. Gustafsson, B.E. Bergström, S., Lindstedt, S., and Norman, A., 1957, Turnover and nature of fecal bile acids in germfree and infected rats fed cholic acid-24-^{14}C. Bile acids and steroids 4I, *Proc. Soc. Exp. Biol. Med.* 94:467-471.
15. Gustafsson, B.E., and Norman, A., 1969, Influence of the diet on the turnover of bile acids in germ-free and conventional rats, *Brit. J. Nutr.* 23:429-442.
16. Hickman, M.A., Rogers, Q.R., and Morris, J.G., 1990, Effect of processing on the fate of dietary [^{14}C]taurine in cats, *J. Nutr.* 120:995.
17. Hickman, M.A., Bruss, M.L., Morris, J.G., and Rogers, Q.R., 1992, Dietary protein source (Soybean vs. casein) and taurine status affect kinetics of the enterohepatic circulation of taurocholic acid in cats, *J. Nutr.* 122:1019-1028.
18. Hickman, M.A., Rogers, Q.R., and Morris, J.G., 1992, Taurine balance is different in cats fed purified and commercial diets, *J. Nutr.* 122:553-559.
19. Hickman. M.A., Morris, J.G., and Rogers, Q.R., 1992, Intestinal taurine and the enterohepatic circulation of taurocholic acid in the cat, *in:* "Taurine: Nutritional Value and Mechanisms of Action", Lombardini, J.B., Schaffer, S.W., and Azuma, J., eds., *Plenum Press, New York,* pp. 45-54.
20. James, M.O., Smith, R.L., Williams, R.T., and Reidenberg, M., 1972, The conjugation of phenylacetic acid in man, sub-human primates and some non-primate species, *Proc. Royal Soc. Lond. B* 182:25-35.
21. Johnston, K., Lamport, A., and Batt, R.M., 1993, An unexpected bacterial flora in the proximal small intestine of normal cats, *Vet. Rec.* 132:362-363.
22. Karhara, M., and Price, J.M., 1961, Quinaldylglycyltaurine: a urinary metabolite of quinaldic acid and kynurenic acid in the cat, *J. Biol. Chem.* 236: 508-511.
23. Kondo, H., Anada, A., Oshsawa, K., and Ishimoto, M., 1971, Formation of sulfoacetaldehyde from taurine in bacterial extracts, *J. Biochem.* 69:621-623.
24. Kritchevsky, D., and Story J.A., 1974, Binding of bile salts in vitro by nonnutritive fiber, *J. Nutr.* 104:458-462.
25. Liddle, R.A., Green, G.M., Conrad, C.K., and Williams, J.A., 1986, Proteins but not amino acids, carbohydrates, or fats stimulate cholecystokinin secretion in the rat, *Am. J. Physiol.* 251:G:243-248.
26. Liddle, R.A., Misukonis, M.A., Pacy, L., and Balber, A.E., 1992, Cholecystokinin cells purified by fluorescent-activated cell sorting respond to monitor peptide with an increase in intracellular calcium, *Proc. National Academy of Sciences,* USA 89:5147-5151.
27. Louie, D.S., May, D., Miller, P., and Owyang C., 1986, Cholecystokinin mediates feedback regulation of pancreatic enzyme secretion in rats, *Amer. J. Physiol.* 250:G252-259.
28. Lo, G.S., 1990, Physiological effects and physio-chemical properties of soy cotyledon fiber, *in:* "New Developments in Dietary Fiber", Furda, I., and Brine, C.J., eds., Plenum Press, New York.
29. Morris, J.G. and Rogers, Q.R., 1992, The metabolic basis for the taurine requirement of cats, *in:* "Taurine: Nutritional Value and Mechanisms of Action", Lombardini, J.B., Schaffer, S.W., and Azuma, J., eds., Plenum Press, New York, pp. 33-44.
30. Morris, J.G., Rogers, Q.R., and Pacioretty, L.M., 1990, Taurine: an essential nutrient for cats, *J. Small Anim. Prac.* 31:502-509.

31. Odle, J., Glass, E.N., Czarnecki-Maulden, G.L., and Baker, D.H., 1992, Urinary excretion of taurine as a function of taurine intake: Potential for estimating taurine bioavailability in the adult cat, *in:* "Taurine: Nutritional Value and Mechanisms of Action", Lombardini, J.B., Schaffer, S.W., and Azuma, J., eds,. *Plenum Press, New York, pp.* 55-62.
32. Pion, P.D., Kittleson, M.D., Rogers, Q.R., and Morris, J.G., 1987, Myocardial failure in cats associated with low plasma taurine: A reversible cardiomyopathy, *Science* 237:764-768.
33. Shimamoto, G., and Berk, R.S., 1990, Taurine catabolism. III Evidence for the participation of the glyoxylate cycle, *Biochim. Biophys. Acta* 632:399-407.
34. Shorey, R.L., Day, P.J., Willis, R.A., Lo, G.S., and Steinke, F.H., 1985, Effect of soybean polysaccharide on plasma lipids, *J. Am. Diet. Assoc.* 85:1461.
35. Skare, K.L., Schnoes, H.K., and DeLuca, H.F., 1982, Bilary metabolites of all-trans-retinoic acid in the rat:isolation and identification of a novel polar metabolite, *Biochemistry* 21:3308-3317.
36. Skare, K.L., Sietsema, W.K., and DeLuca, H.F., 1982, The biological activity of retinotaurine, *J. Nutr.* 112:1626-1630.
37. Sturman J.A., 1993, Taurine in development, *Physiol. Rev.* 73:119-147.
38. Sturman, J.A., Hepner, G.W., Hoffman, A.F., and Thomas, P.J., 1975, Metabolism of [^{35}S] taurine in man, *J. Nutr.* 105:1206-1214.
39. Sturman, J.A., Palachal, T., Imaki, H., Moretz, R.C., French, J., and Wisniewski, H.M., 1987, Nutritional taurine deficiency and feline pregnancy outcome, *in:* "The Biology of Taurine Methods and Mechanisms", Huxtable, R.J., Franconi, F., and Giotti, A., eds., Plenum Press, New York, pp.113-134.
40. Toyama, S., Misono, H., and Soda, K., 1978, Properties of taurine:α-ketoglutarate aminotransferase of *Achromobacter superficialis*. Inactivation and reactivation of the enzyme, *Biochem. Biophys. Acta* 523:75.
41. Yamanishi, R., Kotera, J., Fushiki, T., Soneda, T., Saitoh, T., Oomori, T., Satoh, T., and Sugimoto, E., 1993, A specific binding of the cholecystokinin-releasing peptide (monitor peptide) to isolated rat small-intestine cells, *Biochem. J.* 291:57-63.

REACTIVITY OF TAURINE WITH ALDEHYDES AND ITS PHYSIOLOGICAL ROLE

Mizuho Ogasawara, Tomomi Nakamura,
Ikuo Koyama, Masami Nemoto, and
Tsuguchika Yoshida

Research Center
Taisho Pharmaceutical Co., Ltd.
No. 403. Yoshino-cho
1-Chome Ohmiya-shi
Saitama, 330, Japan

SUMMARY

The chemical reactions of the amino group of taurine with aldehydes were investigated. Glucose, acetaldehyde, and malondialdehyde were used as aldehydes. After taurine had reacted, the amounts of remaining taurine and aldehydes were measured, and the reactivity was evaluated. Amino acids such as glycine, α-alanine, and ß-alanine were compared because of their structural resemblance to each other. Taurine showed a high reactivity with each one of the aldehydes tested.

It is known that protein is altered through reactions of the amino group with various aldehydes. Low density lipoprotein (LDL) was used as a model protein and the inhibiting effect of taurine against the modification of LDL by malondialdehyde was examined. It was shown that the inhibiting effects of taurine correlated with the reactivity of malondialdehyde with amino acids.

Further, the taurine-glucose reaction product showed an antioxidative effect on the peroxidation of liposomes made of yolk phosphatidylcholine as a biomembrane model. The results suggest the possibility of an inhibiting effect of taurine against the modification of protein, as well as an antioxidative effect through the reactions of taurine with aldehydes in vivo.

INTRODUCTION

Taurine is a sulfur-containing amino acid that has an amino group at the ß position and a sulfonic group as the acid radical. Various effects of taurine, such as lowering blood sugar levels in an experimental diabetic model, decreasing serum cholesterol, inhibition of peroxide lipid in serum, and antioxidation (13) have been reported. However, the mechanisms of these effects have not been clarified. We have investigated the chemical

Taurine in Health and Disease, Edited by R. Huxtable and D.V. Michalk, Plenum Press, New York, 1994

reaction of the amino group of taurine in vitro and discuss its potential significance. Generally, the reaction between the amino group and the carbonyl group is called the aminocarbonyl reaction, or the Maillard reaction, and much related research has been conducted in the field of food chemistry. In recent years it has been reported that the amino group of a protein reacts with reducing sugar nonenzymatically in vivo (5,12). Further, it has been reported that compounds containing aldehyde groups also react with the amino group of a protein, causing an alteration of the protein (4,11). Moreover, various aldehydes or oxidants are generated in geriatric disorders, such as diabetes or senility, resulting in the alteration of proteins. Therefore, this reaction has received increased attention in both medical and physiological fields in recent years.

First of all, we examined the chemical reactions of taurine and aldehydes. Next, we investigated whether taurine has an inhibiting effect on protein modification as a result of the amino group competing with amino groups of the protein. Further, the antioxidative effect of the reaction product of taurine and glucose was examined.

MATERIALS AND METHODS

Glycine, α-alanine, ß-alanine, glucose, acetaldehyde, and 2-2'-axobis-(2-amidinopropane) dihydrochloride (AAPH) used were purchased from Wako Pure Chemical Industries, Ltd. Taurine was synthesized in our laboratory, and diacetyl phosphate (DCP) was purchased from Aldrich Chemical Co., Ltd. Malonaldehyde bis(dimethylacetal) was obtained from Tokyo Kasei Kogyo Co., Ltd. Low density lipoprotein (LDL) isolated from human blood plasma was obtained from Protogen AG, and Coatsome NC-10 prepared by Nichiyu Liposome Co., Ltd. was used as yolk phosphatidylcholine. All other reagents were of the highest grade available.

Reactivity of Taurine with Glucose

The amino acids were dissolved in 0.1 M phosphate buffer that had been adjusted to pH 7.4 with 1 N NaOH. Glucose was dissolved in the buffer. Each amino acid solution was mixed with glucose solution at a ratio of 1:1 (v/v). The final concentrations of amino acid and glucose were 250 mM and 250 mM, respectively. After reactions for 34 days at 50°C, the amount of remaining glucose was determined by the HPLC method using a Hitachi L-6000 and a Differential Refractometer RI-8 Electronics Unit (Toyo Soda). An Ionpak KS-801 (Shodex) column was used, which was connected to an Ionpak KS-800P (Shodex), a pre-column. The column temperature was 80°C. Deionized water was used as an eluent.

Reactivity of Taurine with Acetaldehyde

The solutions containing amino acids were prepared according to the method mentioned above. To prevent evaporation, acetaldehyde was dissolved in a phosphate buffer (pH 7.4) below 5°C. Each amino acid solution was mixed with acetaldehyde solution at a ratio of 1:1 (v/v). The final concentrations of amino acid and acetaldehyde were 250 mM and 250 mM, respectively. A vial was filled with the solutions. After reacting for 30 min at 37°C, the remaining acetaldehyde that reached a liquid/vapor equilibrium was determined by head space gas chromatography (GC), using a Shimadzu GC-9A. A G-column (G950, GL Science Industry) column was used. The column temperature was 110°C. Hydrogen gas was used as the carrier gas.

Reactivity of Taurine with Malondialdehyde

The solutions containing amino acids were prepared according to the method mentioned above. Malondialdehyde was prepared from malondialdehyde bis(dimethylacetal) according to the method of Kikugawa, et al. (4), and adjusted to pH 7.4 with 1 N NaOH. Each amino acid solution was mixed with malondialdehyde solution at a ratio of 1:1 (v/v). The final concentrations of amino acid and malondialdehyde were 250 mM and 250 mM, respectively. After reacting for 2 h at 37°C, the amount of each remaining amino acid was determined by the HPLC method. The assay procedures for the amino acid determination were as follows: 0.2 M boric acid/potassium chloride/sodium hydroxide buffer (pH 9.0) was added to each solution. Then an acetonitrile solution containing fluorescamine was added, and the mixture was stirred to derive a fluorescent substance. This was determined by HPLC (Hitachi 655A-12). Fluorescence was monitored with a fluorescence spectrometer (Hitachi F-1000) fitted with 390 and 485 nm filters for excitation and emission, respectively. A TSK gel LS410 (Tosoh) column was used. The column temperature was 50°C. Acetonitrile:H_2O:0.05 M tetra-*n*-butylammonium hydroxide solution (pH 7.4) = 36:54:10 (v/v/v) was used as an eluant.

Inhibiting Effect of Taurine on LDL Modification by Malondialdehyde

Each amino acid was dissolved in 0.01 M phosphate buffered saline adjusted to pH 7.4 with 1 N NaOH. Malondialdehyde was prepared from malonaldehyde bis(dimethylacetal) and diluted with phosphate buffered saline. The solutions containing amino acid, malondialdehyde, and LDL were mixed. The final concentrations of amino acids, malondialdehyde, and LDL were 25 mM, 25 mM, and 1 mg/ml, respectively. They were incubated for 3 h at 37°C. Afterwards, LDL was dialyzed for 20 h at 5°C against 0.01 M phosphate buffered saline containing 0.01% EDTA. The amount of malondialdehyde per solution containing LDL was measured using the thiobarbituric acid method. The results showed the amounts of malondialdehyde that modified LDL.

Preparation of Taurine-Glucose Reaction Product

Taurine (25 mmol) and glucose (50 mmol) were dissolved in 0.1 M phosphate buffered saline. After adjusting the solution to pH 7.4 with 1 N NaOH, a total volume of 60 ml was prepared. It was refluxed for 9 h at 100°C. This solution was diluted by 10 times and was used for the experiment on antioxidative effects.

Peroxidative Reaction of Egg Phosphatidylcholine

A thin membrane was prepared by using a chloroform solution containing 103.4 μmol egg-phosphatidylcholine and 10 μmol DCP, and it was hydrated with 10 ml phosphate buffered saline, to prepare the liposome. A taurine-glucose reaction product was added to this liposome suspension at the ratio of 1:1 (v/v). Phosphate buffered saline was added as a control, and 80 mM glucose and 40 mM taurine were added as a reference, instead of a taurine-glucose reaction product. AAPH (50 mM), with a water-soluble azo-compound as an oxygen radical source, was added to the liposome suspensions at a ratio of 1:1 (v/v). After reacting for 3 h at 37°C, the amount of remaining egg-phosphatidylcholine was determined by HPLC (Hitachi 655A-12). A Lichrosorb Si60 (Merck) column was used at 25°C, and the eluent was methanol:40 mM NaH_2PO_4::96:4 (v/v).

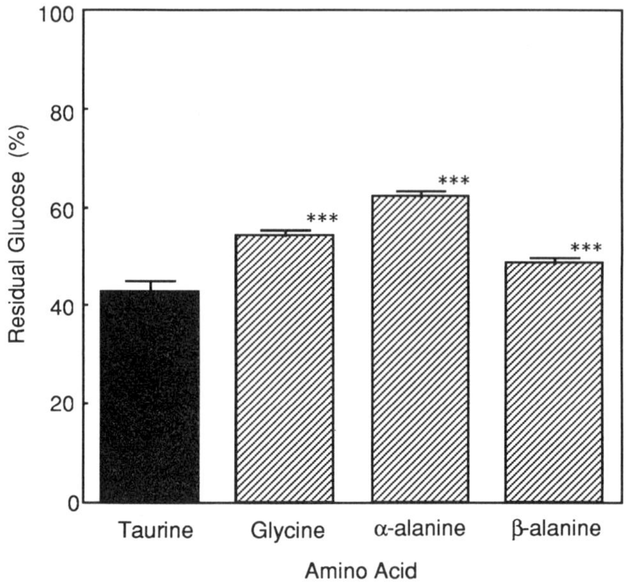

Figure 1. Reaction between amino acids and glucose. Solutions of amino acid (250 mM) and glucose (250 mM) were kept for 34 days at pH 7.4 and 50°C. Residual glucose was determined by HPLC. ***$p < .001$ compared to taurine.

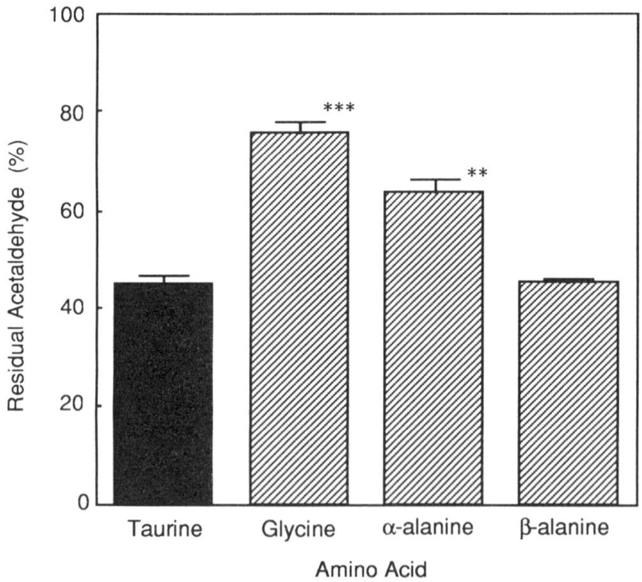

Figure 2. Reaction between amino acids and acetaldehyde. Solutions of amino acid (250 mM) and acetaldehyde (250 mM) were mixed below 5°C at pH 7.4. After 30 min at 37°C, the remaining acetaldehyde was determined by head space gas chromatographic analysis. **$p < .01$; ***$p < .001$ compared to taurine.

RESULTS

Reactivities of Taurine with Aldehydes

The reactivity of taurine and other amino acids with glucose (Fig. 1), acetaldehyde (Fig. 2) and malondialdehyde (Fig. 3) was evaluated. Disappearance of substrate indicated that reaction had occurred in all cases.

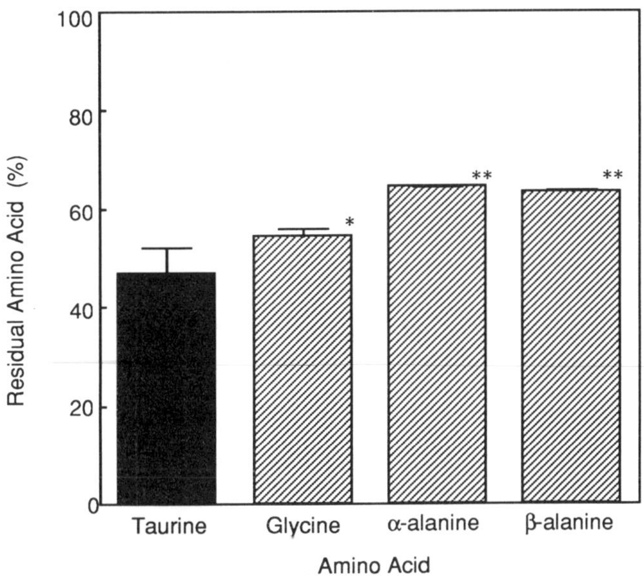

Figure 3. Reactions between amino acids and malondialdehyde. Solutions of amino acid (250 mM) and malondialdehyde (250 mM) were mixed at pH 7.4 for 2 h at 37°C. Remaining amino acid was determined by HPLC. *p < .05; **p < .01 compared to taurine.

Inhibiting Effect of Taurine on LDL Modification by Malondialdehyde

When malondialdehyde is added to an LDL solution and incubated, LDL is modified. We examined the effect of the addition of each amino acid to this solution to investigate whether the modification of LDL was inhibited. The amount of malondialdehyde modified with LDL was 32.2 nmol/ml LDL in the control, without the addition of any amino acids, as shown in Fig. 4. However, when taurine was added to the solution containing LDL, the amount of malondialdehyde reacting was 15.7 nmol/ml LDL, and the modification of LDL was significantly reduced ($P < 0.001$). When the inhibiting effects were compared among the amino acids, taurine was most potent, followed by glycine, ß-alanine, and α-alanine. A tendency for the inhibiting effect to be correlated with the reactivity of amino acids with malondialdehyde was shown.

Antioxidative Effect of Taurine-Glucose Reaction Product

Egg phosphatidylcholine, which was a component of the liposome, was peroxidized by the oxygen radical generated by AAPH, and the percentage of residual egg phosphatidylcholine decreased to about 70% (Fig. 5). At this time, the effect of the addition of the taurine-glucose reaction product on the peroxidation of egg phosphatidylcholine was examined. When taurine and glucose were added, the percentage of residual egg phosphatidylcholine was almost equal to that of the control. On the other hand, when the taurine-glucose reaction product was added, the peroxidation of egg phosphatidylcholine was controlled and the percentage of residual egg-phosphatidylcholine rose to about 90%.

DISCUSSION

Changes in physiological activity due to the alteration of a protein, and its subsequent relationship with various diseases, have received considerable attention in recent years. Rahbar (9) reported that HbAIc, a fraction of hemoglobin, was increased in diabetics. The ß chain N-terminal valine of hemoglobin reacts nonenzymatically with glucose to produce Amadori compounds (12). In diabetes, the glucose concentration in the blood is maintained at a high level. It is also known that glucose reacts with various proteins (glycation), such as albumin, collagen, and myelin in vivo (3) and that this is a cause of complications associated with diabetes. Moreover, acetaldehyde is an intermediate metabolite of ethanol, whose toxicity is problematic. Malondialdehyde is one of the metabolites of lipid

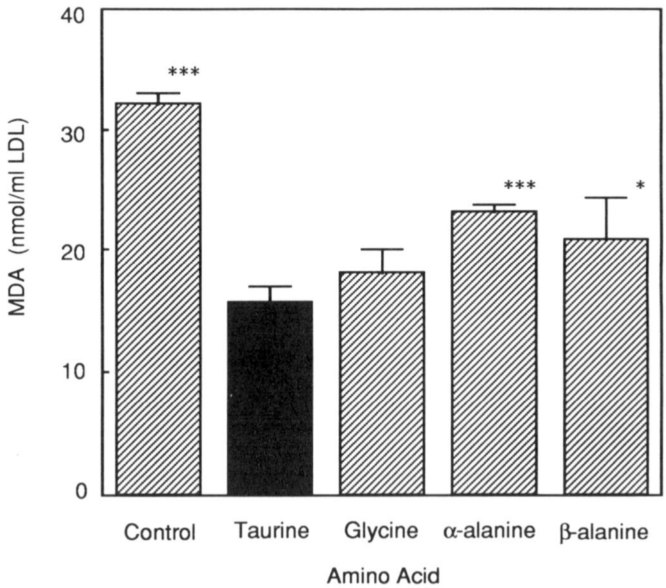

Figure 4. Effect of amino acids (25 mM) on the reaction of malondialdehyde (25 mM) with LDL (1 mg/ml). The mixture was maintained at pH 7.4 for 3 h at 37°C. The LDL was then dialyzed for 20 h at 5°C against 0.01 M phosphate-buffered saline containing 0.1% EDTA. Malondialdehyde bound to LDL was determined by the thiobarbituric acid method. $^{*}p < .05$; $^{***}p < .001$ compared to taurine.

peroxidation. It has been reported that the amount of malondialdehyde in serum increases during renal insufficiency, suggesting a relationship to its pathogenesis.

In our study, we examined the chemical reaction of these aldehydes with taurine. We investigated whether taurine could control protein modification, because the amino group of taurine competed with the amino group of a protein in reactions with aldehydes. Furthermore, the antioxidative effect of the taurine-glucose reaction product was examined.

Taurine showed a high reactivity for glucose, acetaldehyde, and malondialdehyde. As for the lowering effect on blood sugar levels by taurine, in an experimental diabetes model induced by alloxan or dehydroascorbic acid, such an effect might have been partially due to the direct reaction of taurine with glucose. Hobara et al. (*Alcohol Taisya to Kan* 8:101 [1989]) reported that the concentration of acetaldehyde in the blood and liver of a rat after the administration of ethanol to the stomach by gavage (1.5 g/kg weight) significantly decreased within 30 min of the administration of taurine to the stomach by gavage (0.5 g/kg weight). They concluded that this was due to the activation of aldehyde dehydrogenase by taurine. We used a high concentration (250 mM) of taurine and acetaldehyde in our experiment, and incubated them for 30 min at 37°C. A direct reaction between acetaldehyde and taurine was confirmed. The possibility is suggested that taurine scavenges acetaldehyde by a direct reaction in vivo.

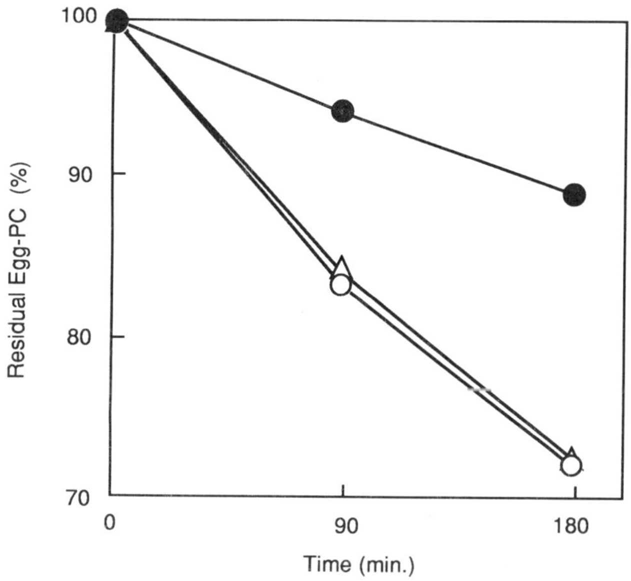

Figure 5. Antioxidative effect of the taurine-glucose reaction product. AAPH was added as an oxygen radical source to a liposome suspension prepared from egg phosphatidylcholine. The mixture was incubated for 3 H at 37°C, and the amount of residual phosphatidylcholine then determined. Open circles: no other addition; Triangles: taurine (40 mM) and glucose (80 mM) added; Closed circles: taurine-glucose reaction product added.

The uptake of LDL by a macrophage is promoted by the alteration of apolipoprotein B. This changes foam cells, causing the deposition of cholesterol and eventually causing arteriosclerosis (2). The inhibiting effect of taurine on LDL modification was evaluated in vitro. Our results demonstrated that the amount of malondialdehyde by which LDL was modified was significantly inhibited by the addition of taurine ($P < 0.001$). Moreover, this inhibiting effect of LDL modification correlated with the reactivity of malondialdehyde and amino acids, In other words, taurine, with its high reactivity, showed a strong inhibiting effect. We speculated that taurine reacted with malondialdehyde, competing with the amino group of the protein of LDL, and that the amounts of malondialdehyde, which modified LDL, were relatively decreased. The serum cholesterol-lowering effect of taurine has already been reported and the inhibiting effect of taurine is considered to be one of the mechanisms explaining the effect. Moreover, it is thought that the protein is modified by acetaldehyde (10). The inhibiting effect of taurine on protein modification due to acetaldehyde may be expected, as well as that due to malondialdehyde.

As for the influence of taurine on lipid peroxidation, it has been reported that a high lipid peroxide value in serum decreased following the administration of taurine (Okamoto et al. *Sulfur Amino Acids* 7:201 [1984]). Taurine is believed to control peroxide formation by stabilizing the cell membrane. We examined the antioxidative effect of the taurine-glucose reaction product. It is generally known that there is an antioxidative effect in the Maillard reaction products of amino acids and a reducing sugar (7). Moreover, glycine-glucose reaction products display the strong scavenging actions of active oxygen species (8). We examined the antioxidative effect of the taurine-glucose reaction product on the peroxidation of egg phosphatidylcholine by active oxygen. An antioxidative effect of the taurine-glucose reaction product was demonstrated after the reaction for 9 h at 100°C. It is though that the reaction temperature influences the generation of Maillard reaction products (1), and that the types of amino acid and the pH of the solution influence their antioxidative effect (6). It is not clear whether the taurine-glucose reaction product exists in vivo. However, glycation in vivo indicates the possibility of the existence of a taurine-glucose reaction product in vivo. Further, Okamoto et al. (8) reported that the protein by which glycation is caused scavenges active oxygen species (8). This suggests the possibility that the taurine-glucose reaction product is produced in vivo and has an antioxidative effect.

From the facts described above, it is suggested that taurine plays various physiological roles by reacting with aldehydes. It is known that taurine is present in high concentrations in such animal tissues as retina and platelets (13). In the future, we plan to investigate the relationship between concentrations of taurine and its reactivity, and to examine the physiological roles of taurine in vivo.

REFERENCES

1. Benzing-Purdie, L.M., Ripmeester, J.A., and Ratcliffe, C.I. 1985, Effects of temperature on Maillard reaction products, *J.Agric.Food Chem.*, 33:31-33.
2. Goldstein, J.L., Ho, Y.K., Basu, S.K., and Brown, M.S. 1979, Binding site on macrophages that mediates uptake and degradation of acetylated low density lipoprotein, producing massive cholesterol deposition, *Proc.Natl.Acad.Sci.USA*, 76:333-337.
3. Harding, J.J. 1985, Nonenzymatic covalent posttranslational modification of proteins *in vivo*, *Adv.Protein Chem.*, 37:247-334.
4. Kikugawa, K., Tsukuda, K., and Kurechi, T. 1980, Studies on peroxidized lipids. I. Interaction of malondialdehyde with secondary-amines and its relevance to nitrosamine formation, *Chem.Pharm.Bull.*, 28:3323-3331.
5. Koenig, R.J., Blobstein, S.H., and Cerami, A. 1977, Structure of carbohydrate of hemoglobin AIc, *J.Biol.Chem.*, 252:2992-2997.
6. Lingnert, H. and Eriksson, C.E. 1980, Antioxidative Maillard reaction products. I. Products from sugars and free amino acids, *J.Food Process.Preserv.*, 4:161-172.
7. Lingnert, H. and Eriksson, C.E. 1981, Antioxidative effect of Maillard reaction products, *Prog.Food Nutr.Sci.*, 5:453-466.
8. Okamoto, G., Hayase, F., and Kato, H. 1992, Scavenging of active oxygen species by glycated proteins, *Biosci.Biotech.Biochem.*, 56:928-931.
9. Rahbar, S. 1968, An abnormal hemoglobin in red cells of diabetics, *Clin.Chim.Acta*, 22:296.
10. Savolainen, M.J., Baraona, E., and Lieber, C.S. 1987, Acetaldehyde binding increases the catabolism of rat serum low-density lipoproteins, *Life Sci.*, 40:841-846.
11. Steinbrecher, U.P., Lougheed, M., Kwan, W.C., and Dirks, M. 1989, Recognition of oxidized low density lipoprotein by the scavenger receptor of macrophages results from derivatization of apolipoprotein B by products of fatty acid peroxidation, *J.Biol.Chem.*, 264:15216-15223.
12. Stevens, V.J., Baraona, E., and Lieber, C.S. 1977, Nonenzymatic glycosylation of hemoglobin, *J.Biol.Chem.*, 252:2998-3002.
13. Wright, C.E., Tallan, H.H., Lin, Y.Y., and Gaull, G.E. 1986, Taurine: biological update, *Ann.Rev.Biochem.*, 55:427-453.

HEPATIC REGULATION OF CYSTEINE UTILIZATION FOR TAURINE SYNTHESIS

M. H. Stipanuk, P. J. Bagley, Y.-C. Hou,
D. L. Bella, M. F. Banks and
L. L. Hirschberger

Division of Nutritional Sciences
Cornell University
Ithaca, New York 14830

INTRODUCTION

Like most other proteinogenic amino acids, cysteine is required both for synthesis of proteins and for synthesis of several other essential organic compounds, which in the case of cysteine include glutathione, taurine and coenzyme A. In addition, cysteine is important as a precursor of inorganic sulfur, both as sulfide and as sulfate. When the sulfur is cleaved from the carbon chain, the carbons enter central pathways of energy metabolism or gluconeogenesis as pyruvate and the nitrogen must be excreted.

Taurine synthesis occurs essentially entirely via the formation and decarboxylation of cysteine sulfinate (Fig. 1), and the liver appears to be the organ in which the largest amount of taurine is synthesized (3,15). The regulation of the conversion of cysteine is taurine is unusual in that the activities of the two enzymes that determine its rate of synthesis in liver are affected in opposite directions by excess dietary protein. Cysteine dioxygenase activity is elevated by an increase in dietary protein (5), whereas cysteine sulfinate decarboxylase activity is decreased by an increase in dietary protein (6).

Intuitively, it is not obvious whether cysteine catabolism should be regulated by the needs for excretion or by the needs for production of essential catabolites. Both taurine and sulfate are important metabolites required by the body for essential functions as well as being products of cysteine catabolism that are excreted in the urine. Both of these compounds can be obtained directly from the diet or can be formed by catabolism of sulfur-containing amino acids. Also, formation of taurine does not contribute to the acid load associated with high intakes of sulfur amino acids or protein, which is due to the formation of SO_4^{2-} and H^+. On the other hand, formation and excretion of sulfur as taurine also involves the excretion of nitrogen and carbon, negating the need for urea synthesis but representing a loss of energy and of a gluconeogenic substrate from the body. Thus, it is reasonable to expect that the cysteine sulfinate-dependent catabolism of cysteine would be highly regulated. The ability to control both total cysteine sulfinate-dependent catabolism and the partitioning of cysteine sulfinate to taurine versus pyruvate and sulfate may be physiologically important.

The purpose of this paper is to summarize several studies conducted in our laboratory in our initial attempts to understand the hepatic regulation of cysteine conversion to taurine. Cysteine catabolism in the isolated rat hepatocyte model used in these studies occurs predominantly via cysteine sulfinate-dependent pathways (2,15).

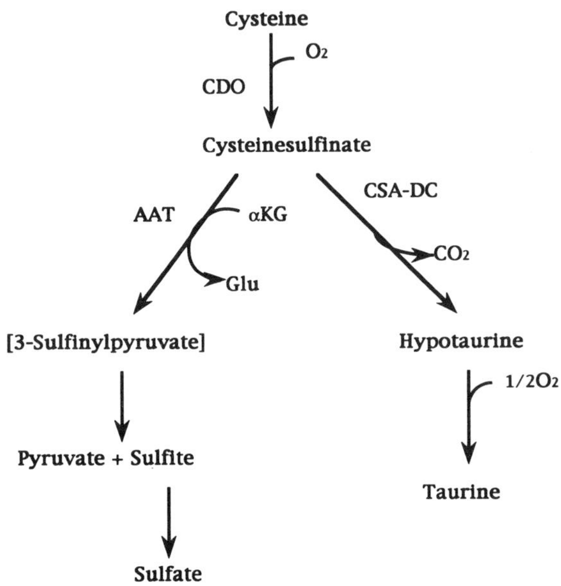

Figure 1. Cysteine sulfinate-dependent metabolism of cysteine in rat liver. AAT: Aspartate (cysteine sulfinate) aminotransferase; CDO: Cysteine dioxygenase; CSA-DC: Cysteine sulfinate decarboxylase; α-KG: α-Ketoglutarate.

METHODS

Studies were conducted using freshly isolated hepatocytes from male adult Sprague-Dawley rats that weighed between 200 and 400 g (Blue Spruce Farms, Altamont, NY). Hepatocytes were isolated as described previously by our laboratory (16). All rats were housed in individual stainless steel mesh cages in a room maintained at 20° C and 60-70% humidity with light from 2000 to 0800 h. Rats had free access to diet and water. In three of the four studies that will be discussed, rats were fed purified diets that contained various levels of casein or sulfur-containing amino acids for two to three weeks prior to the isolation of hepatocytes. The composition of the diets were modifications of the AIN-76A formulation (1). The diets contained in g per kg: 10 to 60 g vitamin-free casein as indicated in the figure or table legends, 0 to 10 g L-methionine or DL-methionine as indicated in the figure or table legends, 50 g cellulose, 50 g corn oil, 10 g AIN-76A vitamin mix, 2 g choline bitartrate, 35 g AIN-76 mineral mix, and equal amounts of sucrose and cornstarch to make 1000 g. In one study, the diet was also modified to contain excess chloride or excess sodium as fixed ions. In the fourth study, hepatocytes were obtained from rats that had been fed a semi-purified diet (Prolab RMH 1000, Agway, Syracuse, NY); the isolated

Table 1. Comparison of [^{35}S]cysteine metabolism in rat hepatocytes incubated in different substrate media

Incubation mixture	[^{35}S]Sulfate	[^{35}S]Glutathione	[^{35}S]Taurine + [^{35}S]hypotaurine	[^{35}S]Metabolite Total
	pmol · min^{-1} · mg cells (wet wt)$^{-1}$			
0.2 mM L-[^{35}S]cysteine plus:				
No Addition	15.0 ± 3.4	4.7 ± 1.2	4.0 ± 1.1	23.7 ± 4.1
Amino Acid Mixture	12.9 ± 2.4	4.9 ± 1.3	2.0 ± 0.9	19.8 ± 3.6
Glucose/Lactate/Pyruvate	15.6 ± 2.7	5.3 ± 1.1	4.4 ± 1.4	25.3 ± 2.5
Amino Acid Mixture + Glucose/Lactate/Pyruvate	16.5 ± 4.6	4.3 ± 1.1	2.8 ± 1.2	23.6 ± 5.0
Methionine	13.1 ± 3.3	5.1 ± 1.4	4.0 ± 1.7	22.2 ± 4.4
Amino Acid Mixture + Methionine	10.6 ± 4.7	5.6 ± 1.5	2.3 ± 0.9	18.5 ± 2.2
1.0 mM L-[^{35}S]cysteine plus:				
No Addition	37.1 ± 7.0	10.6 ± 2.7	11.5 ± 2.8	59 ± 5
Amino Acid Mixture	42.3 ± 8.0	8.4 ± 4.3	14.1 ± 7.8	65 ± 11
Glucose/Lactate/Pyruvate	41.6 ± 10	7.8 ± 1.6	20.5 ± 3.5	70 ± 12
Amino Acid Mixture + Glucose/Lactate/Pyruvate	35.1 ± 9.7	9.0 ± 1.7	12.6 ± 5.6	55 ± 13

Incubations of hepatocytes were carried out in Krebs-Henseleit bicarbonate buffer (pH 7.4) with either 0.2 or 1.0 mM L-[^{35}S]cysteine and various additions of amino acids or energy substrates as indicated. The amino acid mixture contained in μM: 204 Ala, 194 Arg, 64 Asn, 32 Asp, 123 Glu, 643 Gln, 295 Gly, 72 His, 99 Ile, 153 Leu, 477 Lys, 72 Phe, 226 Pro, 61 Trp, 75 Tyr, 184 Val, 303 taurine, 88 citrulline and 72 ornithine. The glucose/lactate/pyruvate mixture contained 10 mM glucose, 2 mM lactate and 0.2 mM pyruvate. L-methionine was added at a concentration of 100 μM.

Values are the means ± SEM (n=5). Values for incubations, with either concentration of cysteine were not significantly different for various additions ($P>0.05$) by analysis of variance and Tukey's ω-procedure.

hepatocytes were preincubated for 30 min with Krebs-Henseleit bicarbonate buffer (pH 7.4) alone; Krebs-Henseleit bicarbonate buffer plus 1 mM L-cysteine; Krebs-Henseleit bicarbonate buffer plus 0.7 mM bromoheptane, which was dissolved in dimethylsulfoxide (DMSO; final concentration 5 ml/liter); or Krebs-Henseleit bicarbonate buffer plus 5 ml/liter DMSO. After the pretreatment, hepatocytes were washed twice with Krebs-Henseleit bicarbonate buffer to remove excess cysteine or bromoheptane.

The utilization of ^{35}S-labeled cysteine for [^{35}S]taurine (plus [^{35}S]hypotaurine) and [^{35}S]sulfate formation was determined by incubating intact hepatocytes with 0.2 mM L-[^{35}S]cysteine plus 0.05 mM bathocuproine disulfonate (a copper chelator) in Krebs-Henseleit bicarbonate buffer (pH 7.4) in an atmosphere of 95% O_2/5% CO_2 for 10 to 30 min as described previously (16). Incubations terminated at zero time served as blanks. Cell viability was confirmed by an initial ATP level greater than 2 μmol/g and by maintenance of an ATP level greater than 2 μmol/g throughout the incubation period (9,16). Incubations were stopped and ^{35}S-labeled metabolites were measured by HPLC methods previously described by our laboratory (16). Activities (V_{max}) of cysteine catabolic enzymes were determined for hepatocytes from each rat or from each pretreatment group using modifications of assays previously described by our laboratory (2). Protein was determined by the method of Lowry et al. (10). Data were analyzed by analysis of variance and Tukey's ω-procedure (14).

Preliminary to these studies, we examined the effect of the presence of amino acids and glucose/pyruvate/lactate in the incubation mixture on the catabolism of cysteine by isolated rat hepatocytes. Hepatocytes (approximately 65 mg wet weight of cells per 2.5 ml incubation volume) from rats fed the semi-purified diet (Prolab RMH 1000) were incubated with 0.2 mM or 1.0 mM L-[^{35}S]cysteine in Krebs-Henseleit bicarbonate buffer (pH 7.4) for 0, 10 or 20 min. Incubation mixtures were supplemented with a mixture of amino acids to resemble rat plasma amino acid concentrations (in μM: 204 Ala, 194 Arg, 64 Asn, 32 Asp, 123 Glu, 643 Gln, 295 Gly, 72 His, 99 Ile, 153 Leu, 477 Lys, 72 Phe, 226 Pro, 61 Trp, 75 Tyr, 184 Val, 303 taurine, 88 citrulline and 72 ornithine); with 10 mM glucose plus 2 mM lactate plus 0.2 mM pyruvate; with both the mixture of amino acids and glucose/lactate/pyruvate; with 0.1 mM L-methionine; or with the amino acid mixture plus 100 μM L-methionine. The production of [^{35}S]sulfate, [^{35}S]taurine plus [^{35}S]hypotaurine, and [^{35}S]glutathione (total GSH + GSSG) were determined using previously described HPLC methods (16). Data were analyzed by one-way analysis of variance and Tukey's ω-procedure (14).

RESULTS AND DISCUSSION

The results of the preliminary study to determine whether supplementation of the standard incubation mixture with amino acids or glucose/lactate/pyruvate would affect the metabolism of cysteine by rat hepatocytes are summarized in Table 1. None of the additions tested affected the rate of production of any of the ^{35}S-labeled metabolites.

In the first study, rats were fed diets that contained either 10, 30 or 60% casein; the casein was supplemented with DL-methionine in a constant ratio of 15 g DL-methionine per kg of casein. As shown in Table 2, hepatic cysteine dioxygenase activity increased markedly as the dietary protein level was increased; the value for the 60% casein group was 7-fold that for the 10% casein group. In contrast to the increase in cysteine dioxygenase activity, cysteine sulfinate decarboxylase activity decreased markedly with increased dietary protein; the value for the 60% casein group was only 7% that for the 10% casein group. As shown in Fig. 2, the rate of taurine production by hepatocytes incubated with ^{35}S-cysteine approximately doubled between 10 and 30% casein, but did not increase further with a further increase in casein level to 60%. The rate of sulfate production from [^{35}S]cysteine increased progressively with the increase in dietary casein, by 3-fold for 30% casein and by 6.6-fold for 60% casein, corresponding to cysteine dioxygenase activity levels that increased similarly by 3.4- and 6-fold, respectively. Total catabolism of cysteine also closely paralled cysteine dioxygenase activity in the hepatocytes with sulfate production accounting for more than 75% of the total catabolism.

Taurine production as a percentage of total catabolism actually decreased as the dietary protein level increased; taurine production as a percentage of total catabolism was 22% for the 10% casein group, 14% for the 30% casein group and 6% for the 60% casein group. This pattern for partitioning of cysteine between taurine and sulfate appeared to result from the combined effects of increased conversion of cysteine to cysteine sulfinate due to increased cysteine dioxygenase activity plus decreased conversion of cysteine sulfinate to

Table 2. Activities of cysteine dioxygenase and cysteine sulfinate decarboxylase in isolated hepatocytes from rats fed diets containing 10%, 30% or 60% casein

Enzyme	Diet (% Casein) 10	30	60
		nmol product · min^{-1} mg protein $^{-1}$	
Cysteine dioxygenase	0.07 ± 0.04[a]	0.30 ± 0.07[b]	0.47 ± 0.03[b]
Cysteine sulfinate decarboxylase	10 ± 6[a]	2.7 ± 1[b]	1.4 ± 0.5[b]

Casein diets were supplemented with DL-methionine at a constant ratio of 15 g DL-methionine per kg casein. Values represent means ± SEM for hepatocyte preparations from 4 or 5 rats. [a] Significantly different ($P<0.05$) from activity found for 10% casein diet (by analysis of variance and Tukey's ω-procedure).

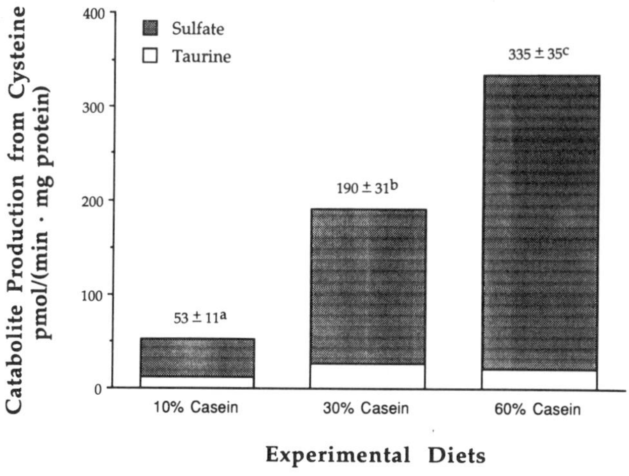

Figure 2. Production of [^{35}S]taurine (plus [^{35}S]hypotaurine) and [^{35}S]sulfate by hepatocytes isolated from rats fed diets that contained 10% casein + 0.15% DL-methionine, 30% casein + 0.45% DL-methionine, or 60% casein + 0.9% DL-methionine. Values are means for sulfate or taurine production by hepatocytes isolated from 4 or 5 rats. Total catabolite production ($\bar{X}$ ± SEM) is printed at the top of each bar; total catabolite production values not followed by the same superscript are significantly different ($P<0.05$) by analysis of variance and Tukey's ω-procedure.

taurine due to decreased activity. It is unlikely that the decreased partitioning of cysteine sulfinate to taurine versus sulfate was due to the increased intracellular cysteine sulfinate concentrations generated by the increased rate of flux of cysteine to cysteine sulfinate because incubation of hepatocytes with increasing concentrations of cysteine sulfinate over the range of tracer only (approximately 0.01 mM) to 0.2 mM resulted in a progressive increase in partitioning to taurine rather than a decrease as observed here (unreported results, Bagley and Stipanuk). Also, intracellular cysteine sulfinate concentrations are much below the K_m of cysteine sulfinate decarboxylase for cysteine sulfinate (4,11) such that one would expect flux to be almost proportionate to concentration. It is possible that an increase in aspartate (cysteine sulfinate) aminotransferase activity may have also occurred and contributed to the increased conversion of cysteine sulfinate to sulfate. Aspartate aminotransferase activity is present in hepatocytes at much higher levels than is cysteine sulfinate decarboxylase, but the K_m of aspartate aminotransferase for cysteine sulfinate is 14 mM (17) compared to the much lower K_m of cysteine sulfinate decarboxylase for cysteine sulfinate, which is 0.05-0.17 mM (11).

In the second study, rats were fed diets with 10% casein alone or supplemented with 0.3% L-methionine or 1.0% L-methionine. In hepatocytes from rats fed the diets supplemented with low or high methionine, cysteine dioxygenase activity increased to 2.8-times and 13-times the basal level, as shown in Table 3. In contrast to the first study, cysteine sulfinate decarboxylase activity was not significantly affected by the addition of methionine to the basal diet. Lower activities may have been observed if higher levels of supplementation had been used as methionine has been shown to affect hepatic cysteine

Table 3. Activities of cysteine dioxygenase and cysteine sulfinate decarboxylase in hepatocytes isolated from rats fed 10% casein diets with or without supplemental L-methionine

	Diet		
Enzyme	10% casein	10% casein + 0.3% L-methionine	10% casein + 1.0% L-methionine
	nmol product · min^{-1} · mg $protein^{-1}$		
Cysteine dioxygenase	0.16 ± 0.6	0.46 ± 0.11[b]	2.0 ± 0.4[b]
Cysteine sulfinate decarboxylase	14.1 ± 0.4	12.1 ± 1.2	11.2 ± 0.3[a]

Values represent means ± SEM for enzyme activities in hepatocyte preparations from 6 or 7 rats.
[a] Significantly different ($P<0.05$) from other dietary treatments (by analysis of variance and Tukey's ω-procedure).

sulfinate decarboxylase activity (7). Nevertheless, in the first study, hepatic cysteine sulfinate decarboxylase activity decreased significantly with a level of protein (30% casein + 0.45% DL-methionine) that provided a level of sulfur amino acids comparable to that in the 10% casein diet supplemented with 1.0% L-methionine (approximately 13 g sulfur amino acids per kg diet). Thus, cysteine sulfinate decarboxylase activity appeared to respond differently to dietary protein than to sulfur-containing amino acids alone.

As shown in Fig. 3, taurine production increased with addition of methionine to the diet. In rats fed the diets supplemented with methionine, the taurine production increased to 4 and 12-times the basal rate of taurine production for diets with 0.3 and 1.0%

L-methionine, respectively. These increases paralleled the increases in cysteine dioxygenase activity to 3 and 13-times the basal levels. Sulfate production also increased in rats fed diets supplemented with methionine, but the fold-increases were much smaller than for taurine. In rats fed diets supplemented with methionine, sulfate production increased to 1.6- and 2.0-times the basal level.

Taurine production as a percentage of total catabolism increased from 22% in hepatocytes from rats fed the basal diet to 42% in those from rats fed the diet supplemented with 0.3% methionine and to 62% in those from rats fed the diet supplemented with 1.0% methionine. This probably represented an increased proportion of total catabolism occurring by cysteine sulfinate-dependent pathways due to the elevation in cysteine dioxygenase

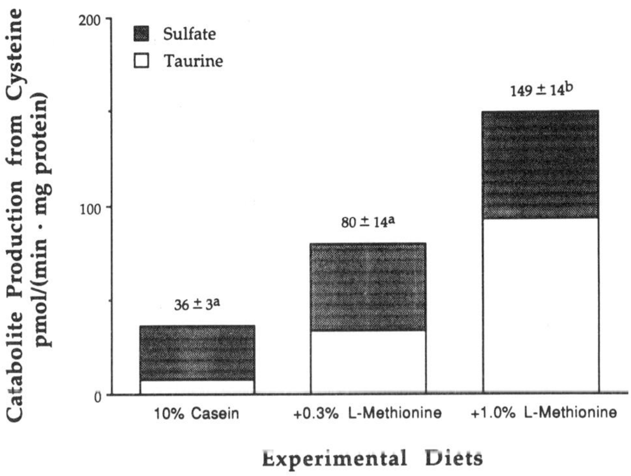

Figure 3. Production of [^{35}S]taurine (plus [^{35}S]hypotaurine) and [^{35}S]sulfate by hepatocytes isolated from rats fed diets that contained 10% casein, 10% casein + 0.3% L-methionine, or 10% casein + 1.0% L-methionine. Values are means for sulfate or taurine production by hepatocytes isolated from 6 or 7 rats. Total catabolite production ($\bar{X} \pm$ SEM) is printed at the top of each bar; total catabolite production values not followed by the same superscript are significantly different ($P<0.05$) by analysis of variance and Tukey's ω-procedure.

activity and/or an increased conversion of cysteine sulfinate to taurine versus sulfate due to a higher intracellular cysteine sulfinate concentration that resulted from increased conversion of cysteine to cysteine sulfinate. The increased cysteine dioxygenase activity plus the sustained high levels of cysteine sulfinate decarboxylase facilitated increased formation of cysteine sulfinate and conversion of that cysteine sulfinate to taurine and allowed taurine to become the major end-product of cysteine catabolism.

Clearly, under conditions where cysteine dioxygenase activity was increased without a reciprocal decrease in cysteine sulfinate decarboxylase activity, taurine production was favored. This was the case in hepatocytes isolated from rats fed a 10% casein diet supplemented with 1.0% methionine. In these hepatocytes, 62% of cysteine catabolism led to taurine production. It is possible that it would be advantageous to the animal to excrete excess nitrogen and sulfur as taurine under conditions of sulfur amino acid excess as this would not generate H^+ or urea. On the other hand, under conditions in which protein is being used for energy and/or for gluconeogenesis (conditions in which we would expect a mixture of amino acids to be available), conversion of sulfur amino acids to sulfate may be favored so that the carbon chain can be utilized for energy or gluconeogenesis.

To evaluate whether acid-base balance plays a role in regulation of the conversion of cysteine to taurine versus sulfate, we fed rats either a 15% or 54% casein diet with no supplemental methionine. In addition, we added excess chloride (by replacing calcium carbonate with calcium chloride; 118 meq Cl^-/kg diet) to the 15% casein diet; the amount of excess chloride added was approximately equal to the predicted equivalents of excess sulfate anion generation in rats fed the 60% casein diet. The intent was to produce acidosis due to an excess of fixed anions rather than to an excess of dietary protein. We also, added excess sodium (118 meq Na^+/kg diet as sodium bicarbonate) to the high protein diet in order to compensate for the sulfur amino acid-generated acid load in the high protein diet.

We saw an effect of protein level on cysteine dioxygenase and cysteine sulfinate decarboxylase activities as had been observed previously, but the addition of excess fixed anion or excess fixed cation had no effect (data not shown). Similarly, the production of both taurine and sulfate was altered by protein level but not by acid-base balance changes in rats fed excess chloride or sodium (data not shown).

The results of this study suggested that increased cysteine dioxygenase activity and increased taurine production in rat liver is a response to increased levels of sulfur amino acids and not a response to metabolic acidosis in general. This is in contrast to changes in glutamine metabolism and ammonium excretion, which respond similarly to acidosis caused by high protein diets or excess fixed anions (13). However, blood pH parameters were not markedly altered in this particular study, and we are presently conducting a study with more extreme treatments that may enable us to make a clearer conclusion.

In the three studies discussed above, the increased production of sulfate and/or taurine in rats fed high levels of methionine or protein was consistently accompanied by a large decrease in incorporation of cysteine into glutathione and a smaller decrease in incorporation of cysteine into protein (data not shown). Total cellular utilization remained relatively constant. Because of this observation that increased catabolism of cysteine to taurine and sulfate is generally accompanied by a reciprocal decrease in incorporation of cysteine into glutathione, the influence of cellular glutathione status on the partitioning of cysteine to taurine was also assessed. It is possible that inhibition of glutathione synthesis could lead to increased catabolic flux in the cell and could have been a contributing factor in the studies in which diets were supplemented with protein or sulfur-containing amino acids. Nonallosteric inhibition of γ-glutamylcysteine synthetase by glutathione has been demonstrated; glutathione was a competitive inhibitor with respect to glutamate with a K_i of 2.3 mM (12).

Pretreatment of hepatocytes to modify intracellular glutathione levels was effective. Control hepatocytes contained 3.0-4.8 μmol glutathione per g and this was not affected by DMSO, the vehicle used to dissolve the bromoheptane. Preincubation of hepatocytes with 1 mM L-cysteine slightly elevated the glutathione level to 5.2 $\pm$ 0.1 μmol per g (mean $\pm$ S.E.M., n=4). Preincubation of cells with 0.7 mM bromoheptane (8) depleted cellular glutathione levels to 0.4 $\pm$ 0.1 μmol per g.

Production of ^{35}S-labeled metabolites from [^{35}S]cysteine by these pretreated hepatocytes is shown in Fig. 4. Glutathione production increased significantly in cells that had been treated with bromoheptane; the rate of [^{35}S]glutathione production in glutathione-depleted hepatocytes was almost double that for the glutathione-repleted hepatocytes (+ cysteine pretreatment group) and about 50% greater than that for the control groups. Taurine production decreased significantly in cells that had been treated with bromoheptane to 27% of the rate for control hepatocytes and to 22% of the rate for glutathione-repleted hepatocytes. Sulfate production also decreased significantly in cells treated with bromoheptane. The activities of cysteine dioxygenase and cysteine sulfinate decarboxylase were not affected by the pretreatment of hepatocytes with cysteine, bromoheptane or DMSO (data not shown).

Thus, the data obtained in this experiment were consistent with the hypothesis that feedback inhibition of glutathione synthesis by glutathione in control and cysteine-pretreated cells may decrease utilization of cysteine for glutathione synthesis and favor its catabolism to taurine and sulfate (especially to taurine in this case because of relatively high cysteine dioxygenase and cysteine sulfinate decarboxylase activities in hepatocytes isolated from rats fed standard 24% protein diets). Additional experiments need to be done to test this hypothesis, especially in view of the substantial decrease in total cysteine metabolism observed in the bromoheptane-treated cells.

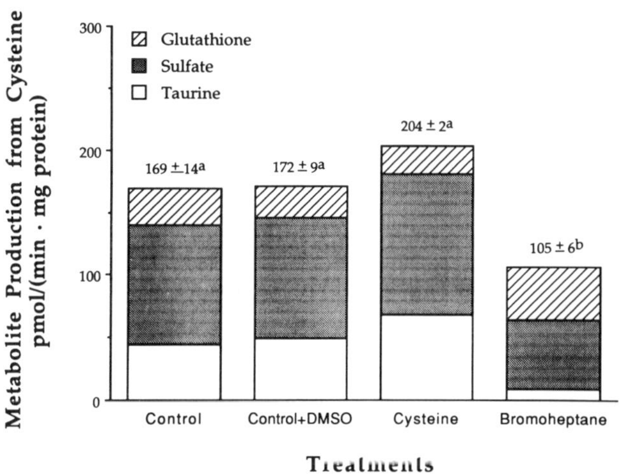

Figure 4. Production of $[^{35}S]$taurine (plus $[^{35}S]$hypotaurine), $[^{35}S]$sulfate, and $[^{35}S]$glutathione (GSH + GSSG) by hepatocytes pretreated to modify cellular glutathione concentrations. Values are means for metabolite production by hepatocytes isolated from 4 rats. Total metabolite production ($\bar{X} \pm$ SEM) is printed at the top of each bar; total metabolite production values not followed by the same superscript are significantly different ($P<0.05$) by analysis of variance and Tukey's ω-procedure. Intracellular glutathione concentration ($\bar{X} \pm$ SEM) at the beginning of the incubation period was 3.6 ± 0.2 μmol/g for control (no pretreatment), 4.1 ± 0.3 μmol/g for the DMSO-control, 5.2 ± 0.1 μmol/g for the hepatocytes repleted by preincubation with 1.0 mM cysteine, and 0.4 ± 0.1 μmol/g for the hepatocytes depleted by preincubation with 0.7 mM bromoheptane.

Our observations seem to indicate consistently that the availability of sulfur amino acids or changes in intracellular concentrations of methionine or cysteine or one of their metabolites is the major signal for changes in cysteine metabolism to taurine. Increased availability of cysteine results in proportionately greater hepatic conversion of cysteine to taurine; this appears to be due to an increase in cysteine dioxygenase activity and the increased intracellular cysteine sulfinate concentrations that are generated by this increased flux. Conversion of cysteine to taurine in rat liver is clearly a process associated with the availability of excess cysteine and appears to provide a convenient means for excretion of both nitrogen and sulfur.

Glutathione and protein synthesis appear to take priority over catabolism of cysteine to taurine and sulfate; the mechanisms involved in this distribution are not yet clear. Preliminary data suggest that regulation of glutathione synthesis could influence the availability of cysteine to catabolic pathways. Our results also suggest that the animal may have regulatory mechanisms that restrict the conversion of cysteine to taurine under conditions in which amino acids are a major source of energy or are required for gluconeogenesis. Decreased activity of hepatic cysteine sulfinate decarboxylase may play an important role in this process.

Much remains to be learned about the regulation of cysteine metabolism in the liver and in other tissues. Although the liver is generally considered to be the organ in which sulfur amino acid catabolism occurs and is regulated, it is important to consider that a large portion of the sulfur-containing amino acids taken up by the liver are incorporated into glutathione which is exported into the plasma (3). This plasma glutathione provides substrate for γ-glutamyltranspeptidase and dipeptidase located on cell membranes (especially in the kidney) which release cysteine back into the circulation. This cysteine has by-passed catabolism by the liver and presumably has a greater chance of being catabolized extra-hepatically.

ACKNOWLEDGMENTS

This research was supported by NRICGP/USDA grant # 92-01219 and by Hatch Project NY(C)-399472.

REFERENCES

1. American Institute of Nutrition, 1977, Report of the American Institute of Nutrition ad hoc committee on standards for nutritional studies, *J. Nutr.* 107:1340-1348.
2. Bagley, P.J., 1993, "Regulation of cysteine catabolism by cysteine dioxygenase and cysteine sulfinate decarboxylase activities in isolated rat hepatocytes", Ph.D. Thesis, Cornell University, Ithaca, New York.
3. Garcia, R.A.G., and Stipanuk, M.H., 1992, The splanchnic organs, liver and kidney have unique roles in the metabolism of sulfur amino acids and their metabolites in rats, *J. Nutr.* 122:1693-1701.
4. Guion-Rain, M.-C., Portemer, C., and Chatagner, F., 1975, Rat liver cysteine sulfinate decarboxylase: purification, new appraisal of the molecular weight and determination of catalytic properties, *Biochim. Biophys. Acta* 384:265-276.
5. Hosokawa, Y., Niizeki, S., Tojo, H., Sato, I., and Yamaguchi, K., 1988, Hepatic cysteine dioxygenase activity and sulfur amino acid metabolism in rats: possible indicators in the evaluation of protein quality, *J. Nutr.* 118:456-461.
6. Jerkins, A.A., Bobnoff, L.E., and Steele, R.D., 1989, Hepatic cysteine sulfinic acid decarboxylase activity in rats fed various levels of dietary casein, *J. Nutr.* 119:1593-1597.
7. Jerkins, A.A., and Steele, R.D., 1991, Dietary sulfur amino acid modulation of cysteine sulfinic acid decarboxylase, *Am. J. Physiol.* 261:E551-E555.
8. Khan, S., and O'Brien, P.J., 1991, 1-Bromoalkanes as new potent nontoxic glutathione depletors in isolated rat hepatocytes, *Biochem. Biophys. Res. Comm.* 179:436-441.
9. Lamprecht, W., and Trautschold, I., 1974, ATP determination with hexokinase and glucose-6-phosphate dehydrogenase, *in:* "Methods of Enzymatic Analysis," H. U. Bergmeyer, ed., Academic Press, New York, vol. 4, pp. 2101-2110.
10. Lowry, O.H., Rosebrough, N.J., Farr, A.L., and Randall, R.J., 1951, Protein measurement with the Folin phenol reagent, *J. Biol. Chem.* 193:265-275.
11. Oertel, W.H., Schmechel, D.E., Weise, V.K., Ransom, D.H., Tappaz, M.L., Krutzsch, H.C., and Kopin, I.J., 1981, Comparison of cysteine sulphinic acid decarboxylase isoenzymes and glutamic acid decarboxylase in rat liver and brain, *Neuroscience* 6:2701-2714.
12. Richman, P.G., and Meister, A., 1975, Regulation of γ-glutamyl-cysteine synthetase by nonallosteric feedback inhibition by glutathione, *J. Biol. Chem.* 250:1422-1426.
13. Schrock, H., and Goldstein, L., 1981, Interorgan relationships for glutamine metabolism in normal and acidotic rats, *Am. J. Physiol.* 240:E519-E525.
14. Steel, R.G.D., and Torrie, J.H., 1960, "Principles and Procedures of Statistics", McGraw-Hill, New York, pp. 99-160.

15. Stipanuk, M.H., Bagley, P.J., Coloso, R.M., and Banks, M.F., 1992, Metabolism of cysteine to taurine by rat hepatocytes, *in:* "Taurine - Nutritional Value and Mechanisms of Action," Lombardini, J.B., Schaffer, S. W., and Azuma, J., eds., Plenum Press, New York, pp. 413-421.
16. Stipanuk, M.H., Coloso, R.M., Garcia, R.A.G., and Banks, M.F., 1992, Cysteine concentration regulates cysteine metabolism to glutathione, sulfate and taurine in rat hepatocytes, *J. Nutr.* 122:420-427.
17. Weinstein, C.L., and Griffith, O.W., 1988, Cysteinesulfonate and β-sulfopyruvate metabolism: partitioning between decarboxylation, transamination and reduction pathways, *J. Biol. Chem.* 263:3735-3743.

IMMUNOHISTOCHEMICAL LOCALIZATION OF TAURINE-CONJUGATED BILE ACIDS IN THE LIVER OF MOUSE, RAT, MONKEY AND HUMAN

W.G. Ding[1], I. Tooyama[2], H. Kimura[2] and K. Kuriyama[3]

[1]Departments of Physiology
[2]Institute of Molecular Neurobiology
Shiga University of Medical Science
Otsu, 520-21
[3]Department of Pharmacology
Kyoto Prefectural University of Medicine
Kyoto, 606 Japan

INTRODUCTION

The liver and brain are organs rich in taurine (14,17). Taurine is present at high concentrations in liver and brain of developing mammals, and, may be important in normal fetal and neonatal development (11,29). In the brain, a role of taurine as an inhibitory neurotransmitter or neuromodulator has been proposed (13,20). A number of immunohistochemical studies have clarified the localization of taurine (3,31) or its synthesizing enzyme, cysteine sulfinic acid decarboxylase (4). However, the localization of taurine in the liver has not been well settled. In our previous immunohistochemical studies using antiserum to taurine, we have found that the antiserum recognizes not only taurine in the brain but also bile acids conjugated with taurine in the liver (15,7,30).

It is well known that bile acids are conjugated with taurine or glycine by two independent enzymatic reactions in hepatocytes (10,23,25). Oral administration of taurine increases the percentage of bile acids conjugated with taurine, although dietary glycine has no effect on the formation of glycine-conjugated bile acids (26). This suggests that the hepatic supply of taurine may be one regulatory factor in the bile acid conjugation pattern. To test this assumption, anatomical details of taurine and its bile conjugates in the liver are required, particularly at microscopic levels of each hepatic lobule.

A role of taurine has also been proposed in the protection of hepatocytes against hepatotoxins such as carbon tetrachloride in the rat (22), and taurine has been suggested as a possible therapeutic agent in the clinical treatment of liver diseases (19).

Taurine in Health and Disease, Edited by R. Huxtable
and D.V. Michalk, Plenum Press, New York, 1994

We will here review our recent study on taurine-conjugated bile acids detected by immunohistochemistry in mouse liver. The results will be compared with our new data for the rat, monkey and human. For further discussion on the functional role(s) of taurine, similar studies in mouse liver during ontogeny and after intoxication with CCl_4 or phenobarbital will be described.

IMMUNOHISTOCHEMICAL PROCEDURES AND ANTIBODY SPECIFICITY

We performed an immunohistochemical study for taurine-like immunoreactivity in liver using the avidin-biotin-peroxidase method (7). Liver specimens were fixed in a mixture of 4% paraformaldehyde and 0.2% picric acid buffered with 0.1 M sodium phosphate, cryoprotected in 15% sucrose in 0.1 M phosphate buffer, and then cut into frozen sections 20 μm-thick in a cryostat. The sections were put through the following steps in a free-floating state: (a) incubate for 2 days with antibody against taurine (diluted 1:20,000-80,000) at 4 °C; (b) incubate for 1 h with biotinylated anti-rabbit IgG (Vector Labs., U.S.A., diluted 1:1000) at room temperature; (c) incubate for 1 h at room temperature with avidin-biotin-peroxidase complex (Vector, diluted 1:4,000); (d) react for 5 min at room temperature with a mixture containing 0.02% 3,3-diaminobenzidine 4HCl, 0.3% nickel ammonium sulfate and 0.0045% H_2O_2 in 0.05 M Tris HCl buffer (pH 7.6).

The immunohistochemical specificity of the taurine antiserum has been described previously (7,15,30). The immuno-spot assay has clearly demonstrated that the taurine antibody reacts not only with taurine itself but also with the three soluble taurine-conjugated bile acids. Immuno-absorption tests have further demonstrated that immuno-positive staining is perfectly abolished when the antiserum has been pre-incubated with taurine (5 mM) or one of hydrophilic taurine-conjugated bile acids (1 mM).

DISTRIBUTION OF LIVER TAURINE-LIKE IMMUNOREACTIVITY IN SOME MAMMALS

Taurine-like immunoreactive granules were deposited in liver cells of mouse, rat, monkey and human. In mouse liver, hepatocytes containing such positive granules were distributed unevenly in each liver lobule (Fig. 1A), being located predominantly in the periphery of the lobule. Hepatocytes around the central vein contained very few immunoreactive granules (Fig. 1A). In thin (4 μm) sections, several clusters of taurine-like immunoreactive granules could be seen to occur in single hepatocytes (Fig. 1B), and these granules often clustered near either the nucleus or cell membrane. In contrast, taurine-like immunoreactivity in rat liver was distributed evenly and diffusely throughout the hepatic lobule; almost all hepatocytes in both central and peripheral regions of lobules contained high taurine-like immunoreactivity. Hepatocytes in the peripheral region tended to contain even more intense taurine-like immunoreactivity (Fig. 1C). In monkey liver, the distribution pattern of taurine-like immunoreactivity was similar to that in rat liver, although the density of positive granules in single hepatocytes was slightly lower (data not shown). In human liver, hepatocytes showing taurine-like immunoreactivity were mainly distributed in the periphery of hepatic lobules (Fig. 2), with a distribution pattern similar to that in mature mice. However, the density of taurine-like immunoreactivity in each cell was generally much lower compared to the other species examined above. These results suggest that the distribution pattern of immunoreactivity differs among mammalian species, which may be supported by many biochemical studies. For instance, the intracellular concentration of taurine in liver is shown to vary considerably among different species (1,8,16). Moreover,

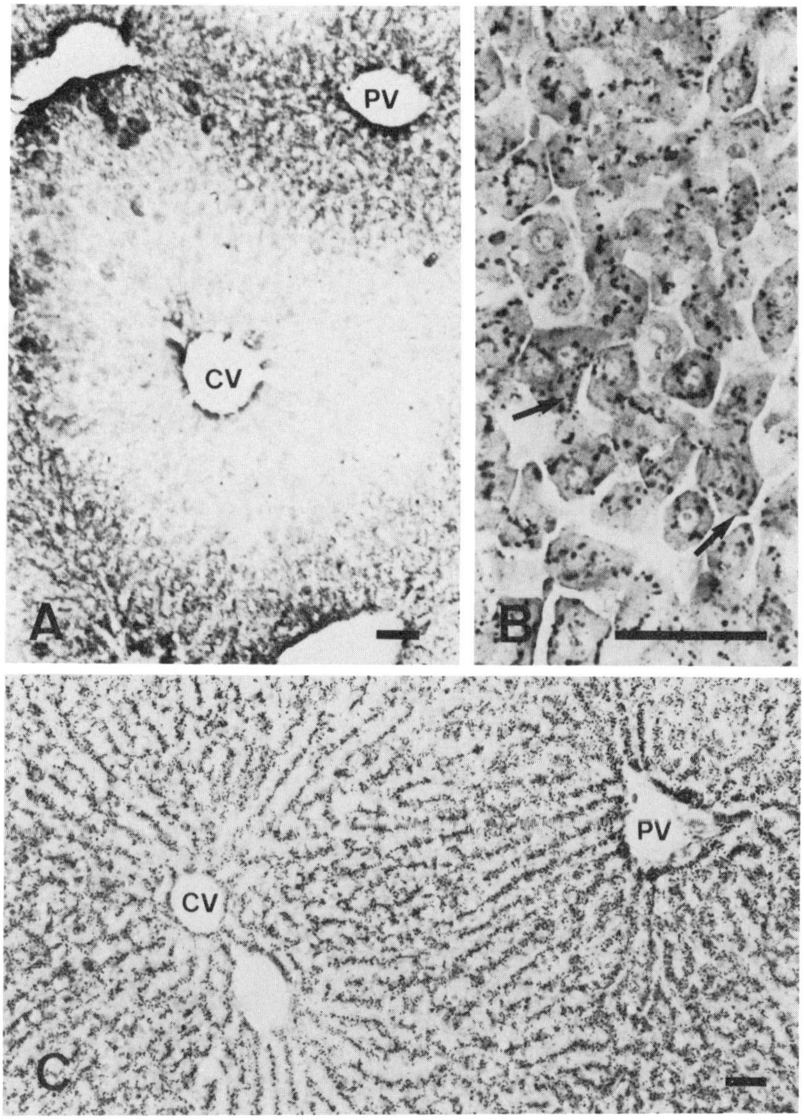

Figure 1. Taurine-like immunoreactivity in mouse (**A-B**) and rat (**C**) liver sections, counterstained with neutral red. **A:** Taurine positive granules are distributed predominantly in the peripheral region of lobules in a mature mouse. **B:** Several taurine positive granules are diffused in each hepatocyte. Arrows point to some granules which seem to contact the membrane of the liver cell. **C:** Taurine-positive granules are distributed both in the peripheral and central regions of the hepatic lobule. CV: central vein; PV: portal vein. Calibration bars: 50 μm

notable differences exist in the manner of bile acids conjugation in different species. This is particularly true in the rat whose bile acids are almost completely conjugated with taurine (2).

At the electron microscopic level, taurine-like immunoreactivity in mature mice was observed in small vesicles, often clustered in the cytoplasm of hepatocytes (Fig. 3A). These positive vesicles were found mainly in the cisternal lumina of smooth-surfaced endoplasmic reticulum suggesting that the taurine-like immunoreactivity substance is synthesized there

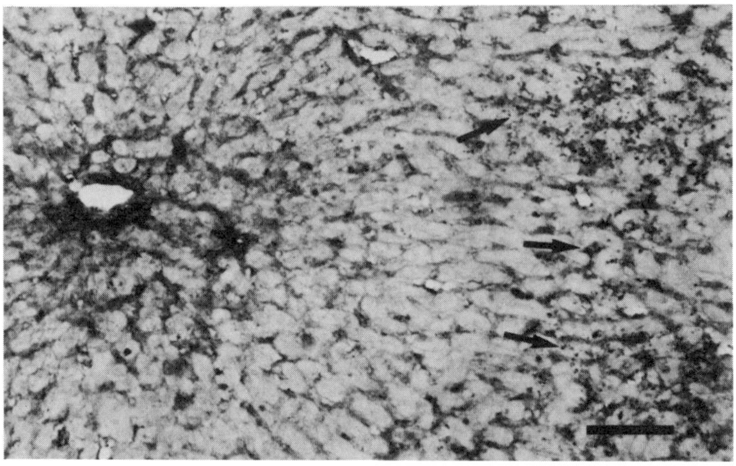

Figure 2. Taurine-like immunoreactivity is observed mainly in the peripheral region of the human hepatic lobule (arrows). Calibration bar: 100 μm

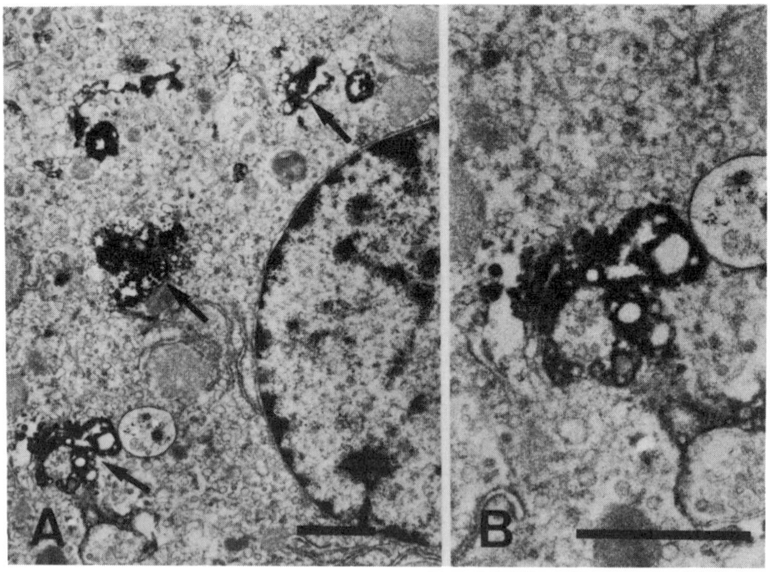

Figure 3. Immunoelectron micrographs of taurine-like immunoreactivity in adult mouse liver. **A:** Taurine-like immunoreactivity is observed in the hepatocyte cytoplasm as small vesicles which often form clusters (arrows). **B:** The positive reaction occurs in the cisternal lumen of smooth endoplasmic reticulum. Calibration bars: 1 μm

(Fig. 3B). More regions of taurine-like immunoreactivity were located nearer bile canaliculi than near capillaries, suggesting that taurine conjugates are excreted mainly to the bile duct.

Taurine is known to be contained in both neuronal and glial cells of the central nervous system. However, positive immunostaining has been observed predominantly in neurons (3,-31). This may be explained by the fact that neuronal taurine exists mainly as particulate forms, probably bound to certain proteins, while glial taurine is present in unbound free form. Free taurine is difficult to fix during tissue preparation so that taurine-like immunoreactivity seen in neurons appears to represent the particulate form of taurine (15). Such a technical difficulty may have prevented the study of the localization in non-neural organs although taurine is contained in many organs of various vertebrates (14,17). Indeed, our preliminary study (data not shown) failed to demonstrate taurine-like immunoreactivity in the kidney which is well known to have a great potential to synthesize free taurine. It is conceivable that taurine-like immunoreactivity in the liver may not sufficiently reflect free taurine, but more preferably reflect particulate forms of taurine, most probably taurine-conjugated bile acids. In fact, our immuno-spot experiment indicated that all soluble taurine-conjugated bile acids, regardless of primary or secondary structure, are immunoreactive with the taurine antiserum. Such a staining property with the taurine antibody in the liver appears to be in harmony with reports indicating that, although taurine is synthesized mainly in perivenous hepatocytes (24), taurine is conjugated with bile acids mainly in periportal lobular periphery (28). Our results showed that higher taurine-like immunoreactivity is presented in periportal hepatocytes of adult mice, rats, monkeys, and humans, again suggesting that taurine-like immunoreactivity in the liver represents taurine-conjugated bile acids. Therefore, taurine synthesized in perivenous zone seems to be rapidly transported to periportal zone via the systemic circulation.

Table 1. Distribution of taurine-like immunoreactivity in mouse liver

Age	Hepatic lobule	
	Peripheral	Central
15-day embryo	-	-
17-day embryo	+	+
19-day embryo	+	+
1-day postnatal	+	+
3-day postnatal	++	++
1-week postnatal	+++	++
2-week postnatal	+++	+++
3-week postnatal	+++	+
6-8-week young	+++	-
3-month adult	+++	-

- = none; + = few; ++ = moderate number; +++ = numerous

ONTOGENIC DEVELOPMENT OF TAURINE-LIKE IMMUNOREACTIVITY IN MICE LIVER

Distribution of taurine-like immunoreactivity was observed in the mouse liver from embryo to adult (Table 1). Taurine-like immunoreactivity appeared first in mouse liver from 17 day-old embryos, where it was found in both the peripheral and central regions of the lobules. In both 19 day-old embryos and one day postnatal mice, the distribution patterns and densities of taurine-like immunoreactivity were similar to those of 17 day-old embryos. Taurine-like immunoreactivity in both the periphery and center of the lobules increased with age until at least three days after birth, and the highest level in the central region were seen in two week-old mice. However, by 3 weeks postnatal, the taurine-like immunoreactivity had notably decreased in some central lobular regions, and in mature mice (6-8 weeks, 3 months), almost no taurine-like immunoreactivity was found in the central region of any lobule, although the density of taurine-like immunoreactivity in the peripheral region of the lobules was similar to that found at 14 days postnatal. These results suggest that the distribution pattern of taurine-like immunoreactivity in mouse liver varies considerably during development, which accords well with previous biochemical studies in mouse liver (11,29), indicating that more taurine-like immunoreactivity is present in neonatal than in adult.

Biochemical studies have shown that more than 50% of bile acid is conjugated with taurine rather than glycine in the embryo and 10-days-old newborn (9); i.e. the ratio of the amount of bile acid conjugated with glycine to that conjugated with taurine is less than 1. This ratio is increased up to 2.4 at 7-12 months of age and to 3 at adulthood. The ratio, however, is also influenced by the diet. Liquid foods rich in fat lead to an increase in the levels of taurine-conjugated bile acid (21). The large amount of taurine seen in the liver of early newborn mice may, therefore, be influenced by the kinds of food.

CARBON TETRACHLORIDE-TREATED MICE

Carbon tetrachloride was chosen as the most widely used reference substance for all hepatotoxic compounds (for review see 18). The hepatotoxicity induced by CCl_4 seems to be mediated by the reaction of free radical metabolites (27). The major histopathological abnormalities include hepatic centrilobular cirrhosis and hepatic steatosis in both periportal

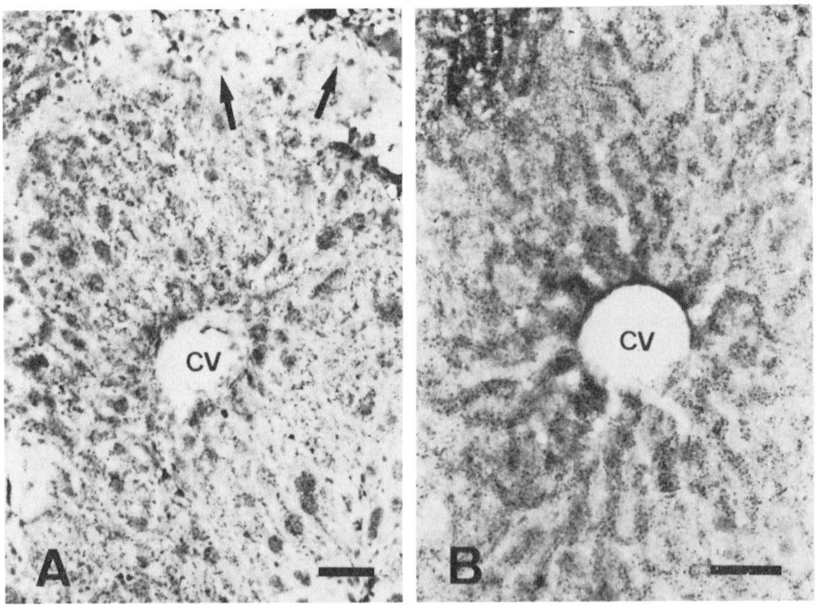

Figure 4. Taurine-like immunoreactivity observed in adult mouse liver following treatment with CCl_4 (**A**) and phenobarbital (**B**). **A:** CCl_4 injection (0.2 ml/100g), twice a week for 14 consecutive weeks. Notable fibrosis in the peripheral region of the lobule can be seen (arrows), and the number of taurine positive granules in the central region is markedly increased. **B:** Phenobarbital injection (100 mg/kg) once a day for two consecutive weeks. Taurine-like immunoreactivity in the central region is markedly increased. Calibration CV: central vein. bars: 50 μm

and perivenous hepatocytes. Mature mice were injected hypodermically with CCl_4 (Nacalai, Japan) and olive oil (1:1 by volume) in a dose of 0.2 ml/100 g, twice a week for 14 consecutive weeks. Control mice were dosed with the same dose of olive oil without CCl_4. In the mice receiving CCl_4 treatment, notable fibrosis was seen in the periphery of lobules and hepatocytes in the central region were found to contain more intense taurine-like immunoreactivity than that observed in control mice (Fig. 4A). At the electron microscopic level, taurine-like immunoreactivity was also detected in smooth endoplasmic reticulum, with a distribution pattern similar to that found in untreated mature mice. Many dilated smooth endoplasmic reticula were observed in hepatocytes, but the smooth endoplasmic reticulum containing taurine-like immunoreactivity seemed to have normal size and shape (data not shown).

These data indicate that CCl_4 intoxication might affect the distribution of taurine-like immunoreactivity in liver. In particular, taurine-like immunoreactivity was markedly increased in the central region of lobules from mice intoxicated by CCl_4. Probably, after injury, hepatic cells in the central region of lobules are activated by an unknown mechanism and start to synthesize taurine. Such an activation occurring in the central hepatocytes, that might have escaped from the CCl_4 suffering, may reflect a compensatory mechanism for degenerative fibrous hepatocytes in the peripheral acinus where taurine synthesis is no longer possible.

PHENOBARBITAL-TREATED MICE

Phenobarbital is frequently used in animal experiments to induce the hepatic microsomal enzyme system such as cytochrome P450-dependent mono-oxygenase (5). This enzyme induction results in an increased metabolic rate of a number of drugs and endogenous substances including bile salts. Mature mice were given intraperitoneally with phenobarbital (Nacalai, Japan, 30 mg/kg or 100 mg/kg), once a day for two consecutive weeks. In mice treated with a repeated dose (30 mg/kg) of phenobarbital, the distribution pattern of taurine-like immunoreactivity did not differ from that in untreated mice. However, in the mice treated with a high and repeated dose (100 mg/kg) of phenobarbital, some cytoplasmic vacuoles, and a slight increase in taurine-like immunoreactivity, were observed in the peripheral region of lobules. The most notable effect, however, was a marked increase in taurine-like immunoreactivity in the central region of lobules (Fig. 4B).

The increase of taurine-like immunoreactivity in phenobarbital-treated mice did not seem to reflect enzyme induction occurring in the liver, because a lower dose of phenobarbital (30 mg/kg) does not affect liver taurine-like immunoreactivity although it is sufficient to induce liver enzymes (6,12).

REFERENCES

1. Awapara, J., 1956, The taurine concentration of organs from fed and fasted rats, *J.Biol. Chem.* 218:571-576.
2. Bremer, J., 1956, Species differences in the conjugation of free bile acids with taurine and glycine, *Biochem. J.* 63:507-513.
3. Campistron, G., Geffard, M., and Buijs, R.M., 1986, Immunological approach to the detection of taurine and immunocytochemical results, *J. Neurochem.* 46:862-868.
4. Chan Palay, V., Lin, C.T., Palay, S., Yamamoto, M., and Wu, J.Y., 1982, Taurine in the mammalian cerebellum: Demonstration by autoradiography with [^{3}H]taurine and immunocytochemistry with antibodies against the taurine-synthesizing enzyme, cysteine-sulfinic acid decarboxylase, *Proc. Natl. Acad. Sci. USA* 79:2695-2699.

5. Conney, A.H., 1967, Pharmacological implications of microsomal enzyme induction, *Pharmacol. Rev.*, 19:317-366.
6. Conney, A.H., Davison, C., Gastel, R., and Burns, J.J., 1960, Adaptive increase in drug-metabolizing enzymes induced by phenobarbital and other drugs, *J. Pharmacol. Exp. Ther.* 130:1-8.
7. Ding, W.G., Tooyama, I., Kimura, H., Kuriyama, K., and Ochi, J., 1993, Distribution of taurine-like immunoreactivity in the mouse liver during ontogeny and after carbon tetrachloride or phenobarbital intoxication, *Histochem. J.* 25:376-383.
8. Dubreuil, R., and Timiras, P.S., 1953, Effect of cortisone on free amino acids in the serum and organs of the rabbits, *Am. J. Physiol.* 174:20-26
9. Encrantz, J.C., and Sjövall, J., 1959, on the bile acids in duodenal contents of infants and children: Bile acids and steroids 72, *Clin. Chim. Acta.* 4:793-799.
10. Garbutt, J.T., Lack, L., and Tyor, M.P., 1971, Physiological basis of alterations in the relative conjugation of bile acids with glycine and taurine, *Am. J. Clin. Nutr.* 24:218-228.
11. Ghisolfi, J., Nguyen, V.B., Thouvenot, J.P., Rolland, M., Putet, G., and Lapalu Traon, C., 1989, Taurine in developing brain, liver and muscle in infants. *Biol. Neonate* 56:186-191.
12. Hart, L.G., Adamson, R.H., Dixon, R.L., and Fouts, J.R., 1962, Stimulation of hepatic microsomal drug metabolism in the newborn and fetal rabbit, *J. Pharmacol. Exp. Ther.* 137:103-106.
13. Haas, H.L., and Hosli, L., 1973, The depression of brain stem neurons by taurine and its interaction with strychnine and bicuculline, *Brain Res.* 52:399-402.
14. Hayes, K.C., and Sturman, J.A., 1981, Taurine in metabolism, *Ann. Rev. Nutr.* 1:401-425.
15. Ida, S., Kuriyama, K., Tomida, Y., and Kimura, H., 1987, Antisera against taurine: Quantitative characterization of the antibody specificity and its application to immunohistochemical study in the rat brain, *J. Neurosci. Res.* 18:626-631.
16. Jacobsen, J.G., and Smith, Jr., L.H., 1963, Comparison of decarboxylation of cysteine sulfinic acid-1-^{14}C and cysteic acid-1-^{14}C by human, dog, and rat liver and brain, *Nature* 200:575-577.
17. Jacobsen, J.G., and Smith, L.H., 1968, Biochemistry and physiology of taurine and taurine derivatives, *Physiol. Rev.* 48:424-511.
18. Kalf, G.F., Post, G.B., and Snyder, R., 1987, Solvent toxicology: recent advances in the toxicology of benzene, the glycol ethers, and carbon tetrachloride, *Annu. Rev. Pharmacol. Toxicol.* 27:399-427.
19. Kroll, J., and Lund, E., 1966, The effect of taurine on serum levels of aminotransferase activity in patients with cirrhosis, *Dan. Med. Bull.* 13:173-174.
20. Kuriyama, K., 1980, Taurine as a neuromodulator, *Fed. Proc.* 39:2680-2684.
21. Lindstedt, S., Avigan, J., Goodman, D.S., Sjövall, J., and Steinberg, D., 1965, The effect of dietary fact on the turnover of cholic acid and on the composition of the biliary bile acids in man, *J. Clin. Invest.* 44:1754-1765.
22. Nakashima, T., Taniko, T., and Kuriyama, K., 1982, Therapeutic effect of taurine administration on carbon tetrachloride-induced hepatic injury, *Jpn. J. Pharmacol.* 32:583-589.
23. Ohkuma, I., Tamura, J., and Kuriyama, K., 1984, Roles of endogenous and exogenous taurine and glycine in the formation of conjugated bile acids: Analyses using freshly isolated and primary cultured rat hepatocytes, *Jpn. J. Pharmacol.* 35:347-358.
24. Penttila, K.E., 1990, Role of cysteine and taurine in regulating glutathione synthesis by periportal and perivenous hepatocytes, *Biochem. J.* 269:659-664.
25. Siperstein, M.D., and Murray, A.W., 1955, Enzymatic synthesis of cholyl-CoA and taurocholic acid, *Science* 123:377-378.
26. Sjövall, J., 1959, Dietary glycine and taurine on bile acid conjugation in man: Bile acids and steroids 75, *Proc. Soc. Exp. Biol. Med.* 100:676-678.
27. Slater, T.F., 1982, Free radicals as reactive intermediates in tissue injury, "Biological Reactive Intermediates 2: Chemical Mechanisms and Biological Effects", Snyder, R., Parke, D.V., Kocsis, J.J., Jollow, D.J., Gibson, G.G., and Witmer, C.M., ed., Plenum Press, New York, pp. 575-589.
28. Strange, R.C., 1984, Hepatic bile flow, *Physiol. Rev.* 64:1055-1102.
29. Sturman, J.A., Rassin, D.K., Gaull, G.E., and Cote, L.J., 1980, Taurine in developing rhesus monkey brain, *J. Neurochem.* 35:304-310.
30. Tomida, Y., and Kimura, H., 1987, Immunohistochemical and biochemical studies of substances with taurine-like immunoreactivity in the brain, *Acta Histochem. Cytochem.* 21:31-40.
31. Yoshida, M., Karasawa, N., Ito, M., and Nagatsu, I., 1986, Demonstration of taurine-like immunoreactive structures in the rat brain, *Neurosci. Res.* 3:356-363.

EFFECTS OF TAURINE ON MICROSOMAL ENZYME ACTIVITIES INVOLVED IN LIVER LIPID METABOLISM OF WISTAR RATS

Alfredo Cantafora, Chong Chao Yan, Yun Sun
and Roberta Masella

Laboratory of Metabolism and
Pathological Biochemistry
Istituto Superiore di Sanità
Roma, Italy

INTRODUCTION

Our previous in vivo studies with guinea pigs, cats and rats showed that changes in the hepatocellular levels of taurine affect the liver lipid composition of these animal species (6,7,30). Specifically, we found that a taurine-supplemented diet lowered the liver concentrations of phospholipid, but had contrasting effects on neutral lipid content. Taurine supplementation increased the liver stores of triglyceride in guinea pigs and cats (species with a low endogenous synthesis of taurine), but depleted these liver stores in rats (species with an active endogenous synthesis of taurine). The changes were not only quantitative but also qualitative. Taurine supplement (but also its depletion) induced significant changes in the cholesterol proportion, in the ratio of phosphatidylethanolamine to phosphatidylcholine and in the fatty acid profile of the major phospholipid classes. The good agreement between our data and previous in vivo (4,18,29) and in vitro (25) studies, indicated that taurine has an important role not only in biliary lipid secretion, but also in liver lipid metabolism.

The enzymatic reactions involved in liver biosynthesis of glycerolipid, glycero-phospholipids and steroids occur principally in the endoplasmic reticulum of hepatocyte. These reactions are modulated by membrane-enzyme interactions which, in turn, depend on both the lipid composition and the chemical environment of membrane itself (27,31). Therefore, slight modifications in the micro-environment of membrane-bound enzymes involved in lipid metabolism may explain a wide spectrum of lipid alterations, such as those induced by taurine variations. This possibility is suggested by the formation of low-affinity charge complexes between taurine and neutral membrane phospholipids (15), that may alter both the membrane physicochemical properties and the regulation of membrane-bound enzymes. Furthermore, it cannot be excluded that long-term modifications in taurine levels may affect enzymatic regulation by inducing lasting alterations in the lipid membrane composition. This has been demonstrated in synaptosomal membrane of developing rat brain in which the age-related taurine decrease is inversely correlated with the molar ratio of phosphatidylethanolamine to phosphatidylcholine (14) and with the rate of phospholipid methylation that converts phosphatidylethanolamine to phosphatidylcholine (13).

In this study we tried to clarify the mechanisms by which taurine exerts its different effects on the liver lipid metabolism of Wistar rats. The choice of this species took into account the fact that most of the knowledge of the liver lipid metabolism derives from studies with rats and that an interesting analogy exists between rat and human: both species have variations in the hepatic synthesis of taurine that depend on the dietary intake, and the availability of taurine in the liver affects the conjugation pattern of bile salts (16). This study showed, in the first place, that the addition of 1% taurine to drinking water for two weeks induced significant changes in the lipid composition of microsomal membrane. The study also demonstrated, by two separate sets of in vitro experiments, that both the membrane lipid changes induced by the in vivo treatment and the taurine levels facing the microsome membrane are factors important to the regulation of membrane-bound enzyme systems.

MATERIALS AND METHODS

All reagents and solvents used were of analytical grade. The radiolabelled substrates [1-^{14}C]oleoyl-CoA (specific activity 52.9 mCi/mmole), cholesteryl [1-^{14}C]oleate (52 mCi/mmole), L-3-phosphatidylcholine,1palmitoyl-2-[1-^{14}C]oleoyl (57 mCi/mmol), [1-^{14}C]arachidonic acid (52 mCi/mmole) and [1-^{14}C]palmitoyl-*sn*-3-phosphorylcholine (52 mCi/mmole) were purchased from Amity PG (Milan, Italy). The unlabelled products were supplied by Sigma Italia (Milan, Italy). Silica gel plates for thin-layer chromatography, without fluorescence indicator, were purchased from E. Merck (Darmstadt, Germany).

Twenty male Wistar rats (by Charles River Italiana, Calco-Como) with body weights ranging from 120 to 150 g, were randomly divided into taurine-treated (10 animals) and control (10 animals) groups. Both the groups were fed ad libitum with the same semisynthetic regimen for rodents containing 18.5% protein, 3.0% fat, 6.0% fiber, 6.0% salt and vitamin mixture (4RF21 Formula for Rodents by Mucedola Srl, Settimo Milanese, Italy). The taurine content in this diet was about 0.02%, as determined by amino acid analyser. Control rats drank tap water. Taurine-treated animals had a 1% addition of taurine to drinking water. After a 2 week treatment, the rats were fasted overnight and then were killed by cervical dislocation under light ether anaesthesia. The liver was immediately excised and homogenised into 4 volumes of a buffer containing 0.25 M sucrose, 20 mM Tris-HCl, 1 mM EDTA and 1% fatty acid-free bovine serum albumin (pH 7.4). The subcellular fraction of endoplasmic reticulum membrane was separated by sequential ultracentrifugation as described by Graham (11). The final microsome preparation was suspended in the homogenisation buffer, divided into various aliquots and stored at -30°C until assayed. Purity of each preparation was checked by measuring the enzymatic activities of 5'-nucleotidase, glucose-6-phosphatase, and succinate dehydrogenase, as markers for plasma membranes, endoplasmic reticulum and mitochondria, respectively (21). The protein concentration was measured by the Bradford's method (5) using bovine serum albumin as calibration standard.

The acyl-CoA:cholesterol acyltransferase, oleoyl-CoA hydrolase and diacyl-glycerol acyltransferase activities were measured by incubating 0.2-0.4 mg of microsome protein with [14C]-oleoyl-CoA, as described by Field et al. (8). The cholesteryl ester hydrolase activity was measured following the method of Khoo et al. (17). The phospholipase A_1 and A_2 activities were measured by incubating for 1 h at 37°C 0.8 mg of microsomal protein in 0.4 ml of 100 mM Tris, 5 mM $CaCl_2$ buffer (pH 7.5) with 100 μM 1-palmitoyl,[2-^{14}C]oleoyl, phosphatidylcholine. The incubation was terminated with addition of chloroform/

methanol mixture (2:1 by volume) and the radioactivity incorporated in free fatty acids (phospholipase A_2 activity) and in lysophosphatidylcholine (phospholipase A_1 activity) was evaluated following thin layer chromatography separation by liquid scintillation counting, as previously described (1). The acyl-CoA:1-acylglycerophosphorylcholine acyltransferase and lysophospholipase activities were carried out as previously described by Masella et al. (20). The determination of arachidonoyl-CoA synthetase activity was measured as described by Bar-Tana et al. (2).

Aliquots of the isolated microsomes were extracted with chloroform:methanol mixture (2:1, by vol.) as described by Folch et al. (9). The lipid classes in these extracts were first isolated by thin-layer chromatography and then analysed by gas-liquid chromatography, as previously described by ourselves (30)

The results are reported as mean $\pm$ standard deviations. The statistical evaluation of differences between control and taurine-treated groups was considered significant at a level of probability > 95%, by the two-sided Student's t-test. The study of correlation was made by the linear regression analysis. The statistical tests were performed with the SOLO package by BMDP Statistical Software, Inc. (Los Angeles, CA).

RESULTS

During the preparatory in vivo treatment with 1% taurine in drinking water for 2 weeks, neither food nor water intakes showed any significant differences between controls and treated groups. Also the growth rates did not show significant variations (7.8 $\pm$ 0.8 and 8.4 $\pm$ 0.9 g/day, in taurine and control group, respectively).

Table 1 shows the microsome membrane composition in control and taurine-treated rats. The treatment induces changes in the lipid content of microsomes but not in their protein content. The level of neutral lipids is significantly decreased and that of polar lipids (i.e., phospholipids) is increased by oral treatment with taurine.

Table 1. Effect of oral supplement with taurine on the gross composition of liver microsome membrane

	Control	Taurine
Protein (mg/g wet tissue)	27.1 $\pm$ 0.5	26.4 $\pm$ 0.4
Neutral Lipids (μmol/g protein)	143.1 $\pm$ 9.3	117.0 $\pm$ 7.6[c]
Polar Lipids (μmol/g protein)	231.9 $\pm$ 9.5	245.3 $\pm$ 6.8[a]

Data are for 10 animals per group. [a] Different from control ($P < 0.05$); [b] Different from control ($P < 0.001$)

The details of microsome membrane lipid composition are shown in the Table 2. It is interesting to note that the above-seen decrease in neutral lipids is entirely due to a fall in the free cholesterol level, the proportion of which is the only one to be markedly decreased. On the other hand, the phospholipids, even if slightly changed in their total concentration, show an increase in the proportion of phosphatidylethanolamine and a symmetrical decrease in the proportion of the other classes.

Most of the variations seen are known to affect the physicochemical properties of the membrane. Therefore, we examined in detail the effect of intracellular taurine levels on three relevant composition parameters: the molar ratios of free cholesterol to total phospholipid (FCH/PL), of phosphatidylethanolamine to phosphatidylcholine (PE/PC) and of sphingomyelin to phosphatidylcholine (SP/PC). The regression analyses, separately made for each of these parameters, are shown in the three panels of Figure 1. Both the FCH/PL and SP/PC molar ratios are inversely related to the taurine liver levels. Conversely, the PE/PC ratio is directly related with the taurine level in the whole organ.

Table 2. Effect of the oral supplement with taurine on the percent distributions of neutral and polar lipid classes in microsome membranes

	Control	Taurine
Neutral Lipids		
Free Cholesterol	62.0 ± 2.5	52.6 ± 2.8^{c}
Cholesteryl Ester	4.9 ± 0.6	5.5 ± 1.0
Triacylglycerol	15.8 ± 1.4	20.9 ± 1.5^{c}
Diacylglycerol	3.9 ± 0.3	4.2 ± 0.3
Free Fatty Acids	13.4 ± 0.7	16.8 ± 0.4^{c}
Polar Lipids		
Phosphatidylcholine	50.4 ± 1.2	41.0 ± 3.4^{c}
Phosphatidylethanolamine	14.9 ± 1.4	39.0 ± 2.9^{c}
Phosphatidylserine	18.0 ± 1.6	9.9 ± 1.3^{c}
Sphingomyelin	12.4 ± 1.0	5.9 ± 0.6^{c}
Phosphatidic Acid	0.5 ± 0.1	1.9 ± 0.4^{c}
Lysophosphatidylcholine	3.8 ± 0.5	2.4 ± 0.3^{c}

Data are means ± SD for 6 samples per group. [c] Different from Control (P < 0.001).

Taurine treatment affects the fatty acid composition of the major glycerophospholipids in rat liver microsomes. The proportion of long-chain polyunsaturated fatty acid is decreased and that of saturated fatty acid is increased (Table 3).

Table 3. Effect of oral supplement with taurine on the fatty acid composition of the major glycerophospholipids in the liver microsome membrane of Wistar rats

	C_{22} polyunsaturated (%)		C_{16}-C_{22} saturated (%)	
	Control	Taurine	Control	Taurine
Phosphatidylcholine	4.8 ± 0.85	2.8 ± 0.81^{c}	50.2 ± 3.80	53.8 ± 2.94
Phosphatidylethanolamine	12.1 ± 0.75	5.9 ± 0.84^{c}	47.3 ± 1.51	51.9 ± 1.93^{b}
Phosphatidylserine	6.9 ± 0.19	4.9 ± 0.22^{c}	56.6 ± 2.00	63.1 ± 0.96^{c}

Data are means ± SD for 6 samples per group. [b]Different from Control (P<0.01); [c]Different from Control (P<0.001)

Table 4 shows the effect of in vivo treatment with taurine on several enzyme activities related to lipid metabolism in rat liver microsomes. These changes indicate that the treatment with taurine may affect the cholesterol storage in liver cells by lowering the acyl-CoA:cholesterol acyltransferase activity (which converts free cholesterol into cholesteryl ester) and augmenting the cholesteryl ester hydrolase activity. The enzymes that regulate the intracellular availability of activated fatty acyls, i.e., the oleoyl-CoA hydrolase and the

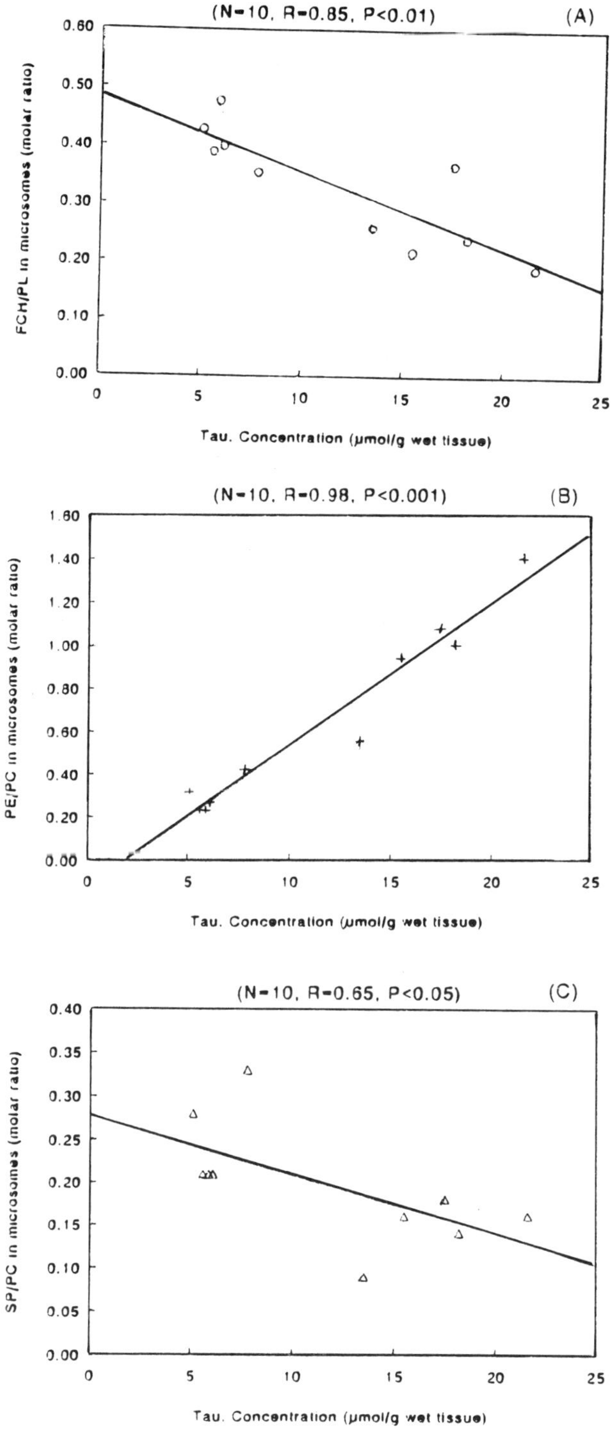

Figure 1. Linear regressions of the molar ratios: (A) free cholesterol to phospholipids (FCH/PL); (B) phosphatidylethanolamine to phosphatidylcholine (PE/PC) and (C) sphingomyelin to phosphatidylcholine (SP/PC) in microsomal membranes of control and taurine-treated animals versus the taurine concentration measured in the respective livers. The results of statistical evaluation are separately reported in each panel.

arachidonoyl-CoA synthetase, are significantly increased by the in vivo treatment with taurine. Conversely, this treatment lowers diacylglycerol acyltransferase activity, the enzyme that regulates the intracellular accumulation of fats by converting diacylglycerol into triacylglycerol.

Table 4. Effect of oral supplement with taurine on enzyme activities related to lipid metabolism in microsomal membranes isolated from rats treated with 1% taurine in drinking water for 2 weeks.

Enzyme Activity	Control	Taurine
Acyl-CoA:cholesterol acyltransferase	16.7 ± 1.1	6.1 ± 0.6[c]
Cholesterol Ester Hydrolase	7.3 ± 1.1	9.1 ± 1.6[a]
Oleoyl-CoA Hydrolase	84.0 ± 3.8	135.6 ± 4.4[c]
Arachidonoyl-CoA Synthetase	64.2 ± 6.9	102.4 ± 9.2[c]
Diacylglycerol acyltransferase	23.8 ± 1.4	15.0 ± 1.3[c]
Phospholipase A_1	0.9 ± 0.1	0.86 ± 0.10
Phospholipase A_2	0.62 ± 0.07	0.83 ± 0.07[c]
Lysophospholipase	3.6 ± 0.6	6.0 ± 1.2[b]
Acyl-CoA:1-acyl-glycerophosphoryl-choline acyltransferase	8.3 ± 1.98	11.6 ± 1.64[b]

The results are the mean ± SD of duplicate determinations conducted on five different microsomal preparations. The values are expressed as pmole/mg protein/min for all enzymes, except for lysophospholipase and Acyl-CoA:1-acyl-glycerophosphorylcholine acyltransferase where the values are expressed as nmole/mg protein/min. [a] Different from Control ($P < 0.05$); [b] Different from Control ($P < 0.01$); [c] Different from Control ($P < 0.001$).

Table 4 shows also some relevant effects of oral supplement with taurine on enzyme activities involved in the retailoring of glycerophospholipids on microsome membrane; i.e., phospholipases, lysophospholipase, and acyl-CoA:1-acyl- glycerophosphoryl-choline acyltransferase. It is interesting to note that while phospholipase A_1 activity (involved in acyl changes at the *sn*-1 position) is not affected by taurine treatment, other activities involved with the remodelling of the *sn*-2 acyl position, i.e., phospholipase A_2, acyl-CoA:1--acyl-glycerophosphorylcholine acyltransferase and lysophospholipase activities, are significantly higher in microsomes of taurine-treated rats than in control microsomes. The pre-treatment with taurine affects the incorporation of both [^{14}C]oleoyl-CoA and [^{14}C]arachidonoyl-CoA into the major neutral lipid and phospholipid classes, as shown in the Figures 2 and 3, respectively. The microsomes isolated from taurine-treated rats show, in general, a higher incorporation of the labelled fatty acyls in both the neutral and polar lipids with the most significant increases in the diacylglycerol and phosphatidylethanolamine classes, respectively.

This study includes a set of experiments made with control microsomes where the enzyme activities are measured in the presence of increasing concentrations of taurine in the incubation medium. Figure 4 shows the effect of large taurine variations in the medium on the incorporation of [^{14}C]oleoyl-CoA into diacylglycerol, free fatty acid, triacylglycerol and cholesteryl ester. Figure 5 shows the effect of taurine levels on the microsomal reacylation of 1-acyl-lysophosphatidylcholine by [^{14}C]oleoyl-CoA. It is interesting to note that the results obtained in these experiments agree with the decreased concentrations of triacylglycerol, cholesteryl ester and phosphatidylcholine found in our previous in vivo study in livers of taurine-treated rats (30).

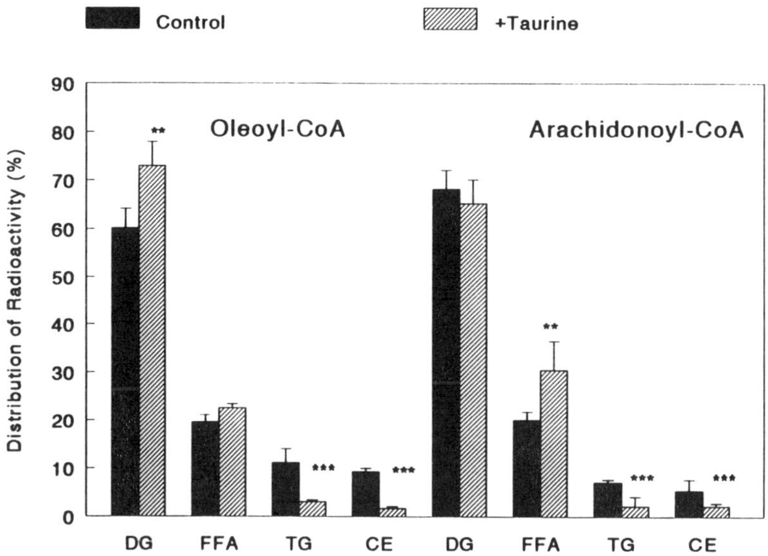

Figure 2. Percent distribution of the radioactivity incorporated in the major neutral lipids of microsomal membrane of control and taurine-treated rats following incubation with radiolabelled oleoyl-CoA and arachidonoyl-CoA

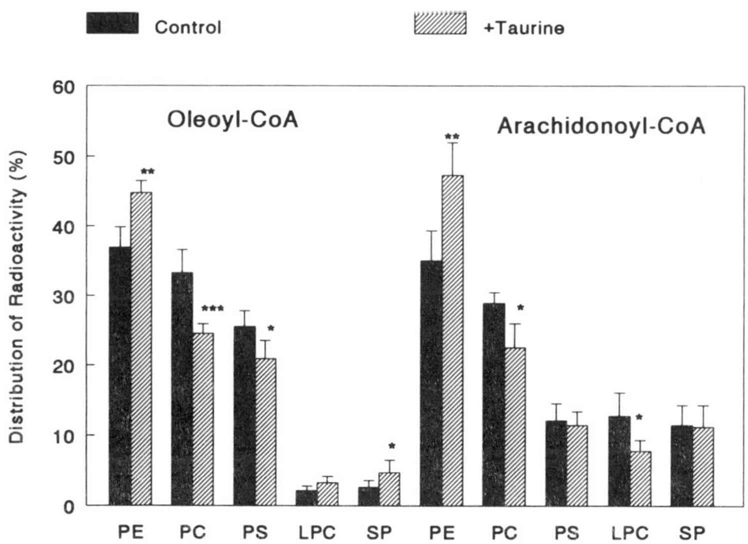

Figure 3. Percent distribution of the radioactivity incorporated in the major phospholipids of microsomal membrane of control and taurine-treated rats following the incubation with radiolabelled oleoyl-CoA and arachidonoyl-CoA

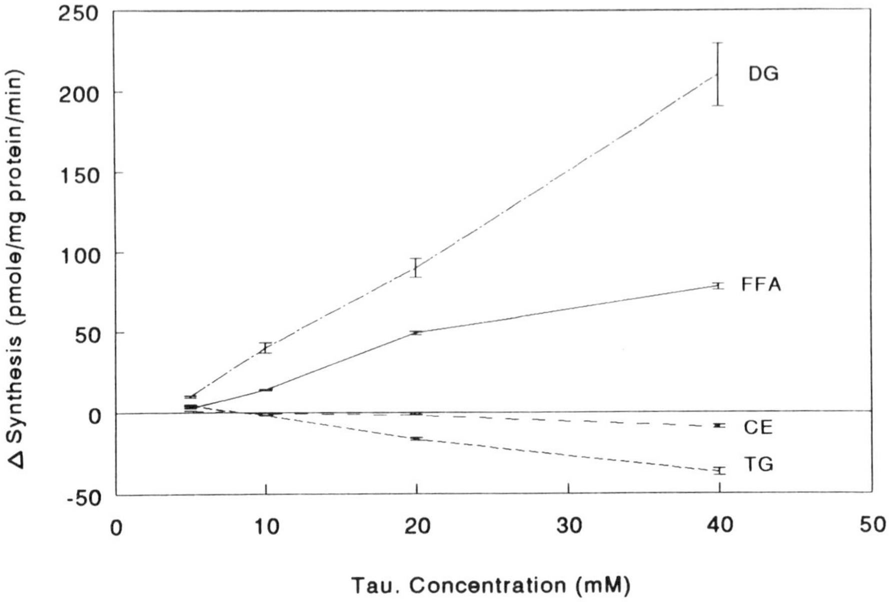

Figure 4. Effect of taurine concentration in incubation media of control microsomes with [^{14}C]oleoyl-CoA on the synthesis of the major neutral lipids

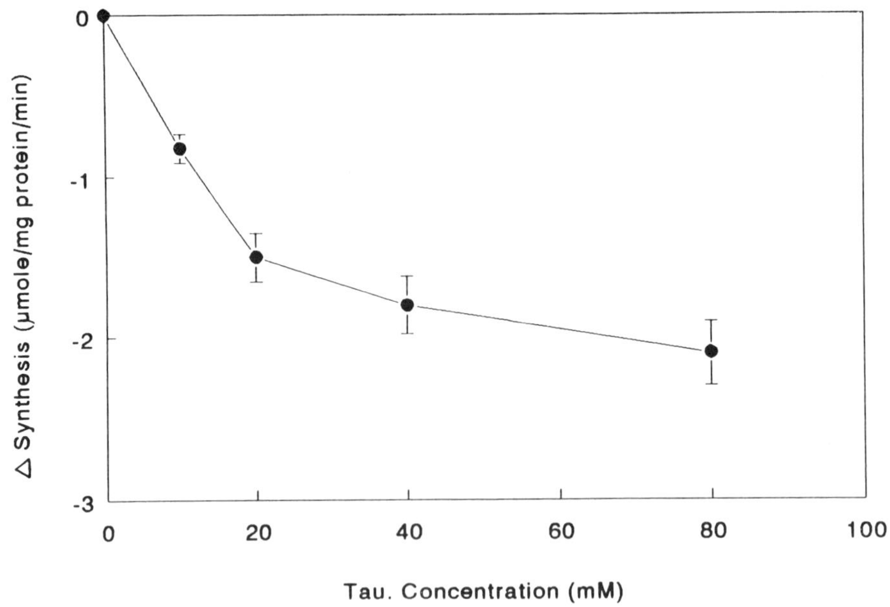

Figure 5. Reacylation of 1-palmitoyl-lysophosphatidylcholine with oleoyl-CoA by liver microsomes of normal rats in the presence of different concentrations of taurine in the incubation medium

DISCUSSION

Cholesterol equilibria between liver and plasma or bile compartments are regulated by hydroxymethylglutaryl-CoA reductase, 7α-cholesterol hydroxylase, acyl-CoA:cholesterol acyltransferase, and cholesteryl ester hydrolase activities. Therefore, the decrease in both cholesterol and cholesteryl ester hepatic content and the increase in bile secretion of taurine conjugates observed in taurine-treated rats (30) are likely to depend on one or more of the above mentioned enzyme activities. Specifically, the stimulation of cholesterol movement from hepatocellular stores (where it is in the esterified form) to exchangeable pools (where it is in the free form) is in accord with the higher activity of 7α-hydroxylase and the lower activity of hydroxymethylglutaryl-CoA reductase induced by taurine administration found in other studies (26,30), and with the decrease in acyl-CoA:cholesterol acyltransferase activity and the increase in cholesteryl ester hydrolase activity found in this study.

In agreement with the decreased cholesterol concentration found in the livers of animals treated with taurine (30), we observe here a significantly lower content of free cholesterol into microsomes of treated animals (89.4 $\pm$ 8.7 and 62.0 $\pm$ 6.6 μmole/mg protein, in control and taurine-treated microsomes, respectively). This decrease in membrane cholesterol content is likely to have an important regulatory role on both acyl-CoA:cholesterol acyltransferase and cholesteryl ester hydrolase activities. However, we cannot exclude that factors other than the long-term variations in membrane composition and the reduced availability of substrate for esterification may play a role in lowering the acyl-CoA:cholesterol acyltransferase activity and the cholesteryl ester content. A possibility, suggested by in vitro experiments with normal microsomes in which the taurine added to incubation medium had a slight, but immediate, reducing effect on cholesterol esterification, is that the acyl-CoA:cholesterol acyltransferase activity may be regulated by the degree of phosphorylation of this enzyme. It is known that this activity is increased in vitro by enzyme phosphorylation (3) and, consequently, not only a reduced availability of substrate (i.e., of membrane free cholesterol), but also a reduced phosphorylation of the enzyme may lower acyl-CoA:cholesterol acyltransferase activity in taurine-treated microsomes. This hypothesis is supported by a recent study in which taurine was found to inhibit protein kinase C-catalyzed phosphorylation of specific proteins in cortical P_2 fraction of rat (19). However, the higher activity of cholesteryl ester hydrolase, another enzyme activated by cyclic AMP-dependent protein kinase and protein kinase C in rat liver (10), observed in the microsomes of taurine-treated animals, suggests that the membrane cholesterol levels more than the phosphorylation- dephosphorylation cycles of acyl-CoA:cholesterol acyltransferase and cholesteryl ester hydrolase enzymes, plays a major role in regulating the enzyme activities after a long-term treatment. The modifications in membrane cholesterol equilibria and the variations in enzyme activities involed in liver cholesterol metabolism due to taurine administration, as result from this and other studies, are outlined in Figure 6.

The lower incorporation of [^{14}C]oleoyl-CoA into triacylglycerol (i.e., the lower diacylglycerol acyltransferase activity measured in microsomes of taurine-treated rats), and its higher incorporation into the diacylglycerol class is in substantial agreement with the previous in vivo observation that taurine administration decreases hepatic triacylglycerol concentration in Wistar rats (30). Conversely, the previous data contrast with results of present experiments on the incorporation of fatty acyl-CoAs into phospholipids. The higher amount of radioactivity incorporated into phospholipids of taurine-treated rat microsomes, following incubation with either [^{14}C]oleoyl-CoA or [14C] arachidonoyl-CoA, suggests that the increased synthesis of diacylglycerol and its decreased conversion into triacylglycerol stimulate phospholipid synthesis (22). However, the taurine-induced increase in the acyl-CoAs hydrolases (demonstrated by the higher oleoyl-CoA hydrolase activity in the microsomes of taurine-treated rats) may lessen the availabilty of acylating substrates for both triacylglycerol and phospholipid synthesis. This, along with the increase in

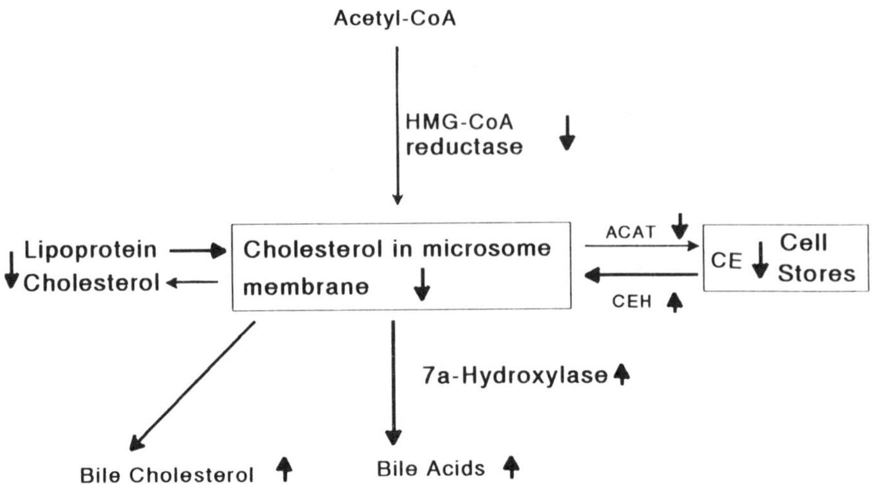

Figure 6. Summary of the effects induced by taurine supplement on liver cholesterol metabolism of rats. Upward arrows indicate an increase and downward arrows a decrease. The increased bile steroid secretion mobilizes the cholesteryl ester cell stores and enhances the uptake of circulating cholesterol. However, this mobilization of intra- and extra-hepatic resources cannot compensate for the reduced biosynthesis and the cholesterol proportion in the microsome membrane is reduced.

phospholipase A and lysophospholipase activities, may result in an enhanced phospholipid catabolism and not in increased levels of phospholipid, as observed in all our experiments.

In this study, the microsomal membrane of taurine-treated animals shows an increased ratio of PE/PC that is in accord with the results of the experiments of incorporation of activated fatty acyls into phospholipid classes of liver microsomes showing that [^{14}C]oleoyl--CoA, and even more the [^{14}C]arachidonoyl-CoA, are better incorporated into phosphatidylethanolamine than into phosphatidylcholine. A further contribution to the increase in phosphatidylethanolamine proportion in liver microsomes of taurine-treated animals may derive from the taurine-induced inhibition of phosphatidylethanolamine methyltransferase activity that causes a lower conversion of phosphatidylethanolamine into phosphatidylcholine, as suggested by Hamaguchi et al. (12). However, the increased proportion of phosphatidylethanolamine in microsomal membrane is paralleled by a decrease, and not an increase, in the phosphatidyl-ethanolamine content of whole liver following in vivo treatment with taurine. The liver cell has relevant lipid exchanges with both plasma and bile compartments that may explain this discrepancy. However, this point deserves further investigation.

Some general considerations may be derived from the present observations on phospholipid composition following treatment with taurine. The long-chain acyl-CoA thioesters are active metabolic intermediates of fatty acids and are required for both mitochondrial and microsomal esterification. The selectivity of the fatty acyl-CoA transferases helps to maintain the positional asymmetry of saturated and unsaturated fatty acids in diacylglycero-phospholipids.There is a good evidence that the acylation of 1-acyl-*sn*-glycero-3-phosphoryl-choline is important for the introduction of polyunsaturated fatty acids (principally arachidonate) into *sn*-2 position. At the same time, hepatic phospholipase A_1 and A_2 and lysophospholipase contribute to diacylglycerophospholipid retailoring by deacylating some specific molecular species of glycero-phospholipids. In this study, the activities of both acyl-CoA:1-acyl-glycero- phosphorylcholine acyltransferase and arachidonoyl-CoA synthetase increased in taurine-treated microsomes. These facts indicate that taurine stimulates phospholipid modification by increasing the deacylation-reacylation cycles of glycero-phospholipids metabolism. However, it is possible that in vivo treatment with taurine stimulates the deacylation more than the reacylation, as suggested by the taurine-induced decrease in polyunsaturated fatty acids of liver lipid classes (30). In accordance with this, we found in in vitro experiments that taurine increased the incorporation of [^{14}C]oleoyl-CoA and decreased that of [^{14}C]arachidonoyl-CoA into microsomal lysophosphatidylcholine (2.1 $\pm$ 0.63% and 3.2 $\pm$ 0.91 %, with oleoyl-CoA, and 12.7 $\pm$ 3.4% and 7.8 $\pm$ 1.5%, with arachidonoyl-CoA, in control and taurine-treated microsomes, respectively).

A higher phospholipase A_2 activity following treatment with taurine, was observed also in the synaptic membrane of guinea pig brain cortex (23). However, the activation of this enzyme was hardly attributable to taurine alone. Many common factors, including calcium, nonenzymatic proteins, cyclic adenosine monophosphate and hormones may affect the phospholipase A_2 enzyme activity (28). The effect of taurine on phospholipase activities observed in this study is probably mediated by interaction of taurine with membrane phospholipids. This would be in agreement with a previously proposed hypothesis that the interaction of taurine and phospholipids may alter both the conformation and charge of phospholipids and, consequently, the accessibility of these substrates to the enzymes (15).

In conclusion, the in vivo administration of taurine affects the free cholesterol content, the distribution of lipid classes, and their fatty acyl composition in the liver microsome membrane. These changes are known to affect the physical and the biochemical properties of microsome membrane and, therefore, may explain the wide spectrum of effects exerted by taurine on liver lipid metabolism. Furthermore, the immediate effect of variations in taurine on enzymatic processes occurring at the microsomal membrane interface suggests that once taurine triggers some slight change in the membrane microenvironment (probably by formation of low-affinity charge complexes) this primary change may induce, in the long term, marked changes in membrane composition and in membrane-bound enzyme activities. This study has not recognized the enzyme activity that directly affected by taurine intracellular levels to trigger the further changes in lipid environment of microsome membrane. Changes in either cholesterol or phosphatidylethanolamine proportion, due to their relevant effects on fluidity and function of hepatocyte membrane (24), are good candidates for this role. However, more investigations are required to understand the sequence and the details of these events in the living cell.

ACKNOWLEDGEMENTS

We thank Dr. Giorgio Pinto and Mrs. Tiziana Marinelli for technical support. This work was partially supported by the Progetto Finalizzato Invecchiamento of CNR, grant No. 93.00499.PF40 (Inv. 93.3.000).

REFERENCES

1. Angelico, M., Alvaro, D., Masella, R., Ginanni Corradini, S., and Cantafora, A. 1987, Transport, utilization and biliary secretion of lysophosphatidylcholine in the rat liver, *Biochim.Biophys.Acta*, 905:91-99.
2. Bar-Tana, J., Rose, G., and Shapiro, B. 1975, Long-chain fatty acyl-CoA synthetase from rat liver microsomes, *Methods Enzymol.*, 35:117-122.
3. Beg, Z.H., Stonik, J.A., and Brewer, H.B. 1987, Modulation of the enzymic activity of 3-hydroxy-3-methylglutaryl Coenzyme A reductase by multiple kinase systems involving reversible phosphorylation: a review, *Metabolism*, 36:900-917.

4. Bellentani, S., Pecorari, M., Cordana, P., Marchegiano, P., Manenti, F., Bosisio, E., Fabiani, E.D., and Galli, G. 1987, Taurine increases bile acid pool size and reduces bil saturation index in the hamster, *J.Lipid Res.*, 28:1021-1027.
5. Bradford, M. 1976, A rapid and sensitive method for the quantitation of microgram quantities of protein utilizing the priciple of protein-dye binding, *Anal.Biochem.*, 72:248-254.
6. Cantafora, A., Blotta, I., Rossi, S.S., Hofmann, A.F., and Sturman, J.A. 1991, Dietary taurine content changes liver lipids in cats, *J.Nutr.*, 121:1522-1528.
7. Cantafora, A., Mantovani, A., Masella, R., Mechelli, L., and Alvaro, D. 1986, Effect of taurine administration on liver lipids in guinea pigs, *Experientia*, 42:407-408.
8. Field, F.J., Albright, E., and Mathur, S. 1991, Inhibition of acylcoenzyme A: cholesterol acyltransferase activity by PD128042: effect on cholesterol metabolism and secretion in CaCO-2 cells, *Lipids*, 26:1-8.
9. Folch, T., Lees, B., and Sloane-Stanley, G.J. 1957, A simple method for the isolation and purification of total lipids from animal tissues, *Biol.Chem.*, 226:497-509.
10. Gosh, S. and McLean Grogan, W. 1989, Activation of rat liver cholesterol ester hydrolase by cAMP-dependent protein kinase and protein kinase C, *Lipids*, 24:733-736.
11. Graham, J. 1984, "Centrifugation", IRL Press, Washington.
12. Hamaguchi, T., Azuma, J., and Schaffer, S. 1991, Interaction of taurine with methionine: Inhibition of myocardial phospholipid methyltransferase, *J.Cardiovasc.Pharmacol.*, 18:224-230.
13. Huxtable, R.J. 1992, Physiological actions of taurine, *Physiol.Rev.*, 72:101-163.
14. Huxtable, R.J., Crosswell, S., and Parker, D. 1989, Phospholipid composition and taurine content of synaptosomes in developing rat brain, *Neurochem.International*, 15:233-238.
15. Huxtable, R.J. and Sebring, L.A. 1986, Towards a unifying theory for the action of taurine, *Trends Pharmacol.Sci.*, 7:481-485.
16. Ide, T., Horii, M., Yamamoto, T., and Kawashima, K. 1990, Contrasting effects of water-soluble and water-insoluble dietary fibers on bile acid conjugation and taurine metabolism in the rat, *Lipids*, 25:335-340.
17. Khoo, J.C., Steinberg, D., Huang, J.J., and Vagelos, R.P. 1976, Triglyceride, diglyceride, monoglyceride, and cholesterol ester hydrolases in chicken adipose tissue activated by adenosine 3',5'-monophosphate-dependent protein kinase. Chromatographic resolution and immunochemical differentiation from lipoprotein lipase, *J.Biol.Chem.*, 251:2882-2890.
18. Kibe, C., Wake, C., Kuramoto, T., and Hoshita, T. 1980, Effect of dietary taurine on bile acid metabolism in guinea pigs, *Lipids*, 15:224-229.
19. Li, Y.-P. and Lombardini, J.B. 1991, Taurine inhibits protein kinase C-catalyzed phosphorylation of specific proteins in a rat cortical P_2 fraction, *J.Neurochem.*, 56:1747-1753.
20. Masella, R. and Cantafora, A. 1991, Effect of physiological membrane-damaging detergents bile salt and lysolecithin on selectivity of reacylation and deacylation of phospholipids. A study with rat liver microsomes, *Acta Toxicol.Therap.*, 12:197-209.
21. Morre, J.D. 1975, Isolation of Golgi apparatus, *Methods Enzymol.*, 22:130-148.
22. Muir, J.G. and Murray, A.W. 1986, Mimicry of phorbol ester responses by diacylglycerols. Differential effects on phosphatidylcholine biosynthesis, cell-cell communication and epidermal growth factor binding, *Biochim.Biophys.Acta*, 855:176-184.
23. Ricci, G., Federici, G., Lucente, G., Achilli, M., and Cavallini, D. 1982, L-Lanthionine oxidation by snake venom L-amino acid oxidase, *Physiol.Chem.Phys.*, 14:193-199.
24. Schachter, D. 1984, Fluidity and function of hepatocyte plasma membranes, *Hepatology*, 4:140-151.
25. Stephan, Z.F., Lindsey, S., and Hayes, K.C. 1987, Taurine enhances low density lipoprotein binding, *J.Biol.Chem.*, 262:6069-6073.
26. Sugiyama, K., Ohishi, A., Ohnuma, Y., and Maramatsu, K. 1989, Comparison between the plasma cholesterol-lowering effect of glycine and taurine in rats fed on high cholesterol diets, *Agric.Biol. Chem.*, 53:1647-1652.
27. Taylor, R. 1993, Superoxide dismutase dysfunction: a radical route to familial ALS, *J.NIH Res.*, 5:64-67.
28. van den Bosch, H. 1980, Intracellular phospholipases, *Biochim.Biophys.Acta*, 604:191-246.
29. Yamanaka, Y., Tsuji, K., Ichikawa, T., Nakagawa, Y., and Kawamura, M. 1985, Effect of dietary taurine on cholesterol gallstone formation and tissue cholesterol contents in mice, *J.Nutr.Sci.& Vitaminology*, 31:225-232.
30. Yan, C.C., Bravo, E., and Cantafora, A. 1993, Effect of taurine levels on liver lipid metabolism: an in vivo study in the rat, *Proc.Soc.Exp.Biol.Med.*, 202:88-96.
31. Yeagle, P.L. 1989, Lipid regulation of cell membrane structure and function, *FASEB Journal*, 3:1833-1842.

INHIBITORS OF ANION EXCHANGER ACTIVITY REDUCE SODIUM CHLORIDE-DEPENDENT TAURINE TRANSPORT BY BRUSH BORDER VESICLES

Russell W. Chesney and Andrea M. Budreau

Department of Pediatrics
University of Tennessee
Memphis, Tennessee 38163

ABSTRACT

The erythrocyte anion exchanger, or band 3 protein, catalyzes the exchange of chloride and bicarbonate in many cells. Anion exchanger activity is predominantly found in the basolateral membrane of the cortical collecting duct. Because proximal tubular renal taurine uptake across the brush border surface is dependent upon sodium and chloride, the effect of inhibitors of anion exchanger activity on renal taurine accumulation by rat renal brush border membrane vesicles was examined. The anion exchanger probes, DIDS (4,4-'diisothiosulfonic acid), and NBD-taurine (N-[7-nitrobenz-2-oxa-1,3-diazol-4-yl]-aminoethane sulfonic acid), and NBD-taurine (N-[7-nitrobenz-2-oxa-1,3-diazol-4-yl]-aminoethane sulfonic acid), were used as inhibitors of anion exchanger activity. DIDS, NAP-, and NBD-taurine all markedly inhibit the initial rate of NaCl-dependent accumulation of taurine by brush border membrane vesicles. NAP- and NBD-taurine accumulation is chloride dependent because inhibition was not found when uptake was performed in the presence of $NaNO_3$ in place of chloride. In the presence of maximal inhibition of taurine uptake by NAP- or NBD-taurine, no additional inhibition was evident after incubation with DIDS. On the other hand, when a competitive inhibitor of taurine uptake, β-alanine, was used, additional inhibition of taurine accumulation was found in the presence of NAP- or NBD-taurine ($p < .01$ respectively). These results suggest some interaction of the anion exchanger and the taurine accumulation process at the apical surface of the proximal tubule. Because NBD-taurine is a fluorescent probe, it may be possible to isolate membrane peptides that bind NBD-taurine and demonstrate fluorescence. Preliminary isoelectric focusing experiments of brush border protein incubated with NBD-taurine show fluorescence localized to several particular fractions. Hence this observation that inhibitors of the anion exchanger system block NaCl-dependent taurine uptake can potentially serve as a means of isolating the taurine transporter protein.

Taurine in Health and Disease, Edited by R. Huxtable
and D.V. Michalk, Plenum Press, New York, 1994

INTRODUCTION

The erythrocyte anion exchanger catalyzes the exchange of chloride and bicarbonate in many cells. anion exchanger activity is predominantly found in the basolateral membrane of the cortical collecting duct. NAP-taurine (N-[4-azido-2-nitrophenyl]-2-aminoethane sulfonate, dihydrate), a photoreactive probe which is transported by the renal proximal tubular organic anion transport system (10,18), inhibits human erythrocyte exchange by binding to a site near the Cl^- transport site, locking the transport protein in the outward-facing form and preventing the conformational changes necessary for Cl^- transport (11-13). NBD-taurine (N-(7-nitrobenz-2-oxa-1,3-diazol-4-yl)-aminoethane sulfonic acid), is a fluorescent substrate-analogue also transported by the erythrocyte anion exchange system that has been shown to inhibit Cl^- exchange in a competitive manner from both sides of the membrane (7,9). Because proximal tubular renal taurine uptake across the brush border surface is dependent upon sodium and chloride, the effect of inhibitors of anion exchanger activity on renal taurine accumulation by rat renal brush border membrane vesicles (brush border membrane vesicles) was examined. The specific inhibitors characterized in these studies were selected because of their potential for use as probes to aid in the isolation of the taurine transporter protein.

METHODS

Animals

Male Sprague-Dawley rats (Harlan Sprague-Dawley Inc., Indianapolis, Indiana), aged 56-60 days and weighing 200-250 g each, were used in all studies. Experiments were conducted according to the National Institutes of Health Guide for Care and Use of Laboratory Animals.

Membrane Vesicle Preparation

Renal cortex brush border membrane vesicles were prepared by a modification of the method of Booth and Kenny (1). Rats were decapitated and exsanguinated, the kidneys removed, decapsulated, and placed in a solution of 0.05 mM mannitol, 1 mM $MgSO_4$, 2 mM tris (hydroxyethyl) aminoethane (Tris), and 2 mM N-2-hydroxyethylpiperazine-N'-2-ethane sulfonic acid (HEPES), pH 7.1. Renal cortex samples weighing 1.5 to 6.0 g each (wet weight) were homogenized in 20 volumes of the same buffer for 1 min with a Polytron (Kinematica, Switzerland) at setting no. 4 for 45 s and setting no. 7 for 15 s. Magnesium chloride was added to the homogenate to a final concentration of 10 mM and stirred on ice for 15 min to aggregate the intracellular and basolateral membranes. Repeated centrifugations and an additional $MgCl_2$ precipitation step were performed, as described elsewhere (4). The final membrane preparation was suspended in 396 mM mannitol, 1 mM $MgSO_4$, 2 mM HEPES-Tris (pH 7.35), and contained approximately 10 mg protein/ml. In all experiments membrane vesicles were used for uptake studies immediately after preparation.

Enzyme and Protein Determinations

Membrane purity was routinely assessed from the enrichment of various membrane marker enzymes relative to the homogenate. Enzymes assayed included γ-glutamyl transferase, a brush border membrane marker (16), 5'-nucleotidase, a marker for plasma membranes (8), ouabain-inhibitable Na^+/K^+-ATPase as a basolateral membrane marker

(17), malate dehydrogenase to indicate microsomal membranes (15), succinic cytochrome c reductase as a mitochondrial marker (19), and N-acetyl-β-D-glucosaminidase to indicate lysosomes (14). Protein was determined by the method of Bradford (2) using the BioRad Coomassie Brilliant Blue Dye reagent.

Uptake Studies

Uptake of radiolabelled taurine by vesicles in the presence and absence of various inhibitors was assayed by a Millipore rapid filtration technique (5). In general, incubation of the membranes was initiated by adding freshly prepared vesicles to media containing various salts and known amounts of labelled and unlabelled taurine to achieve the desired final concentration. Experimental uptake media contained varying concentrations of [^{3}H]taurine, 2 mM HEPES-Tris (pH 7.35), 1 mM $MgSO_4$, 100 mM NaCl or $NaNO_3$, and mannitol to balance osmolarity with that of the intravesicular medium (300 mOsm). At specific time intervals, 50 μl samples of the incubation mixture were placed onto pre-wetted 0.45 μm Millipore filters (HAWP) and washed with 3 volumes (3.0 ml each) of an ice-cold "stop" solution, which contained 2 mM HEPES-Tris, 1 mM $MgSO_4$, and 150 mM NaCl. Experiments using the light-sensitive NAP-taurine were performed in the presence of a red safe-light only. In some experiments, vesicles were loaded with NAP-taurine by performing the last 3 centrifugations of the membrane preparation in solutions containing 1.0 mM NAP-taurine. All incubations were performed in triplicate and each experiment was performed a minimum of three times. Data are expressed as picomoles taurine/mg protein retained by the vesicles on the filters. Filters were dried overnight in scintillation vials, then dissolved in a toluene-based cocktail and measured for radioactivity in a liquid scintillation counter (Packard Tri-Carb 2000CA). Nonspecific retention of radioactivity to detergent-lysed (0.1% SDS) membranes never exceeded 1-1.5% of the total amount found.

Electrophoresis Experiments

Brush border membrane vesicles were incubated with 100 mM NaCl and 1 mM NBD-taurine for 45 min, washed with approximately 10 volumes of buffer, then solubilized on ice for 1 h in buffer containing the non-denaturing zwitterionic detergent CHAPS. After centrifugation at 100,000 x g for 1 h and desalting by overnight dialysis at 4°C, the solubilized protein was subjected to preparative isoelectric focusing using a Rotofor (BioRad). The harvested fractions were scanned for fluorescence with a spectrofluorometer. Those fractions demonstrating fluorescence were then placed on a 12% polyacrylamide gel and separated according to molecular weight, as were various controls.

Analytical

Data comparisons were made with Statview 512$^{+\text{TM}}$ software package using Student's t-test for grouped, independent data (6). Data are expressed as mean $\pm$ SE.

Materials

[^{3}H]Taurine (25.6 Ci/mmole) was obtained from Du Pont, New England Nuclear Corp. (Boston, MA). NAP-taurine was custom made by Pierce Chemicals (Rockford, IL). NBD-taurine (N-[7-nitrobenz-2-oxa-1,3-diazol-4-yl]-aminoethane sulfonic acid) was purchased from Molecular Probes, Inc. (Eugene, OR). DIDS (4,4'-diisothiocyana-tostilbene-2,2-disulfonic acid) was obtained from Sigma Chemical Co. (St. Louis, MO). All other chemicals used to prepare media were reagent grade.

RESULTS

Analyses of brush border membrane preparations showed an enrichment ratio of γ-glutamyl transferase of 9.57 $\pm$ 0.52 relative to starting homogenate. The enrichment ratio of 5'-nucleotidase was 7.70 $\pm$ 0.76. Enrichment of other marker enzymes was not evident.

A number of preliminary experiments were performed in which the concentrations of NAP-taurine and NBD-taurine were varied to determine optimum levels with which to achieve maximal inhibition of taurine uptake (data not shown). At 35 μM and 100 μM, NAP-taurine only slightly inhibited uptake of 10 μM taurine; the degree of inhibition was not significant. Inhibition of taurine uptake by NBD-taurine was dose-dependent, as evidenced by increased inhibition with increasing NBD-taurine concentration. Both NAP-and NBD-taurine, at 1.0 mM each, significantly inhibited the secondary active, NaCl-dependent component of 10 μM taurine uptake by rat brush border membrane vesicles (Figs. 1, 2). The concentration of taurine was then varied (10, 25, 50, 75,100 μM) and the effect of 1.0

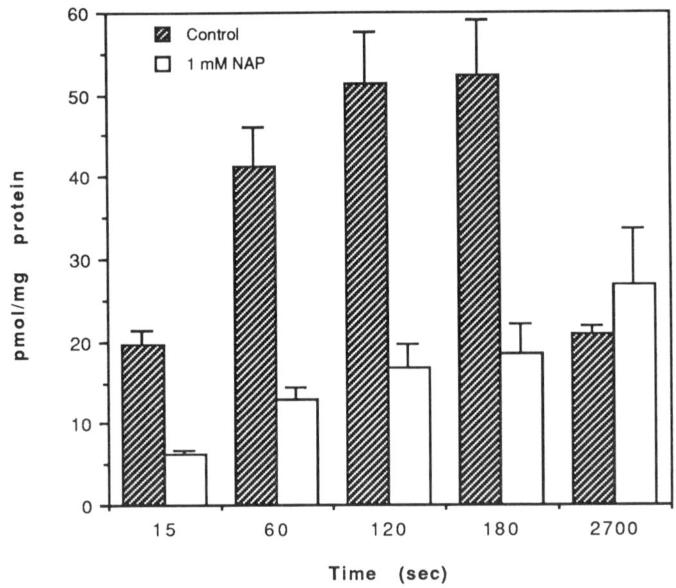

Figure 1. 10 μM taurine uptake in the presence and absence of 1 mM NAP-taurine. Data are the mean of 3 experiments, each performed in triplicate with different membrane preparations. Conditions are as described in the Methods section.

mM NAP-taurine on uptake was measured at 15 s, a time point accepted as representative of initial rate of uptake (20). The effect of NAP-taurine decreased with increasing taurine concentration. Uptake of both 10 and 25 μM taurine was significantly inhibited, ($p < 0.02$ and 0.01, respectively). Uptake of 50 μM taurine was affected, but did not reach statistical significance ($p < 0.07$). The NaCl-independent diffusional component of taurine transport was not affected, as uptake of taurine at concentrations greater than 50 μM was not significantly inhibited (data not shown). In some experiments brush border membranes were loaded with 1.0 mM NAP-taurine during vesicle preparation, then taurine uptake was measured in the presence and absence of external NAP-taurine. In the presence of 1.0 mM internal NAP-taurine, uptake of 10 μM taurine was significantly inhibited regardless of whether NAP-taurine was present in the external medium.

DIDS, a specific, nonpenetrating inhibitor of anion permeability, also significantly inhibited taurine uptake when used at a final concentration of 0.5 mM, as demonstrated previously (20). However, these inhibitory effects were not additive: When brush border

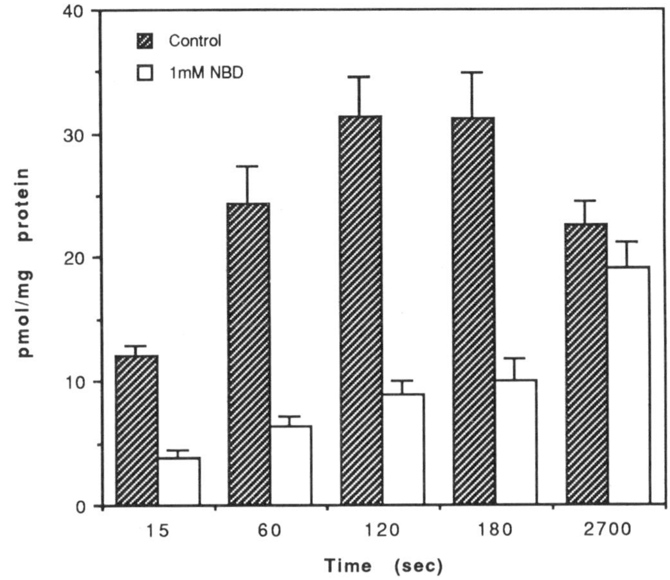

Figure 2. 10 μM taurine uptake in the presence and absence of 1 mM NBD-taurine. Data are the mean of 3 experiments, each performed in triplicate with different membrane preparations. Conditions are as described in the Methods section.

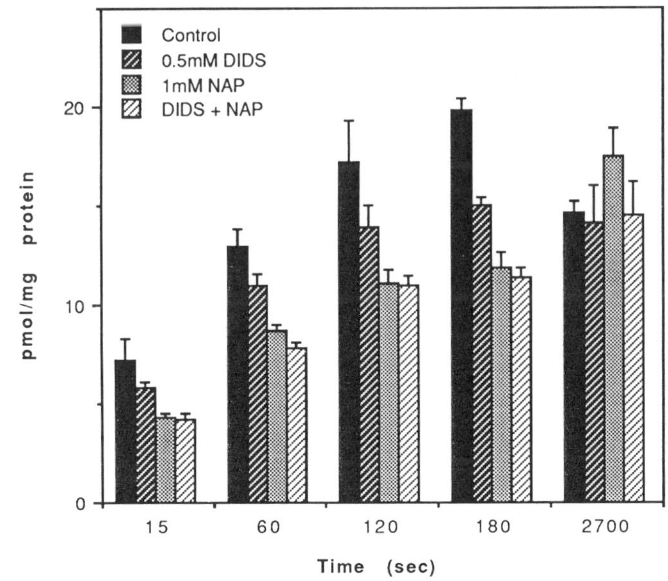

Figure 3. Inhibition of 10 μM taurine uptake by 1 mM NAP-taurine $\pm$ 0.5 mM DIDS. Data are the mean of 3 experiments, each performed in triplicate with different membrane preparations. Conditions are as described in the Methods section.

membrane vesicles were incubated with either taurine analogue in conjunction with DIDS, inhibition was no greater than that achieved with NAP- or NBD-taurine alone (Figs. 3, 4).

The Cl^--dependence of the observed inhibition was examined by substituting an equimolar amount of $NaNO_3$ for the NaCl in the incubation mixture. As shown in Figures 7 and 8, taurine uptake in the presence of NO_3^- was not affected by either analogue,

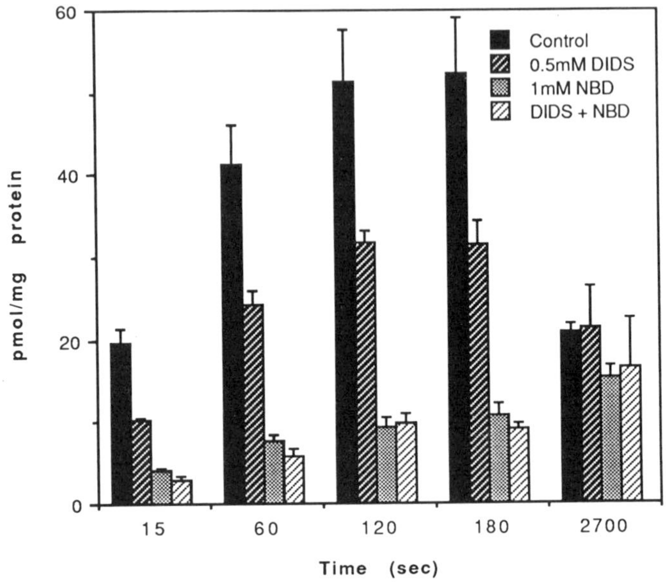

Figure 4. Inhibition of 10 μM taurine uptake by 1 mM NBD-taurine $\pm$ 0.5 mM DIDS. Data are the mean of 3 experiments, each performed in triplicate with different membrane preparations. Conditions are as described in the Methods section.

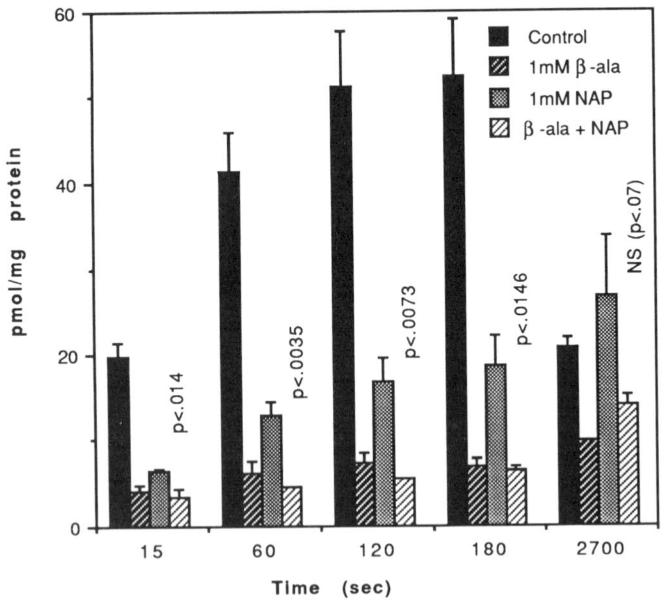

Figure 5. Inhibition of 10 μM taurine uptake by 1 mM NAP-taurine $\pm$ 1.0 mM β-alanine. Data are the mean of 3 experiments, each performed in triplicate with different membrane preparations. Conditions are as described in the Methods section.

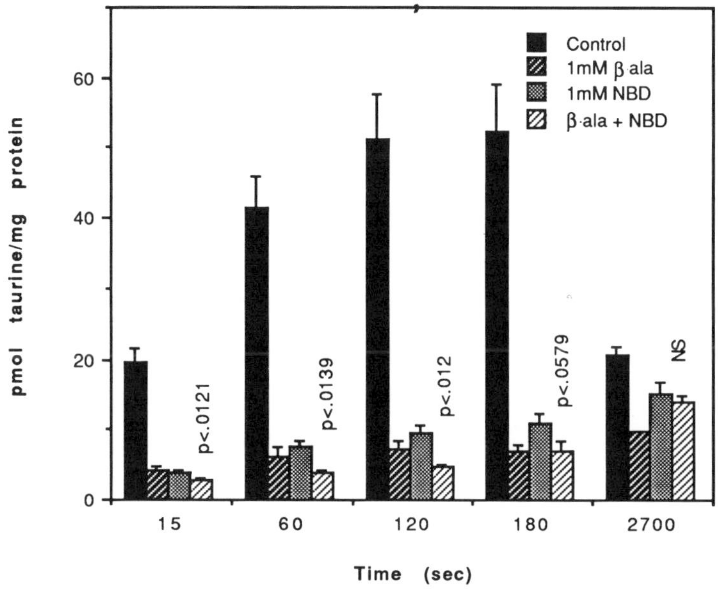

Figure 6. Inhibition of 10 μM taurine uptake by 1 mM NBD-taurine $\pm$ 1.0 mM β-alanine. Data are the mean of 3 experiments, each performed in triplicate with different membrane preparations. Conditions are as described in the Methods section.

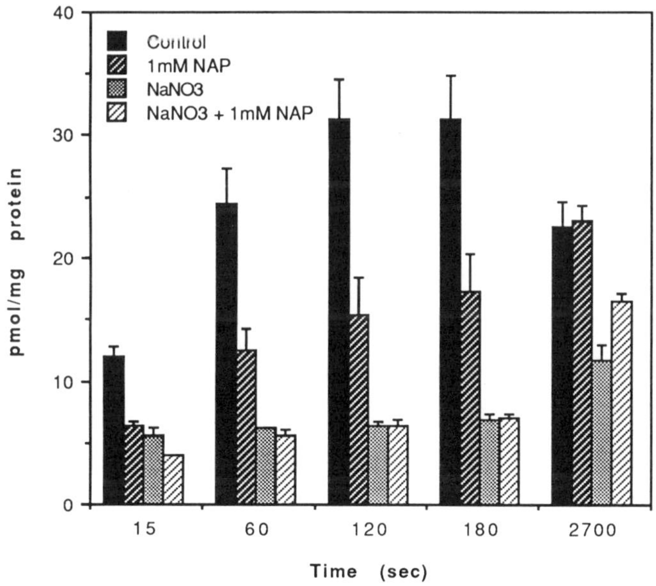

Figure 7. 10 μM taurine uptake in the presence and absence of 1 mM NAP-taurine with 100 mM $NaNO_3$ replacing NaCl in the incubation mixture. Data are the mean of 3 experiments, each performed in triplicate on different membrane preparations. Media are as described in Methods.

indicating that the inhibition is indeed Cl^--dependent. These experiments served also to confirm that the effects seen were not simply the result of nonspecific binding of the analogues to the membranes. On the other hand, when a known competitor for taurine binding sites, such as β-alanine, was used, inhibition of taurine uptake was increased significantly (Figs. 5, 6).

Isoelectric focusing of solubilized brush border membranes after incubation with 1 mM NBD-taurine yielded fractions with peaks of fluorescence which may indicate binding of the probe to certain membrane peptides. Polyacrylamide gel electrophoresis of those fractions showed distinct bands at approximately 80 kd. Further study is required before any conclusions may be drawn from these preliminary experiments.

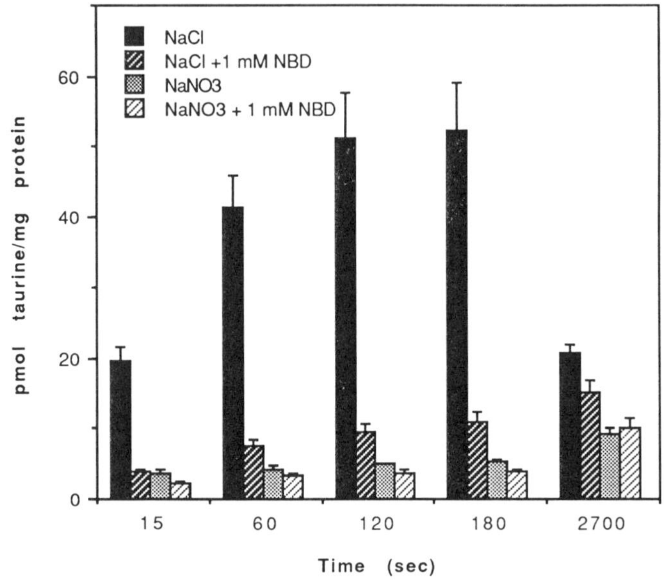

Figure 8. 10 μM taurine uptake in the presence and absence of 1 mM NBD-taurine with 100 mM $NaNO_3$ replacing NaCl in the incubation mixture. Data are the mean of 3 experiments, each performed in triplicate on different membrane preparations. Media are as described in Methods.

DISCUSSION

This study suggests that there is a strong interaction between the anion exchanger and the taurine accumulation process at the apical surface of the proximal tubule of the rat kidney. The role of various anion exchange inhibitors was investigated because the proximal tubule renal process is so highly dependent upon sodium and chloride. The disulfonic stilbenes have been shown in the past to block the transport of several organic solutes (9), yet little is known about the interaction of a NaCl-dependent amino acid transport process with anion exchanger activity. Perhaps the most significant finding of the present study is that all of the anion exchanger probes, DIDS, NAP-taurine, and NBD-taurine, appear to inhibit markedly the initial rate of NaCl-dependent taurine uptake. As a further characteristic, it was shown that both NAP- and NBD-taurine inhibition of taurine accumulation was Cl^- dependent, because the inhibition was absent when uptake was performed under the condition of external $NaNO_3$ rather than NaCl. In contrast to studies done by Knauf et al. in erythrocytes (12), where sidedness plays an important role in the inhibition of anion transport, NAP-taurine inhibited taurine uptake in proximal brush border membrane vesicles regardless of whether it was present inside or outside the vesicles, though the inhibition was not additive when it was present on both sides of the membrane at once. Hence, these classical anion exchange inhibitors appear to be particularly active against the taurine transport process.

In 1976, Cabantchik stated that the fact that NAP-taurine and DIDS compete for binding sites suggests that inhibitory and transport sites may be closely related (3). He further suggested that these findings implicated a 95,000 MW protein as an element of the anion transport system, since it appeared that this protein contained that binding site. The results presented here are somewhat at variance with those obtained with erythrocytes, because in the presence of maximal inhibition of taurine by NAP- or NBD-taurine, no additional inhibition was evident after incubation with DIDS. The results can further be interpreted to indicate that β-alanine inhibits taurine accumulation at some site different from the anion inhibitors. These results strongly suggest some interaction of the anion exchanger and the taurine accumulation process at the apical surface of the proximal tubule. Because these inhibitors are clearly important in Cl^--Cl^- or Cl^--HCO_3^- exchange, it is possible that they block the Na^+ and Cl^- binding sites on the taurine transport molecule. This, however, remains a speculation until the primary and secondary structures of the transport molecule are known, as well as the interactions with binding proteins. At the present time, the details of cytoskeletal fixation, membrane-lipid interactions, binding site of the amino acid, Na^+ binding site, rank order binding of ions and solute, regulation by second messengers or tyrosine kinases and other protein kinases are not understood nearly as well as those of the Na^+-H^+ antiporter or the Na^+K^+- ATPase systems. Hence, the clues provided by the use of anion inhibitors have the potential for generating useful insights into the nature of the taurine transport process in renal proximal tubule.

Preliminary studies using a nondenaturing technique of isoelectric focusing have shown that NBD-taurine fluorescence is localized to several particular fractions, which suggests that this methodology can potentially serve as a means of isolating the taurine transporter protein. This laboratory is currently involved in using molecular approaches to the isolation of the taurine transporter, but the utilization of a fluorescent probe may provide a means of further clarifying the nature and physicochemical characteristics of the transporter protein.

ACKNOWLEDGMENTS

This study was supported in part by the the LeBonheur Chair of Excellence of the Department of Pediatrics of the University of Tennessee, and NIH grant DK37221-08.

REFERENCES

1. Booth, A.G, and Kenny, A.J., 1974, A rapid method for the preparation of microvilli from rabbit kidney, *Biochem. J.* 142:575-581.
2. Bradford, M., 1976, A rapid and sensitive method for the quantitation of microgram quantities of protein utilizing the principle of protein-dye binding, *Anal. Biochem.* 72:248-253.
3. Cabantchik, Z.I., Knauf, P.A., Ostwald, T., Markus, H., Davidson, L., Breuer, W., and Rothstein, A., 1976, The interaction of a photoreactive probe with the anion transport system of the human red blood cell, *Biochim. Biophys. Acta* 455:526-537.
4. Chesney, R.W., Gusowski, N., and Friedman, A.L., 1983, Renal adaptation to altered dietary sulfur amino acid intake occurs at the luminal brush border membrane, *Kidney Int.* 24:588-594.
5. Chesney, R.W., Sacktor, B., and Rowen, R.R., 1973, The binding of D-glucose to the isolated rabbit renal brush border, *J. Biol. Chem.* 218:2182-2191.
6. Daniel, W.D., 1983, Biostatistics: A foundation for analysis in the health sciences. Wiley, New York, pp 177-180.

7. Darmon, A., Eidelman, O., and Cabantchik, Z.I., 1982, A method for measuring anion transfer across membranes of hemoglobin-free cells and vesicles by continuous monitoring of fluoresence, *Anal. Biochem.* 119:313-321.
8. Dixon, T.F., and Purdom, M., 1954, Serum 5-nucleotidase, *J. Clin. Path.* 7:341-343.
9. Eidelman, O., Zangvill, M., Razin, M., Ginsburg, H., and Cabantchik, Z.I., 1981, The anion transfer system of erythrocyte membranes, *Biochem. J.* 195:503-513.
10. Goldinger, J.M., Khalsa, B.D.S., and Hong, S.K., 1984, Photoaffinity labeling of organic anion transport system in proximal tubule, *Am. J. Physiol.* 247 (Cell Physiol. 16):C217-227.
11. Knauf, P.A., Breuer, W., McCulloch, L., and Rothstein, A., 1978, N-(4-azido-2-nitrophenyl)-2-aminoethylsulfonate (NAP-taurine) as a photoaffinity probe for identifying membrane components containing the modifier site of the human red blood cell anion exchange system, *J. Gen. Physiol.* 72:631-649.
12. Knauf, P.A., Law, F-Y., Tarshis, T., and Furuya, W., 1984, Effects of the transport site conformation on the binding of external NAP-taurine to the human erythrocyte exchange system, *J. Gen. Physiol.* 83:683-701.
13. Knauf, P.A., Mann, N.A., Kalwas, J.E., Spinelli, L.J., and Ramjeesingh, M., 1987, Interactions of NIP-taurine, NAP-taurine, and Cl^- with the human erythrocyte anion exhange system, *Am. J. Physiol.* 253 (Cell Physiol. 22):C652-C661.
14. Leaback, D.H., and Walker, P.G., 1961, Studies on glucosaminidase IV. The fluorimetric assay of N-acetyl-β-D-glucosaminidase, *Biochem. J.* 78:151-156.
15. Lowry, O.H., and Passonneau, J.V., 1972, A flexible system of enzymatic analysis, Academic Press, New York.
16 Pillion, D.J,. Jeske, A.H., and Leibach, F.H., 1976, Gamma-glutamyl transpeptidase in the urine from an isolated rabbit kidney perfused with and without DMSO, *Biochem. Pharmacol.* 25:913-918.
17. Post, R.L., and Sen, A.K., 1967, Sodium and potassium-stimulated ATPase, energy-linked reactions, *Methods Enzymol.* X:762-768.
18. Stokols, M.F., Koschier, F.J., Goldinger, J.M., and Hong, S.K., 1981, Renal transport of NAP-taurine, *Am. J. Physiol.* 241 (Renal Fluid Electrolyte Physiol. 10):F9-F13.
19. Tisdale, H.D., 1967, Preparation and properties of succinic cytochrome c reductase (Complex II-III), *Methods Enzymol.* X:213-217.
20. Zelikovic, I,. Stejskal-Lorenz, E., Lohstroh, P., Budreau, A., and Chesney, R.W., 1989, Anion dependence of taurine transport by rat renal brush border membrane vesicles, *Am. J. Physiol.* 256:F646-F655.

EXPRESSION OF TAURINE TRANSPORTER AND ITS REGULATION BY DIET IN XENOPUS LAEVIS OOCYTES FOLLOWING INJECTION OF RAT KIDNEY CORTEX mRNA

Xiaobin Han and Russell W. Chesney

Department of Pediatrics
University of Tennessee
Memphis, Tennessee 38163

ABSTRACT

NaCl-dependent taurine transport adapts to changes in the dietary intake of sulfur amino acids. The renal adaptive response is expressed by enhanced NaCl-dependent taurine cotransport by brush border membrane vesicles after a low taurine diet and reduced transport after a high taurine diet as compared to a normal taurine diet. In order to determine if this adaptive regulation is dependent on new protein synthesis, the *Xenopus laevis* oocyte expression system was utilized to define the translational regulation of taurine transporter activity. Poly(A)$^+$ RNA was isolated from kidney cortex of Sprague Dawley rats fed either a low, normal or high taurine diet for 28 days. Injection of poly(A)$^+$ RNA resulted in a time- and dose-dependent increase in NaCl-taurine co-transport. Taurine uptake was stimulated about 2-10-fold after injection of poly(A)$^+$ RNA (10-40 ng) as compared to H_2O-injected oocytes. Taurine uptake by oocytes was sodium- and anion-dependent ($Cl^- > Br^- > SCN^- > I^-$). The K_m and V_{max} of the taurine transporter were 22.5 μM and 8.35 pmol/h/oocyte respectively, similar to the K_m of 17.0 μM found in rat brush border membrane vesicles. Because the adaptive response involves an augmented or reduced V_{max} of the transporter, taurine uptake by oocytes injected with poly(A)$^+$ RNA from rats fed each diet was examined. Poly(A)$^+$ RNA from rats fed a low taurine diet elicited twice the taurine uptake elicited from rats fed a normal taurine diet and more than three times the uptake from high taurine-fed rats. Northern blot analysis after hybridization with an RNA probe for the taurine transporter cDNA from MDCK cells (obtained from Dr. Uchida) indicated that the molecular size of taurine transporter mRNA is about 1.9 kb and is regulated by diet. Expression of taurine transporter by the oocytes injected with 30 ng of capped transcript from pNCT was significantly reduced by taurine in the medium. In conclusion, taurine uptake by oocytes after injection of mRNA is similar to brush border membrane vesicles taurine transport. The long-term adaptive response is regulated at the level of mRNA, and the short-term adaptive response is regulated at the level of protein synthesis or secretion. We speculate that the renal adaptive response to altered dietary sulfur amino acid intake is both transcriptionally and translationally regulated.

Taurine in Health and Disease, Edited by R. Huxtable and D.V. Michalk, Plenum Press, New York, 1994

INTRODUCTION

Studies have shown that the renal tubular epithelium can adapt to alterations in the sulfur amino acid composition of the diet, particularly in terms of reabsorption of the sulfur-containing β-amino acid taurine (1-3,6). Taurine is a compound whose body pool size is regulated by the kidney (3). The renal adaptive response is expressed by enhanced NaCl-dependent taurine co-transport by brush border membrane vesicles after a low taurine diet as compared to a normal taurine diet. These changes in tubule and brush border membrane vesicle accumulation of taurine after dietary manipulation are associated with a change in the rate of taurine uptake (V_{max}) rather than in the affinity of the uptake system for taurine (K_m) (2,7,10). In order to determine if this adaptive regulation is dependent on new protein synthesis, the *Xenopus laevis* oocyte expression system was utilized to define the dietary regulation of taurine transporter activity. Poly(A)$^+$ RNA was isolated from the kidney cortex of Sprague-Dawley rats fed either low, normal or high taurine diets for 28 days. It was found that injection of poly(A)$^+$ RNA resulted in a time- and dose-dependent increase in NaCl-taurine co-transport with kinetic and substrate characteristics like those in brush border membrane vesicles, and that the expression was increased in oocytes injected with poly(A)$^+$ RNA from rats fed a low taurine diet and decreased in oocytes injected with poly(A)$^+$ RNA from rats fed a high taurine diet, as compared to oocytes injected with poly(A)$^+$ RNA from normal taurine-fed rats.

METHODS

Animals

Male Sprague-Dawley rats (Harlan Sprague-Dawley Inc., Indianapolis, Indiana), aged 56-60 days and weighing 200-250 g each, were used in all studies. Rats were fed low taurine diet, normal taurine diet, or high taurine diet for 28 days prior to sacrifice. The high taurine diet is supplemented with 3% taurine and 0.5% methionine (weight/weight), the normal taurine diet is supplemented only with 0.5% methionine, and the low taurine diet is supplemented with neither amino acid. The composition of each diet has been described in detail (4).

RNA Preparation

Total RNA was prepared as previously described (12). Briefly, guanidinium thiocyanate was used to disrupt rat kidney cortex tissues, and the resulting homogenate was layered on a cushion of a dense solution of CsCl (5.7 M, 0.01 M EDTA, pH 7.5). Total RNA was pelleted from the lysate after centrifugation at 32,000 x g in a Beckman ultracentrifuge for 18 h at 25°C. Poly(A)$^+$ RNA was isolated by oligo(dT)-cellulose chromatography, reprecipitated in ethanol, and then finally dissolved in DEPC-treated water to a final concentration of 1.0 μg/μl. Capped mRNA transcript was made by SP_6 polymerase using Sal I-cut pNCT (taurine transporter from MDCK cells) as a template.

Microinjection of *Xenopus laevis* Oocytes

Ovarian lobes were dissected from anesthetized frogs by incubation of ovarian fragments for 40 min with 2 mg/ml collagenase type II in calcium-free buffer at room temperature. Defolliculated stage V-VI oocytes were selected and incubated overnight at 18°C in media (50% L-15 [Leibovitz, Sigma], 1 mM L-glutamine, 15 mM HEPES, pH 7.6, 100 μg/ml gentamycin sulfate) before injection (5,14,15). Water (40 nl) containing 0-40 ng of poly(A)$^+$ RNA and 30 ng of capped transcript was injected into each oocyte. Injected oocytes were maintained at 18°C in the above media for 1-3 days.

Measurement of Taurine Uptake

Oocytes were transferred to Na^+-containing uptake solution (2 mM KCl, 1 mM $MgCl_2$, 100 mM NaCl) or Na^+-free uptake solution containing 10 μM unlabelled taurine and 0.5 μCi/ml ^{14}C-taurine (Du Pont, NEN). After incubation for an appropriate time at room temperature, oocytes were transferred to a 24-well cell culture plate and washed rapidly 5 times with 1 ml of ice-cold Na^+-free uptake solution. Finally, individual oocytes were transferred to miniscintillation vials, solubilized in 100 μl of 10 % SDS, and counted in 2 ml of scintillation cocktail (Aquasol, NEN).

Northern Blot Analysis

10 μg of poly(A)$^+$ RNA isolated from kidney cortex from each group of rats was separated in a 1.2% agarose gel containing 17.6% formaldehyde and transferred to a nylon membrane (Genescreen plus, NEN) by overnight capillary blotting in 10x SSC. Before hybridization, the membrane was incubated for 1 h at 42°C in a prehybridization solution containing 50% formamide, 1 M NaCl, 5% dextran sulfate, and 1% SDS. The Northern blot was hybridized overnight at 42°C with ^{32}P-labelled pNCT subclone probe (randomly primed Hind III fragment of pNCT) in the prehybridization solution. The blot was washed successively in 2x SSC/0.1% SDS, 1x SSC/0.1% SDS, and 0.1x SSC/0.1% SDS at 65°C for 30 minutes respectively, and was exposed to Kodak film at -80° C for one week.

Materials

[^{14}C]-taurine (92.1 mCi/mmol) and [^{32}P]-dCTP (3000 Ci/mmol) were purchased from Du Pont, NEN. pNCT was a generous gift from Dr. S. Uchida.

Data Analysis

All experiments were performed in duplicate. The data presented are the mean $\pm$ SE for 6 oocytes. Student's t test for paired data was applied to determine significant differences in the means.

RESULTS

Basic Observations

Oocytes injected with water exhibited a low rate of taurine uptake, usually indistinguishable from the background. Injection of poly(A)$^+$ RNA from rat kidney cortex resulted in an increased rate of NaCl-taurine co-transport. Taurine uptake was stimulated about 2-10-fold after injection of poly(A)$^+$ RNA (10-40 ng) as compared to water-injected oocytes.

Because the adaptive response involves an augmented or reduced V_{max} of initial rate transport activity, taurine uptake by oocytes injected with poly(A)$^{+}$ RNA from rats fed each diet was examined.

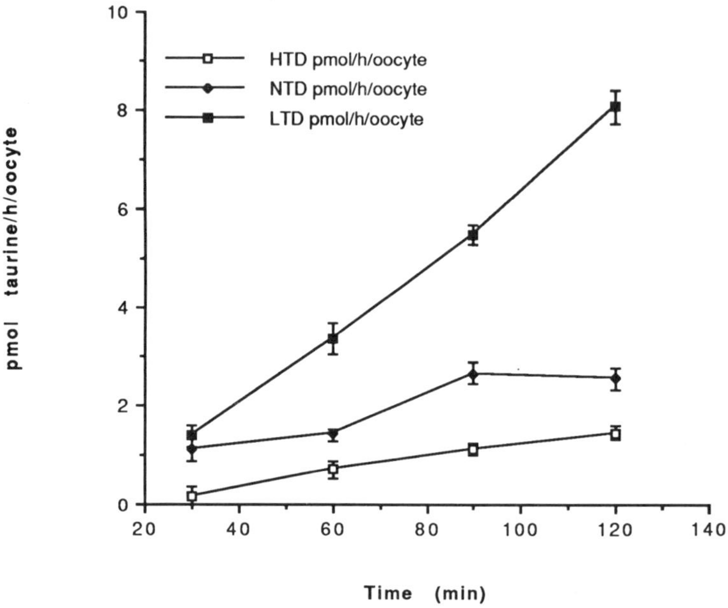

Figure 1. Time course of taurine uptake by mRNA-injected oocytes. Oocytes were injected with 40 nl of water containing 40 ng of mRNA, and uptake was assessed at 3 days post-injection.
HTD: high taurine diet; NTD: normal taurine diet; LTD: low taurine diet.

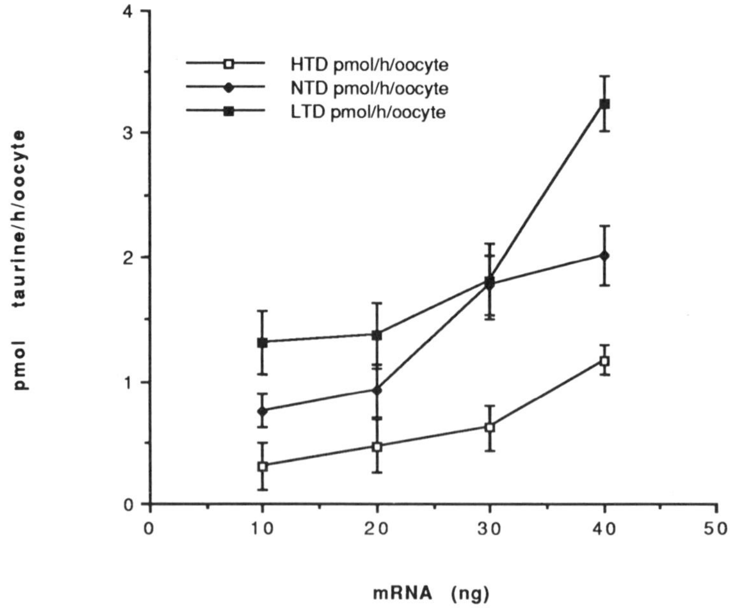

Figure 2. Dose-dependent expression of taurine uptake. Oocytes were injected with 10-40 ng mRNA and uptake was assessed 3 days later. Abbreviations: see Fig. 1.

Time Course of Taurine Uptake by Poly(A)$^+$ RNA-injected Oocytes

As illustrated in Fig. 1, the rate of uptake of taurine by poly(A)+ RNA-injected oocytes was linear. Taurine uptake by the oocytes injected with poly(A)$^+$ RNA from the kidney cortex of rats fed a low taurine diet was significantly higher than that from rats fed a normal taurine diet, and taurine uptake was significantly decreased when the oocytes were injected with poly(A)$^+$ RNA from the kidney cortex of rats fed a high taurine diet.

Dose-dependent Expression of Taurine Uptake

Expression of the taurine transporter was related to the amount of poly(A)$^+$ RNA injected (between 10-40 ng) (Fig. 2). Although the linearity is similar, higher transport rates were obtained by injecting poly(A)$^+$ RNA from kidneys of rats fed a low taurine diet. Based on these results, 40 ng of poly(A)$^+$ RNA were injected into oocytes in all subsequent experiments.

Time-dependent Expression of Taurine Uptake

Fig. 3 depicts the time-dependent expression of taurine uptake after poly(A)$^+$ RNA (40 ng/40 nl) injection into oocytes. Marked expression of taurine uptake was observed after 2 days, and this increased further to reach a plateau after 3 days. Poly(A)$^+$ RNA from low taurine-rats elicited twice the taurine uptake of normal taurine-fed rats, and more than three times the uptake of high taurine-fed rats.

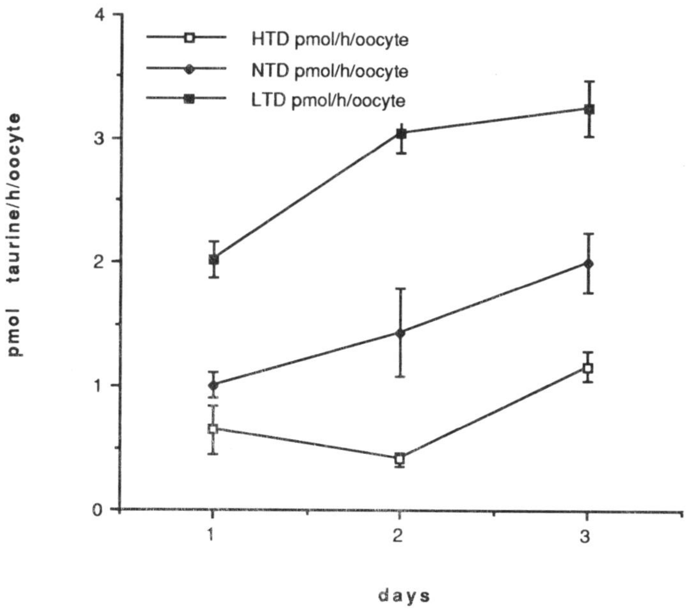

Figure 3. Time-dependent expression of taurine uptake. 40 ng of mRNA from rat kidney cortex were injected into oocytes, and uptake was assessed 3 days later in the presence of NaCl. For abbreviations, see Fig. 1.

Characterization of Expressed Taurine Uptake

The transporter expressed after injection of poly(A)$^+$ RNA from low taurine-fed rat kidney cortex was characterized to confirm that the expressed transporter was the same as that found in brush border membrane vesicles. The expressed taurine uptake obtained by subtracting the uptake in poly(A)$^+$ RNA-injected oocytes showed Michaelis-Menton kinetics (Fig. 4). Eadie-Hofstee transformation yielded an apparent K_m and V_{max} of 22.5 μM and 8.35 pmol/h/oocyte, respectively.

Anion Dependence of Taurine Transporter

Oocytes were injected with 40 ng of mRNA from low taurine-fed rat kidney cortex. Experiments were performed in which Cl^- in the uptake media was replaced with various anions. Oocytes were preincubated for 30 min in Na^+- and taurine-free uptake solution, then switched into Na^+-containing taurine uptake solution. Taurine uptake by oocytes injected with poly(A)$^+$ RNA from kidney cortex of low taurine-fed rats was sodium- and anion-dependent. As shown in Fig. 5, taurine uptake was reduced markedly when Cl^- was replaced with I^-, or SCN^-. Bromide, however, restored in part the taurine transport by oocytes.

Regulation of Taurine Transporter mRNA by Dietary Taurine

As shown in Fig. 6, the abundance of taurine transporter mRNA was increased in renal cortex from rats fed a low taurine diet and reduced in the tissue from rats fed a high taurine diet, as compared to rats fed a normal taurine diet. The average size of mRNA for the rat kidney taurine transporter is about 1.9 kb (1.4-2.4 kb).

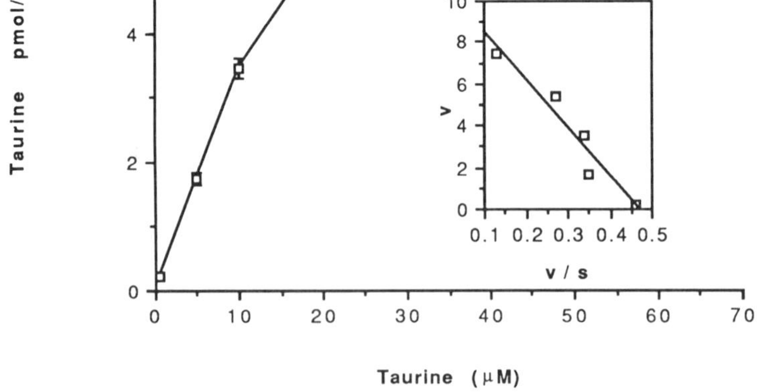

Figure 4. Kinetics of taurine uptake induced in oocytes by injection of 40 ng mRNA from kidney cortex of rats fed a low taurine diet. Uptake of different levels of taurine was assessed at 3 days post-injection.

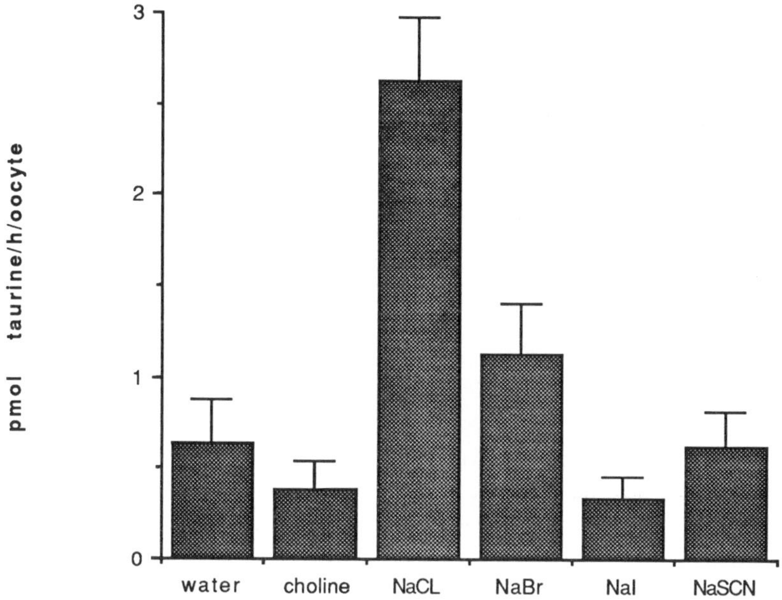

Figure 5. Taurine uptake by oocytes in the presence of various anions. Oocytes were injected with 40 ng of mRNA from low taurine-fed rats, and experiments were performed in which Cl^- in the uptake medium was replaced with various anions.

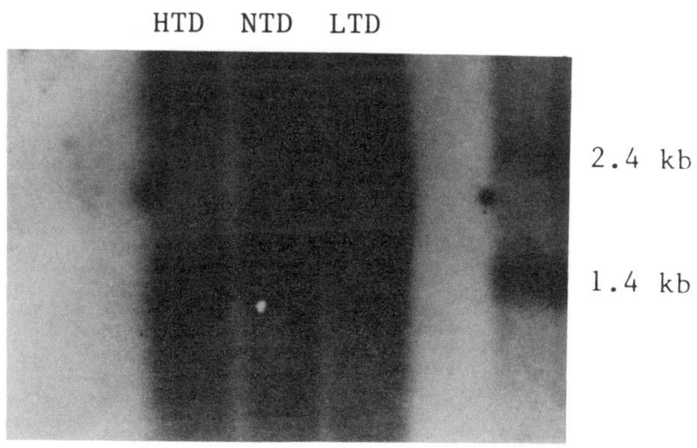

Figure 6. Expression and regulation of taurine transporter mRNA by diet. 10 μg of poly$(A)^+$ RNA from rats fed each diet was loaded per well in a standard formaldehyde/agarose gel. The RNA probe was made by T_7 RNA polymerase using Not I-cut pNCT as a template.

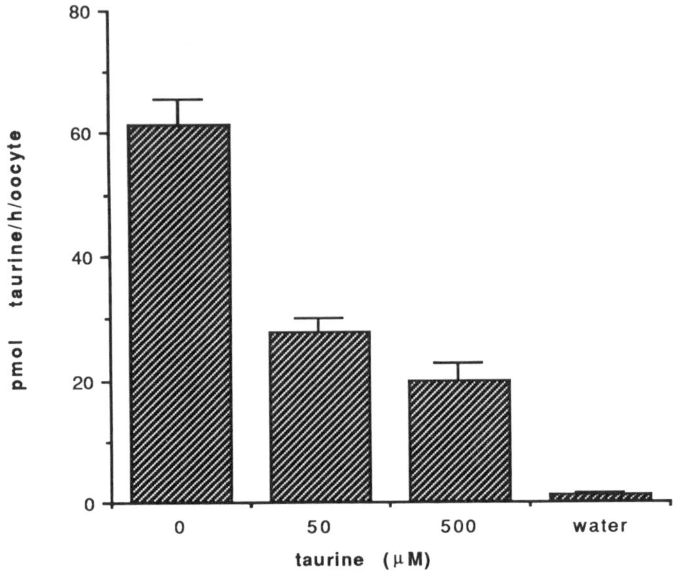

Figure 7. Oocytes were incubated in media containing different levels of taurine (0, 50, 500 μM) for 3 days, then taurine uptake was measured.

Regulation of Expression of Taurine Transporter by Medium Taurine

Oocytes were injected with 30 ng of capped transcript and incubated in media containing 0, 50 or 500 μM taurine for 3 days, then taurine uptake was measured. Results show that expression of the taurine transporter was significantly reduced by the presence of 50 μM taurine in the medium as compared to the control without taurine (Fig. 7).

Twenty oocytes from each group, with and without capped mRNA injection, were incubated in the above media for 8 h. [^{35}S]methionine (120 μCi) was added to the media and the oocytes were further incubated for 24 h. The labelled protein samples were prepared and analyzed by SDS-polyacrylamide gel electrophoresis (12%), then exposed to film for 7 days. An [^{35}S]methionine-labelled translation product was found (MW about 97.5 kd) from the oocytes incubated in the media without taurine (data not shown).

DISCUSSION

The renal adaptive response to alterations in dietary sulfur amino acid intake involve increased or decreased initial rate activity of the NaCl-dependent taurine transporter at the brush border membrane surface of the proximal tubule (1-3,6). This study was conducted to determine at what level the renal adaptive response to altered dietary sulfur amino acid intake is regulated. The results showed that the endogenous Na^+-dependent taurine uptake rate in *Xenopus laevis* oocytes was small and was not Cl^--dependent. Injection of poly(A)$^+$ RNA resulted in a time- and dose-dependent increase in NaCl-taurine co-transport. Taurine uptake was stimulated about 2-10-fold after injection of poly(A)$^+$ RNA (10-40 ng) as compared to water-injected oocytes. Poly(A)$^+$ RNA-injected oocytes showed saturable taurine transport, with a K_m and V_{max} of the taurine transporter of 22.5 μM and 8.35 pmol/h/oocyte respectively, similar to the K_m of 17.0 μM found in rat brush border membrane vesicles (3). Other characteristics of the transporter expressed in poly(A)$^+$ RNA-injected oocytes also resembled those of brush border membrane vesicles. Taurine uptake by oocytes injected with poly(A)$^+$ RNA was sodium- and anion-dependent ($Cl^- > Br^- > SCN^- > I^-$), as is taurine transport in brush border membrane vesicles (16).

Because the adaptive response involves an augmented or reduced V_{max} of initial rate transporter activity, taurine uptake by oocytes injected with poly(A)$^+$ RNA from rats fed each diet was examined. Poly(A)$^+$ RNA from low taurine-fed rats elicited twice the taurine uptake elicited from normal taurine-fed rats and > 3-fold the uptake from high taurine-fed rats. Northern blot analysis showed that the size of mRNA for the rat kidney taurine transporter is about 1.9 kb. A low taurine diet increased the abundance of taurine transporter mRNA isolated from rat kidney cortex while a high taurine diet decreased the abundance of mRNA, as compared to a normal taurine diet. These results indicate that regulation of the taurine transporter by diet occurs at the level of mRNA.

Regulation of expression of the taurine transporter by medium taurine was also studied by injection of capped transcript into the oocytes (Fig. 7). Results showed that expression of the taurine transporter was significantly reduced by medium taurine (50 μM) as compared to the taurine-free control. Western analysis showed a band for the taurine transporter in the control, but not in the taurine-supplemented groups (data not shown). It suggested that down-regulation of the taurine transporter by oocytes in the presence of medium taurine occurs at the level of protein synthesis or secretion.

These study results suggest that the rat kidney taurine transporter is encoded for by a different gene than those encoding for the taurine transporters which have been cloned, because the size of mRNA and the K_m for each taurine transporter are different (8,9,11,13). In order to understand better the renal adaptive response to dietary taurine, efforts to clone the rat kidney taurine transporter and to raise antibodies to the transporter are underway.

In conclusion, taurine uptake by oocytes after injection of mRNA is similar to taurine transport in brush border membrane vesicles. The long-term adaptive response is regulated at the level of mRNA, and the short-term adaptive response is regulated at the level of protein synthesis or secretion. We speculate that the renal adaptive response to altered dietary sulfur amino acid intake is both transcriptionally and translationally regulated.

ACKNOWLEDGMENTS

We greatly appreciate the gift of the pNCT clone from Dr. S. Uchida. We are grateful to Dr. L. Albritton for technical advice, and to Andrea Budreau for help with preparation of the manuscript. This study was supported in part by the LeBonheur Chair of Excellence of the Department of Pediatrics of the University of Tennessee, and NIH grant DK37221-08.

REFERENCES

1. Chesney, R.W., Gusowski, N., and Dabbagh, S., 1987, Studies on renal adaptation to altered dietary amino acid intake: reduced renal taurine cortex content increases the V_{max} of taurine uptake by brush border membrane vesicles. *Pediatr. Nephrol.* 1:9-15.
2. Chesney, R.W., Gusowski, N., and Friedman, A.L., 1983, Renal adaptation to altered dietary sulfur amino acid intake occurs at the luminal brush border membrane, *Kidney Int.* 24:588-594.
3. Chesney, R.W., Gusowski, N., and Dabbagh, S., 1985, Renal cortex taurine content regulates renal adaptive response to altered dietary intake of sulfur amino acids, *J. Clin. Invest.* 76:2213-2221.
4. Chesney, R.W., Lippincott, S., Gusowski, N., Padilla, M., and Zeikovic, I., 1986, Studies on renal adaptation to altered dietary amino acid intake: tissue taurine responses in nursing and adult rats. *J. Nutr.* 116: 1965-1976.
5. Dumont, J.N., 1972, Oogenesis in Xenopus laevis (Daudin), *J. Morphol.*, 136:153-180.
6. Friedman, A.L., Albright, P.W., Gusowski, N., Padilla, M., and Chesney, R.W., 1983, Renal adaptation to alteration in dietary amino acid intake, *Am. J. Physiol.* 245:F159-F166.

7. Friedman, A., Albright, P.W., and Chesney, R.W., 1981, Dietary adaptation of taurine transport by rat renal epithelium, *Life Sci.* 29:2415-2419.
8. Jhiang, S.M., Fithian, L., Smanik, P., McGill, J., Tong, Q., and Mazzaferri, E.L., 1993, Cloning of the human taurine transporter and characterization of taurine uptake in thyroid cells, *FEBS* 318(2):139-144.
9. Liu, Q.R., Lopez-Corcuera, B., Nelson, H., Mandiyan, S., and Nelson, N., 1992, Cloning and expression of a cDNA encoding the transporter of taurine and β-alanine in mouse brain, *Proc. Natl. Acad. Sci. USA* 89:12145-12149.
10. Rozen, R., and Scriver, C.R., 1982, Renal transport of taurine adapts to perturbed taurine homeostasis, *Proc. Natl. Acad. Sci. USA* 79:2101-2105.
11. Smith, K.E., Borden, L.A., Wang, C.H.D., Hartig, P.R., Branchek, T.A., and Weinshank, R.L., 1992, Cloning and expression of a high affinity taurine transporter from rat brain, *Molec. Pharmacol.* 42:563-569.
12. Sambrook, J., Fritsch, E.F., and Maniatis, T., 1989, *in:* "Molecular Cloning: A Laboratory Manual", second edition, Cold Spring Harbor Laboratory Press, pp 7.3-7.52, 1989.
13. Uchida, S., Kwon, H.M., Yamauchi, A., Preston, A.S., Marumo, F., and Handler, J.S., 1992, Molecular cloning of the cDNA for MDCK cell Na^+- and Cl^--dependent taurine transport that is regulated by hypertonicity, *Proc. Natl. Acad. Sci. USA* 89:8230-8224.
14. Wallace, R.A., Jared, D.W., Dumont, J.N., and Sega, M.W., 1973, Protein incorporation by isolated amphibian oocytes. III. Optimum incubation conditions, *J. Exp. Zool.* 184:321-334.
15. Wallace, R.A., and Misulovin, Z., 1978, Long-term growth and differentiation of Xenopus oocytes in a defined growth medium, *Proc. Natl. Acad. Sci. USA* 75:5534-5538.
16. Zelikovic, I., and Chesney, R.W., 1989, Ionic requirements for amino acid transport, *Am. J. Kidney Dis.* 14(4):313-316.

REGULATION OF TAURINE TRANSPORT BY EXTERNAL TAURINE CONCENTRATION AND MEDIUM OSMOLALITY IN RENAL TUBULAR CELLS IN CULTURE

Deborah P. Jones, Bin Jiang, Russell W. Chesney

LeBonheur Children's Medical Center
University of Tennessee
Department of Pediatrics
956 Court Avenue, Room B310
Memphis, TN 38163

INTRODUCTION

Taurine is actively transported into the renal tubular cell by a Na^+- and Cl^--dependent system specific for this ß-amino acid and structural analogs (14). The proximal epithelial transport system regulates the urinary excretion of taurine. Taurine transport is regulated by the dietary intake of taurine in the rat (1), by extracellular taurine concentration in kidney cells in culture (7), and by medium osmolality in MDCK cells (11). We have previously demonstrated that taurine transport in both MDCK and LLC-PK1 cells is increased following incubation of cell monolayers in taurine-free medium and in decreased following incubation of cells in high extracellular taurine (8). The LLC-PK1 is a porcine cell line of proximal tubular origin and the MDCK (Madin Darby Canine Kidney) cell line is from the collecting duct of the dog.

The adaptive response to medium taurine concentration was expressed by both the proximal (LLC-PK1) and the distal (MDCK) cell lines within 8 h following incubation in defined medium and was prevented when protein synthesis and protein export were inhibited (6). Inhibition of RNA transcription had no effect on the ability of the cells respond to high or low medium taurine concentration (6). The ability of cells to adapt to changes in substrate availability is not unique to taurine and is observed for many vital cell nutrients such as neutral amino acids, phosphate and sulfate. Although the role of taurine in the renal tubular cell is not completely elucidated, taurine is postulated to have several key functions in the kidney; among these are that of a intracellular antioxidant, as well as an osmolyte (5). We postulate that there are two distinct mechanisms for regulation of the ß-amino acid transport system depending upon the stimulus and the cell type studied. The following

Taurine in Health and Disease, Edited by R. Huxtable
and D.V. Michalk, Plenum Press, New York, 1994

studies were performed to further examine the mechanisms by which medium hypertonicity and medium taurine concentration regulate taurine transport in cultured renal epithelial cells.

In many systems, the ß-amino acid transporter also appears to accept the substrate GABA, a neurotransmitter. However, recent studies in the MDCK cell lines indicate that GABA also shares a transport system with the osmolyte, betaine, which is transported by a separate system. Our studies of taurine transport reveal interaction between both betaine and GABA and the activity of taurine transport. Therefore, we will present preliminary studies of GABA transport in these cells for comparison with taurine.

METHODS

LLC-PK1 and MDCK cell lines were obtained from American Type Culture Collection (Rockville, MA) and were maintained as previously described (7). Cells were subcultured by trypsinization and seeded onto 0.4 μm polycarbonate filter supports (Costar, transwell). For experiments, medium was replaced with a hormonally defined, serum-free formulation.

Uptake studies were performed on confluent monolayers 10-14 days after seeding. Briefly, cells were washed with Earle's Balanced Salt Solution (EBSS) at 37°C. Uptake was initiated by the addition of EBSS with or without sodium, pH 7.4, with 50 μM [^{3}H]taurine or 10 μM [^{3}H]GABA at 37°C (7). Uptake was terminated by the removal of uptake solution followed by three rapid washes with cold EBSS. The uptake solution contained [^{14}C]inulin (0.1 μCi/ml). The contralateral solution was sampled for contamination with [^{14}C]inulin as a measure of monolayer integrity. Monolayers were considered to be intact if the inulin leakage was less than 1%. Cells were solubilized in 1% SDS in 0.2 N NaOH. An aliquot was dispersed into Optifluor, and then radioactivity counted in a Packard Tricarb 2000-CA Liquid Scintillation Analyzer (Packard, Downer's Grove, IL). Total cell protein was measured by the Lowry method. Uptake was expressed as Mol taurine/mg cell protein.

RESULTS

Effect of Hypertonicity and Medium Taurine Concentration on Taurine Uptake in LLC-PK1 and MDCK Cells

LLC-PK1 and MDCK cells were incubated in either standard, serum-free, hormonally-defined medium, or medium to which raffinose was added to achieve a final medium osmolality of 500 mOsm. In addition, the medium taurine concentration was varied so that the concentration of taurine was either 0, 50, or 500 μM. Taurine uptake was measured in the presence of sodium following the 24 h incubation (Figure 1). Taurine uptake by the apical surface of MDCK cells was unchanged following exposure to hypertonic medium except in the absence of taurine (Figure 1); taurine accumulation by the basolateral surface of MDCK cells was increased in the presence of both 0 and 50 μM taurine. Conversely, taurine uptake by either the apical or the basolateral surface of LLC-PK1 cells was not changed following incubation in hypertonic medium, however, the adaptive response following manipulation of medium taurine concentration was present on both surfaces, with relatively greater on the apical surface as previously shown (7). Taurine uptake by the basolateral cell surface was significantly increased by 4 h and was fully expressed by 12 h (data not shown).

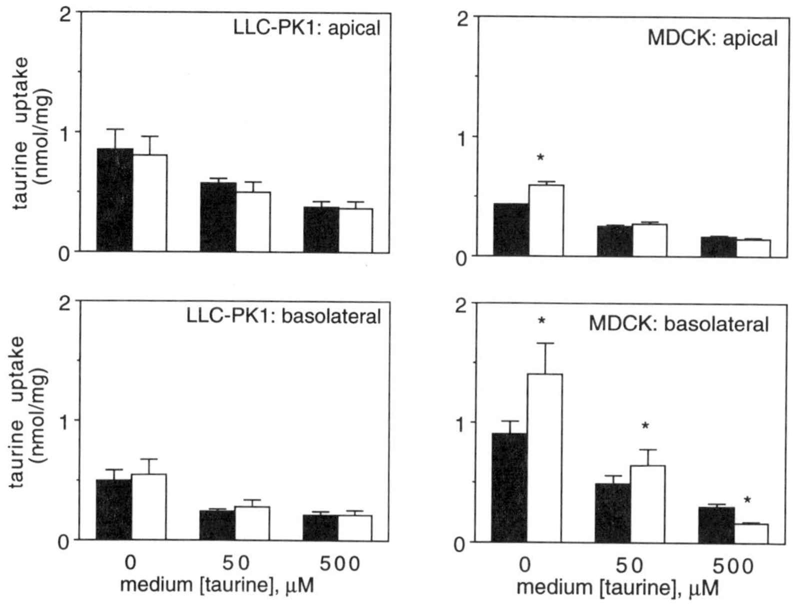

Figure 1. Polarity of taurine transport in LLC-PK1 and MDCK cells in response to variation in medium taurine concentration and medium tonicity. Apical and basolateral taurine uptake was measured for 15 min in LLC-PK1 and assessed at 20 min in MDCK cells. Values represent the mean ± SEM of 6 filters. * $p < .001$.

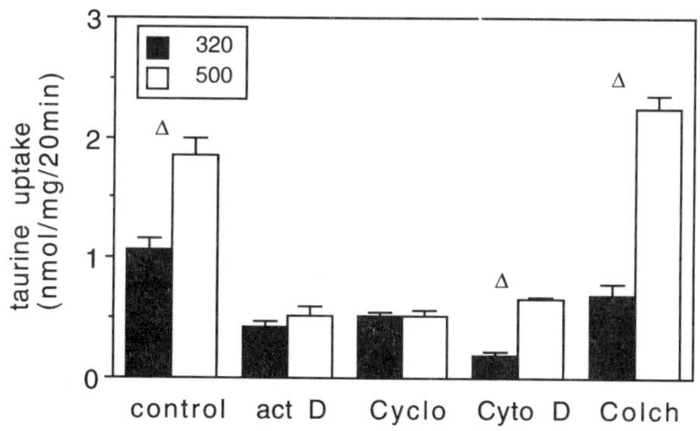

Figure 2. Effect of inhibitors on basolateral taurine uptake in MDCK cells. Values represent the mean ± SEM for 4 filters.

Role of Protein Synthesis, RNA Transcription, and Microtubular Function on Adaptation to Medium Hypertonicity

MDCK monolayers were incubated in the presence of either actinomycin D, (0.5 μg/ml), which inhibits RNA transcription, cycloheximide (50 μM), which inhibits protein synthesis, and two inhibitors of microtubular depolymerization, cytochalasin D (10 μM) or colchicine (70 μM) for 2 h prior to and during a 12 h incubation in either standard or hypertonic medium. Taurine accumulation by the basolateral cell surface was 95% of baseline (monolayers in isotonic medium) in the cycloheximide-treated group and 103% of baseline in the actinomycin D-treated group compared to an increase in taurine uptake of 161% of baseline in the control group (Figure 2). Exposure to hypertonicity in the presence of cytochalasin D and colchicine increased taurine accumulation to 319% and 330%, of baseline values respectively (see Figure 2). To summarize, incubation in the presence of either cycloheximide or actinomycin D prevented the adaptation of MDCK cells to hypertonicity while incubation with cytochalasin D or colchicine did not prevent the observed increase in basolateral taurine accumulation. Thus, expression of the adaptive response to hypertonicity requires RNA transcrip- tion and protein synthesis.

The Role of Medium Taurine and Other Substrates in Adaptation of MDCK Cells to Hypertonic Medium

MDCK cells were incubated in the presence of standard, serum-free, hormonally defined medium containing 500 μM taurine, GABA, betaine, L-alanine or ß-alanine for 24 h and then bathing medium was changed to either standard or hypertonic medium with the same added amino acid for an additional 24 h. Basolateral taurine accumulation was measured and compared to control (0 μM taurine). The increase in taurine accumulation was markedly blunted if cells had been previously exposed to high medium taurine as compared to those incubated in the absence of taurine as uptake was reduced in both groups exposed

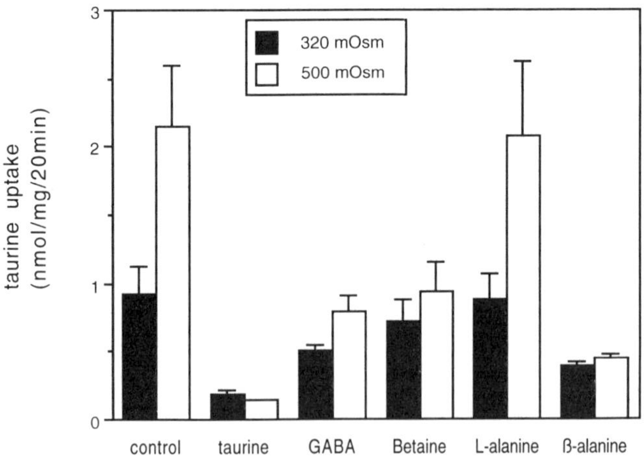

Figure 3. Effect of taurine and other substrates on taurine uptake in hyperosmolar medium. Confluent MDCK cells were incubated in isotonic medium containing 500 μM taurine, GABA, betaine, L-alanine or ß-alanine. Following the first 24 h incubation, medium was changed to either isotonic or hypertonic medium containing the same amino acid. Basolateral taurine uptake was measured at 20 min. Values represent the mean $\pm$ SEM for four filters.

to 500 μM taurine (Figure 3). The adaptive response of MDCK cells to hypertonicity was prevented by ß-alanine, but not by L-alanine (Figure 3). In addition, both GABA and betaine prevented the increase in taurine accumulation following exposure to hypertonic medium.

Effect of Medium Solute Concentration on GABA Uptake by LLC-PK1 Cells

Both apical and basolateral uptake of GABA by LLC-PK1 cell monolayers was decreased when cells were incubated for 24 h in the presence of 500 μM GABA. Incubation with taurine significantly decreased GABA accumulation by the basolateral but not the apical surface, while betaine had no appreciable effect (Figure 4).

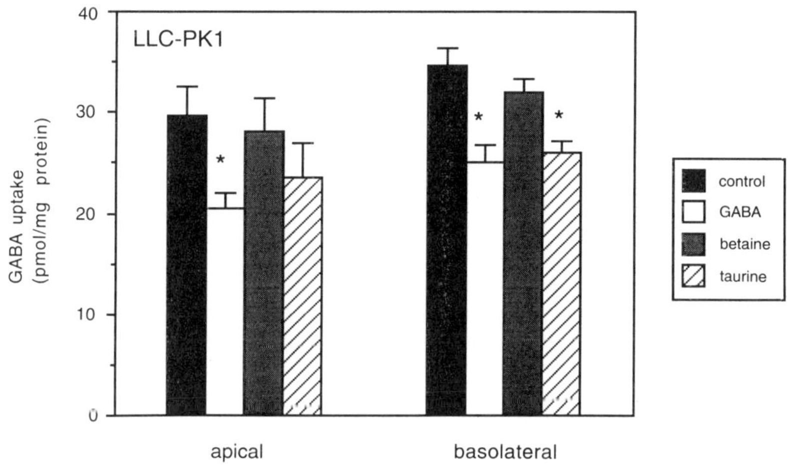

Figure 4. Uptake of GABA (10 μM) by LLC-PK1 cell monolayers following 24 h in medium containing 500 μM taurine, betaine, or GABA.

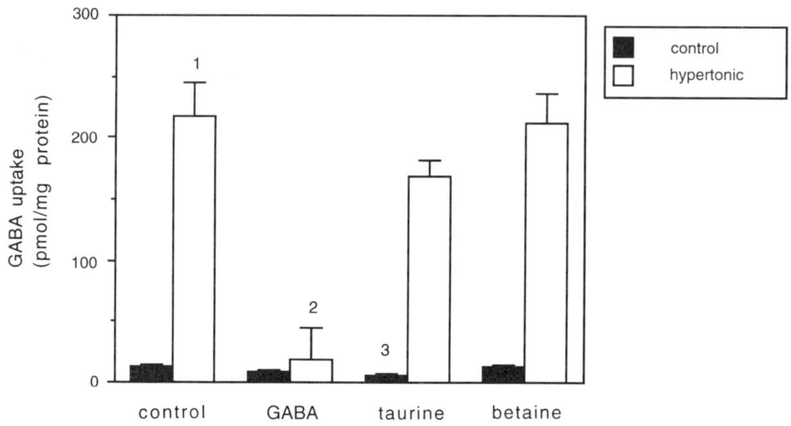

Figure 5. Uptake of GABA (10 μM) by MDCK cell monolayers following 24 h in medium containing either no additional osmolyte, or with 500 μM taurine, betaine, or GABA. Medium was made hypertonic to 500 mOsm as described before. 1: p=.0001, compared to control,2: p=.Ol compared to hypertonic control, and 3: p=.0006 compared to control.

Effect of Medium Osmolality and Solute Concentration on GABA Uptake by MDCK Cells

Uptake of GABA by MDCK cells was increased following a 24 h incubation in hypertonic medium. This response was significant on both the apical and the basolateral surfaces. If cell monolayers are co-incubated in hypertonic medium and high external GABA concentration, the increase in GABA uptake is prevented in a manner similar to that observed for taurine. Taurine reduces GABA uptake by cells in normotonic medium. However, neither betaine nor taurine prevent the increase in GABA uptake by MDCK cells in response to hypertonicity.

DISCUSSION

MDCK cells increase basolateral taurine accumulation in response to incubation in hyperosmolar medium while LLC-PK1 cells do not exhibit increased taurine accumulation following exposure to hypertonic medium. In contrast, both cell lines respond to changes in external taurine concentration with reciprocal changes in taurine accumulation (7). The enhanced taurine uptake which occurs in response to hyperosmolality is primarily localized to the basolateral surface of MDCK cells, whereas the adaptive response to medium taurine concentration is expressed on both the apical and the basolateral surfaces of LLC-PK1 and MDCK cell lines (7, 11). The response of taurine transporter activity to increased external medium tonicity is dissimilar in these two cultured renal epithelial cell lines, and may be related to the renal tubular site of origin. However, one must be cautious in drawing conclusions about the intact renal tubule from data in continuous cell lines.

In comparing the renal tubular handling of taurine and betaine, the accumulation of betaine is enhanced following long-term exposure of MDCK and PAP-HT25 cells (of papillary origin) to hyperosmolar medium (6, 15). In addition, the cellular content increased by 24 h in hypertonic medium and continued to increase over the next 5 days (9). Like taurine, betaine accumulation by MDCK cells is primarily localized to the basolateral surface (13). Kinetic analysis reveals the presence of two distinct transport systems for betaine, a high affinity system and a low affinity system. Following exposure of MDCK cells to medium made hypertonic by the addition of NaCl, the V_{max} of both the low-affinity and the high-affinity systems is increased (13). The increased V_{max} was noted to peak at 24 h and then to decline if betaine was present in the medium, presumably after sufficient osmolyte was accumulated within the cell (10, 13). Uchida and coworkers reported that exposure of MDCK cells to medium made hypertonic by the addition of raffinose increases the V_{max} for basolateral taurine uptake in MDCK cells, and this is accompanied by increased expression of a messenger RNA which imparts taurine transporter activity when injected into Xenopus oocytes (11, 12). Therefore, basolateral taurine transport and betaine transport increase when MDCK cells are exposed to a hypertonic environment.

Medium taurine concentration modifies the response of MDCK cells to hypertonic medium. As previously shown (11), incubation of MDCK cells in taurine-free medium further enhanced the adaptive response to hypertonicity, so that the two stimuli seemed additive. Our studies confirm their data. In addition, high medium taurine concentration in the face of hypertonicity blunts the enhanced basolateral uptake of taurine. One might speculate that in the presence of high medium taurine, adequate taurine is accumulated in the cell without the need to increase taurine transport. One stimulus for regulation of taurine transport, at least in response to medium taurine concentration or substrate supply, is the intracellular content of taurine. Uchida reported that cell taurine was unchanged in MDCK cells following incubation for 48 h in either isotonic or hypertonic taurine-free medium. However, the cellular taurine content was two-fold greater in monolayers incubated in hypertonic medium containing 50 μM taurine compared to isotonic medium containing 50 μM taurine (18). Cellular taurine content is probably not the only stimulus that increases taurine transport since taurine accumulation by MDCK cells was enhanced following exposure to hypertonic medium despite intracellular taurine levels which were comparable to control.

Medium taurine regulates taurine transport by a mechanism which requires functional protein synthesis but not RNA transcription (6). In contrast, the response of MDCK cells to hypertonic medium is prevented by actinomycin D, an inhibitor of RNA transcription, as well as cycloheximide, an inhibitor of protein synthesis. As previously noted, injection of mRNA from MDCK cells exposed to hypertonic medium into *Xenopus* oocytes has demonstrated increased taurine transporter activity due to increased message. Thus, the mechanisms involved in increased taurine transport induced by changes in medium osmolality appear to be distinct from those involved in the response to changes in medium taurine concentration. RNA transcription is required for the osmolar response but not the adaptive response to changes in substrate concentration.

Structural analogs of taurine such as ß-alanine also blunted the adaptive response to hypertonic medium. Interestingly, the presence of 500 μM GABA, which is not known to function as an osmolyte, also blunted the response to hypertonic medium as well as reduced taurine uptake by the monolayers exposed to isotonic medium. GABA has been shown to inhibit taurine uptake competitively and has been proposed to be a potential substrate for the ß-amino acid transporter (2). In addition GABA is postulated to be transported by the betaine transporter (12). Betaine also reduced the increase in taurine accumulation typically expressed following exposure to hypertonicity. The ability of betaine to alter taurine transport may be related to its known function as an osmolyte rather than as a substrate of the ß-amino acid transport system. There may be regulatory mechanisms which enable one osmotically active substance to affect the accumulation of another.

Indeed, the presence of one osmolyte has been shown to alter the accumulation of another (3). Renal medullary cells incubated in high extracellular osmolality (700 mOsm) in the presence of 0, 0.5 and 15 mM betaine exhibit reciprocal changes in the cell sorbitol content and very small, yet significant changes in cell GPC and inositol content (3, 8). This was not observed when inositol concentration was varied between 0, 18 and 15 mM. In addition, the intracellular concentration of betaine increases when sorbital is decreased; likewise, the accumulation of sorbitol is increased with low betaine (3, 8). Potential mechanisms by which the accumulation of one osmolyte can affect the accumulation of another include changes in the transport of the substrate, alterations in the availability of substrate for synthesis, and inhibition of enzymes involved in the synthesis of the osmolyte (8).

In contrast to the inhibitory action of GABA on taurine uptake during exposure to hypertonicity, taurine had little effect on GABA accumulation following hypertonic medium and only a slight effect in normal osmolality. More studies are required to explain the distinct features involved in the regulation of the taurine and GABA transport system, however it would seem from these data that accumulation of neither betaine nor taurine into the cell down-regulates the accumulation of GABA even though GABA shares a common transport system with betaine and possibly taurine.

In conclusion, there are at least two distinct mechanisms involved in the regulation of taurine transport: external taurine and medium osmolality. The response to medium osmolality requires protein synthesis and RNA transcription, and remains intact in the presence of microtubular toxins. The increase in taurine transport which occurs following increased external osmolality is expressed by the basolateral surface of MDCK cells but not by LLC- PK1 cells. The concentration of taurine in the external medium may modify the change in taurine uptake following hypertonic medium. That is taurine-replete cells behave differently than taurine-depleted cells. In addition, taurine supply has a more generalized effect on taurine uptake in that it induces changes in accumulation by both surfaces of LLC-PK1 and MDCK cells. The adaptation which follows manipulation of taurine concentration does not appear to require RNA synthesis for expression yet is absent when protein synthesis or exocytosis is inhibited. Thus, the cell processes leading to changes in taurine transport at the membrane are divergent depending upon the physiologic stimulus as well as the cell studied.

ACKNOWLEDGMENTS

Supported in part by grants from the National Institutes of Health (DK 37223), The National Kidney Foundation of West Tennessee, LeBonheur Children's Medical Center, and The American Heart Association (91-004470, Clinician Scientist Award).

REFERENCES

1. Chesney, R.W., Gusowski N., and Dabbaugh, S., 1985, Renal cortex taurine regulates the adaptive response to altered dietary intake of sulfur amino acids. *J. Clin. Invest.* 76:2213- 2221.
2. Danzler, W.H., and Silbernagl, S., 1976, Renal tubular reabsorption of taurine, g-amino butyric acid (GABA) and ß-alanine studied by continuous microperfusion. *Pfluegers Arch.* 367:123-128.
3. Garcia-Perez, A., Burg, M.B., 1991, Renal medullary organic osmolytes. *Physiol. Rev.* 71:1081-1115.
4. Heilig, C.W., Brenner, R.M., Yu, A.S.L., Kine, B.C., and Gullans, S.R., 1990, Modulation of osmolytes in MDCK cells by solutes, inhibitors and vasopressin. *Am. J. Physiol.* 259:F653- F659.
5. Huxtable RJ, 1992, Physiological actions of Taurine. *Physiol. Rev.* 72:101-163.
6. Jones, D.P., Miller, L.A., Dowling, C., Chesney, R.W., 1991, Regulation of Taurine transporter activity in LLC-PK1 cells: Role of protein synthesis and protein kinase C activation. *J. Am. Soc. Nephrol.* 2:1021-1029.
7. Jones, D.P., Miller, L.A., Chesney, R.W., 1993, Polarity of taurine transport in cultured renal epithelial cell lines: LLC-PK1 and MDCK. *Am. J. Physiol.* 265:F137-F145.
8. Moriyama, T., Garcia-Perez, A., Burg, M.B., 1990, Factors affecting the ratio of different organic osmolytes in renal medullary cells. *Am. J. Physiol.* 259:F847-F858.
9. Nakanishi, T., Balaban, R.S., Burg, M.B., 1988, Survey of osmolytes in renal cell lines. Am J Physiol 255: C181-C191.
10. Nakanishi, T., Turner, R.J., Burg, M.B., 1990, Osmoregulation of betaine transport in mammalian renal medullary cells. *Am. J. Physiol.* 258: F1061-1067.
11. Uchida, S., Kwon, H.M., Preston, A.S., and Handler, J.S., 1991, Taurine behaves as an osmolyte in MDCK cells: Protection by polarized, regulated transport of taurine. *J. Clin. Invest.* 88:656-662.
12. Uchida, S., Nakanishi, T., Kwon, H., Preston, A., Handler, J.S., 1991, Expression of MDCK cell Na and Cl dependent taurine transporter in *Xenopus laevis* oocytes. *J. Biol. Chem.* 266:9605-9609.
13. Yamamauchi A, Kwon, H.M., Uchida, S., Preston, A.S., Handler, J.S., 1991, Myo-inositol and betaine transporters regulated by tonicity are basolateral in MDCK cells. *Am. J. Physiol.* 261:F197-F202.
14. Zelikovic, I., Stejskal-Lorenz, E., Lohstroh, P. Budreau, A., and Chesney, R.W., 1989, Anion dependence of taurine transport by rat brush border membrane vesicles. *Am .J. Physiol.* 256:F646-F6-55.

ROLE OF TAURINE IN THE KIDNEY: OSMOREGULATORY TAURINE ACCUMULATION IN RENAL MEDULLA

Takeshi Nakanishi[1], Yoshihiro Takamitsu[2], and Minoru Sugita[1]

[1]Fifth Department of Internal Medicine
[2]Department of Nephrology and Dialysis
Hyogo College of Medicine
Nishinomiya, 663 Japan

INTRODUCTION

Comparative studies of a wide variety of organisms indicate that almost all cells that are able to adapt to a high-salt environment do so by balancing the increased extracellular osmolality with high intracellular concentrations of certain organic solutes ("organic osmolytes"). The organic osmolytes fall mainly into three groups: polyols (such as sorbitol and myo-inositol), methylamines (such as glycerophosphorylcholine and betaine), and amino acids (such as taurine, glycine and proline). Cells apparently use these organic compounds to adjust their intracellular osmolality because, unlike "perturbing" solutes such as NaCl, KCl, and urea, the organic osmolytes inhibit enzymes and other cellular processes relatively little even at high concentrations (17).

Using X-ray microprobe analysis, Beck et al. firstly observed the apparent osmolality gap between the intracellular and extracellular electrolyte concentration in the renal papilla under antidiuretic conditions (2). Bagnasco and Burg identified polyols and methylamines in the renal inner medulla that maintain the osmotic balance between inside and outside of the cells (1).

Taurine is widely distributed in the animal kingdom and is the most abundant free amino acid in the kidney. Taurine is well established as playing an important role in the maintenance of intracellular osmolal concentration in marine invertebrates, teleosts, and amphibians. As for mammalians, considerable evidence suggests that taurine also plays a salient role in osmoregulation in brain and heart (10, 11). In the kidney, however, the role of taurine in osmoregulation has remained controversial.

To clarify the osmoregulation of taurine accumulation in the kidney under various hyperosmotic conditions, we directly measured taurine as well as sorbitol, myo-inositol, glycerophosphorylcholine and betaine content along the corticomedullary axis of kidneys from control, water deprived, sodium-loaded and urea-loaded rats. Sodium infusion and urea infusion help to determine the effects of hyperosmolality due to sodium compared to urea on the osmolyte content of the inner medullary cells separately, while the combination of sodium and urea has an effect on osmolyte accumulation in water-deprived animals.

METHODS

Male Wistar rats, weighing approximately 280-380 g, were cannulated in the inferior vena cava via insertion of a polyethylene catheter (PE 10 tubing) in the femoral vein 72 h prior to sacrifice. The animals were deprived of food from the time of cannulation until sacrifice for a 72 h period in order to eliminate any effect of food intake on osmolyte accumulation in renal inner medulla.

The rats were divided into the following four groups: a) Control rats underwent sham operation of femoral vein cannulation; b) Dehydrated rats were deprived of water for 72 h prior to sacrifice and underwent sham operation of femoral vein cannulation; c) Sodium-loaded rats were infused with hypertonic salt solution for 72 h continuously (5.45% NaCl solution; 18 ml/day) prior to sacrifice; d) Urea-loaded rats were infused with hyperosmolar urea solution for 72 h continuously (15 % urea solution; 18 ml/day) prior to sacrifice.

Tissue extracts were prepared as reported previously (3, 16). Kidneys were rapidly removed. They were sliced with scissors into three segments along the corticopapillary axis: cortex, outer medulla, and inner medulla. These sections were frozen in liquid nitrogen, lyophilized over 24 h and weighed to obtain the dry weight. The tissue water content was calculated by the difference between the wet and dry weights. The dried sections were homogenized in 3 ml cold 7% perchloric acid with a glass homogenizer. The homogenates were centrifuged for 20 min at cold temperature. The pellets were then dissolved in 0.25 N NaOH and tested for protein content using the Bio-Rad protein assay (Bio-Rad Labs, Richmond, Ca). Tissue sodium and potassium were determined in the supernatants of perchloric acid extracts using a two-wavelength atomic absorption spectrophotometer (Hitachi, TYPE-308, Tokyo, Japan).

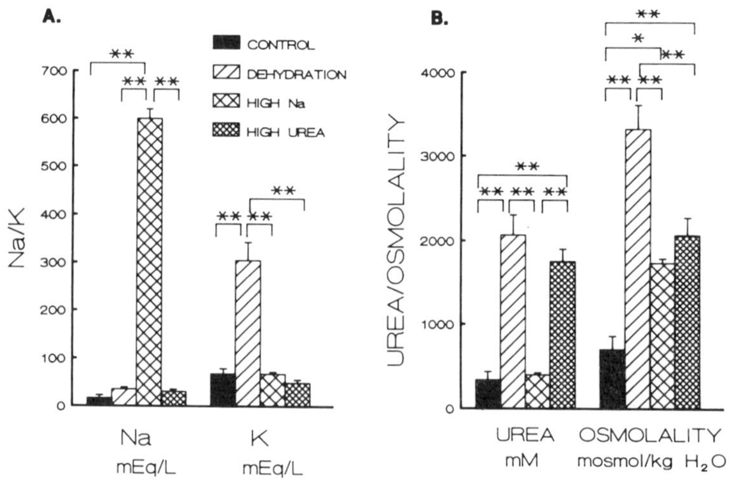

Figure 1. Effect on urine parameters of several hyperosmotic conditions (* $p<0.05$, ** $p<0.01$)

Betaine, glycerophosphorylcholine, sorbitol, myo-inositol and amino acid osmolytes were measured by high-performance liquid chromatography (HPLC) of perchloric acid extracts of tissue (14). The tissue supernatants from the centrifugation step were then neutralized with 2.0 N KOH to approximately pH 7, and passed through a Sep-Pak C18 cartridge. In addition, extracts used to measure betaine and myo-inositol (but not glycerophosphorylcholine, sorbitol or amino acids) were also placed on a 1 ml AG1-X4 (hydroxyl form) column, and betaine and myo-inositol were eluted with 4 ml of water (6). The organic osmolytes were measured in the processed extracts by HPLC with the use of a Sugar-Pak I (Waters, Milford, MA) or Sugar-Pak Ca (Nihon Waters, Tokyo, Japan) column. Amino acids were measured with a amino acid analyzer (Hitachi L-8500, Tokyo, Japan) using the same samples diluted with the same volume of 0.04 N HCl.

Results are shown as mean $\pm$ SEM. Statistical significance of differences was determined using the Bonferroni method (13) for multiple comparisons or Student's t test for paired comparisons. A probability of $p<0.05$ was taken to indicate a statistically significant difference. Linear correlations were determined by least-squares regression analysis of individual data points.

RESULTS

Metabolic Parameters

Urinary sodium concentration was over ten times higher in sodium-loaded rats than in rats of any other group. However, the concentrations in control, dehydrated and urea-loaded rats did not differ significantly (Figure 1). Urinary potassium concentration was four to six times higher in dehydrated rats than in rats of any other group, while those of the control, sodium-loaded and urea-loaded rats did not differ significantly. Urine urea concentration was significantly higher in dehydrated and urea-loaded rats than in control and sodium-loaded rats. Urine osmolality was the highest in dehydrated rats and was significantly higher in dehydrated, sodium-loaded and urea-loaded rats than in controls.

Tissue Sodium, Potassium, Urea and Water Content

Figure 2 shows the tissue Na^+, K^+, urea and water content along the corticopapillary axis. Significant differences are indicated only for the inner medullary content of each solute or water among the 4 groups. Significant differences in other areas are discussed in the text.

Along the corticomedullary axis, tissue Na^+, K^+, urea and water content increased, and were highest in renal inner medulla in all groups. Among the 4 groups, cortical or outer medullary content of Na^+, K^+ and water did not differ significantly. Inner medullary Na^+ content was highest in sodium-loaded rats, and was significantly higher than in controls. It was also higher in dehydrated rats than in controls, but not significantly so (Figure 2).

There were no significant differences in inner medullary potassium content or water content (water-to-protein ratio: liter/kg protein) among the groups tested.

Inner medullary urea content was significantly higher in dehydrated and urea-loaded rats than in control and sodium-loaded rats. The outer medullary urea content was highest in urea-loaded rats.

Tissue Osmolyte Content Along the Corticomedullary Axis

In all groups, glycerophosphorylcholine and sorbitol were low throughout the cortex and outer medulla, but rose steeply in the inner medulla (Figure 3).

Betaine content rose from the cortex to the outer medulla by approximately 10 times and then increased slightly in the inner medulla from control and sodium-loaded groups or

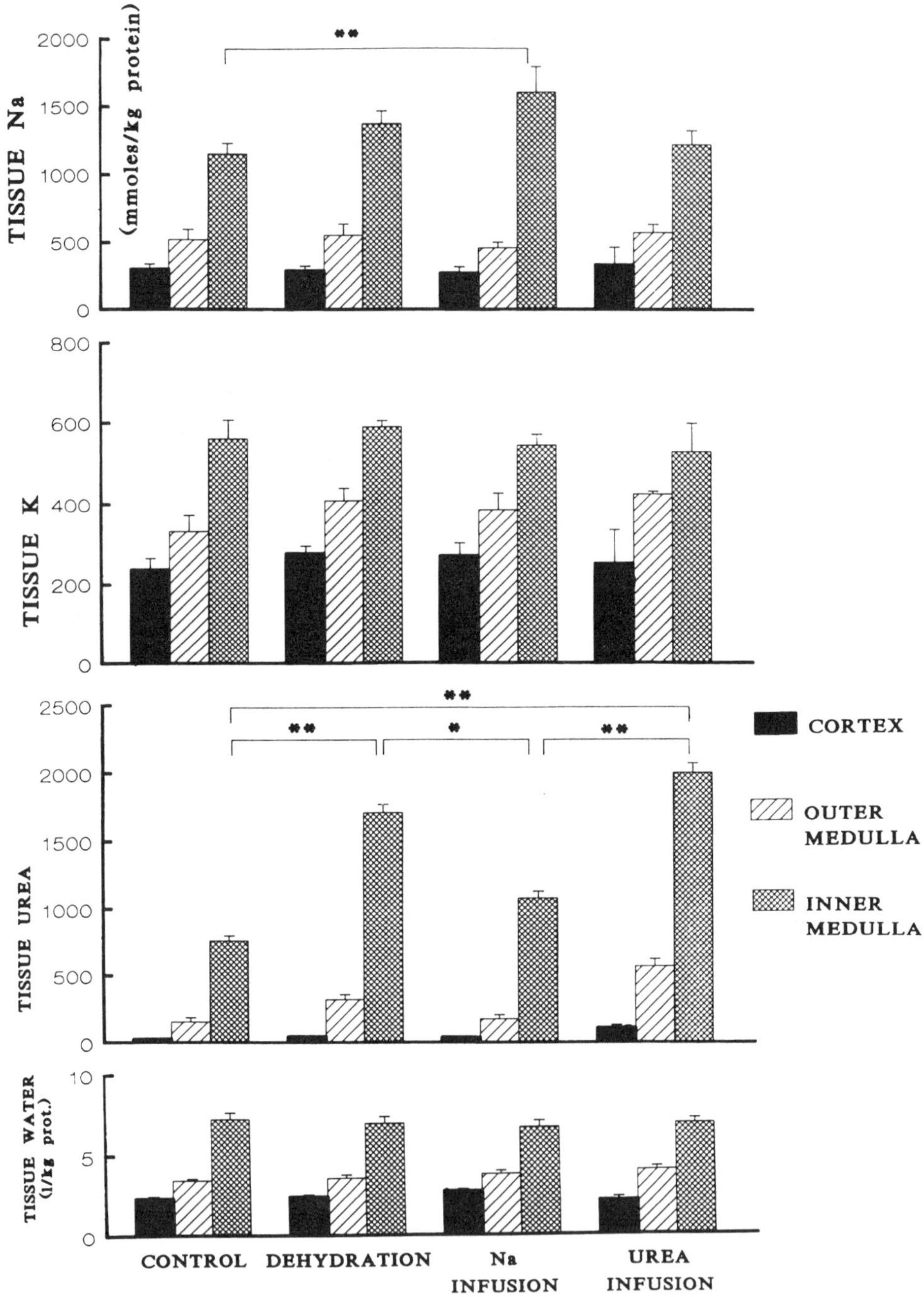

Figure 2. Effect of several hyperosmotic conditions on renal Na^+, K^+, urea and water content along the corticopapillary axis (* $p<0.05$, ** $p<0.01$). The units, per Kg protein, are mEq (K^+, Na^+), mmol (urea) or liters (water).

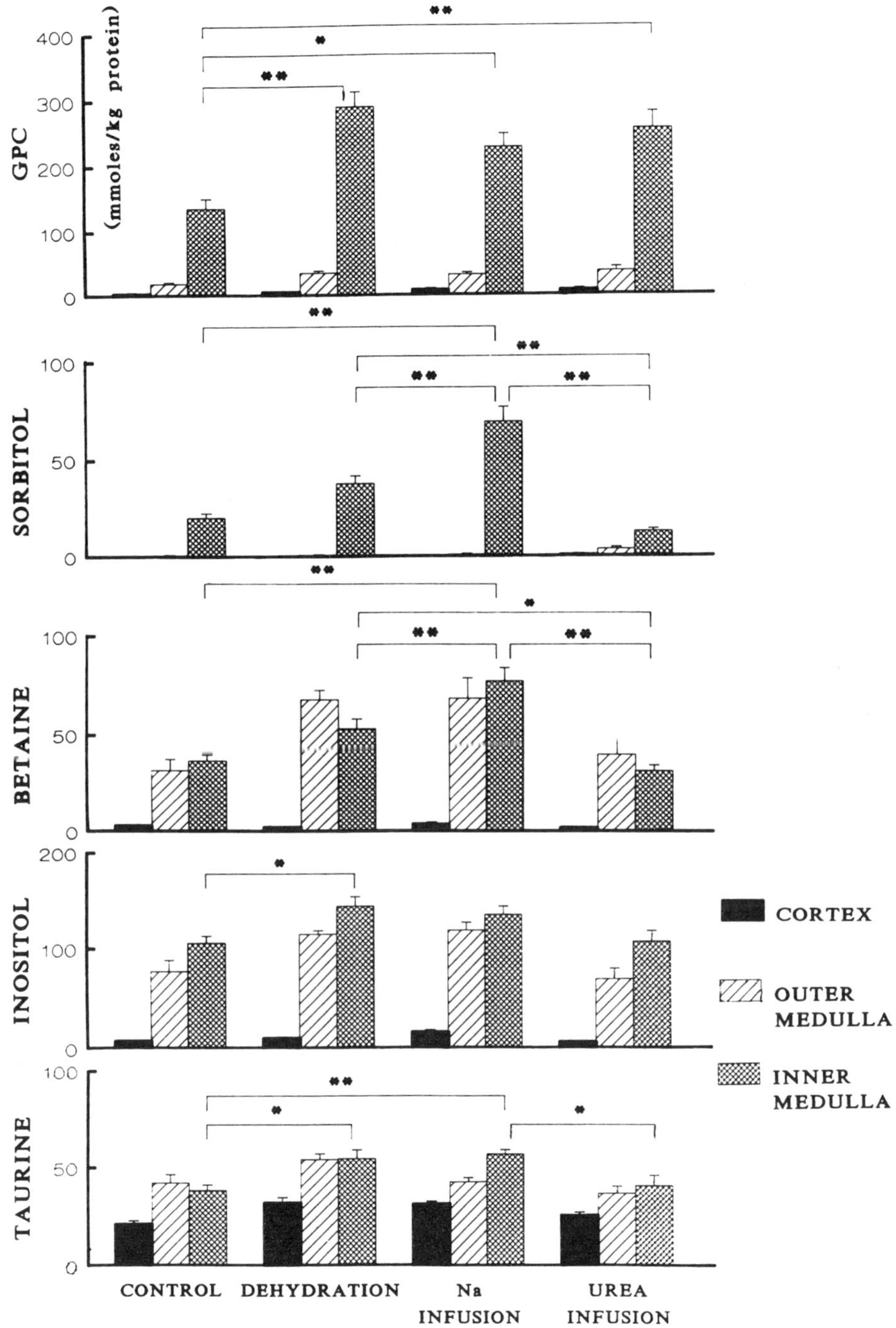

Figure 3. Effect of several hyperosmotic conditions on renal osmolyte contents along the corticopapillary axis (* $p<0.05$, ** $p<0.01$). Units are mmol/Kg protein

decreased in rats from dehydrated and urea-loaded groups. Myo-inositol increased from the cortex to the inner medulla monotonically among all groups tested.

Cortical taurine content was relatively higher than the contents of other osmolytes in all groups. Taurine content rose from the cortex to the outer medulla by approximately 2 times or less and then maintained an identical level in the inner medulla or increased slightly.

Cortical Content of Each Osmolytes

Cortical inositol content was highest in salt-loaded rats. Cortical taurine contents in dehydrated and salt-loaded rats were significantly higher than in control.

Outer Medullary Content of Each Osmolyte

Outer medullary betaine and inositol contents in dehydrated and salt-loaded rats were significantly higher than in control and urea-loaded rats. Outer medullary taurine content in dehydrated rats was significantly higher than in control and urea-loaded rats.

Inner Medullary Content of Each Osmolyte

Glycerophosphorylcholine content in dehydrated, sodium-loaded and urea-loaded rats did not differ significantly, but was highest in dehydrated and lowest in salt-loaded rats.

Myo-inositol content was significantly higher in dehydrated rats than in controls. It was also higher in salt-loaded rats than controls, but not significantly so.

Taurine as well as sorbitol and betaine content was highest in salt-loaded rats and was significantly higher than in control and urea-loaded rats. Taurine content was also higher in dehydrated rats than in control. Taurine content in the inner medulla was comparable to sorbitol or betaine content.

Correlation Between Osmolyte Content in Inner Medulla and Other Parameters

Tables 1 and 2 show the coefficients of linear correlation between each osmolyte and urine or tissue parameter; i.e., urea, sodium, potassium, sodium plus potassium, and osmolality (for urine only). The coefficients of correlation that were significantly different from zero are displayed in the tables.

Table 1. Coefficients of correlation between each osmolyte in inner medulla and urine parameters

Urine	Inositol	Betaine	Taurine	Sorbitol	GPC	Total
Urea	NS	NS	NS	NS	4=0.57[c]	NS
Na^{+} [1]	NS	NS	NS	NS	4=0.78[c]	NS
Na^{+} [2]	NS	NS	NS	r=0.82[a]	NS	NS
Na^{+} + K^{+}	r=0.60[c]	r=0.73[c]	r=0.53[b]	r=0.86[c]	r=0.36[a]	r=0.65[c]
K^{+}	NS	NS	NS	NS	r=0.50[b]	r=0.45[b]
Osmolality	r=0.54[c]	NS	NS	NS	r=0.73[c]	r=0.63[c]

[1]All data points except for the sodium-loaded group;

[2]The data from the sodium-loaded group only; GPC: Glycerophosphorylcholine

NS, not significant; [a] $p<0.05$; [b] $p<.01$; [c] $p<0.001$

As for the correlation between osmolyte content and urine sodium concentration, the sodium-loaded group was clearly different from other groups because urinary sodium concentrations were very much higher than those of the other groups. The linear correlations were determined by applying separately two sets of data points: all data points except those of the sodium-loaded group and the data from the sodium-loaded group only.

Inner medullary content of each osmolyte, including taurine, was well correlated with the urine sodium plus potassium concentrations (Figure 4A) and tissue sodium contents (Figure 4B).

The sum of the tissue osmolyte contents correlated well with urine sodium, urine potassium, urine sodium plus potassium, urine osmolality, tissue urea, tissue sodium, and tissue sodium plus potassium contents.

Table 2. Coefficients of correlation between each osmolyte and tissue parameters in inner medulla

Tissue	Inositol	Betaine	Taurine	Sorbitol	GPC	Total
Urea	$r=0.46^b$	NS	NS	NS	$r=0.69^c$	$r=0.54^c$
Na^+	$r=0.50^b$	$r=0.72^c$	$r=0.44^c$	$r=0.72^c$	$r=0.44^a$	$r=0.65^c$
$Na^+ + K^+$	$r=0.49^a$	$r=0.68^c$	NS	$r=0.64^c$	$r=0.43^a$	$r=0.61^c$
K^+	NS	NS	NS	NS	NS	NS

NS, not significant; [a] $p<0.05$; [b] $p<0.01$; [c] $p<0.001$

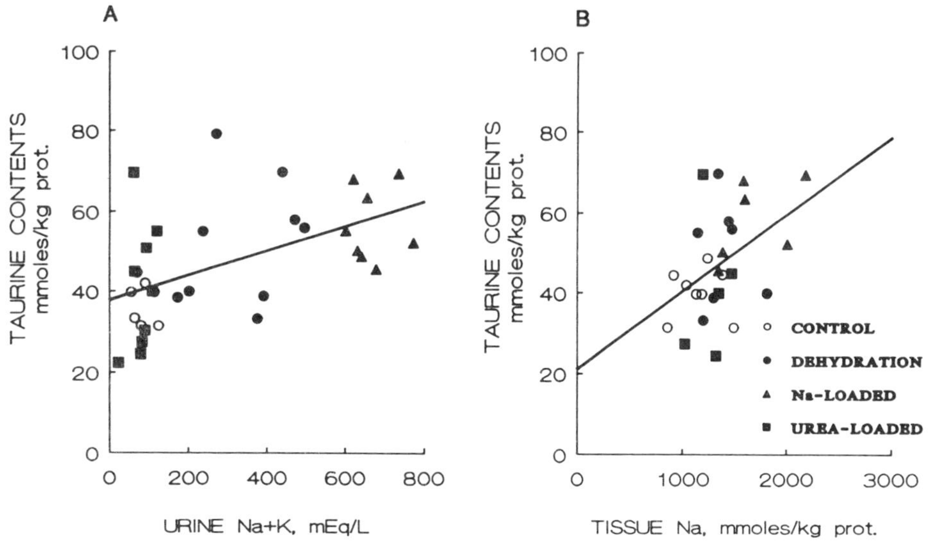

Figure 4. A: Relation between inner medullary contents of taurine and urinary sodium plus potassium concentrations. [Taurine content]=[Urine $Na^+ + K^+$]*0.0308+37.9 ($r=0.53$, $p<0.01$); **B:** Relation between inner medullary contents of taurine and tissue sodium contents. [Taurine content]=[Tissue Na^+]*0.0192+21.2 ($r=0.44$, $p<0.05$)

DISCUSSION

This study examined the changes in renal taurine, and methylamines and polyols along the corticopapillary axis, occurring in response to antidiuresis, salt loading and urea loading.

Taurine Accumulation in the Kidney

Taurine content was the highest in the inner medulla for all hyperosmotic groups, while the extracellular osmolality was highest along the corticopapillary axis. Cortical taurine contents in dehydrated and sodium-loaded rats were significantly larger than in control. Outer medullary taurine content in dehydrated rats was significantly larger than in control and urea-loaded rats. Taurine contents in the inner medulla from the dehydrated and sodium-loaded rats were significantly larger than those from control and urea-loaded rats.

In the brain and heart of the mammal, the role of taurine in osmoregulation has been established (10,11); that is, the intracellular accumulation of taurine might partly balance the extracellular hyperosmolality.

In a previous study, no gradients for taurine concentration along the corticopapillary axis was observed in the ruminant kidney, although taurine was a major amino acid component in the renal inner medulla (9). The reason for the difference between the present and previous studies might be related to the species differences or to the method of determination. The separation of amino acids was earlier performed only by sulfonated cation exchange resin and not by HPLC (4).

Mechanism of Taurine Accumulation in Inner Medulla

Inner medullary taurine contents correlated well with the sum of urine sodium plus potassium concentrations or tissue sodium content, but not with urine urea concentrations or tissue urea contents. Similar correlations were observed in inositol and betaine contents. These data suggest that the accumulation of taurine as well as myo-inositol and betaine in the renal inner medulla may also be attributed to effects of extracellular electrolytes.

In MDCK cells, taurine as well as betaine and myo-inositol increased when medium osmolality was increased by the addition of NaCl (7, 8, 12). These additional osmolytes, which appear after switching to a hyperosmotic medium, are transported into MDCK cells by a secondary, sodium-dependent active transport mechanism. Hypertonicity is associated with an increase in the V_{max} for each transporter, but not with any change in K_m (7, 8, 12). There is little direct evidence that active transport of taurine is osmoregulated in renal medullary cells in vivo.

Assuming that tissue sodium is retained principally in the extracellular space of papillary tissues and that the sodium-dependent transport of these solutes is primarily a mechanism for osmolyte accumulation as is true for MDCK cells (12, 15), it seems quite likely that the extracellular electrolytes, principally sodium, accelerate the transport of these osmolytes into the renal inner medullary cells.

NaCl and Urea Differ in Their Effects on Osmolyte Accumulation

In inner medulla, high sodium content is associated with increases in all of the osmolytes except myo-inositol. Urea, however, induces an increase only in glycerophosphorylcholine, and may in fact induce a decrease in sorbitol and betaine contents.

Along the corticopapillary axis, an increasing content of inositol, sorbitol, and glycerophosphorylcholine from cortex to inner medulla was observed, but not of betaine in dehydrated and urea-loaded rats. Failure of the betaine to increase from outer medulla to inner medulla might be related to a high urea concentration in inner medulla, since high concentrations of urea inhibit the betaine accumulation in MDCK cells (8).

Thus, NaCl and urea differ in their effects on the osmolyte accumulation in renal medulla.

Estimation of Osmolyte Concentration in Renal Tissues

In this study we measured osmolyte contents and not the concentrations. The latter cannot be estimated without knowledge of the intra-and extracellular spaces and distributions. The water content in the inner medulla of both groups was significantly larger than that in the cortex. If the osmolyte concentrations in the tissue were calculated using these values, the osmolyte content gradient along the corticopapillary axis could be canceled by the water content.

However, several reasonable assumptions and estimation of intracellular space in the renal tissues can be made to evaluate the intracellular concentrations of organic osmolytes. First, Law demonstrated the ratio of the cellular water-to-total tissue water in the cortex (63%), medulla (55%) and papilla (40%) respectively (5). Secondly, the non-urea organic osmolytes are likely to be largely intracellular, as suggested by the low blood and urine concentrations found in this and other studies (1, 16).

Figure 5 shows the estimated concentration of several organic osmolytes; i.e., polyols, methylamines and taurine, in control and sodium-loaded rats. The estimated concentration of each osmolyte in inner medulla from sodium-loaded rats was higher than controls. The pattern of the estimated concentration gradient of each osmolyte along the corticomedullary axis was very similar to that of the contents. In both control and sodium-loaded rats, the estimated concentration of betaine, inositol and taurine in outer medulla is higher than in inner medulla.

In conclusion, the rat renal inner medulla contains not only high concentrations of polyols and methylamines but also taurine as a concomitant of the renal concentrating mechanism. The taurine content especially in the inner medulla is highly regulated by

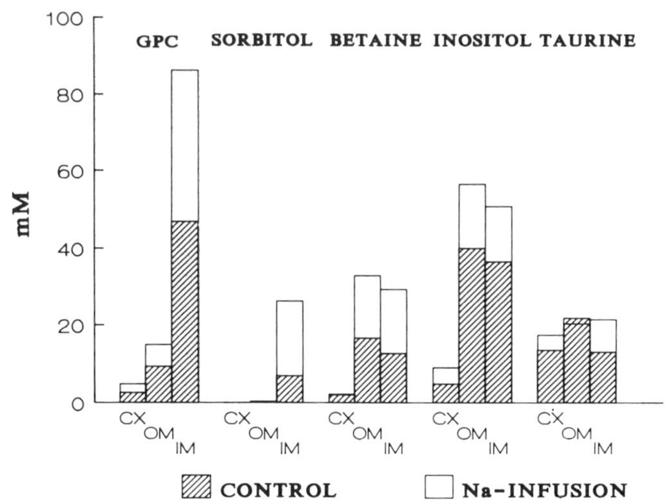

Figure 5. Estimated intracellular concentration of each osmolyte in control and sodium-loaded rats

hyperosmolality with sodium, consistent with previous observations in brain and heart. Kidney cells can use taurine for the purpose of adjusting the intracellular osmolality in the same manner as in bacteria, plants, and other organs (17).

ACKNOWLEDGEMENTS

The authors are grateful to Y. Sato of the Central Research Laboratory at Hyogo College of Medicine for operating the high-speed amino acid analyzer. We also thank A. Takiguchi and K. Obata for expert technical assistance. This work was partly supported by Grant-in-Aid for Scientific Research 03670046 from the Japan Ministry of Education.

Part of the data discussed have been published in: Nakanishi, T., Uyama, O., Nakahama, H., Takamitsu, Y., and Sugita, M., 1993, Determinants of relative amounts of medullary organic osmolytes: effect of NaCl and urea differ, *Am. J. Physiol.* 264:F472-479.

REFERENCES

1. Bagnasco, S., Balaban, R., Fales, H.M., Yang, Y., and Burg, M., 1986, Predominant osmotically active organic solutes in rat and rabbit renal medullas, *J. Biol. Chem.* 261:5872-5877.
2. Beck, F., Dorge, A., Rick, R., and Thurau, K., 1984, Intra- and extracellular element concentrations of rat renal papilla in antidiuresis, *Kidney Int.* 25:397-403.
3. Gullans, S.R., Blumenfeld, J.D., Balschi, J.A., Kaleta, M., Brenner, R.M., Heilig, C.W., and Hebert, S.C., 1988, Accumulation of major organic osmolytes in rat renal inner medulla in dehydration, *Am. J. Physiol.* 255:F626-634.
4. Joseph, M.H., and Marsden, C.A., 1986, Amino acids in small peptides, "HPLC of Small Molecules", Lim, C.K., ed., IRL Press, Oxford, pp. 13-28.
5. Law, R.O., 1987, The volume and ioniccomposition of cells in incubated slices of rat renal cortex, medulla and papilla, *Biochim. Biophys. Acta* 931:276-285.
6. Nakanishi, T., and Burg, M.B., 1989, Osmoregulatory fluxes of myo-inositol and betaine in renal cells, *Am. J. Physiol.* 257:C964-970.
7. Nakanishi, T., Turner, R.J., and Burg, M.B., 1989, Osmoregulatory changes in myo-inositol transport by renal cells, *Proc. Natl. Acad. Sci. USA* 86:6002-6007.
8. Nakanishi, T., Turner, R.J., and Burg, M.B., 1990, Osmoregulation of betaine transport in mammalian renal medullary cells, *Am. J. Physiol.* 258:F1061-1067.
9. Robinson, R.R., Owen, E.E., and Schmidt-Nielsen, B., 1966, Intra-renal distribution of amino acids in antidiuretic ruminants *Comp. Biochem. Physiol. [A]* 19:187-195.
10. Thurston, J.H., Hauhart, R.E., and Dirgo, J.A., 1980, Taurin: a role in osmotic regulation of mammalian brain and possible clinical significance, *Life Science* 26:1561-1568.
11. Thurston, J.H., Hauhart, E.E., and Naccarato, E.F., 1981, Taurine: possible role in osmotic regulation of mammalian heart, *Science* 214:1373-1374.
12. Uchida, S., Nakanishi, T., Kwon, H.M., Preston, A.S., and Handler, J.S., 1991, Taurine behaves as an osmolyte in Madin-Darby canine kidney cells: protection by polarized, regulated transport of taurine, *J. Clin. Invest.* 88:656-662.
13. Wallenstein, S., Zucker, C.L., and Fleiss, J.L., 1980, Some statistical methods useful in circulation research, *Circ. Res.* 47:1-9.
14. Wolff, S.D., Yancey, P.H., Stanton, T.S., and Balaban, R.S., 1989, A simple HPLC method for quantitating major organic solutes of renal medulla, *Am. J. Physiol.* 256:F954-956.
15. Yamauchi, A., Kwon, H.M., Uchida, S., Preston, A.P., and Handler, J.S., 1991, Myo-inositol and betaine transporters regulated by tonicity are basolateral in MDCK cells, *Am. J. Physiol.* 261:F197-202.
16. Yancey, P.H., and Burg, M.B., 1989, Distribution of major organic osmolytes in rabbit kidneys in diuresis and antidiuresis, *Am. J. Physiol.* 257:F602-607.
17. Yancey, P.H., Clark, M.E., Hand, S.C., Bowlus, R.D., and Somero, G.N., 1982, Living with water stress: evolution of osmolyte system, *Science* 217:1214-1222.

TAURINE AND EXPERIMENTAL KIDNEY DISEASE

Howard Trachtman[1] and John A. Sturman[2]

[1]Department of Pediatrics, Division of Nephrology,
Schneider Children's Hospital, Long Island Campus
for the Albert Einstein College of Medicine,
New Hyde Park, NY 11042
[2]The Institute for Basic Research in Developmental Disabilities
Department of Developmental Biochemistry
Staten Island, NY 10314

INTRODUCTION

There is abundant evidence that taurine functions as an antioxidant in a variety of biological systems. Taurine reacts with HOCl, a potent oxidizing agent generated by activated neutrophils, to form the less toxic product, N-chlorotaurine (20). This reaction retards damage to ocular surfaces exposed to HOCl (13). Taurine protects cultured lymphoblastoid cells against iron-ascorbate or retinol-induced oxidant damage and enhances cell viability (16,17). Taurine supplementation protects the lung from oxidant-induced damage following exposure to ozone (2) or nitrogen dioxide (7). Taurine alone or in combination with niacin reduces pulmonary fibrosis following intratracheal installation of the oxidants, amiodarone (26) or bleomycin (27). Finally, taurine attenuates adriamycin-induced cardiotoxicity in perfused chick hearts by decreasing myocardial lipid peroxidation (9).

Reactive oxygen molecules are involved in progressive organ damage in several experimental models of kidney disease. In ischemic acute renal failure caused by bilateral renal artery clamping, intra-renal infusion of free radical scavengers such as catalase and superoxide dismutase lessens the reperfusion injury following restoration of renal blood flow (15). The hydroxyl radical contributes to the nephrotoxicity of the aminoglycoside antibiotic, gentamicin (25). Administration of vitamin E reduces proteinuria, maintains tissue levels of sulfhydryl proteins and lessens structural injury in nephrotoxic serum nephritis (5).

Oxygen free radicals are implicated in the pathogenesis of the glomerulopathies induced by puromycin aminonucleoside and adriamycin. Administration of allopurinol, catalase, and the iron chelating agent, desferioxamine, reduces the severity of acute phase puromycin aminonucleoside-induced injury (4,19). Provision of tungsten, an inhibitor of renal xanthine oxidase/xanthine dehydrogenase, alleviates adriamycin nephropathy (6). In diabetic nephropathy, hyperglycemia directly promotes lipid peroxidation of cell membranes

Taurine in Health and Disease, Edited by R. Huxtable
and D.V. Michalk, Plenum Press, New York, 1994

and causes renal cell injury (3). Glycosylation of proteins initiates a series of autoxidative chemical reactions that result in the accumulation of advanced glycosylation end products in the kidney, compounds that contribute to the development of diabetic nephropathy (10,18).

In view of this evidence, we tested whether the administration of taurine could modify the extent of kidney damage under circumstances in which reactive oxygen molecules have been implicated in the deteriorating course of renal disease.

TAURINE AND EXPERIMENTAL RENAL DISEASE

Chronic Puromycin Aminonucleoside Nephropathy

We first studied chronic puromycin aminonucleoside nephropathy, a model of the clinical disorder, focal segmental glomerulo-sclerosis (21). Puromycin aminonucleoside is an agent that is toxic to the glomerular epithelial cell. Nephropathy was induced in male Sprague-Dawley rats by 7 subcutaneous injections of puromycin aminonucleoside, 2 mg/kg body weight, over a 12 week period (8). Experimental animals received a 1% taurine drinking water and control rats were given untreated tap water. Both groups were fed standard rodent chow containing 22% protein.

Taurine supplementation did not alter body weight, blood pressure, or hematocrit. Moreover, it did not reduce total urinary protein excretion nor did it affect on the serum albumin, cholesterol, or triglyceride concentrations. However, long term provision of taurine-containing water decreased the urinary albumin:creatinine excretion ratio from 44 to 26 (Table 1).

Table 1. Taurine and puromycin aminonucleoside nephropathy

	Control	Taurine-treated
Creatinine clearance (ml/min/100 g body weight)	0.17 ± 0.03	0.33 ± 0.03
Urinary albumin:creatinine ratio	44 ± 4	26 ± 4*
Segmental glomerulosclerosis (%)	15.9 ± 2.1	9.9 ± 1.9*
Tubulointerstitial injury score (0-4+)	2.7 ± 0.3	1.5 ± 0.3*
Malondialdehyde (nmol/g wet wt)	231 ± 17	92 ± 11*

The control group consisted of 24 rats treated with puromycin as described in the text. The taurine-treated group (23 rats) received a similar treatment with puromycin, but were maintained on drinking water containing 1% taurine. Data are means ± SEM. * $p<0.1$ compared to control group.

After 12 weeks, the inulin clearance, an index of the glomerular filtration rate, was two fold higher in taurine-treated rats (0.26±0.04 versus 0.13±0.03 ml/min/100 g body weight, $P<0.05$). These functional changes were associated with a 50% reduction in the number of segmentally sclerosed glomeruli and an attenuation in the amount of tubulointerstitial scarring. Renal cortical malondialdehyde content, assayed as thiobarbituric acid-reactive substances (14), was significantly reduced by long term administration of taurine (Table 1). In addition, there was a decreased peak height of an oxidizing metallo-protein complex that was detected in the renal parenchyma using electron paramagnetic resonance spectroscopy.

Streptozocin-induced Diabetes

We next examined the effect of taurine on the development of diabetic nephropathy. Diabetes was induced in male Sprague-Dawley rats by an intravenous injection of streptozocin, 60 mg/kg body weight. Experimental animals were provided 1% taurine drinking water, while control animals were given tap water. Both groups of rats were maintained for 40 weeks without insulin treatment.

Taurine did not alter the water intake or urine output in the streptozocin-diabetic rats. The severity of the hyperglycemia and glycosylated hemoglobin levels, 13.4±0.7 versus 13.8±0.6%, was comparable in the two study groups (Table 2). Taurine had no effect on weight gain, blood pressure or hematocrit and did not reduce the rate of cataractogenesis. There was a modest reduction in serum cholesterol and triglyceride levels in taurine-treated streptozocin-diabetic rats; renal tissue taurine content was increased from 6.8±1.0 to 13.6±0.9 mmol/kg wet weight.

Table 2. Taurine and streptozocin-induced diabetes

	Weight (g)	Plasma glucose (mg/dl)	Creatinine clearance (ml/min/100 g body weight)
Diabetic, untreated			
Initial	256 ± 6	136 ± 2	0.45 ± 0.02
10 months	273 ± 19	570 ± 16	0.48 ± 0.04
Diabetic, taurine			
Initial	253 ± 4	143 ± 6	0.43 ± 0.03
10 months	281 ± 24	567 ± 21	0.70 ± 0.06*

Data are means ± SEM for 8 untreated and 5 taurine-treated rats. * $p<.01$ compared to corresponding untreated group.

Oral taurine supplementation reduced urinary protein:creatinine excretion ratio and daily urinary protein excretion by 50%. There was no change in serum albumin concentration. The creatinine clearance was persistently elevated in the taurine-supplemented streptozocin-rats. These findings were associated with a modest decrease in the percentage of sclerosed glomeruli from 4.0±1.8 to 0.8±0.3%, and the degree of tubulointerstitial fibrosis from 0.8±0.2 to 0.2±0.1 on a 0-3+ scale.

Taurine treatment did not alter the content of the initial glycosylation product, fructoselysine, or the Amadori rearrangement compound, carboxymethyllysine, in skin collagen of streptozocin-diabetic rats (Table 3). However, taurine supplementation inhibited the late autoxidative reactions that lead to the chronic renal accumulation of advanced glycosylation end products. The level of fluorescence products detected by excitation at 370 nm and emission at 440 nm was reduced by 20%. The levels of pentosidine, an advanced glycosylation end product that is formed by the crosslinking of L-lysine and L-arginine levels by pentoses and hexoses, in rat skin collagen were unaffected by taurine treatment (measurements were performed with the assistance of John Baynes).

Table 3. Collagen and glyco-oxidation in streptozocin diabetes

	FL	CML	Pentosidine	Fluorescence
Diabetes, untreated	19.3 ± 1.6	0.84 ± 0.07	8.5 ± 0.30	767 ± 52
Diabetes, +taurine	15.8 ± 1.6**	0.82 ± 0.18	8.1 ± 0.60	623 ± 28*
Normal, untreated	2.2 ± 0.3	0.38 ± 0.02	1.4 ± 0.07	181 ± 2
Normal, +taurine	1.0 ± 0.1*	0.35 ± 0.07	0.4 ± 0.03*	103 ± 25*

Data are means ± SEM. * $p<0.05$, ** $p<0.06$ compared to corresponding untreated group. Abbreviations: FL: fructose lysine; CML: carboxymethyl lysine.

Acute and Chronic Renal Failure

The effect of taurine was evaluated in rats with acute renal failure induced by bilateral renal artery clamping for 60 min. Pretreatment for 2 weeks with 1% taurine drinking water, increasing the renal cortical content of taurine by nearly twofold, did not alleviate the severity of renal dysfunction. In addition, rats subjected to a 5/6th nephrectomy to induce chronic renal failure were examined to ascertain whether taurine could attenuate the loss of renal function. Provision of a 1% taurine solution in the drinking water had no demonstrable effect on kidney function over a 36 week observation period. The lack of a beneficial effect of taurine on the course of chronic renal failure in the remnant kidney model is consistent with our inability to detect increased levels of oxygen free radicals in plasma or renal tissue using the spin trap reagent, α-phenyl-N-butyl-nitrone and an HPLC assay (24).

Immunohistochemical Localization Studies

We explored whether the different impact of taurine in experimental renal disease was paralleled by changes in the distribution of the amino acid within the kidney (23). Renal tissue was stained with a specific polyclonal antibody raised against a taurine-bovine serum albumin conjugate. The antiserum was pre-absorbed against the structurally similar amino acids, GABA and ϐ-alanine, to eliminate nonspecific staining of the renal tissue. In normal kidney, taurine staining was most evident in the cytoplasm of collecting tubule cells, with considerably less staining in proximal tubular epithelium and glomeruli. In untreated rats with chronic puromycin or streptozocin-diabetic nephropathy, taurine staining was enhanced in medullary collecting ducts and glomeruli, with less pronounced alterations in other renal structures. All changes were exaggerated by oral taurine supplementation. In both disease states, the immunohistochemical localization pattern was associated with increased renal content of taurine. In order to determine the specificity of these changes in taurine staining within the kidney, immunohistochemical localization of GABA was examined in the same two models of renal disease. In normal rats, the pattern of GABA staining was similar to taurine, with predominance of the amino acid in the tubules. However, in contrast to taurine, GABA staining in glomeruli and the tubulointerstitium was reduced in animals with chronic puromycin nephropathy or streptozocin-diabetes. These changes were paralleled by a 75% decrease in renal cortical content of GABA. In rats with acute or chronic renal failure, in which taurine treatment had no beneficial impact on kidney function, the intensity of taurine or GABA staining was unchanged in the renal parenchyma.

TAURINE AND CULTURED RENAL CELLS

Experiments were then performed with cultured renal cells to clarify the mechanism of action of taurine within the kidney.

Cell Taurine Content

To determine whether the in vivo localization was paralleled by differences in cell taurine content, taurine levels were measured in rat mesangial cells, LLC-PK_1 and opossum kidney cells (analogues of the proximal tubule) and Madin Darby canine kidney (MDCK) cells an analogue of the distal nephron.

Taurine content differed in the various cell types and was highest in the opossum kidney cells 6.53$\pm$0.91 μmol/mg protein and lowest in the MDCK cells, 3.07$\pm$0.73 μmol/mg protein. The value was intermediate in LLC-PK_1 cells, 3.67$\pm$0.21 μmol/mg protein. Thus, taurine content varies in renal epithelial cells, even in cell lines that are derived from the same nephron segment, such as opossum kidney and LLC-PK_1 cells.

Taurine content was 6.3$\pm$0.66 μmol/mg protein in the mesangial cells. The effect of an altered ambient taurine or glucose concentration on mesangial cell taurine content was examined by raising the taurine or glucose level to 500 μM or 33.3 mM, respectively. Raising the taurine content alone had no effect on mesangial cell taurine content, 6.6$\pm$0.33 μmol/mg protein. However, the combination of an elevated glucose and taurine level in the medium resulted in a six-fold increase in mesangial cell taurine content to 37.5$\pm$4.0 μmol/mg protein. Additional studies, using mannitol and raffinose as nonmetabolized extracellular solutes, are under way to ascertain whether this change in mesangial cell taurine content represented a cell volume regulatory response to high glucose-induced hyperosmolality or was a specific adaptive response to attenuate glucose-mediated cell injury.

Taurine and Collagen Synthesis

Scarring of glomeruli and the tubulointerstitium in chronic renal disease reflects net accumulation in these structures of extracellular matrix materials, primarily collagen (1). Therefore, we measured collagen production in rat mesangial cells, opossum kidney, LLC-PK_1 and MDCK cells in vitro, focusing on the effect of exposure to high glucose medium (22). Cells were grown to confluence and then assigned to control Dulbecco's modified Eagle's medium or media supplemented either with glucose (27.8 mM), taurine (500 μM), glucose (27.8 mM) and taurine (500 μM), or mannitol (27.8 mM). Collagen production was assayed after 7 days based upon the incorporation of ^{3}[H]proline followed by collagenase digestion.

Collagen production nearly doubled in the mesangial cells exposed to high ambient glucose, an effect that was reversed by the concomitant addition of taurine (Table 4). Glucose did not promote increased collagen production by renal tubular epithelial cell lines. In addition, glucose had no effect on collagen production by human fetal lung fibroblasts. Thus, the effect of high glucose on extracellular matrix protein synthesis is cell-specific, even within the same organ.

Taurine and lipid peroxidation

We next examined whether glucose enhances lipid peroxidation in cultured renal cell and whether taurine attenuates this process (22). Once again, mesangial cells, LLC-PK_1, opossum kidney and MDCK cells were grown to confluence and studied under the test

Table 4. Collagen production in renal cells

	Mesangial cells	Epithelial cells	
		MDCK	$LLC\text{-}PK_1$
DME, 5.6 mM glucose	33.5 ± 5.9	3.8 ± 0.2	5.9 ± 0.1
+ 27.8 mM glucose	54.2 ± 3.7□	3.5 ± 0.3	5.3 ± 0.2
+ 500 μM taurine	25.9 ± 2.7	4.3 ± 0.3	5.7 ± 0.4
+ glucose + taurine	31.2 ± 2.7	3.7 ± 0.2	6.5 ± 0.3
+ 27.8 mM mannitol	30.8 ± 3.0	4.6 ± 0.7	6.1 ± 0.1

Cells were maintained in Dulbecco's modified Eagle's medium (DME) with the additions indicated. Data are expressed as percentage of total protein synthesis (means ± SEM for 4 samples). * $p < .001$ versus all other groups.

conditions described above. Lipid peroxidation was evaluated after 1 week by measuring malondialdehyde content based on an assay of thiobarbituric acid-reactive substances (14).

High glucose increased lipid peroxidation in mesangial cells but not in $LLC\text{-}PK_1$, opossum kidney or MDCK cells (Table 5). Taurine reversed the pro-oxidant effect of high glucose on the rat mesangial cells. These results were confirmed by determination of conjugated dienes in an extract of membrane lipids (11).

Table 5. Lipid peroxidation in renal cells: malondialdehyde assay

	Mesangial cells	Epithelial cells	
		MDCK	$LLC\text{-}PK_1$
DME, 5.6 mM glucose	2.73 ± 0.29 (8)	1.45 ± 0.19 (8)	1.02 ± 0.08 (3)
+ 27.8 mM glucose	4.12 ± 0.42 (8)†	1.96 ± 0.48 (8)	1.06 ± 0.07 (3)
+ 500 μM taurine	2.97 ± 0.42 (6)	1.79 ± 0.26 (8)	1.08 ± 0.07 (3)
+ glucose + taurine	2.62 ± 0.33 (6)	1.49 ± 0.18 (7)	1.15 ± 0.05 (3)
+ 27.8 mM mannitol	2.38 ± 0.60 (6)	1.63 ± 0.13 (7)	0.81 ± 0.03 (3)

Cells were maintained in Dulbecco's modified Eagle's medium (DME) with the additions indicated. Data are nmol/mg protein (means ± SEM for 4 samples). † $p < 0.025$ versus 5.6 mM glucose, and glucose plus taurine groups.

Taurine and Cell Proliferation

High glucose concentrations exert an antiproliferative effect on cultured human (12) and rat mesangial cells (28). We tested whether taurine could reverse this process by measuring ^{3}[H]thymidine uptake. Cells were plated in 24 well plates at an initial density of 50,000 cells/well. After 24 h, cells were serum-deprived for 24 h. Cells were then maintained in one of the following test media for one week; control, high glucose, high taurine or combined high glucose-high taurine. On the final day, 1 μCi [^{3}H]thymidine was added to each well for 6 h.

High glucose concentrations reduced thymidine uptake by nearly 33% (Table 6). The addition of taurine significantly restored the rate of mesangial cell proliferation compared to cells grown in high glucose alone. We confirmed these data by trypsinizing cells and counting them after they had been maintained in the test conditions outlined above for 1 week. Again, there was a significant decrease in the number of mesangial cells that were grown in the presence of an elevated ambient glucose that was partially reversed by the addition of taurine. Addition of mannitol to raise the osmolality of the medium to the same degree as glucose had no effect on tritiated thymidine uptake or cell number. Finally, the percentage of cells stained with 5'-bromo-2'-deoxyuridine, used as a marker of dividing cells, was reduced by high glucose from 11.7 to 4.7% and restored 10.2% by the concomitant addition of taurine to the incubation media.

Table 6. Taurine and mesangial cell proliferation: [^{3}H]thymidine incorporation

	5.5 mM glucose	33.3 mM glucose
- Taurine	9350 ± 645	6048 ± 411[a]
+ Taurine	11278 ± 763	7785 ± 217[b]

Values are cpm (means ± SEM for 6 samples per group). [a] $p < .01$ versus control; [b] $p < .05$ versus no taurine group.

Table 7. Effect of advanced glycosylation end products on [^{3}H]thymidine incorporation in mesangial cells

	Control	AGEs 25 μg/ml	AGEs 250 μg/ml
- Taurine	6403 ± 426	5726 ± 394	2988 ± 436
+ Taurine	5774 ± 731	9505 ± 816	6087 ± 454

Values are cpm (means ± SEM for determinations in quadruplicate). [a] $p < .01$ versus control; [b] $p < .01$ versus no taurine group.

It has also been established that advanced glycosylation end products inhibit proliferation of mesangial cells (18). Therefore, we examined whether taurine affected this adverse interaction. There was a dose-dependent reduction in thymidine uptake by the mesangial cells when the concentration of advanced glycosylation end products was increased from 25-250 μg/ml. Taurine significantly increased proliferation at the two doses of advanced glycosylation end products that were tested. (Table 7). This finding was confirmed by determining cells counts which were increased at both levels of advanced glycosylation end products in media containing taurine. Taurine increased the number of bromodeoxyuridine-stained cells from 3.5 to 8.8% in 25 μg/ml glycosylation end products-containing medium and from 2.3 to 3.8% in 250 μg/ml glycosylation end products-containing medium ($P<0.05$). The effects of taurine to reduce lipid peroxidation, normalize collagen production and restore mesangial cell proliferation may represent an interaction between endogenous antioxidants and altered bioactivity of TGF-β in diabetes (28).

In conclusion, the experimental findings that have been compiled to date indicate that in select experimental renal disease states, namely chronic puromycin nephropathy and streptozocin diabetes, taurine acts as an endogenous antioxidant to preserve kidney function and retard progressive structural damage. Immunohistochemical localization studies indicate that in these glomerulopathies, there is increased taurine content in medullary tubules and secondary changes in glomeruli, two regions of oxidant stress. Cell culture experiments indicate that taurine is present in glomerular mesangial and renal tubular epithelial cells. Taurine prevents high glucose-induced lipid peroxidation and enhanced collagen production by mesangial cells. It normalizes the rate of proliferation of mesangial cells exposed to high glucose or advanced glycosylation end products. Clinical studies are needed to determine the role of dietary taurine supplementation as an adjunctive therapy in patients with primary glomerular disease or incipient diabetic nephropathy.

REFERENCES

1. Ayo, S.H., Radnick, R.A., Glass, W.A.,II, Garoni, J.A., Rampt, E.R., Appling, D.R., and Kreisberg, J.I. 1991, Increased extracellular matrix synthesis and mRNA in mesangial cells grown in high-glucose medium, *Am.J.Physiol.*, 260:F185-F191.
2. Banks, M.A., Porter, D.W., Martin, W.G., and Castranova, V. 1990, Effects of *in vitro* ozone exposure on peroxidative damage, membrane leakage, and taurine content of rat alveolar macrophages, *Toxicol.Appl.Pharmacol.*, 105:55-65.
3. Baynes, J.W. 1991, Role of oxidative stress in development of complications in diabetes, *Diabetes*, 40:405-412.
4. Diamond, J.R., Bonventre, J.V., and Karnovsky, M.J. 1986, A role for oxygen free radicals in aminonucleoside nephrosis, *Kidney Int.*, 29:478-483.
5. Endreffy, E., Turi, S., Laszik, Z., Bereczki, C., and Kasa, K. 1991, The effects of vitamin E on tissue oxidation in nephrotoxic (anti-glomerular basement membrane) nephritis, *Pediatr.Nephrol.*, 5:312-317.
6. Ginevri, F., Gusmano, R., Oleggini, R., Acerbo, S., Bertelli, R., Perfumo, F., Cercignani, G., Allegrini, S., D'Allegri, F., and Ghiggeri, G. 1990, Renal purine efflux and xanthine oxidase activity during experimental nephrosis in rats: difference between puromycin aminonucleoside and adriamycin nephrosis, *Clin.Sci.*, 78:283-293.
7. Gordon, R.E., Shaker, A.A., and Solano, D.F. 1986, Taurine protects hamster bronchioles from acute NO_2-induced alterations, *Am.J.Pathol.*, 125:585-600.
8. Grond, J., Weening, J.J., and Elema, J.D. 1984, Glomerular sclerosis in nephrotic rats: comparison of the long-term effects of adriamycin and aminonucleoside, *Lab.Invest.*, 51:277-284.
9. Hamaguchi, T., Azuma, J., Harada, H., Takahashi, K., Kishimoto, S., and Schaffer, S.W. 1989, Protective effect of taurine against doxorubicin-induced cardiotoxicity in perfused chick hearts, *Pharmacol.Res.*, 21:729-734.
10. Lorenzi, M. 1992, Glucose toxicity in the vascular complications of diabetes: the cellular perspective, *Diab.Metab.Rev.*, 8:85-103.
11. Massey, K.D. and Burton, K.P. 1990, Free radical damage in neonatal rat cardiac myocyte cultures: effects of α-tocopherol, trolox and phytol, *Free Radic.Biol.Med.*, 8:449-458.
12. Nahman, N.S.,Jr., Leonhart, K.L., Cosio, F.G., and Hebert, C.L. 1992, Effects of high glucose on cellular proliferation and fibronectin production by cultured human mesangial cells, *Kidney Int.*, 41:396-402.
13. Nakamori, K., Koyama, I., Nakamura, T., Nemoto, M., Yoshida, T., Umeda, M., and Inoue, K. 1993, Quantitative evaluation of the effectiveness of taurine in protecting the ocular surface against oxidant, *Chem.Pharm.Bull.(Tokyo)*, 41:335-338.
14. Ohkawa, H., Ohishi, N., and Yagi, K. 1979, Assay for lipid peroxides in animal tissues by thiobarbituric acid reaction, *Anal.Biochem.*, 94:351-358.
15. Paller, M.S., Hoidal, J.R., and Ferris, T.F. 1984, Oxygen free radicals in ischemic acute renal failure in the rat, *J.Clin.Invest.*, 74:1156-1164.
16. Pasantes-Morales, H., Wright, C.E., and Gaull, G.E. 1984, Protective effect of taurine, zinc and tocopherol on retinol-induced damage in human lymphoblastoid cells, *J.Nutr.*, 114:2256-2261.
17. Pasantes-Morales, H., Wright, C.E., and Gaull, G.E. 1985, Taurine protection of lymphoblastoid cells from iron-ascorbate induced damage, *Biochem.Pharmacol.*, 34:2205-2206.
18. Skolnik, E.Y., Yang, Z., Makita, Z., Radoff, S., Kirstein, M., and Vlassara, H. 1991, Human and rat mesangial cell receptors for glucose-modified proteins: Potential role in kidney tissue remodelling and diabetic nephropathy, *J.Exp.Med.*, 174:931-939.
19. Thakur, V., Walker, P.D., and Shah, S.V. 1988, Evidence suggesting a role for hydroxyl radical in puromycin aminonucleoside-induced proteinuria, *Kidney Int.*, 34:494-499.
20. Thomas, E.L., Grisham, M.B., Melton, D.F., and Jefferson, M.M. 1985, Evidence for a role of taurine in the *in vitro* oxidative toxicity of neutrophils toward erythrocytes, *J.Biol.Chem.*, 260:3321-3329.
21. Trachtman, H., Del Pizzo, R., Futterweit, S., Levine, D., Rao, P.S., Valderrama, E., and Sturman, J.A. 1992, Taurine attenuates renal disease in chronic puromycin aminonucleoside nephropathy, *Am.J.Physiol.Renal,Fluid Electrolyte Physiol.*, 262:F117-F123.

22. Trachtman, H., Futterweit, S., and Bienkowski, R.S. 1993, Taurine prevents glucose-induced lipid peroxidation and increased collagen production in cultured rat mesangial cells, *Biochem.Biophys.Res.-Commun.*, 191:759-765.
23. Trachtman, H., Lu, P., and Sturman, J.A. 1993, Immunohistochemical localization of taurine in rat renal tissue: Studies in experimental disease states, *J.Histochem.Cytochem.*, 41:1209-1216.
24. Trachtman, H., Wilson, D., and Rao, P.S. 1992, The role of oxygen free radicals in the development of chronic renal failure, *Life Sci.*, 50:1877-1883.
25. Walker, P.D. and Shah, S.V. 1988, Evidence suggesting a role for hydroxyl radical in gentamicin-induced acute renal failure in rats, *J.Clin.Invest.*, 81:334-341.
26. Wang, Q., Hollinger, M.A., and Giri, S.N. 1992, Attenuation of amiodarone-induced lung fibrosis and phospholipidosis in hamsters by taurine and/or niacin treatment, *J.Pharmacol.Exp.Ther.*, 262:127-132.
27. Wang, Q., Hyde, D.M., and Giri, S.N. 1992, Abatement of bleomycin-induced increases in vascular permeability, inflammatory cell infiltration, and fibrotic lesions in hamster lungs by combined treatment with taurine and niacin, *Lab.Invest.*, 67:234-242.
28. Wolf, G., Sharma, K., Chen, Y., Ericksen, M., and Ziyadeh, F.N. 1992, High glucose-induced proliferation in mesangial cells is reversed by autocrine TGF-β, *Kidney Int.*, 42:647-656.

PROTECTING EFFECT OF TAURINE AGAINST HYPOXIC CELL DAMAGE IN RENAL TUBULAR CELLS CULTURED IN DIFFERENT TRANSPLANT PRESERVATION SOLUTIONS

Peter Wingenfeld,[1] Stefan Strübind,[1]
Ute Gehrmann,[1] Thomas Minor,[2]
Wolf Isselhard,[2] and Dietrich Michalk[1]

[1]Department of Pediatrics
[2]Institute for Experimental Medicine
University of Cologne
Cologne, Federal Republic of Germany

INTRODUCTION

The transplantation of human organs is an important field in current medicine and medical research. One of the limiting steps in organ transplantation is the preservation of the graft. Frequently the hypoxic storage of the removed organ limits the possibility of sucessful transplantation. The hypoxic cell damage due to insufficient organ preservation is a major cause for the loss or the dysfunction of transplanted organs like hearts and kidneys (6,9,13,17,20,24).

Methods of preservation and suitable solutions for the adequate preservation of the graft are still under investigation (2,6,7,13-16,18,19,22,26) and many of the pathophysiological mechanisms of cell damage during and subsequent to hypoxia are still unknown or subjects of controversy (2,3,13,25,26).

Some of the determinating factors of the extent of hypoxic and posthypoxic deterioration are the energy deficiency of the organ (3), intracellular calcium overload (31), osmoregulatory dysfunction (26,27), the oxygen paradox, and membrane oxidation induced by toxic metabolites and free oxygen radicals (3).

Different transplant preservation solutions are conventionally used for the preservation of the explanted kidney. Two different types of a kidney preservation solution are represented by the Euro Collins solution (6) and the University of Wisconsin Solution (2,7). The first one is based on components and constituents of the intra- and extracellular environment whereas the University of Wisconsin solution was developed considering intracellular fluid components and metabolite concentrations and was therefore supplemented

Taurine in Health and Disease, Edited by R. Huxtable
and D.V. Michalk, Plenum Press, New York, 1994

with agents acting as osmoregulators, antioxidants, radical scavengers and membrane stabilizators.

Because of the unique functions of taurine as modulator of transmembrane calcium transport (11), as osmoregulator (27,28), and as free radical scavenger (21), the purpose of our study was to investigate in a cell culture model if regularly used transplant preservation solutions like the University of Wisconsin or the Euro Collins solution can be improved by supplementation with taurine. Furthermore, we wanted to see if supplementation with taurine during hypoxic preservation is able to improve posthypoxic outcome. Finally, we wanted to compare the efficiency of the two storage solutions within this experimental model.

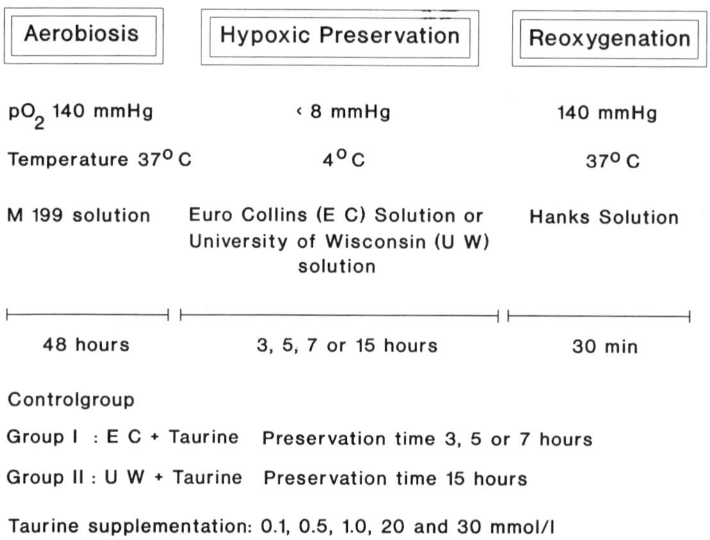

Figure 1. Experimental course and groups

MATERIALS AND METHODS

The experiment was divided in three different time periods (Fig. 1).

Aerobiosis

Proximal renal tubular cells of the pig from the available cell strain LLC PK 1 (Kasza pig kidney; ICN Biomedical Meckenheim FRG) were cultured aerobically as monolayers in flasks (Easy access, 25 cm^2, canted neck; Falcon # 3813) for 48 h in M 199 medium with an admixture of fetal calf serum (50 ml/l, 20%), penicillin (100 I.U./ml), and streptomycin (100 μg/ml). The cultures were incubated with 21% oxygen and 5% carbon dioxide. The temperature was kept constantly at 37°C and the oxygen partial pressure was about 140 mm Hg. The cultures were fed three times a week. After 48 h of subculturing, each monolayer contained between 1 and 1 1/2 million cells.

Hypoxia

At the end of this period the culture medium was removed and replaced by 5 ml preservation solution; either Euro Collins or University of Wisconsin solution (Fig. 1). The renewed solutions were equilibrated beforehand with nitrogen or nitrogen plus 5% carbon dioxide resulting in oxygen partial pressures below 8 mm Hg, determined by means of a acid base laboratory.

With a new and simple method, the flasks were plugged with a screw cap, described elsewhere, which allowed the filling and flushing of the flasks with a nearly oxygen-free atmosphere and to realize oxygen partial pressures below 8 mm Hg in the supernatant of the culture. Additionally this technique enabled microphotographic studies of the cultures.

With constant low flow gas flushing and defined supernatant volume, the flasks were placed in a water bath at 4°C for different time courses. The gas persufflation was performed with saturated nitrogen (group II) or nitrogen plus 5 % carbon dioxide (group I) through the modified seal screw cap.

Reoxygenation

After 3, 5, 7 or 15 h, the preservation medium was removed and replaced by a cristalloid perfusion solution without taurine. For this we used Hank's balanced salt solution (Sigma chemicals, St. Louis, USA). Thereafter the cell cultures were stored aerobically for 30 min at a temperature of 37°C and an oxygen partial pressure of 140 mm Hg.

Experimental Groups

Group I was cultured 48 h aerobically and afterwards preserved hypoxically with Euro Collins solution (6) for time courses of 3, 5 or 7 h before reoxygenation (Fig. 1). Group II was preserved hypoxically 15 h with University of Wisconsin solution (2). The further treatment was equivalent to group I (Fig. 1). During hypoxic storage, taurine (Sigma chemicals, St. Louis USA) was added to the preservation solution resulting in final concentrations of 0.1, 0.5, 1.0, 20 (group I) and 1, 10, 20 or 30 mM (Group II). Controls were preserved without addition of taurine.

Measurements

Before each hypoxic preservation and reoxygenation, 2-3 cultures were rinsed from the monolayer surface with Trypsin-EDTA (0.5% Trypsin - 0.025% EDTA, Fa. Gibco) and the cells were counted in a Neubauer chamber to get basic values of cell numbers and viability rates were determined by trypan blue exclusion test (12).

After each experimental period, 1 ml of the supernatant was removed for the determination of enzyme leakage and the cells were rinsed from the bottom of the flask by treatment with Trypsin-EDTA solution (2.5% Trypsin - 0.025% EDTA, 1 ml, Fa. Gibco) added to the supernatant. After centrifugation (300 g, 5 min and 30°C), the supernatant was removed and the cells were resuspended. In a small aliquot, we determined the percentage of remaining cells by cell counting with a Neubauer chamber and measured the viability of these cells by the trypan blue exclusion test.

The product of both parameters, regarded as survival index (SI), was calculated (SI = [remaining cells/cells at the beginning] x 100 x viability of the remaining cells [% x %]), and served as an index for the survival of the cultures. This parameter reflected differences in the number of remaining cells after experimental treatment or effects like cell lysis or detachment from the monolayer surface during the experimental course (10,23). The measurement of viability alone is not sufficient to consider these effects adequately.

Furthermore, the survival index allowed a better comparison of different preservation conditions.

Additionally, the leakage of lactate dehydrogenase was measured with standard biochemical assays (LDH Monotest, Boehringer Mannheim) after each experimental period in an aliquod of the supernatant of the cultures.

As parameters of the metabolic status of the whole culture after reoxygenation, the content of high energy phosphates and adenine nucleotides was measured by means of high pressure liquid chromatography of neutralized perchloric acid extracts (HPLC Merck-Hitachi D 6000 detector L 4250, autosampler AS 2000, carrier potassium dihydrogen phosphate and tetrabutylammonium phosphate, gradient 80% carrier and 20% pure methanol, peaks detected at 259 nm compared with defined nucleotide standards). Furthermore the energy charge potential was calculated (1):

$$\text{Energy charge} = \frac{[ATP] + 0.5\,[ADP]}{[ATP] + [ADP] + [AMP]}$$

Before denaturation of proteins with perchloric acid (ice bath, 0.25 ml 1 M) for chromatography, the protein content of each culture was measured in an aliquot of the rinsed culture as described by Bradford et al. (4). Analyses of 5 to 8 cell cultures per group were used for the determination of each parameter.

Statistical analyses were done with Students t-test in comparison to the controls. Different levels of significance were accepted as significant (a: $p < 0.05$; b $p < 0.01$; c $p < 0.001$). The data are presented as mean ± standard deviations.

RESULTS AND DISCUSSION

Group I (Preservation with Euro-Collins Solution)

After 3 h of hypoxic preservation with Euro Collins solution, taurine supplementation in the concentration range of 0.1-1.0 mM significantly increased the number of remaining cells, the viability rate and the survival index (Table 1). During this short preservation period, a 20 mM taurine had an opposite effect, reducing vitality rates and survival indices. This effect must have been reversible since it was not observed after 5 and 7 h, respectively.

After 5 h of cold storage with Euro Collins solution comparable effects occured. Viability and survival index were markedly improved in the experimental groups treated with taurine both physiological and pharmacological concentrations (Table 1).

Subsequent to 7 h of hypoxic storage, groups preserved with pharmacological taurine concentrations showed markedly higher viability rates and significantly increased survival indices (Table 1). With longer preservation times, higher concentrations of taurine were needed for the same protective effect. A possible explanation for the dose- and preservation time-dependent action of taurine is an osmotic swelling of the cells during longer preservation periods, which may be counteracted by the osmoregulatory function of higher taurine doses (11,26,27).

During hypoxic preservation, considerable amounts of lactate dehydrogenase were released (Fig. 2). As suspected, higher amounts of lactate dehydrogenase were released into the supernatant of the cultures after longer preservation times (Fig. 2), indicating a major cellular deterioration during hypoxic storage. The patterns of leakage of lactate dehydrogenase were inverse to the measured viability rates and survival indices (Tables 1 and 2 and Fig. 2). Significantly higher viability rates and survival indices were associated with reduced enzyme release during hypoxic preservation in the cultures preserved with taurine-supplemented solutions.

The additional leakage of lactate dehydrogenase after reoxygenation compared to that during hypoxic preservation indicates further cell damage due to the metabolic events of reoxygenation.

The overall cell damage was mainly caused during the period of oxygen deficiency, and was only partly determined by the pathophysiological events occurring with renewed oxygenation.

Table 1. Percentage of remaining cells, viability rates and survival index of different groups after hypoxic preservation with Euro Collins solution

Preservation time (h)	Taurine (mm)	Remaining cells (%)	Viability (%)	SI [% x %] x 10^3
3	0	78 ± 27	72 ± 5	5.48 ± 1.66
	0.1	116 ± 25 a	77 ± 5 a	7.60 ± 1.60 a
	0.5	105 ± 20 a	84 ± 2 b	8.82 ± 1.61 b
	1	101 ± 24	70 ± 5	7.08 ± 2.25
	20	63 ± 20	50 ± 7 c	3.15 ± 1.22 a
5	0	58 ± 20	61 ± 5	3.44 ± 0.86
	0.1	68 ± 19	62 ± 5	3.70 ± 1.18
	0.5	71 ± 15	76 ± 7 a	5.33 ± 0.84 b
	1	56 ± 5	65 ± 2	3.64 ± 0.42
	20	77 ± 20	81 ± 3 c	6.24 ± 1.73 b
7	0	57 ± 0	46 ± 6	2.78 ± 0.17
	0.1	42 ± 8 a	53 ± 4 a	2.24 ± 0.50
	0.5	37 ± 23	60 ± 6 a	2.16 ± 1.21
	1	60 ± 7	63 ± 3 b	3.85 ± 0.78 b
	20	93 ± 21 a	62 ± 8 b	5.84 ± 1.27 c

SI: Survival index; n = 5 - 8 cultures per experiment, mean ± SD; a $p \leq 0.05$, b $p \leq 0.01$, c $p \leq 0.001$

Taurine supplementation reduced the overall leakage markedly, especially after longer preservation times (Fig. 2).

The viability rates of the remaining cells after reoxygenation were markedly higher in comparison to the posthypoxic values (Fig. 3), which suggest a posthypoxic recovery of the cell cultures. The number of remaining cells, the survival indices after hypoxia and reoxygenation, as well as the lactate dehydrogenase leakage indicated a further growing of the cell cultures after oxygen deficiency in spite of a deterioration of the cultures due to reoxygenation (Figs. 3, 4). Some taurine-treated groups showed slightly higher viability rates than controls after short preservation periods and 30 min reoxygenation (Fig. 3).

Table 2. Leakage of lactate dehydrogenase before and during hypoxic preservation with Euro Collins solution and during 30 min reoxygenation

Taurine during hypoxia	Lactate Dehydrogenase Leakage during hypoxia	during reoxygenation
[mmol/l]	[units LDH /liter supernatant]	
	3 hours	30 min
0	19.4 ± 7.8	13.2 ± 1.9
0.1	12.6 ± 5.0 a	22.5 ± 4.6 b
0.5	13.2 ± 4.2	21.2 ± 3.7 b
1.0	13.3 ± 1.8	18.2 ± 0 c
20	20.9 ± 12.4	22.4 ± 3.5 c
	5 hours	30 min
0	24.9 ± 8.6	21.8 ± 4.0
0.1	26.9 ± 3.1	16.0 ± 1.5 a
0.5	11.1 ± 0 b	20.0 ± 2.7
1.0	13.5 ± 2.1 a	18.9 ± 2.7
20	11.6 ± 3.1 b	15.9 ± 3.0 a
	7 hours	30 min
0	47.3 ± 10.0	15.2 ± 2.3
0.1	41.9 ± 8.2	17.0 ± 4.7
0.5	35.0 ± 3.8 a	21.9 ± 2.5 b
1.0	30.6 ± 4.2 b	15.1 ± 3.2
20	30.7 ± 4.6 b	19.4 ± 3.4 a

n = 5 cultures per experiment, mean ± SD; a $p \leq 0.05$, b $p \leq 0.01$, c $p \leq 0.001$

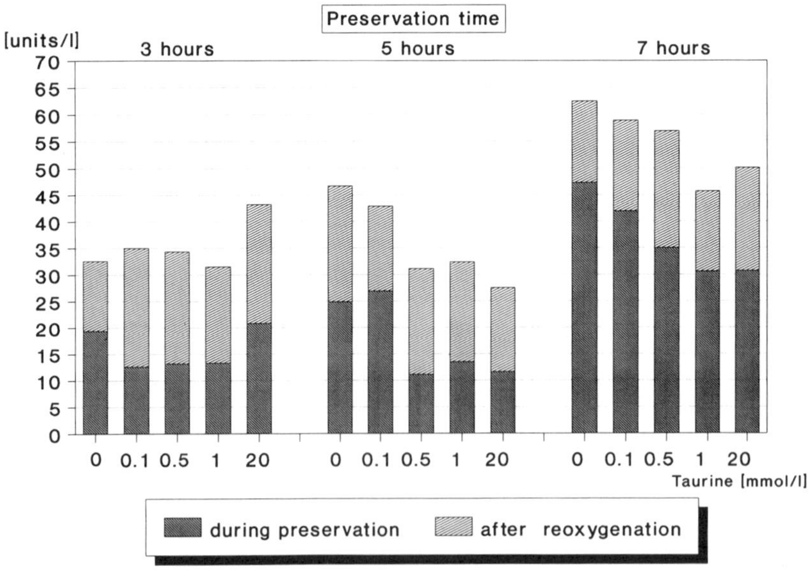

Figure 2. Lactate dehydrogenase leakage during hypoxic preservation with Euro Collins solution and durin 30 min reoxygenation (Group I). For standard deviations and significance see Table 2

Although viability was only slightly improved, taurine supplementation during hypoxic preservation resulted in significantly higher survival rates 30 min after reoxygenation. The same doses of taurine, which improved posthypoxic viability and survival indices, exerted beneficial effects during hypoxia and reoxygenation. The data show that taurine administered only during hypoxia seems to be able to protect the cells against the hypoxic cell damage and to improve the posthypoxic recovery.

Table 3. Viability rates and survival indices after different times of hypoxic preservation with Euro Collins solution and 30 min reoxygenation

Time [h]	Taurine [mmol/l]	Viability [%]	Survival index [% x %] x 10^3
3 + 0.5	0	80.8 ± 6.94	5.37 ± 1.09
	0.1	86.6 ± 2.30	5.13 ± 1.23
	0.5	83.4 ± 4.83	9.99 ± 2.23 b
	1.0	88.8 ± 3.76 a	9.35 ± 3.41 a
	20	84.8 ± 5.98	6.51 ± 1.17
5 + 0.5	0	93.6 ± 5.18	8.59 ± 1.64
	0.1	88.3 ± 2.99	3.86 ± 0.51 c
	0.5	91.6 ± 2.88	5.28 ± 1.43 b
	1.0	95.5 ± 1.00	12.52 ± 4.59
	20	92.5 ± 3.00	12.07 ± 3.43 a
7 + 0.5	0	90.8 ± 2.17	7.58 ± 1.33
	0.1	92.0 ± 3.16	8.61 ± 2.73
	0.5	94.6 ± 1.34 b	11.40 ± 1.50 b
	1.0	93.6 + 1.52 a	11.27 ± 2.04 b
	20	91.0 ± 4.53	11.80 ± 2.18 b

n = 5 cultures per experiment, mean ± SD; a $p \leq 0.05$, b $p \leq 0.01$ c $p \leq 0.001$

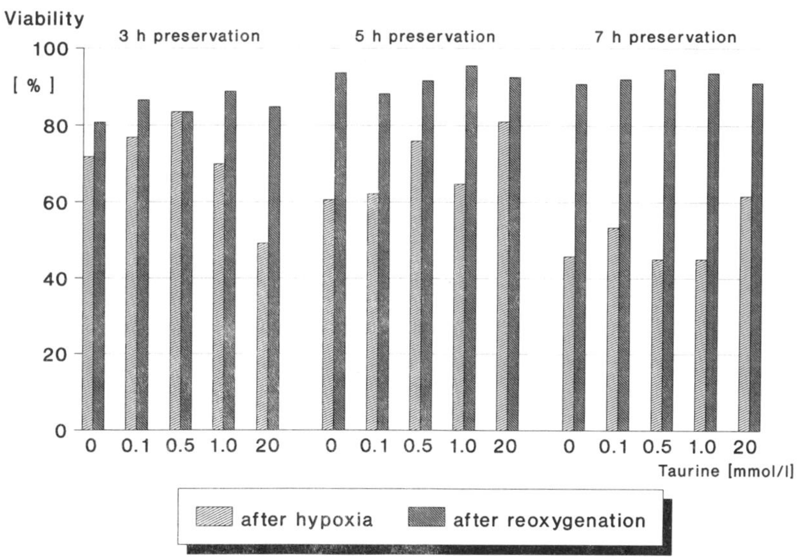

Figure 3. Viability rates after 3, 5, and 7 h hypoxic preservation with Euro Collins solution and 30 min reoxygenation (Group I). For standard deviations and significances see Tables 1 and 3.

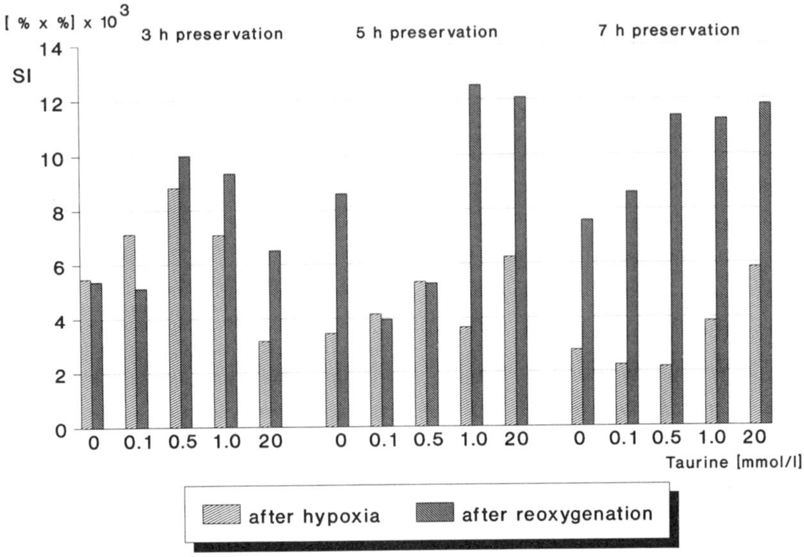

Figure 4. Survival index (SI) after 3, 5, and 7 h hypoxic preservation with Euro Collins solution and 30 min reoxygenation (Group I). For standard deviations and significances see Tables 1 and 3.

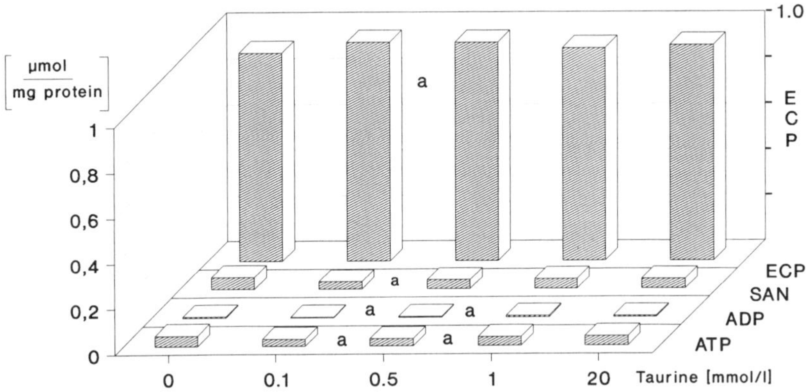

	0	0.1		0.5		1	20	
ECP	0,913	0,958	a	0,956		0,931	0,944	[ ]
SAN	0,054	0,034	a	0,041		0,045	0,045	[µmol/mg protein]
ADP	0,005	0,003	a	0,003	a	0,006	0,005	[µmol/mg protein]
ATP	0,046	0,031	a	0,032	a	0,039	0,041	[µmol/mg protein]

Figure 5. Sum of adenine nucleotides (SAN), ATP, ADP and energy charge potential (ECP) after 3 h hypoxic preservation and 30 min reoxygenation. **a:** $p < .05$ versus control group (Student's t test).

The measurements of metabolic parameters after 3 h hypoxic storage and reoxygenation confirmed the posthypoxic recovery of the cells (Fig. 5). The content of high energy phosphates argues for viable cell cultures and the resulting energy charge potentials of more than 0.9 confirm adequate metabolic conditions. The measured metabolite concentrations of adenine nucleotides were slightly higher, as described by other authors (8,23,30). The differences in metabolite concentrations between the several groups may be due to the variable content of high energy phosphates of the cell cultures (10) and different methods of determination (8,23,30). Therefore the energy charge potential was considered as a further metabolic parameter with a smaller deviation range (1).

Group II (Preservation with University of Wisconsin Solution)

Preservation with the University of Wisconsin solution enabled 3 fold longer hypoxic preservation. The results of viability and survival index measurements after 15 h of hypoxic preservation were similar to groups preserved between 3 and 5 h with Euro Collins solution (Table 1).

Viability as well as survival index were significantly increased in cultures preserved with a taurine concentration of 20 mM in the preservation solution (Table 4). Similar to the long time hypoxic preservation with Euro Collins solution, higher taurine concentrations were needed to exert protective effects during hypoxia.

The amounts of released enzyme activities during hypoxic preservation confirm the protective effects of the taurine supplementation. Comparable to the results after hypoxic preservation with Euro Collins solution, the leakage of lactate dehydrogenase into the supernatant was markedly reduced in groups preserved with taurine in pharmacological concentrations (14.6 $\pm$ 1.62 U/liter [control] versus 7.2 $\pm$ 2.05 $p \leq 0.01$ [10 mM taurine] and 11.1 $\pm$ 1.99 $p \leq 0.05$ [20 mM taurine]).

Table 4. Remaining cells, viability rates and survival indices after aerobiosis and 15 hours hypoxic preservation with University of Wisconsin Solution

Taurine mmol/l	Remaining cells [%]	Viability [%]	Survival index [% x %] x 103
0	55.9 $\pm$ 21.52	86.8 $\pm$ 1.89	4.85 $\pm$ 1.90
1	34.9 $\pm$ 9.28	91.2 $\pm$ 3.70 a	3.19 $\pm$ 0.89
10	88.2 $\pm$ 26.58 a	84.8 $\pm$ 3.70	7.49 $\pm$ 2.27
20	72.3 $\pm$ 8.25 a	94.0 $\pm$ 2.16 b	6.70 $\pm$ 0.82 a
30	67.0 $\pm$ 26.21	76.4 $\pm$ 5.94 a	5.09 $\pm$ 1.96

n = 5 cultures per experiment, mean $\pm$ SD; a $p \leq 0.05$, b $p \leq 0.01$

This experimental procedure imitated the conditions of common kidney transplantation, where the organ is removed from a well-oxygenated organism, then stored under hypoxic conditions and at last reoxygenized after transplantation.

In this experimental model, taurine administered during hypoxia markedly reduced the cell damage due to oxygen deficiency. The beneficial effect outlasted the period of reoxygenation. The mechanisms for the improved survival rates are unclear. One factor may be the osmoregulatory effect of taurine, which counteracted the osmotic cell swelling during hypoxia (27,28). With longer preservation times, higher taurine concentrations are needed to reduce the osmotic changes. Microphotographs of the cultures during hypoxia and subsequent to reoxygenation are in accordance with this hypothesis.

Further studies are needed to determine if processes are involved which influenced membrane functions (29), intracellular electrolyte concentrations or metabolic processes (24). Especially a modulation of the intracellular calcium content and a reduction of free oxygen radicals after reoxygenation could be responsible for the cell protective effects of taurine (5,11,21,31).

The changes of viability and survival rates of the cell cultures during and after hypoxic preservation with Euro Collins solution suggest a taurine-induced stimulation of protein synthesis and cell growth in spite of oxygen deprivation and considerable deterioration of cellular integrity during hypoxia.

In face of the overall generation time of the cell strain LLC PK 1 of about 30 h and generation times during the growth period of 7.5 to 9 h (10), accelerated growth processes have to take place during the observed short reoxygenation period. It is not evident if taurine administered during hypoxia accelerated cell generation by itself or if different mediators are involved.

SUMMARY

We conclude that, within this experimental model and under these experimental conditions, taurine supplementation of standard kidney preservation solutions improves survival of kidney cells during hypoxic preservation. The protective effect depends on the taurine concentration, the hypoxic preservation time and the used preservation solution. Physiological taurine concentrations are effective during short hypoxic periods, whereas pharmacological taurine concentrations seem to be needed for longer periods of hypoxia. Within this experimental model University of Wisconsin solution seems to be more effective than Euro collins solution.

ACKNOWLEDGMENTS

We gratefully acknowledge the skilful technical assistance of Mr. Hans Georg Herb and Mr. Norbert July from the Institute for Experimental Medicine of the University of Cologne.

REFERENCES

1. Atkinson, D.E. 1968, Energy charge of adenylate pool as a regulatory parameter. Interaction with feedback modifiers, *Biochemistry*, 7:4030-4034.
2. Belzer, F.O. and Southhard, J.H. 1988, Principles of solid-organ preservation by cold storage, *Transplantation*, 45:673-676.
3. Bonventre, J.V. 1993, Mechanisms of ischemic acute renal failure, *Kidney Int.*, 43:1160-1178.
4. Bradford, M. 1976, A rapid and sensitive method for the quantitation of microgram quantities of protein utilizing the priciple of protein-dye binding, *Anal.Biochem.*, 72:248-254.
5. Cantiello, H.F., Scott, J.A., and Rabito, C.A. 1986, Polarized distribution of the Na+/H+ exchange system in a renal cell line (LLC-PK1) with characteristics of proximal tubular cells, *J.Biol.Chem.*, 261:3252-3258.
6. Collins, G.M., Bravo-Shugarman, M., and Terasaki, P.I. 1969, Kidney preservation for transportation, *Lancet*, 2:1219-1222.
7. Collins, G.M. and Wicomb, W.N. 1992, New organ preservation solutions, *Kidney Int.*, 42 (Suppl. 38):197-202.
8. Gstraunthaler, G., Pfaller, W., and Kotanko, P. 1985, Biochemical characterization of renal epithelial cell cultures (LLC-PK1 and MDCK), *Am.J.Physiol.*, 248:F536-F544.
9. Hefty, T., Fraser, S., Nelson, K., Barry, J., and Bennett, W. 1991, Efficacy of UW vs. Euro-Collins solution in paired cadaveric kidneys: a prospective study, *Transplantation Proc.*, 23:2370-2371.
10. Hull, R.N., Cherry, W.R., and Weaver, G.W. 1976, The origin and characteristics of a pig kidney cell strain, LLC PK1, *In Vitro*, 12:670-677.
11. Huxtable, R.J. 1992, Physiological actions of taurine, *Physiol.Rev.*, 72:101-163.

12. Jauregul, H.O., Hayner, N.T., and Driscoll, J.L. 1981, Trypan blue dye uptake and lactate dehydrogenase in adult rat hepatocytes - freshly isolated cells, cell suspensions, and primary monolayer cultures, *In Vitro*, 17:1100-1110.
13. Kumar, M.S.A., Samhan, M., Al Sabawi, N., Al Abdullah, I.H., Silva, O.S.G., White, A.G., and Abouna, G.M. 1991, Preservation of cadaveric kidneys longer than 48 hours: comparison between Euro-Collins solution, UW solution and machine perfusion, *Transplantation Proc.*, 23:2392-2393.
14. Lindell, S.L., Gandolph, D., Southard, J.H., and Belzer, F.O. 1991, Comparison of PBS, HTK, and UW solutions for kidney preservation, *Transplantation Proc.*, 23:2399-2401.
15. Marshall, V.C., Biguzas, M., Jablonski, P., Scott, D.F., Howden, B.O., Thomas, A.C., Cham, C.W., and Walls, K. 1990, UW solution for kidney preservation, *Transplantation Proc.*, 22:496-497.
16. Marshall, V.C., Jablonski, P., Biguzas, M., Howden, B.O., and Walls, K. 1990, Kidney preservation with UW solution: the nature of the impermeant, *Transplantation Proc.*, 22:2131-2132.
17. McAnulty, J.F., Ploeg, R.J., Southard, J.H., and Belzer, F.O. 1989, Successful five-day perfusion preservation of the canine kidney, *Transplantation*, 47:37-41.
18. McAnulty, J.F., Vreugdenhil, P.K., Southard, J.H., and Belzer, F.O. 1990, Use of UW cold storage solution for machine perfusion of kidneys, *Transplantation Proc.*, 22:458-459.
19. McCabe, R., Kimmelstiel, F., Cooper, J., and Ramey, W. 1990, Rescue perfusion of the cold-stored kidney, *Transplantation Proc.*, 22:2219-2220.
20. Mohacsi, P.J., Herbertt, K.L., and Thompson, J.F. 1992, Human kidney preservation with University of Wisconsin solution: an initial report of the Australian experience, *Transplantation Proc.*, 24:256-257.
21. Nakamori, K., Koyama, I., Yoshida, T., Umeda, M., and Inoue, K. 1990, Effectivness of taurine in protecting biomembrane against oxidant, *Chem.Pharm.Bull.*, 38:3116-3119.
22. Ploeg, R.J., van Bockel, J.H., Langendijk, P.T.H., Groenewegen, M., van der Woude, F.J., Persijn, G.G., Thoregood, J., and Hermans, J. 1992, Effect of preservation solution on results of cadaveric kidney transplantation, *Lancet*, 340:129-136.
23. Sahai, A., Cole, L.A., Clarke, D.L., and Tannen, R.L. 1989, Rocking promotes differentiated properties in LLC-PK1 cells by improved oxygenation, *Am.J.Physiol.*, 315:C1064-C1069.
24. Sepulveda, F. and Pearson, J.D. 1982, Characterization of neutral amino acid uptake by cultured epithelial cells from pig kidney, *J.Cell.Physiol.*, 112:1822-1828.
25. Tisone, G., Elli, M., Pisani, F., Utzeri, G., De Paolis, P., Torlone, N., Manca da Villahermosa, S., Famulari, A., and Casciani, C.U. 1991, Use of UW solution versus Euro-Collins solution in kidney transplnation from cadaveric donors, *Transplantation Proc.*, 23:2402-2403.
26. Torras, J., Bordalba, J.R., Seron, D., Carrera, M., Castelao, A.M., Poveda, R., Alsina, J., and Grino, J. 1992, An experimental comparison between isotonic saline solution, Euro-Collins, and a flush solution with mannitol in the prevention of renal damage due to warm ischemia, *Transplantation Proc.*, 24:54-55.
27. Trachtman, H. 1991, Cell volume regulation: a review of cerebral adaptive mechanisms and implications for clinical treatment of osmolal disturbances: I, *Pediatr.Nephrol.*, 5:743-750.
28. Trachtman, H. 1992, Cell volume regulation: a review of cerebral adaptive mechanisms and implications for clinical treatment of osmolal disturbances: II, *Pediatr.Nephrol.*, 6:104-112.
29. Troyer, D.A., Kreisberg, J.I., and Venkatachalam, M.A. 1986, Lipid alterations in LLC-PK1 cells exposed to mercuric chloride, *Kidney Int.*, 29:530-538.
30. Venkatachalam, M.A., Patel, Y.J., Kreisberg, J.I., and Weinberg, J.M. 1988, Energy thresholds that determine membrane integrity and injury in a renal epithelial cell line (LLC-PK1), *J.Clin.Invest.*, 81:745-758.
31. Weinberg, J.M., Davis, J.A., Roeser, N.F., and Venkatachalam, M.A. 1990, Role of increased cytosolic free calcium in the pathogenesis of rabbit proximal cell injury and protection by glycine or acidosis, *J.Clin.Invest.*, 87:581-590.

MECHANISMS UNDERLYING PHYSIOLOGICAL AND PHARMACOLOGICAL ACTIONS OF TAURINE ON MYOCARDIAL CALCIUM TRANSPORT

Stephen W. Schaffer[1], Cherry Ballard[1] and Junichi Azuma[2]

[1]Department of Pharmacology, School of Medicine
University of South Alabama
Mobile, Alabama, USA
[2]The Third Department of Internal Medicine
Osaka University Medical School
1-1-50 Fukushima, Osaka 533, Japan

INTRODUCTION

One of the major functions of taurine in the heart is the modulation of calcium transport (21). Depending on the status of the myocardium, taurine is capable of either increasing or decreasing intracellular calcium levels (21). This complicated biphasic pattern has been difficult to resolve mechanistically. The only rational explanation for this phenomenon assumes that taurine exerts multiple effects on calcium transport, some which promote myocyte calcium accumulation while others enhance cellular calcium loss.

Since the magnitude of the cytosolic calcium pool is directly dependent upon calcium entry, calcium extrusion from the cell and calcium release from the sarcoplasmic reticulum, modulation of cytosolic calcium levels by taurine must involve changes in the amount of calcium transported in and out of both the cell and calcium storage sites (Figure 1). Four transporters play dominant roles in the maintenance of normal calcium homeostasis. First, the major transporter regulating calcium entry is the sarcolemmal L-type calcium channel, although there is some evidence that calcium also can enter the cell via the Na^+-Ca^{2+} exchanger. Secondly, the sarcolemmal Na^+-Ca^{2+} exchanger has a high capacity to transport calcium and is considered the primary transporter extruding calcium from the cell. Third, calcium release from the sarcoplasmic reticulum occurs preferentially via the ryanodine-sensitive calcium channel. Although other mechanisms for sarcoplasmic reticular calcium release have been described, they probably contribute minimally to overall calcium release. Fourth, the sarcoplasmic reticular calcium pump functions to maintain sarcoplasmic reticular calcium stores. Depletion of these stores will reduce the amount of calcium available for calcium release, causing a reduction in cytosolic calcium levels. Thus, the effects of taurine

Taurine in Health and Disease, Edited by R. Huxtable and D.V. Michalk, Plenum Press, New York, 1994

on cytosolic calcium content must involve modulation of one or more of the following transport processes: (1) sarcoplasmic reticular calcium pump activity, (2) sarcoplasmic reticular ryanodine-sensitive calcium channel activity, (3) sarcolemmal Na^+-Ca^{2+} exchanger activity or (4) sarcolemmal calcium channel activity. The aim of this report is to clarify the effects of taurine on these four transport processes.

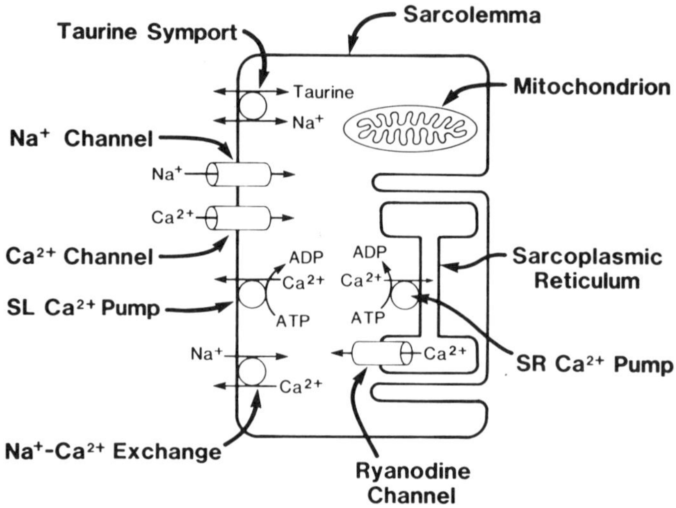

Figure 1. Key sodium and calcium transporters of the heart.

EFFECT OF TAURINE ON SARCOPLASMIC RETICULAR CALCIUM PUMP ACTIVITY

Two approaches have been undertaken to examine the effect of taurine on sarcoplasmic reticular calcium transport. Some investigators have utilized a skinned muscle fiber preparation in which the permeability of the cell membrane has been altered, allowing free flow of small ions, such as calcium, across the modified membrane. Since the sarcoplasmic reticulum of this preparation remains intact, skinned fibers allow the study of calcium movement into and out of intracellular calcium stores. The other approach relies on the isolation of sarcoplasmic reticular vesicles, which can be used to study ATP-dependent calcium transport. Both procedures have provided useful information on the actions of taurine.

Using skinned muscle fiber preparations, Steele and coworkers (25,26) and Galler et al. (4) found that taurine increased the size of caffeine-induced calcium and tension transients. Since the taurine effect was observed at concentrations of caffeine (10 mM) which yielded maximal calcium release from the sarcoplasmic reticulum, Steele and Smith (25) concluded that taurine must enhance the accumulation of calcium by the sarcoplasmic reticulum. However, this conclusion appears to be inconsistent with a series of isolated membrane studies describing the ineffectiveness of taurine as a stimulator of sarcoplasmic reticular calcium pump activity (3,23,30). Although it is possible that taurine might indirectly modulate the activity of the sarcoplasmic reticular calcium pump, alternative mechanisms of enhancing activity seem unlikely. It is known that sarcoplasmic reticular Ca^{2+} ATPase activity is regulated by both protein phosphorylation and changes in phospholipid N-methylation (5,17). However, taurine reduces the phosphorylation of a number of cardiac phosphoproteins (12,19), an effect which if extended to the sarcoplasmic reticular calcium pump regulator, phospholamban, would decrease calcium pump activity. Similarly, the inhibition of phospholipid N-methylation by taurine should cause a reduction, rather than an increase, in sarcoplasmic reticular Ca^{2+} ATPase activity (5). Thus, most data are incompatible with direct stimulation of sarcoplasmic reticular calcium pump activity by taurine.

Nevertheless, these data do not exclude the possibility that flux of calcium into the sarcoplasmic reticulum may be changed with altering taurine levels. This could occur if taurine affected cytosolic calcium content. Since the sarcoplasmic reticular calcium pump is a Ca^{2+}-dependent ATPase, which in normal myocytes does not operate at V_{max}, exposure of the pump to elevated concentrations of calcium would elevate substrate levels, thereby increasing the rate of calcium transport into the sarcoplasmic reticular storage vesicles. This notion is supported by isolated myocyte studies showing that pharmacological concentrations of taurine cause $[Ca^{2+}]_i$ to rise (24,28). Thus, elevations in taurine are capable of increasing the amount of calcium available for transport by the sarcoplasmic reticular calcium pump. It is important to point out, however, that while this mechanism will increase the rate of sarcoplasmic reticular calcium filling, the taurine effect does not involve a change in sarcoplasmic reticular calcium pump activity per se, but rather a change in velocity secondary to a rise in the concentration of one of the ATPase substrates. The attractiveness of this theory is that it is consistent with both the isolated sarcoplasmic reticular studies and the conclusions of Steele and coworkers (25,26).

EFFECT OF TAURINE ON SARCOPLASMIC RETICULAR RYANODINE-SENSITIVE CALCIUM CHANNEL FUNCTION

Taurine has a dual effect on sarcoplasmic reticular ryanodine-sensitive calcium channel function. In the presence, but not the absence of medium calcium, taurine enhances the release of $^{45}Ca^{2+}$ from $^{45}Ca^{2+}$-loaded junctional sarcoplasmic reticular vesicles (16). Since this calcium release mechanism is calcium dependent and is blocked by ruthenium red and FLA 365, the calcium transport process affected by taurine seems to represent Ca^{2+}-induced Ca^{2+} release. In the mammalian myocardium, Ca^{2+}-induced Ca^{2+} release plays a central role in excitation-contraction coupling (31). During excitation, relatively small amounts of calcium enter the cell via the sarcolemmal calcium channel. This source of calcium is insufficient by itself to yield maximal contractile force. However, the elevation in cytosolic calcium levels caused by the slow inward calcium current, triggers the release of calcium from sarcoplasmic reticular calcium stores. This Ca^{2+}-induced Ca^{2+} release mechanism elevates intracellular Ca^{2+} concentration sufficiently to produce maximal interaction between myosin and actin. Since some inotropic agents influence the sarcoplasmic reticular calcium release process (31), promotion of Ca^{2+}-induced Ca^{2+} release may represent one mechanism by which taurine enhances contractile function.

The other effect of taurine on sarcoplasmic reticular calcium release is tied to its inhibition of the enzyme phospholipid methyltransferase (16). This enzyme catalyzes the conversion of phosphatidylethanolamine to phosphatidylcholine, a reaction pathway involving three methylation steps. In accordance with several other calcium transporters, the status of the ryanodine-sensitive calcium channel is highly dependent upon the phospholipid content of the sarcoplasmic reticulum. Perfusion of isolated rat heart with buffer containing methionine enhances phospholipid N-methylation and causes a reduction in Ca^{2+}-induced Ca^{2+} release (16). Since the sarcoplasmic reticulum is located in an environment rich in taurine, activity of sarcoplasmic reticular phospholipid methyltransferase normally will be maintained in a partially inhibited state and Ca^{2+}-induced Ca^{2+} release will consequently be elevated. Thus, taurine ensures effective operation of the Ca^{2+}-induced Ca^{2+} release process, both by directly promoting Ca^{2+}-induced Ca^{2+} release and by modulating ryanodine-sensitive calcium channel activity through changes in the phospholipid environment of the channel.

EFFECT OF TAURINE ON SODIUM-CALCIUM EXCHANGER ACTIVITY

In contrast to the ryanodine-sensitive calcium channel, taurine exerts no direct effect on the sarcolemmal Na^+-Ca^{2+} exchanger (23). However, it is capable of indirectly modulating Na^+-Ca^{2+} exchanger activity. This indirect effect of taurine is related to its ability to alter the phospholipid microenvironment of the Na^+-Ca^{2+} exchanger. Panagia et al. (15) have shown that the Na^+-Ca^{2+} exchanger is very sensitive to the phosphatidylethanolamine and phosphatidylcholine content of the sarcolemmal membrane. Elevations in the methylation state of the membrane by exposure of the heart to the methylating agent, L-methionine, causes a dramatic decline in Na^+-Ca^{2+} exchanger activity. Taurine serves as an inhibitor of phospholipid N-methyltransferase, thereby reducing the conversion of phosphatidylethanolamine to phosphatidylcholine and maintaining the Na^+-Ca^{2+} exchanger in a more active state (7). In this regard, it is interesting that rats made taurine deficient by treatment with the taurine transport inhibitor, ß-alanine, exhibit a reduction in the capacity of the Na^+-Ca^{2+} exchanger, an effect consistent with the regulation of sarcolemmal phospholipid N-methylation and Na^+-Ca^{2+} exchanger activity by taurine (8).

In addition to modulation of Na^+-Ca^{2+} exchanger activity, taurine also affects flux through the exchanger by altering intracellular levels of sodium. Sperelakis et al. (24) have reported that taurine stimulates the fast component of the inward Na^+ current, thereby elevating intracellular Na^+ concentration and promoting Ca^{2+} influx via the Na^+-Ca^{2+} exchanger. Another mechanism for modulation of intracellular sodium and Na^+-Ca^{2+} exchanger flux by taurine has been proposed by Suleiman et al. (27). It is known that taurine transport across the cell membrane occurs by a taurine/Na^+ symport mechanism (22). Consequently, rapid rates of taurine efflux are associated with sodium loss from the cell, which enhance Ca^{2+} efflux via the Na^+-Ca^{2+} exchanger.

EFFECT OF TAURINE ON SARCOLEMMAL CALCIUM CHANNEL ACTIVITY

The effect of taurine on sarcolemmal calcium channel flux is highly dependent on extracellular calcium concentration. Cardiomyocytes incubated in medium containing either normal or slightly elevated concentrations of calcium experience a significant decrease in L-type calcium current and an increase in T-type calcium current following exposure to millimolar concentrations of taurine (18). On the other hand, a slight stimulation in flux through the L-type calcium channel is observed at lower medium calcium concentrations (18). Although the effect of taurine on L-type channel flux is calcium dependent, the increase in calcium current observed in cardiomyocytes incubated in medium containing a low calcium concentration is quite small and virtually disappears when the medium calcium concentration is increased to the normal range. This pattern contrasts with fura-2 studies showing elevations in intracellular calcium concentration of heart cells following addition of taurine to normal incubation medium (24,28). These data suggest that taurine-mediated elevations in L-type calcium current probably play a minor role in modulation of intracellular calcium levels. On the other hand, it has been proposed that reductions in L-type calcium current may contribute to taurine-induced decreases in intracellular calcium levels of the calcium overloaded myocardium (18). This interesting idea deserves further investigation.

TAURINE-MEDIATED ELEVATIONS IN $[Ca^{2+}]_i$ IN THE HYPODYNAMIC HEART

Taurine has little effect on calcium transport of the normal heart. However, hearts made hypodynamic by reductions in perfusate calcium content or exposure to a calcium antagonist are sensitive to the effects of taurine (21). In the untreated, hypodynamic heart, calcium entry through the L-type calcium channel is depressed. Consequently, Ca^{2+}-induced Ca^{2+} release is suppressed, as is the size of the sarcoplasmic reticular calcium pool (Figure 2). Although a balance is maintained between calcium influx and efflux in these hearts, the amount of calcium handled by individual sarcolemmal and sarcoplasmic reticular transporters is small. Therefore, cytosolic calcium content is insufficient to support normal contractile function.

Exposure of the hypodynamic heart or isolated myocytes to medium containing millimolar concentrations of taurine leads to an elevation in intracellular calcium concentration (21,24,28). Several mechanisms could contribute to this rise in cytoplasmic calcium levels (Figure 3). First, high extracellular taurine causes the accumulation of sodium by the cell. Sperelakis et al. (24) have shown that taurine can promote rapid sodium influx via the tetrodotoxin-insensitive sodium channel. Similarly, it is known that taurine

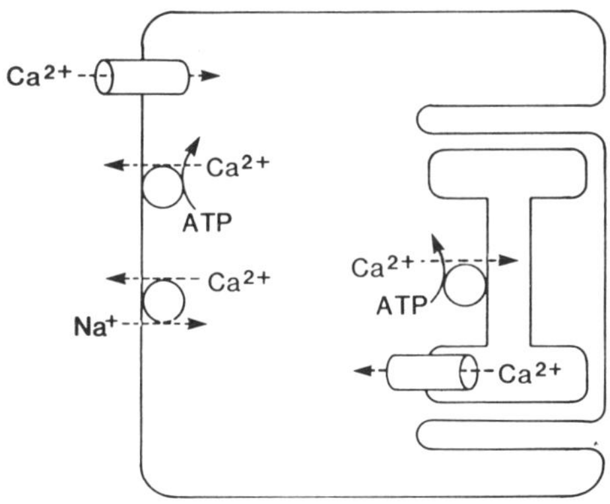

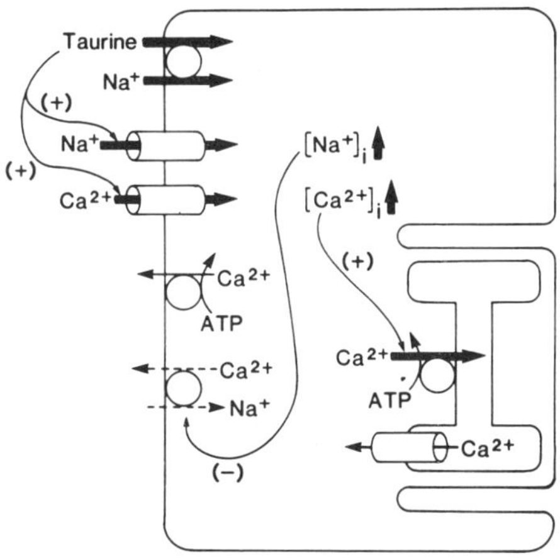

Figure 2 (left). Calcium transport in the hypodynamic heart. There is a general decrease in flux through various calcium transporters in the hypodynamic heart. Broken lines represent a decrease in flux relative to the normal heart. **Figure 3 (right).** Effect of elevated extracellular taurine on calcium transport of the hypodynamic heart. The major effect of taurine is an increase in intracellular sodium and calcium levels. Taurine acts directly on sarcolemmal sodium and calcium channels to enhance ion influx. It also stimulate taurine influx via the taurine/Na^+ symporter, causing a slow rise in intracellular sodium levels. The increase in intracellular sodium and calcium content stimulates calcium uptake by the sarcoplasmic reticulum and decreases calcium efflux via the Na^+-Ca^{2+} exchanger. Broken and bold lines represent decreases and increases in taurine-linked ion flux, respectively.

uptake occurs via a taurine/Na^+ symport mechanism (22). Significant taurine influx, therefore, could theoretically be accompanied by an elevation in intracellular sodium content. However, this latter mechanism seems rather remote as a cause of the taurine-mediated positive inotropic effect. Since the V_{max} for taurine influx via the symporter is 32 nmol/g dry weight/min (10) while myocardial sodium content is approximately 15 μmol/g dry weight (29), a significant elevation in intracellular sodium levels by this mechanism could only be achieved after prolonged perfusion; the positive inotropic effect occurs within seconds after exposure to buffer containing elevated taurine levels. Thus, more important than the effect of extracellular taurine on taurine influx would be its effect on taurine efflux. By reducing the taurine gradient across the membrane, leakage of taurine and sodium from the cell would be slowed and intracellular sodium levels would be preserved. This taurine-mediated elevation in intracellular sodium would be accompanied by altered Na^+-Ca^{2+} exchanger flux, leading to an increase in intracellular calcium concentration. Secondly, taurine has been reported to stimulate T-type calcium current. Although this mechanism may contribute to taurine-mediated cellular calcium accumulation (24), this seems unlikely since T-type calcium current appears to be more closely tied to pacemaker activity during diastole than excitation-contraction coupling. Thirdly, high intracellular levels of taurine promote Ca^{2+}-induced Ca^{2+} release from the sarcoplasmic reticulum. This effect assumes importance following significant elevations in intracellular taurine levels, as occurs in certain pathological states. However, significant elevations in intracellular taurine content only occur after prolonged exposure to medium rich in taurine. Thus, the most logical explanation for the rise in intracellular calcium following exposure to external taurine is the elevation in intracellular sodium concentration.

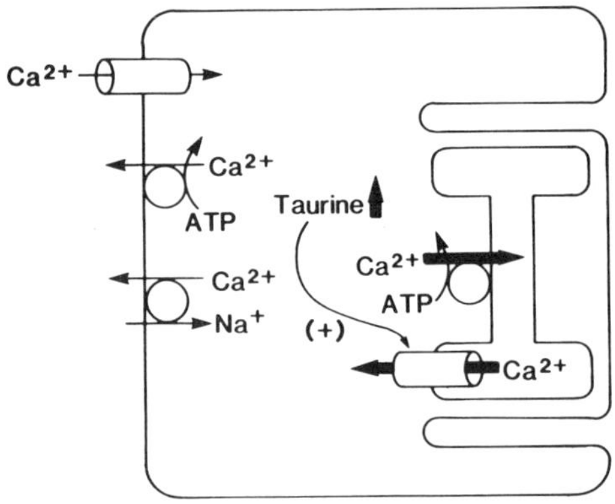

Figure 4. Effect of elevated intracellular taurine on calcium transport of the hypodynamic heart. Taurine directly promotes calcium release from the sarcoplasmic reticulum. In the hypodynamic heart, the released calcium will be resequestered by the sarcoplasmic reticulum and the net effect will be enhanced uptake and release of calcium from sarcoplasmic reticular calcium stores. Bold lines represent increases in taurine-linked ion flux.

In contrast to elevated extracellular taurine content, changes in intracellular taurine concentration will have the greatest effect on Ca^{2+}-induced Ca^{2+} release (21). Dramatic elevations in intracellular taurine should lead to enhanced Ca^{2+}-induced Ca^{2+} release, thereby increasing trigger calcium and stimulating contractile function (Figure 4). Since rat heart contains a very active sarcoplasmic reticular calcium pump, the released calcium should be readily resequestered by the sarcoplasmic reticulum during diastole. Thus, the size of the sarcoplasmic reticular calcium stores should not change and the only apparent effect of elevated intracellular taurine would be an increase in the flux of calcium into and out of the sarcoplasmic reticulum.

TAURINE-MEDIATED REDUCTIONS IN $[Ca^{2+}]_i$ IN THE CALCIUM OVER-LOADED HEART

Excessive calcium accumulation is extremely toxic to the myocyte; therefore, considerable interest has been directed to agents, such as taurine, which reduce intracellular calcium stores. One of the major causes of excessive calcium accumulation in the heart is intracellular sodium overload (Figure 5). Increases in intracellular sodium reduce the extent of calcium extruded from the cell by the Na^+-Ca^{2+} exchanger. Moreover, if the degree of sodium retention is large, the direction of Na^+-Ca^{2+} exchange can be reversed and calcium will actually enter the cell in exchange for intracellular sodium.

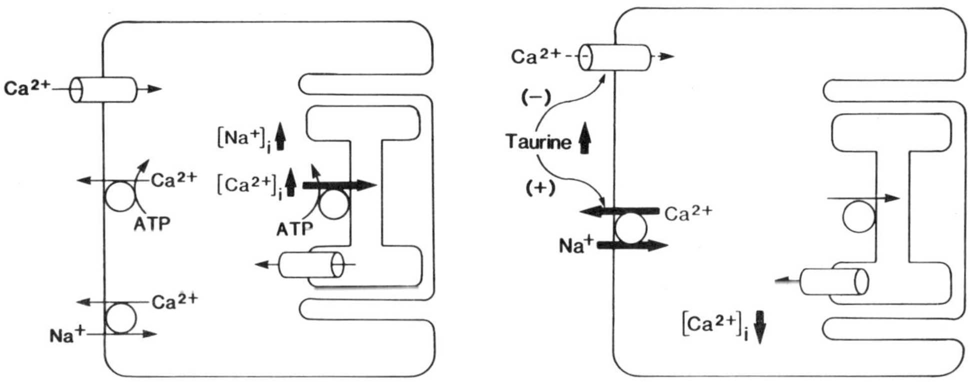

Figure 5 (left). Calcium transport in the calcium overloaded heart. Because of the elevation in intracellular calcium levels, calcium uptake by the sarcoplasmic reticulum is enhanced. Bold lines represent an increase in flux relative to the normal heart. **Figure 6 (right).** Effect of elevated extracellular taurine on calcium transport of the calcium overloaded heart. Taurine maintains the Na^+-Ca^{2+} exchanger in an active state by inhibiting phospholipid N-methylation. It also suppresses L-channel calcium flux. Bold and broken lines represent increases and decreases in taurine-linked ion flux, respectively.

Elevated extracellular taurine has been reported to prevent calcium overload in a series of heart failure models (6,13,14) as well as in myocytes subjected to the calcium paradox (28) or exposed to medium containing low Na^+ and high Ca^{2+} (11). The mechanism responsible for these effects has not been established. Two possible candidates for this phenomenon are shown in Figure 6. Sawamura et al. (18) have reported that pharmacological concentrations of taurine can suppress calcium influx via the L-type calcium channel, but only if extracellular calcium content is high. This might decrease intracellular calcium content, although the effect of calcium overload on this response is uncertain. Another mechanism contributing to calcium regulation is phospholipid N-methylation (7). Through modulation of sarcolemmal phospholipid N-methyltransferase activity, taurine maintains the phospholipid microenvironment of the Na^+-Ca^{2+} exchanger in a less methylated state, thereby ensuring a more active transporter. Since the Na^+-Ca^{2+}

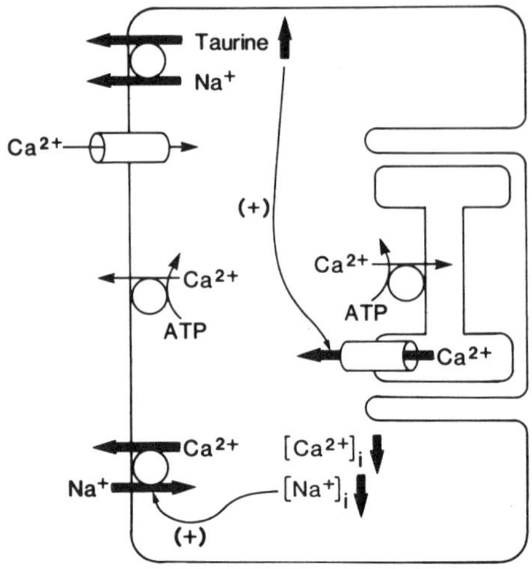

Figure 7. Effect of elevated intracellular taurine on calcium transport of calcium overloaded heart. Taurine indirectly facilitates calcium loss from the cell by promoting sodium loss via the taurine/Na^+ symporter. The decrease in intracellular sodium levels enhances calcium efflux via the Na^+-Ca^{2+} exchanger. Less calcium is stored within the sarcoplasmic reticulum because taurine stimulates release of calcium from the internal calcium storage sites.

exchanger is the major mechanism for the extrusion of calcium from the cell, reductions in Na^+-Ca^{2+} exchanger activity can result in intracellular calcium accumulation.

Studies by Schaffer and coworkers (9,20) have shown that drug-induced taurine depletion renders the heart susceptible to adriamycin toxicity and ischemic injury. Conversely, elevated intracellular taurine is reported to be cardioprotective (2). According to the hypothesis advanced by Chapman et al. (2), the cardioprotective activity of taurine is linked to taurine efflux via the taurine/Na^+ symporter. Rapid loss of taurine from the myocyte via this transporter will result in a corresponding reduction in intracellular sodium content. The resulting change in the sodium gradient across the cell membrane will favor calcium efflux from the cell, leaving less calcium to be transported into the sarcoplasmic reticulum (Figure 7). Also contributing to the redistribution of calcium is the taurine-mediated inhibition of both calmodulin-linked phosphorylation and sarcoplasmic reticular phospholipid methyltransferase activity, effects which mediate reductions in sarcoplasmic reticular calcium pump activity (16,19). At the same time, taurine-mediated facilitation of Ca^{2+}-induced Ca^{2+} release from the sarcoplasmic reticulum will cause intravesicular calcium stores to fall (16).

REFERENCES

1. Azari, J., and Huxtable, R.J., 1980, The mechanism of the adrenergic stimulation of taurine influx in the heart, *Eur. J. Pharmacol.*, 61: 217-223.
2. Chapman, R.A., Suleiman, M-S, and Earm, Y.E., 1993, Taurine and the heart, *Cardiovasc. Res.*, 27: 358-363.
3. Entman, M.L., Bornet, E.P., and Bressler, R., 1977, The effect of taurine on cardiac sarcoplasmic reticulum, *Life Sci.*, 21: 543-550.

4. Galler, S., Hutzler, C., and Haller, T., 1990, Effects of taurine on Ca^{2+}-dependent force development of skinned muscle fibre preparations, *J. Exp. Biol.*, 152: 255-264.
5. Ganguly, P.K., Panagia, V., Okumura, K., and Dhalla, N.S., 1985, Activation of Ca^{2+}-stimulated ATPase by phospholipid N-methylation in cardiac sarcoplasmic reticulum, *Biochem. Biophys. Res. Commun.*, 130: 472-478.
6. Hamaguchi, T., Azuma, J., Awata, N., Ohta, H., Takihara, K., Harada, H., Kishimoto, S., and Sperelakis, N., 1988, Reduction of doxorubicin-induced cardiotoxicity in mice by taurine, *Res. Commun. Chem. Pathol. Pharmacol.*, 59: 21-30.
7. Hamaguchi, T., Azuma, J., and Schaffer, S., 1991, Interaction of taurine with methionine: Inhibition of myocardial phospholipid methyltransferase, *J. Cardiovasc. Pharmacol.*, 18: 224-230.
8. Harada, H., Allo, S., Viyuoh, N., Azuma, J., Takahashi, K., and Schaffer, S.W., 1988, Regulation of calcium transport in drug-induced taurine-depleted hearts, *Biochim. Biophys. Acta*, 944: 273-278.
9. Harada, H., Cusack, B.J., Olson, R.D., Stroo, W., Azuma, J., Hamaguchi, T., and Schaffer, S.W., 1990, Taurine deficiency and doxorubicin: Interaction with the cardiac sarcolemmal calcium pump, *Biochem. Pharmacol.*, 39: 745-751.
10. Huxtable, R.J., Chubb, J., and Azari, J., 1980, Physiological and experimental regulation of taurine content in the heart, *Federation Proc.,* 39: 2685-2690.
11. Ihara, Y., Takahashi, K., Harada, H., Sawamura, A., Schaffer, S., and Azuma, J., 1992, Taurine attenuates contracture induced by perfusion with low sodium, high calcium medium in chick hearts, *in:* "Taurine: Nutritional Value and Mechanisms", J.B. Lombardini, S.W. Schaffer and J. Azuma, eds., Plenum Press, New York, pp.
12. Lombardini, J.B., and Liebowitz, S.M., 1989, Taurine modifies calcium ion uptake and protein phosphorylation in rat heart, *in:* "Taurine and the Heart", H. Iwata, J.B. Lombardini and T. Segawa, eds., Kluwer Academic Publishers, Boston/Dordrecht/London, pp. 117-137.
13. McBroom, M.J., and Welty, J.D., 1977, Effects of taurine on heart calcium in the cardiomyopathic hamster, *J. Mol. Cell. Cardiol.*, 9: 853-858.
14. Ohta, H., Azuma, J., Onishi, S., Awata, N., Takihara, K., and Kishimoto, S., 1986, Protective effect of taurine against isoprenaline-induced myocardial damage, *Basic Res. Cardiol.*, 81: 473-481.
15. Panagia, V., Makino, N., Ganguly, P.K., and Dhalla, N.S., 1987, Inhibition of Na^{+}-Ca^{2+} exchange in heart sarcolemmal vesicles by phosphatidylethanolamine N-methylation, *Eur. J. Biochem.*, 166: 597-603.
16. Punna, S., Hamaguchi, T., Azuma, J., and Schaffer, S., Effect of taurine and methionine on sarcoplasmic reticular Ca^{2+} transport and phospholipid methyltransferase activity, *J. Cardiovasc. Res.* In press.
17. Sasaki, T., Inui, M., Kimura, Y., Kuzuya, T., and Tada, M., 1992, Molecular mechanism of regulation of Ca^{2+} pump ATPase by phospholamban in cardiac sarcoplasmic reticulum, *J. Biol. Chem.*, 267: 1674-1679.
18. Sawamura, A., Sada, H., Azuma, J., Kishimoto, S., and Sperelakis, N., 1990, Taurine modulates ion influx through cardiac Ca^{2+} channels, *Cell Calcium* 11: 251-259.
19. Schaffer, S.W., Allo, S., Harada, H., and Azuma, J., 1990, Regulation of calcium homeostasis by taurine: Role of calmodulin, *in:* " Taurine: Functional Neurochemistry, Physiology and Cardiology", H. Pasantes-Morales, D.L. Martin, W. Shain and R.M. DelRio, eds., Wiley-Liss Inc., New York, pp. 217-225.
20. Schaffer, S.W., Allo, S., and Mozaffari, M., 1987, Potentiation of myocardial ischemic injury by drug-induced taurine depletion, *in:* " The Biology of Taurine", R.J. Huxtable, F. Franconi and A. Giotti, eds., Plenum Press, New York/London, pp. 151-158.
21. Schaffer, S.W., and Azuma, J., 1992, Myocardial physiological effects of taurine and their significance, *in:* " Taurine: Nutritional Value and Mechanisms", J.B. Lombardini, S.W. Schaffer and J. Azuma, eds., Plenum Press, New York, pp. 105-120.
22. Schaffer, S.W., Kulakowski, E.C., and Kramer, J.H., 1985, Taurine transport by reconstituted membrane vesicles, *in:* " Taurine in Nutrition and Neurology", R.J. Huxtable and H. Pasantes-Morales, eds., Plenum Press, New York/London, pp. 143-160.
23. Schaffer, S.W., Punna, S., Duan, J., Harada, H., Hamaguchi, T., and Azuma, J., 1992, Mechanism underlying physiological modulation of myocardial contraction by taurine, *in:* " Taurine: Nutritional Value and Mechanisms", J.B. Lombardini, S.W. Schaffer and J. Azuma, eds., Plenum Press, New York, pp. 193-198.
24. Sperelakis, N., Satoh, H., and Bkaily, G., 1992, Taurine effects on ionic currents in myocardial cells, *in:* " Taurine: Nutritional Value and Mechanisms", J.B. Lombardini, S.W. Schaffer and J. Azuma, eds., Plenum Press, New York, pp. 129-143.

25. Steele, D.S., and Smith, G.L., 1992, Intracellular effects of taurine: Studies on skinned cardiac preparations, *in:* " Taurine: Nutritional Value and Mechanisms", J.B. Lombardini, S.W. Schaffer and J. Azuma, eds., Plenum Press, New York, pp. 163-172.
26. Steele, D.S., Smith, G.L., and Miller, D.J., 1990, The effects of taurine on Ca^{2+} uptake by the sarcoplasmic reticulum and Ca^{2+} sensitivity of chemically skinned rat heart, *J. Physiol.*, 422: 499-511.
27. Suleiman, M-S, Rodrigo, G.C., and Chapman, R.A., 1992, Interdependence of intracellular taurine and sodium in guinea pig heart, *Cardiovasc. Res.* 26: 897-905.
28. Takahashi, K., Harada, H., Schaffer, S.W., and Azuma, J., 1992, Effect of taurine on intracellular calcium dynamics of cultured myocardial cells during the calcium paradox, *in:* " Taurine: Nutritional Value and Mechanisms", J.B. Lombardini, S.W. Schaffer and J. Azuma, eds., Plenum Press, New York, pp. 153-161.
29. Tani, M., and Neely, J.R., Na^+ accumulation increases Ca^{2+} overload and impairs function in anoxic rat heart, *J. Mol. Cell. Cardiol.*, 22: 57-72.
30. Welty, J.D., and Welty, M.C., 1981, Effects of taurine on subcellular calcium dynamics in the normal and cardiomyopathic hamster heart, *in:* " The Effects of Taurine on Excitable Tissues", S.W. Schaffer, S.I. Baskin and J.J. Kocsis, eds., Spectrum Publications, New York, pp. 295-311.
31. Williams, A.J., 1992, Ion conductance and discrimination in the sarcoplasmic reticulum ryanodine receptor/calcium release channel, *J. Muscle Res. Cell Motility*, 13: 7-26.

CARDIOPROTECTIVE ACTIONS OF TAURINE AGAINST INTRACELLULAR AND EXTRACELLULAR CALCIUM-INDUCED EFFECTS

Hiroyasu Satoh

Department of Pharmacology, Nara Medical University
Kashihara, Nara 634, Japan

INTRODUCTION

Taurine, a sulfur-containing amino acid, is present in myocardial cells at high concentration (around 10 mM) (13). Taurine produces many functions on the ion fluxes across cell membrane and the contractile force (13,31,33,34). It has already been reported that taurine produces a positive inotropic effect in heart muscles (3,7,8,10), that it has beneficial effects in treatment of congestive heart failure (2,36), and that is is protective against Ca^{2+} overload and cardiomyopathy (1,19,37). Taurine levels in the cells are lowered in ischemic heart failure (17,18).

It has recently been shown that the actions of taurine are modulated by both intracellular and extracellular Ca^{2+} concentrations ($[Ca^{2+}]_i$ and $[Ca^{2+}]_o$) (10,15,28,35). In ischemia and cardiac failure (at high $[Ca^{2+}]_i$ level), taurine causes a negative inotropic effect. In contrast, at low $[Ca^{2+}]_o$, taurine causes a positive inotropic effect. These findings suggest that taurine may possess a normalizing action on the $[Ca^{2+}]_i$ level. In the present experiments, therefore, the effects of taurine on electrophysiological functions of heart muscle at different $[Ca^{2+}]_i$ or $[Ca^{2+}]_o$ levels has been examined. Whole cell voltage- and current-clamp experiments have been performed to examine ionic currents in single embryonic chick ventricular cells, and conventional microelectrode experiments have been performed to examine the action potential configuration in guinea pig ventricular muscle and the spontaneous activity in rabbit sinoatrial node cells.

MATERIALS AND METHODS

Cell Isolation and Preparation

Single heart cells were isolated from embryonic chick hearts (3- to 17-day-old) by standard techniques (24,25). The single cells were round-shaped and cultured for 1 or 2

Taurine in Health and Disease, Edited by R. Huxtable and D.V. Michalk, Plenum Press, New York, 1994

days. The small preparations were also isolated from guinea pig ventricular muscle and rabbit sinoatrial node.

Whole-cell Voltage- and Current-clamp and Action Potential Recordings

Whole-cell voltage- and current clamp recordings were made using an Axopatch patch-clamp amplifier and standard techniques (12), as reported previously (21,24,25). Patch pipettes were fabricated using a two-stage puller, and they had resistances of 3-5 MΩ. Experiments were carried out at room temperature (22°C). The data were stored and analyzed on an IBM-AT microcomputer using the PCLAM analysis program. The current traces were filtered with a cut-off frequency of 1 KHz for plotting. Also, the action potential recordings were obtained by conventional microelectrode technique (22). All values given are means ± SEM.

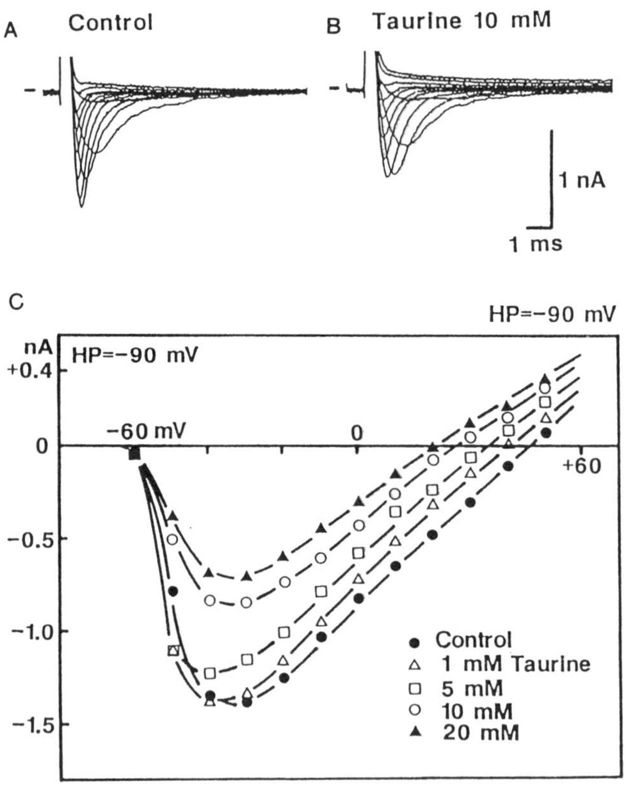

Figure 1. Inhibition of I_{Na} in 17-day-old embryonic chick cardiomy ocytes. **A and B:** Family of current recordings taken from a myocyte in the absence and presence of taurine (10 mM). Test pulses were applied to between -60 and +50 mV from a holding potential of -90 mV. **C:** Current-voltage relationships for I_{Na} in the presence and absence of taurine. The $[Ca]_i$ was pCa 10. To avoid the interference of other currents, $MgCl_2$ (2 mM) and 4-aminopyridine (0.3 mM) were added to the bath solution, and Cs^+ was substituted for K^+ in pipette solution. The experiments were performed at room temperature (22°C). Symbols used are control (filled circles), 1 mM (open triangles), 5 mM (squares), 10 mM (open circles), and 20 mM taurine (filled triangles). Values represent mean ± SEM (the bars are less than the thickness of the symbols).

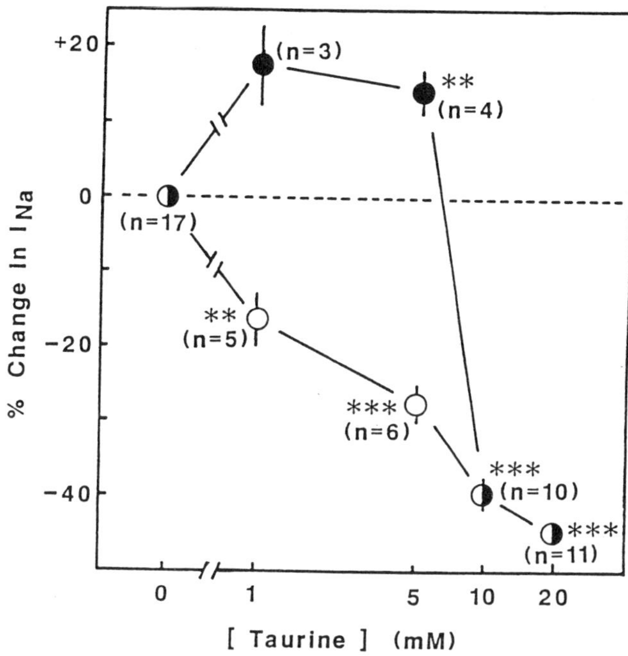

Figure 2. Dose-response curve for changes in I_{Na} produced by exposure to different concentrations of taurine. In some cells, taurine (1 and 5 mM) enhanced I_{Na}, whereas in other cells, the same concentration produced inhibition of I_{Na}. The data were separated in two groups. This graph plots cells in which I_{Na} was enhanced (filled circles) and cells in which I_{Na} was only inhibited (unfilled circles) by taurine (1 and 5 mM). Taurine (10 and 20 mM) produced only inhibition of the I_{Na}. The half-filled circles are comprised of both set of data. Values represent mean + SEM. **:P < 0.01, ***:P < 0.001, with respect to control value.

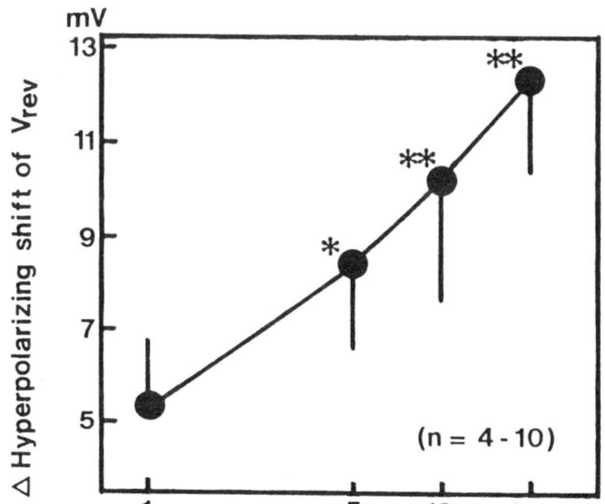

Figure 3. Hyperpolarizing shift of the reversal potential for I_{Na} in embryonic chick cells (17-day-old). Test pulses were applied between -60 and +50 mV from a holding potential of -90 mV. The reversal potentials (V_{rev}) were measured from the data in Figure 1. Values represent mean ± SEM. *: $P < 0.05$, **: $P < 0.01$, with respect to control value (+35.2 ± 3.4 mV, n = 17).

Experimental Solutions

The cells, on glass coverslips, and the small preparations were placed in the chamber, and superfused with a modified Tyrode solution. The composition of this solution was (in mM): NaCl 137, KCl 5.4, $CaCl_2$ 0.1, $MgCl_2$ 1.0, NaH_2PO_4 0.3, glucose 5.0, HEPES 5.0 (adjusted to pH 7.4). To measure each ionic current, tetrodotoxin (10 μM), $CoCl_2$ (3 mM), $MnCl_2$ (2 mM), 4-aminopyridine (3 mM) and/or tetraethylammonium (15 mM) were added to the bath solution. The internal (pipette) solution contained (in mM): KOH 110, KCl 20, $MgCl_2$, EGTA 10, MgATP 5, creatine phosphate 5, aspartic acid 150, and HEPES 5 (adjusted to pH 7.2). To examine the inward current (I_{Na} and I_{Ca}), K^+ was displaced by Cs^+. The $[Ca^{2+}]_i$ level was adjusted to pCa 10 or pCa 7 (9,38). In conventional microelectrode experiments, the glass electrode was filled with 3 M KCl (ca. 10-20 MΩ). Taurine, kindly supplied by Taisho Pharmaceutical Co. (Tokyo, Japan), was dissolved in the bath or the pipette solutions.

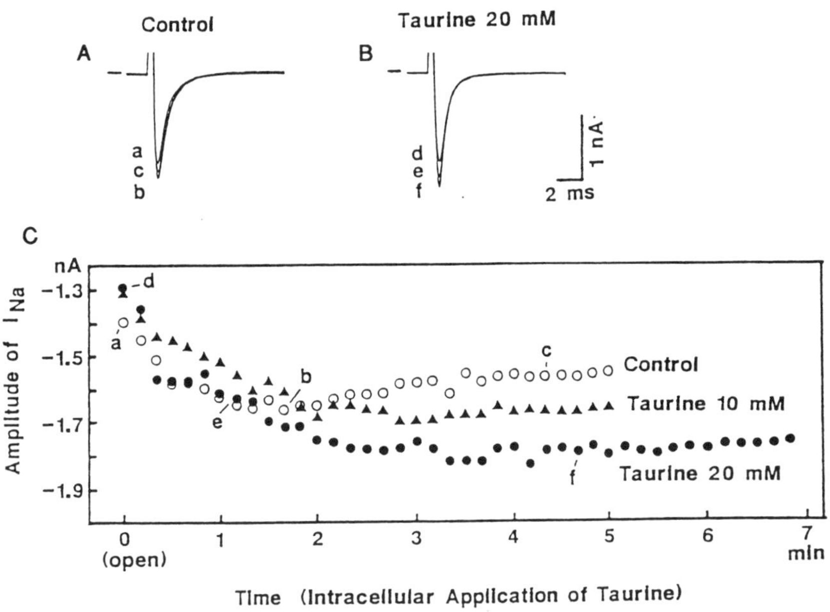

Figure 4. Effects on I_{Na} of intracellular application of taurine. **A and B:** Time-dependent change in peak I_{Na} in control and in taurine (20 mM). **C:** Comparison of time-dependent changes of peak I_{Na} in the absence and presence of taurine. Symbols: control (open circles), 10 mM taurine (triangles), and 20 mM taurine (filled circles).

RESULTS

Effects on Ionic Currents

In single embryonic chick ventricular cells (3- to 17-day old), the effects of taurine on the ionic currents were examined by whole-cell voltage-clamp method.

Sodium Current (Tetrodotoxin-sensitive). In single 17-day-old embryonic chick cells (capacitance ca. 9 pF), the effects on the Na^+ current (I_{Na}) were examined (Fig. 1). At 1 to 5 mM, taurine enhanced the fast Na^+ current (I_{Na}) in some cells (Fig. 2). At 10 to 20 mM, taurine inhibited I_{Na} in all cells by 38.2 ± 4.3% (n = 10) and 44.3 ± 1.8%, respectively. The time decay of the inactivation phase (a single component) was unaffected. The time constant was 1.1 ± 0.5 ms (n = 13) in control. Taurine (10 mM) shifted the reversal potential in the hyperpolarizing direction by 10.2 ± 3.1 mV (n = 10)(Fig. 1C). The shift was concentration-dependent (Fig. 3). The reversal potential shift was not modified by Ni^{2+} (2 mM), a Na^+-Ca^{2+} exchange blocker. Taurine did not affect the kinetics of either the inactivation (h_∞, $V_{1/2}$ was -86.3 mV) or activation (m_∞, $V_{1/2}$ was -49.8 mV). Intracellular application of 10 mM taurine enhanced I_{Na} by 7.8 ± 3.5% (n = 8, P < 0.05) while 20 mM enhanced I_{Na} 14.6 ± 5.6% (n = 5, P < 0.01), as compared to the steady state control value (Fig. 4).

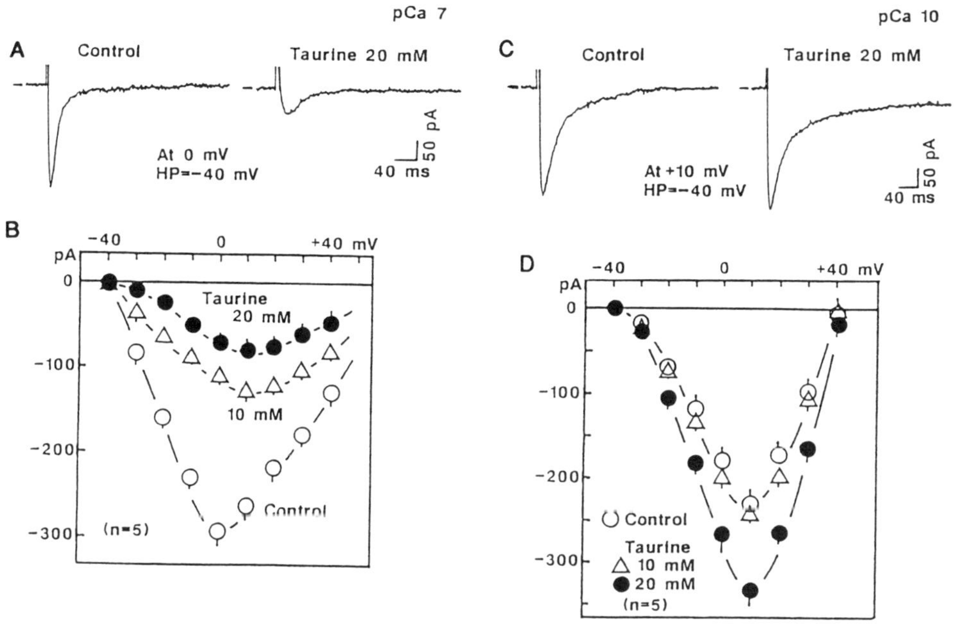

Figure 5. Inhibition of I_{Ca} induced by taurine at different $[Ca]_i$ levels. **A:** Current traces at pCa 7 in a 3-day-old cell in control and in 20 mM taurine. Test pulse was applied to 0 mV from a holding potential of -40 mV. **B:** Current-voltage relationship for I_{Ca} at pCa 7. **C:** Current traces at pCa 10 in a 3-day-old cell in control and 20 mM taurine. Test pulse was applied to +10 mV from a holding potential of -40 mV. **D:** Current-voltage relationship for I_{Ca} at pCa 10. Symbols: control (open circles), 10 mM taurine (triangles), and 20 mM taurine (filled circles). Values are means ± SEM.

Calcium Current. Taurine (10 and 20 mM) at pCa 7 inhibited the L-type Ca^{2+} current (I_{Ca}) at 0 mV by 56.3 ± 2.8% (n = 5, P < 0.01) and 73.2 ± 2.6% (n = 5, P < 0.001) in 3-day-old embryonic chick cardiomyocytes (capacitance was ca. 12 pF), respectively (Fig. 5A-B). In contrast, at pCa 10, 10 mM taurine enhanced I_{Ca} at +10 mV by 7.2 ± 3.1% (n = 5, P>0.05) and 20 mM taurine enhanced I_{Ca} 47.3 ± 2.4% (n = 5, P < 0.01) (Fig. 5C-D). The inactivation curve (f_∞) for I_{Ca} was not modified by taurine. The inactivation time course had two components. The fast and slow time constants (5.8 ± 2.8 ms and 28.8 ± 1.8 ms in control) at pCa 7 were unaffected by taurine. At pCa 10, the fast time constant (15.8 ± 1.0 ms in control) was not affected, but the slow time constant (67.7 + 3.8 ms in control) was slowed by 14.5 + 2.9% (n = 4, P < 0.05) in 10 mM taurine and by 66.2 + 2.8% (n = 4, P < 0.001) in 20 mM taurine. Intracellular application of taurine (20 mM) inhibited I_{Ca} by 63.0 ± 2.8% (n = 4, P < 0.01), even at pCa 10 (Fig. 6). In addition, taurine (20 mM) caused a dormant T-type Ca^{2+} current to appear, although it inhibited the L-type current (Fig. 7).

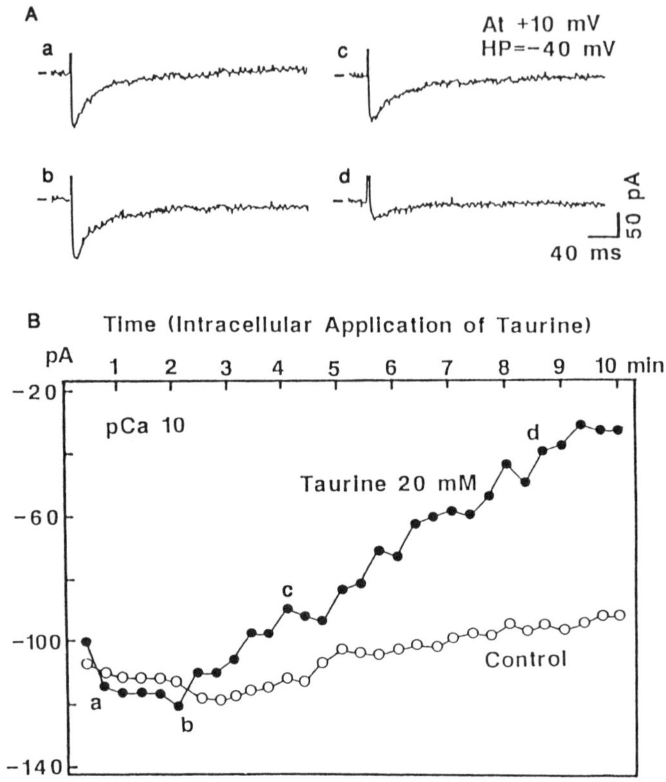

Figure 6. Inhibition of Ca^{2+} current by intracellular application of taurine. **A:** Current traces in a 3-day-old cell after application of 20 mM taurine. Test pulse was applied to +10 mV from a holding potential of -40 mV. **B:** Time-dependent change in peak amplitude of the Ca^{2+} current elicited by intracellularly applied taurine. Symbols: control (open circles) and 20 mM taurine (filled circles).

Delayed Rectifying Potassium Current. The delayed rectifying K^+ current (I_K) in 10-day-old embryonic chick myocytes (capacitance was ca. 10 pF) was also dependent on the $[Ca^{2+}]_i$ level (Fig. 8). In general, the I_K current is activated at high $[Ca^{2+}]_i$, whereas I_K becomes smaller at low $[Ca^{2+}]_i$ (Figs. 8A and C), as previously described (14,20). At pCa 7, taurine (10 and 20 mM) inhibited I_K at +90 mV by 17 ± 3% (n = 5, P < 0.01) and by 26 ± 3%, respectively (n = 5, P < 0.001) (Fig. 8B). At pCa 10, the same taurine concentrations enhanced I_K at +90 mV by 19 ± 3% (n = 7, P < 0.01) and 29 ± 3% (n = 7, P < 0.001), respectively (Fig. 8D). The actual activation curves (P_∞) at both pCa 7 and 10 were depressed by taurine. The half-maximum activation tended to shift in the hyperpolarizing direction, but not significantly. Taurine (20 mM) shifted it from +0.2 ± 2.2 mV to -10.6 ± 1.8 mV (n = 5) at pCa 7, and from +18.5 ± 4.6 mV to +6.6 ± 5.2 mV (n = 5) at pCa 10.

Transient Outward Current. The effect of taurine on the transient outward current (I_{to}) was examined in single embryonic chick cells (10-day-old). Test pulses were applied between -20 and +70 mV from a holding potential of -70 mV. The I_{to} was sensitive to 4-aminopyridine, and its amplitude (at pCa 10) was 327 ± 5.1 pA (n = 6) at +70 mV. Taurine at 10 mM enhanced I_{to} at +70 mV by 15.1 ± 1.3% (n = 6, P < 0.05), and at 20 mM by 37.2 ± 2.1% (n = 6, P < 0.001).

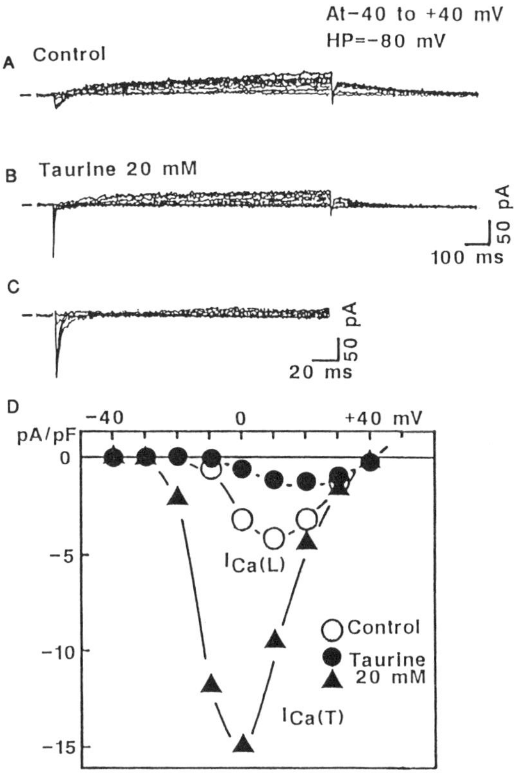

Figure 7. Appearance of T-type Ca^{2+} current induced by taurine at pCa 7 in a 3-day-old cell. The K^+ in pipette solution was not replaced by Cs^+, and tetraethylammonium was not added to the bath solution. **A and B:** Current traces in the absence and presence of 20 mM taurine. Test pulses were applied between -40 and +40 mV from a holding potential of -80 mV. **C:** High speed trace of B. **D:** Current-voltage relationship for the L- and T-type Ca^{2+} currents. Symbols: control (open circles), L-type Ca^{2+} current (filled circles), and T-type Ca^{2+} current (triangles) in the presence of taurine (20 mM).

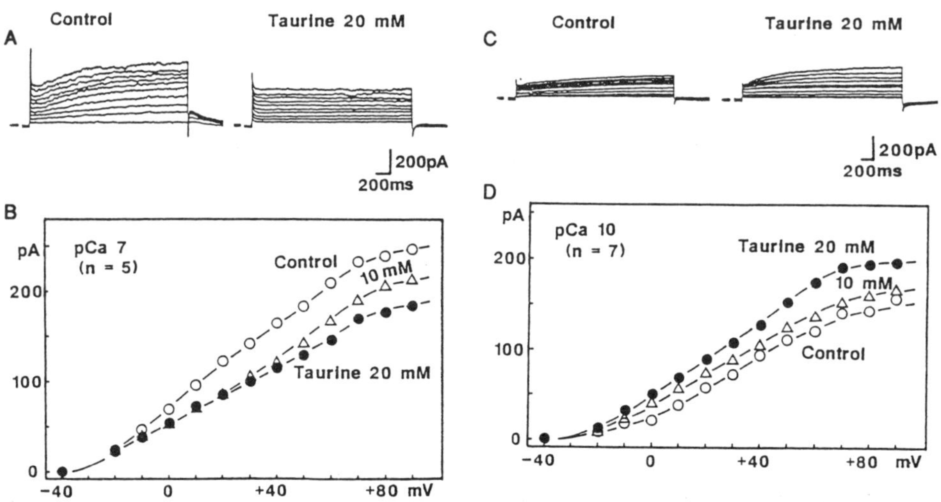

Figure 8. Inhibition of I_K induced by taurine at different $[Ca]_i$ levels. Test pulses were applied between -20 to +90 mV from a holding potential of -40 mV. **A:** Current traces in a 10-day-old cell at pCa 7 in the absence and presence of 20 mM taurine. **B:** Current-voltage relationship for I_K at pCa 7 (n = 5). **C:** Current traces at pCa 10 in the absence and presence of 20 mM taurine. **D:** Current-voltage relationship for I_K at pCa 10 (n = 7). Symbols: control (open circles); 10 mM taurine (triangles); and 20 mM taurine (filled circles). Values are represented as mean ± SEM.

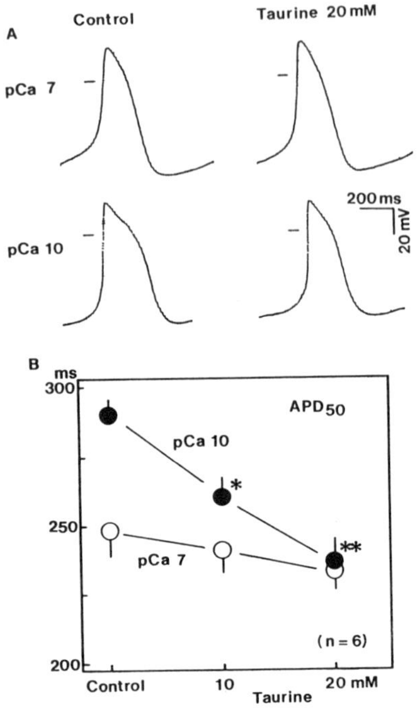

Figure 9. Effects of taurine on the action potential duration from a 10-day-old embryonic chick ventricular cell. **A:** Action poten tials at pCa 7 and pCa 10 in the absence and presence of taurine (20 mM). **B:** Summarized effects of taurine on the action potential duration at 50% repolarization (APD_{50}). Values represent mean ± SEM. *: $P < 0.05$, **: $P < 0.01$, with respect to control value.

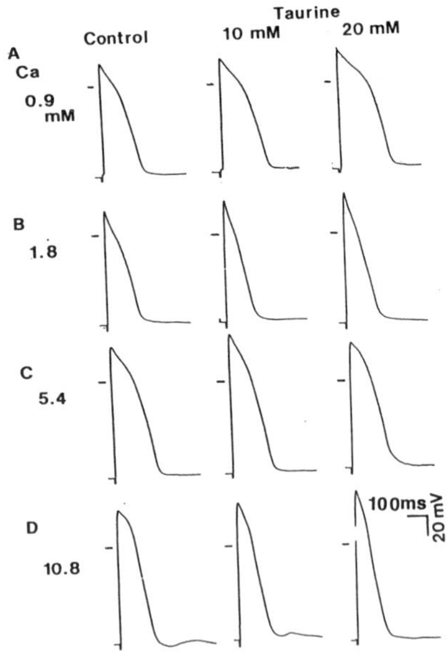

Figure 10. Effects of taurine on the action potential of guinea pig ventricular muscle. The preparations were driven at 1 Hz. **A:** Extracellular Ca^{2+} concentration ($[Ca]_o$) 0.9 mM. **B:** Normal Tyrode solution ($[Ca]_o$ = 1.8 mM). **C:** 5.4 mM $[Ca]_o$. **D:** 10.8 mM $[Ca]_o$. Taurine (10 and 20 mM) antagonized the alteration of the action potential durations induced by the different $[Ca]_o$. The action potential recordings were obtained from different preparations. The short line represents zero mV.

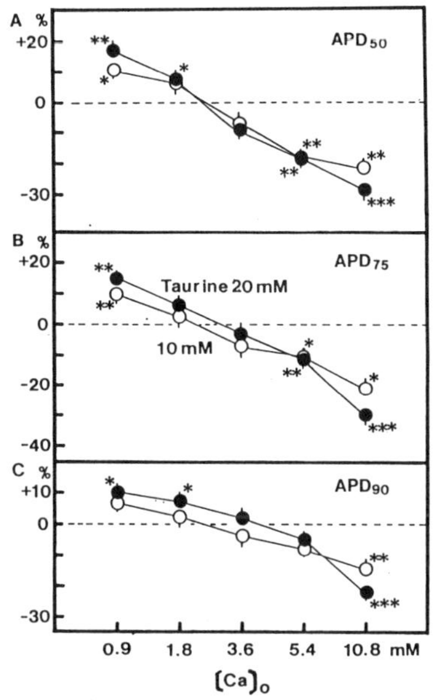

Figure 11. Effects of taurine on the action potential duration at different $[Ca]_o$ solutions. The preparations were constantly driven at 1 Hz. The action potential duration was shortened with an increase in $[Ca]_o$ (0.9 to 10.8 mM). **A:** 50% repolarization of the action potential (APD_{50}). **B:** 75% repolarization of the action potential (APD_{75}). **C:** 90% action potential duration (APD_{90}). Symbols used are 10 mM taurine (open circles) and 20 mM taurine (filled circles). Values represent mean + SEM. *: $P < 0.05$, **: $P < 0.01$, ***: $P < 0.001$, with respect to control value.

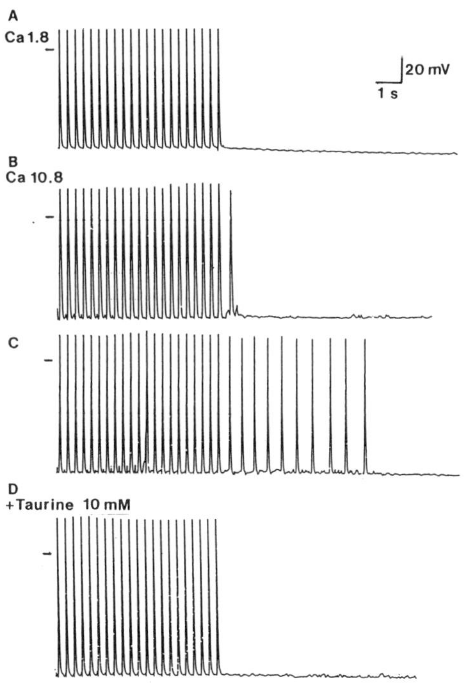

Figure 12. Abolishment by taurine of a delayed after-depolarization induced by high $[Ca]_o$. Action potentials are represented on a slow time scale. The preparation was stimulated at 3 Hz. **A:** Normal Tyrode solution ($[Ca]_o$ = 1.8 mM). **B and C:** Appearance of delayed after-depolarization and spontaneous action potentials. Abnormal electrical activities was completely blocked by addition of taurine (10 mM). Short line represents zero mV.

Effects on the Action Potential

In single 10-day-old embryonic chick cells, current-clamp experiments were performed at different $[Ca]_i$ levels (Fig. 9A). Taurine (10 and 20 mM) at pCa 10 shortened the 50% action potential duration (APD_{50}) by 8.2 ± 2.5% ($n = 6$, $P < 0.05$) and by 16.0 ± 2.0% ($n = 6$, $P < 0.01$), respectively (Fig. 9B). Taurine also tended to decrease the APD_{50} at pCa 7, but not to a significant extent.

Effects on the action potential in guinea pig papillary muscle (its diameter was 2-3 mm) were examined at different $[Ca]_o$ levels (Fig. 10). Generally, low $[Ca]_o$ increased action potential duration, whereas high $[Ca]_o$ decreased it. Taurine (10 and 20 mM) potentiated prolongation of the action potential at low $[Ca^{2+}]_o$ (0.9 mM), and shortened the action potential duration at high $[Ca^{2+}]_o$ (5.4 and 10.8 mM) (Figs. 10 and 11). The change was produced in a $[Ca^{2+}]_o$-dependent manner. Under Ca^{2+} overload conditions (induced by high $[Ca^{2+}]_o$ and/or high stimulation frequencies), the preparations produce abnormal spontaneous activity and elicit a delayed afterdepolarization once stimulation was stopped (Figs. 10D and 12A-C). As shown in Fig. 12D, taurine (10 mM) completely abolished the delayed after-depolarization and the spontaneous action potentials induced by both 10.8 mM $[Ca^{2+}]_o$ and 3 Hz stimulation.

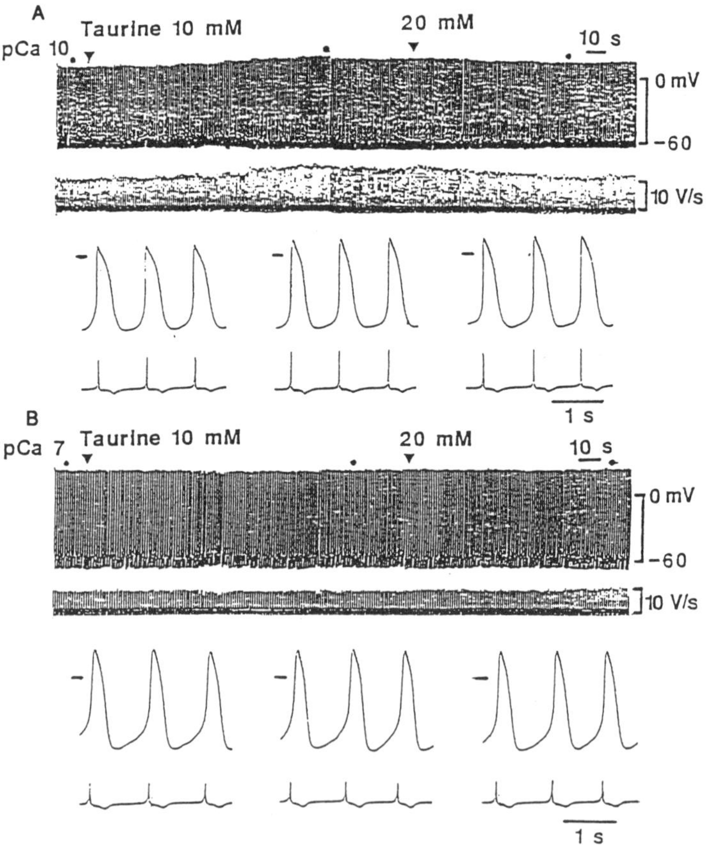

Figure 13. Effects of taurine on the spontaneously beating embryonic chick ventricular myocytes (3-day-old). **A:** The action potentials and the maximum rate of depolarization at pCa 10. **B:** The action potentials and the maximum rate of depolarization at pCa 7. The short line at the left of the recordings represents the zero mV.

Effects on Automaticity

The spontaneous activity in young embryonlc chick heart myocytes (3-day-old) was examined using current-clamp mode. At pCa 10 (the cycle length in control was 980 ± 12 ms, n = 12), taurine at 10 mM had little or no effect, but at 20 mM caused a positive chronotropic effect by 10.1 ± 2.1% (n = 7, $P < 0.05$) (Fig. 13A). The action potential amplitude and the maximum rate of depolarization (V_{max}) were enhanced. On the other hand, at pCa 7 (the cycle length in control was 880 ± 22 ms, n = 6), taurine at 10 and 20 mM caused a negative chronotropic effect of 10.3 ± 1.7% (n = 6, $P < 0.05$) and 15.0 + 1.8% (n = 6, $P < 0.05$), respectively (Fig. 13B).

In spontaneously beating rabbit sino-atrial node preparations (size 0.5x0.5 mm), taurine (5 to 20 mM) caused a positive chronotropic effect at 0.9 mM $[Ca^{2+}]_o$ that was not significantly (Fig. 14). In contrast, at 5.4 to 10.8 mM $[Ca^{2+}]_o$, taurine (1 to 5 mM) caused a negative chronotropic effect. Taurine (10 mM) caused a negative chronotropic effect at 5.4 mM $[Ca^{2+}]_o$ and a positive chronotropic effect at 10.8 mM $[Ca^{2+}]_o$. Taurine (20 mM) also caused a positive chronotropic effect at 10.8 mM $[Ca^{2+}]_o$, presumably due to tachycardia by which taurine abolished irregular rhythms. Taurine at low $[Ca^{2+}]_o$ (0.9 mM) also increased the V_{max} and APD_{50}, but at high $[Ca^{2+}]$n (5.4 to 10.8 mM) decreased them.

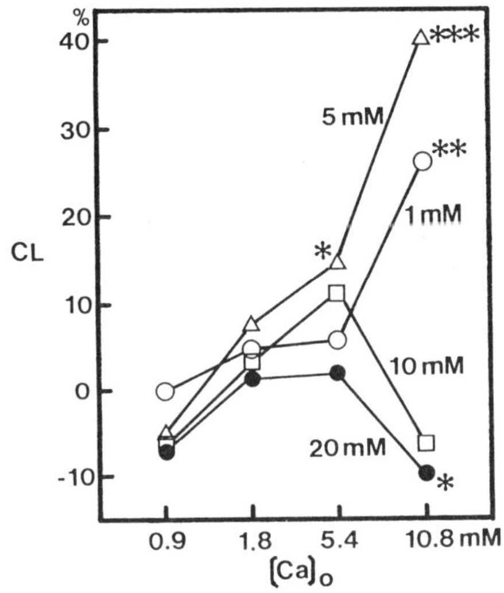

Figure 14. Chronotropic changes by taurine in rabbit sinoatrial node at different extracellular Ca^{2+} concentrations ($[Ca]_o$). The cycle length (CL) was measured, and the percentage changes were plotted. The control value was 480 ± 10 ms (n = 7) at 0.9 mM $[Ca]_o$, 452 ± 9 ms (n = 18) at 1.8 mM $[Ca]_o$, 433 ± 9 ms (n = 8) at 5.4 mM $[Ca]_o$, and 424 ± 8 ms (n = 6) at 10.8 mM $[Ca]_o$. Symbols used are 1 mM (open circles), 5 mM (triangles); 10 mM (squares); and 20 mM (filled circles) taurine. Values represent mean + SEM. The SEM bars are less than the thickness of the symbols. *: $P < 0.05$, **: $P < 0.01$, ***: with respect to control value.

DISCUSSION

On the Normalizing Actions on $[Ca^{2+}]_i$

Taurine inhibited I_{Ca} at pCa 7, whereas it stimulated I_{Ca} at pCa 10, suggesting that taurine may exert a normalizing action on the $[Ca^{2+}]_i$ level by regulating I_{Ca}. When taurine enhances I_{Ca}, there is an elevation in $[Ca^{2+}]_i$, and vice versa. It is unknown whether taurine acts directly on the Ca^{2+} channels. One regulator of Ca^{2+} channels is $[Ca^{2+}]_i$ level, inhibition occurring at high $[Ca^{2+}]_i$. Therefore, these results indicate that taurine possesses an ability to alter Ca^{2+} movement.

Also, the effects on total outward K^+ current was dependent on $[Ca^{2+}]_i$. Taurine enhanced the I_K current at pCa 10, and inhibited it at pCa 7, although the I_K was activated with an increase in $[Ca^{2+}]_i$ level. These results can explain the action of taurine on the action potential: prolongation occurs at high $[Ca^{2+}]_i$ and shortening at low $[Ca^{2+}]_i$. However, taurine prolonged the action potential at low $[Ca^{2+}]_o$ while shortening it at high $[Ca^{2+}]_o$ in rabbit sino-atrial node and guinea pig ventricular muscles. This difference might result from a difference between embryonic and adult hearts. It seems that the actions of taurine are antagonism of Ca^{2+}-induced effects.

In addition, the modulation of I_{Na} regulates the $[Ca^{2+}]_i$ level. The I_{Na} inhibition leads to reduction of $[Ca^{2+}]_i$ via Na^+-Ca^{2+} exchange. Taurine exerted a dual action on I_{Na}, with inhibition or stimulation at lower concentrations (1 to 5 mM) and inhibition only at higher concentrations (10 and 20 mM). The dual action of taurine may reflect two sites of action: a direct stimulation of the fast Na^+ channels; and an indirect inhibition of the channels. Another possibility is that taurine acts only on the fast Na^+ channels: stimulating them from the inner surface and inhibiting them from the outer surface.

These dual actions of taurine are consistent with previous reports (15,28,29). Taurine antagonizes the depressions in contraction, Ca^{2+}-dependent slow action potential and Ca^{2+} influx of cardiac muscle under conditions of ischemia and high $[Ca^{2+}]_o$. The dual action on these ionic currents could normalize the $[Ca^{2+}]_i$ level by direct modulation of the ionic channels and indirectly mediate Na^+-Ca^{2+} exchange. Under Ca^{2+} overload conditions (see Fig. 12), the abolishment of abnormal spontaneous activity would be due to the $[Ca^{2+}]_i$ reduction by taurine application. The spontaneous firing rate in the presence of taurine was decreased at high pCa or $[Ca^{2+}]_o$, and increased at low pCa or $[Ca^{2+}]_o$. Since the spontaneous activitly is dependent on $[Ca^{2+}]_i$ or/and $[Ca^{2+}]_o$ levels, taurine modulates $[Ca^{2+}]_i$ level and thus would regulate the spontaneous firing rate. Therefore, these results suggest that taurine may play an important role in 'Ca^{2+} homeostasis' in maintaining cell functions and protecting against cell damage.

On Cardioprotective Actions

In the present experiments, taurine inhibited I_{Na} and I_{Ca} at pCa 7. These inhibitions may protect cardiac muscles by lowering $[Ca^{2+}]_i$. A reduction in $[Ca^{2+}]_i$ should lead to a reduction in the degree of Ca^{2+} overload accompanying ischemia and cardiac failure. The I_{Na} inhibition exerts a class I antiarrhythmic action, and the I_{Ca} inhibition exerts a class IV antiarrhythmic and antihypertensive action. Actually, taurine abolished the delayed after-depolarization (triggered activity). At low pCa, the enhancement of I_{Ca} induced by taurine may also be a protective effect on the cardiac muscle cells (cardiotonic action). On the other hand, taurine inhibited I_K at pCa 7, whereas it enhanced I_K at pCa 10. The I_K current is greatly regulated by $[Ca^{2+}]_i$ level, resulting in the regulation of action potential duration (see Figs. 9, 10 and 11). The dual action is also an antagonism against the $[Ca^{2+}]_i$-induced effects, and a protection against cell damage (27).

The spontaneous activity is generally decreased at low $[Ca^{2+}]_o$ and is increased at high $[Ca^{2+}]_o$. Taurine caused a positive chronotropic effect at low $[Ca^{2+}]_o$ or pCa, whereas it caused a negative chronotropic effect at high $[Ca^{2+}]_o$ or pCa (see Figs. 13 and 14). The effects are to maintain and protect the cell functions. However, at high concentrations (20 mM), taurine caused a positive chronotropic effect (tachycardia) at 10.8 mM $[Ca^{2+}]_o$. This might be due to occurrence of cellular Ca^{2+} overload (triggered activity) by high $[Ca^{2+}]_o$ and taurine application, presumably resulting from the limitation of the capacity of protective actions of taurine. This might also result from inhibition of irregular ionic currents causing arrhythmias, since the regular rhythm was recovered after taurine application. Taurine caused the sudden emergence of a dormant T-type Ca^{2+} current. This current contributes to the generation of pacemaker activity during diastole (11). The automaticity is depressed under the Ca^{2+} overload conditions (23,26). Since the stimulation of the T-type current activates the pacemaker current, this is also one of the protective actions of taurine.

The hyperpolarizing shift of the reversal potential induced by taurine is not due to modulation of I_{Na} by taurine. Since the shift was elicited not only when I_{Na} was inhibited but also even when I_{Na} was enhanced, it appears to be produced by an additional effect of taurine. The shift was not modified by the Na^+-Ca^{2+} exchange blocker, Ni^{2+}. Recently, it has been reported that taurine elevates $[Na]_i$ by taurine-Na^+ co-transport in guinea pig cardiomyocytes (4,5). Therefore, the $[Na]_i$ elevation by the co-transport would cause the shift of reversal potential, and might lead to a positive inotropic effect.

Table 1. Effects of taurine on the ionic currents of cardiac muscle cells.

		pCa 10	pCa 7
Fast Na^+ current (TTX-sensitive)			
Extracellular application	at 1-5 mM	Enhancement or Inhibition	-
	at 10-20 mM	Inhibition	-
Intracellular application		Inhibition	-
Slow Na^+ current (TTX-insensitive)		Enhancement	-
L-type Ca^{2+} current			
Extracellular application		Enhancement	Inhibition
Intracellular application		Inhibition	-
T-type Ca^{2+} current		Enhancement (Appearance)	-
Delayed rectifying K^+ current		Enhancement	Inhibition
Transient outward current		Enhancement	-

Usually, taurine was added to the bath solution. TTX: Tetrodotoxin

The modulation of the inotropic effect by taurine has already been shown. Sperelakis and colleagues (35) suggest that the enhancement of the slow, tetrodotoxin-insensitive, Na^+ current induced by taurine may cause a positive inotropic effect via Na^+-Ca^{2+} exchange. Taurine does not change cyclic nucleotide levels (29,30). Taurine has been reported to inhibit phosphatidylinositol turnover in heart, and protect against myocardial damage associated with inositol triphosphate accumulation (16). In addition, taurine can displace Ca^{2+} from calmodulin binding sites (32) and thereby inhibit the calmodulin-dependent protein kinase implicated in the regulation of both transmitter release and ion conductance (6). Therefore, taurine would cause the ionotropic effect by modulation of ionic currents and intracellular signal transmissions.

Table 2. Effect of taurine on action potentials

	pCa 10	pCa 7
Embryonic chick cardiomyocytes		
Spontaneous activity	Enhancement	Inhibition
Action potential duration	Shortening	No effect
	Low $[Ca]_o$	**High $[Ca]_o$**
Rabbit sino-atrial node		
Spontaneous activity	Enhancement	Inhibition
Action potential duration	Prolongation	Shortening
Guinea-pig ventricular muscle		
Action potential duration	Prolongation	Shortening

Taurine was applied to the bath solution.

SUMMARY

The effects of taurine were modulated by the $[Ca^{2+}]_i$ or/and $[Ca^{2+}]_o$ levels, consistent with recent reports (10,28,31). Taurine may directly and indirectly regulate the $[Ca^{2+}]_i$ level by modulating Ca^{2+} channels (dependent on $[Ca^{2+}]_i/[Ca^{2+}]_o$) and Na^+ channel (via Na^+-Ca^{2+} exchange). Thus, taurine antagonizes Ca^{2+} ($[Ca^{2+}]_o$ or $[Ca^{2+}]_i$)-induced cardiac functions. The data for the effects of taurine on the ionic currents and action potentials (automaticity) are summarized in Tables 1 and 2. These results indicate that taurine exerts potent cardioprotective actions under the conditions induced by low Ca^{2+} level as well as by calcium overload. In conclusion, the effects of taurine are complex, there being a number of actions on cardiac muscle which may show the possible therapeutic use of this sulfur amino acid.

ACKNOWLEDGEMENT

Part of these data were obtained in Professor Sperelakis' laboratory in the Department of Physiology and Biophysics, University of Cincinnati. The author appreciates his helpful criticisms.

REFERENCES

1. Azari, J., Brumbaugh, P., and Huxtable, R.J. 1980, Prophylaxis by taurine in the hearts of cardiomyopathic hamsters, *J.Mol.Cell.Cardiol.*, 12:1353-1366.
2. Azuma, J., Sawamura, A., Awata, N., Ohta, H., Hamaguchi, T., Harada, H., Takihara, K., Hasegawa, H., Yamagami, T., Ishiyama, T., Iwata, H., and Kishimoto, S. 1985, Therapeutic effect of taurine in congestive heart failure: a double-blind crossover trial, *Clin.Cardiol.*, 8:276-282.
3. Baskin, S.I. and Finney, C.M. 1979, Effects on taurine and taurine analogues on the cardiovascular system, *Sulfur-Containing Amino Acids*, 2:1-18.
4. Benyajati, S. and Bay, S.M. 1992, Sodium-taurine cotransport in reptilian renal brush-border membrane vesicles, *Pflugers Arch.*, 421:168-175.
5. Chapman, R.A., Suleiman, M.-S., and Earm, Y.E. 1993, Taurine and the heart, *Cardiovasc.Res.*, 27:358-363.
6. Cheung, W.Y. 1980, Calmodulin plays a pivotal role in cellular regulation, *Science*, 207:19-27.

7. Dietrich, J. and Diacono, J. 1971, Comparison between ouabain and taurine effects on isolated rat and guinea pig hearts in low calcium medium, *Life Sci.*, 10:499-507.
8. Dolara, P., Ledda, F., Mugelli, A., Mantelli, L., Zilletti, L., Franconi, F., and Giotti, A. 1978, Effect of taurine on calcium, inotropism and electrical activity of the heart, *in:* "Taurine and Neurological Disorders", Barbeau, A. and Huxtable, R.J., eds., Raven Press, New York, pp. 151-159.
9. Fabiato, A. and Fabiato, F. 1979, Calculator programs for computing the composition of the solutions containing multiple metals and ligands used for experiments in skinned muscle cells, *J.Physiol.Paris*, 75:463-505.
10. Franconi, F., Martini, F., Stendardi, I., Matucci, R., Zilletti, L., and Giotti, A. 1982, Effect of taurine on calcium levels and contractility in guinea pig ventricular strips, *Biochem.Pharmacol.*, 31:3181-3-186.
11. Hagiwara, N., Irisawa, H., and Kameyama, M. 1994, Contribution of two types of calcium currents to the pacemaker potentials of rabbit sino-atrial node cells, *J.Physiol.(Lond)*,
12. Hamill, O.P., Marty, A., Neher, E., Sakmann, B., and Sigworth, F.J. 1981, Improved patch-clamp techniques for high-resolution current recording from the cells and cell-free membrane patches, *Pflügers Archiv.*, 391:85-100.
13. Huxtable, R.J. and Sebring, L.A. 1983, Cardiovascular actions of taurine, *in:* "Sulfur Amino Acids: Biochemical and Clinical Aspects", Kuriyama, K., Huxtable, R.J. and Iwata, H., eds., Alan R. Liss, New York, pp. 5-38.
14. Isenberg, G. 1975, Is potassium conductance of cardiac Purkinje fibers controlled by $[Ca^{2+}]_i$? *Nature*, 253:273-274.
15. Kramer, J.H., Chovan, J.P., and Schaffer, S.W. 1981, Effect of taurine on calcium paradox and ischemic heart failure, *Am.J.Physiol.*, 240:H238-H246.
16. Kuriyama, K., Hashimoto, T., Kimori, M., Nakamura, Y., and Yamamoto, S. 1989, Taurine and receptor mechanisms in the heart: Possible correlates with the occurrence of ischemic myocardial damages, *in:* "Taurine and Heart", Iwata, H., Lombardini, J.B. and Segawa, T., eds., Kluwer Academic Publishers, Boston, pp. 139-158.
17. Lombardini, J.B. 1980, Effects of ischemia on taurine levels, *in:* "Natural Sulfur Compounds", Cavallini, D., Gaull, G.T. and Zappia, V., eds., Plenum Press New York, pp. 295-306.
18. Loriette, C., Pierre, Y., and Chatagner, F. 1980, Effect of dietary cholic acid and cholesterol on liver and kidney cystathionase and cysteine sulfinate decarboxylase activities and taurine concentrations in the rat, *Experientia*, 36:285-286
19. McBroom, M.J. and Welty, J.D. 1977, Effect of taurine on heart calcium in the cardiomyopathic hamster, *J.Mol.Cell.Cardiol.*, 9:853-859.
20. Meech, R.W. 1983, The sensitivity of Helix aspersa injected calcium ions, *J.Physiol.(Lond)*, 305:583-588.
21. Satoh, H. 1992, Inhibition in L-type Ca^{2+} channel by stimulation of protein kinase C in isolated guinea-pig ventricular cardiomyocytes, *Gen.Pharmacol.*, 23:1097-1102.
22. Satoh, H. 1993, Comparative electrophysiological and actions of ATP-sensitive potassium channel openers in canine Purkinje fibers, *Gen.Pharmacol.*, 24:565-575.
23. Satoh, H. and Hashimoto, K. 1988, On electrophysiological responses to phorbol esters which stimulate protein kinase C in rabbit sino-atrialnode cells, *Naunyn-Schmiedebergs Arch.Pharmacol.*, 337:308-315.
24. Satoh, H. and Sperelakis, N. 1992, Taurine inhibition of fast Na^+ current in embryonic chick ventricular myocytes, *Eur.J.Pharmacol.*, 218:83-89.
25. Satoh, H. and Sperelakis, N. 1993, Effects of taurine on Ca^{2+} currents in young embryonic chick cardiomyocytes, *Eur.J.Pharmacol.*, 231:443-449.
26. Satoh, H., Tsuchida, K., and Hashimoto, K. 1989, Electrophysiological actions of A23187 and X-537A in spontaneously beating and in voltage-clamped rabbit sino-atrial node preparations, *Naunyn-Schmiedebergs Arch.Pharmacol.*, 339:320-326.
27. Satoh, H. and Uchida, T. 1993, Morphological and electrophysiological changes induced by calcium ionophores (A23187 and X 537A) in spontaneously beating rabbit sino-atrial node mechanical cells, *Gen.Pharmacol.*, 24:49-57.
28. Sawamura, A., Sada, H., Azuma, J., Kishimoto, S., and Sperelakis, N. 1990, Taurine modulates ion influx through cardiac Ca^{2+} channels, *Cell Calcium*, 11:251-259.
29. Sawamura, A., Sperelakis, N., and Azuma, J. 1986, Protective effect of taurine against decline of cardiac slow action potentials during hypoxia, *Eur.J.Pharmacol.*, 120:235-240.
30. Schaffer, S.W., Allo, S., Harada, H., and Azuma, J. 1990, Regulation of calcium homeostasis by taurine: Role of calmodulin, *in:* "Taurine: Functional Neurochemistry, Physiology, and Cardiology", Pasantes-Morales, H., Martin, D.L., Shain, W. and del Río, R.M., eds., Wiley-Liss, New York, pp. 217-225.

31. Schaffer, S.W., Kramer, J., and Chovan, J.P. 1980, Regulation of calcium homeostasis in the heart by taurine, *Federation Proceedings*, 39:2691-2694.
32. Segawa, T., Nomura, Y., and Shimazaki, I. 1985, Possible involvement of calmodulin in modulatory role of taurine in rat cerebral β-adrenergic neurons, *in:* "Taurine: Biological Actions and Clinical Perspectives", Oja, S.S., Ahtee, L., Kontro, P. and Paasonen, M.K., eds., Alan R. Liss, New York, pp. 321-341.
33. Sperelakis, N. and Satoh, H. 1993, Taurine effects on ion channels of cardiac muscle, *in:* "Ionic Channels and Effect of Taurine on the Heart", Noble, D. and Earn, Y., eds., Kluwer Academic Publishers, Boston, pp. 93-118.
34. Sperelakis, N., Satoh, H., and Bkaily, G. 1992, Taurine effects on ionic currents in myocardial cells, *Adv.Exp.Med.Biol.*, 315:129-143.
35. Sperelakis, N., Yamamoto, T., Bkaily, G., Sada, H., Sawamura, A., and Azuma, J. 1989, Taurine effects on action potentials and ionic currents in chick myocardial cells, *in:* "Taurine and the Heart", Iwata, H., Lombardini, J.B. and Segawa, T., eds., Kluwer Academic Publishers, Boston, pp. 1-19.
36. Takihara, K., Azuma, J., Awata, N., Ohta, H., Hamaguchi, T., Sawamura, A., Tanaka, Y., Kishimoto, S., and Sperelakis, N. 1986, Beneficial effect of taurine in rabbits with chronic congestive heart failure, *Am.Heart J.*, 112:1278-1284.
37. Takihara, K., Azuma, J., Awata, N., Ohta, H., Sawamura, A., Kishimoto, S., and Sperelakis, N. 1985, Taurine's possible protective role in age-dependent response to calcium paradox, *Life Sci.*, 37:1705-1710.
38. Tsien, R.Y. and Rink, T.J. 1980, Neutral carrier ion-selective microelectrodes for measurements of intracellular-free calcium, *Biochim.Biophys.Acta*, 559:623-638.

ANTIHYPERTENSIVE EFFECT OF TAURINE ON SALT-INDUCED HYPERTENSION

Masao Nakagawa, Kazuo Takeda, Tatsuo Yoshitomi, Hiroshi Itoh, Tetsuo Nakata and Susumu Sasaki

Second Department of Medicine
Kyoto Prefectural University of Medicine
Kyoto 602, Japan

INTRODUCTION

The antihypertensive effect of taurine has been demonstrated in essential hypertension (5) and experimental hypertension (6,15). In DOCA/salt hypertensive rats, administration of taurine in drinking water restored the blood pressure to normotensive levels and inhibited the development of hypertension. In terms of antihypertensive mechanism, taurine is reported to inhibit sympathetic hyperactivity in DOCA/salt hypertensive rats (18). Moreover, oral administration of taurine is known to normalize the delayed turnover of norepinephrine in the hypothalamus (7). In spontaneously hypertensive rats, long-term administration of taurine results in an increase of taurine content in the brain with a reduction of blood pressure (15). Orally administered taurine is absorbed from the gastrointestinal tract and accumulates in the brain after passing through the blood brain barrier (23). Aside from oral administration, intracerebroventricular (i.c.v.) administration of taurine elicited augmented depressor responses in DOCA/salt hypertensive rats (11,14). These findings suggest that orally administered taurine acts centrally to inhibit the sympathetic nervous system, resulting in a reduction of blood pressure in hypertension. On the other hand, it is suggested that sympathetic hyperactivity contributes to the development of hypertension in salt-induced hypertension in Dahl salt sensitive rats (13), as in DOCA/salt hypertension (20) and the spontaneously hypertensive rat (19). It is possible that taurine inhibits the development of the hypertension induced by salt loading. However, the regulatory mechanism of taurine on salt-induced hypertension is not defined.

In the present study, taurine was orally administered in a drinking water to Dahl salt sensitive and salt resistant rats to elucidate whether taurine prevents the salt-induced hypertension. In addition, to explore the central antihypertensive mechanism of taurine, the blood pressure responses to central administration of hypertonic NaCl and taurine were evaluated.

Taurine in Health and Disease, Edited by R. Huxtable and D.V. Michalk, Plenum Press, New York, 1994

MATERIAL AND METHODS

Sixty male rats of the Dahl strain, half of which were Dahl salt-resistant rat (DR) while the rest was Dahl salt-sensitive rat (DS), were purchased together from Eisai Experimental Animal Center (Tsukuba, Japan). All rats were 4 weeks old upon arrival and were fed with a ordinary sodium diet (0.6%) until they were 6 weeks old, after which two separate studies were performed. For the first study, 18 DR and 16 DS rats were divided into four subgroups, depending on the water they were fed: (i) nine DR rats with tap water; (ii) nine DR rats with 2% taurine solution (iii) eight DS rats with tap water; and (iv) eight DS rats with 2% taurine solution. All rats were fed with 8% NaCl diet and with drinking water ad libitum. The rats were observed under these conditions for 6 weeks and then terminal experiments were done to measure blood pressure directly. When rats were 12 weeks old, the pressor responses to intracisternal (i.c.) injections of hypertonic saline were measured. For the second study, blood pressure responses to i.c.v. injections of taurine were recorded in DS and DR rats on normal or high sodium diets. Systolic blood pressure and heart rate were measured weekly by tail cuff methods with using a programmable sphygmomanometer (PS-100, Riken Kaihatsu). After systolic blood pressure had been measured at the 12th week, each rat was transferred into an individual metabolic cage for collection of urine and measurement of water intake. Electrolytes in urine were measured by flame photometry. Then each rat was anesthetized with urethane (120mg/100g, i.p.) and a catheter inserted into the left femoral artery for recording blood pressure, and into the left femoral vein for drug injections. Blood pressure was measured by connecting the femoral artery cannula to a small-volume transducer (P23XL, Statham) and heart rate was counted by counting pulsatile pressure using a tachometer (1332, NEC-Sanei). For i.c. injection, rats were mounted on a stereotaxic apparatus (David Kopf) and the occipito-atlanta membrane was exposed carefully. A 27 gauge needle was inserted into the cisternal space and connected to a microsyringe (Terumo). Ten μl of 10% NaCl solution were injected slowly. For i.c.v. injection, rats were also mounted on a stereotaxic apparatus and a 23 gauge stainless steel needle connected to a micro-syringe was inserted into the right lateral ventricle.

After these experiments, the whole brain was removed and stored in 10% formalin solution to verify the success of i.c. or i.c.v. injections, and to measure taurine concentration. For the latter, brain was homogenized with 0.4 N perchloric acid, then centrifuged at 20,000 g for 10 min. The pellet was resuspended in 0.4 N perchloric acid, and centrifuged at 20,000 g for 10 min. The supernatant was applied to an amino acid analyzer (L-8500, Hitachi). Drugs used were norepinephrine bitartrate (Nakarai Tesk), 100 and 200 ng/100g; hexamethonium (Nakarai Tesk) 40mg/kg. Taurine (Nakarai Tesk) was dissolved in saline and 10 μl (150 μg), confirmed to reduce blood pressure in urethane-anesthetized rats (10), was injected i.c.v.

Data are expressed as means $\pm$ SEM. Analysis of variance was used to compare the difference among four groups. Newman-Keul's test was applied as a post-hoc-test (Multiple range test) to F ratios significant at 5% or less to determine significance of differences between pairs of means.

RESULTS

Antihypertensive Effect of Taurine in Dahl Salt-Sensitive Rats

After the animals were given the 8% salt diet, the systolic blood pressures obtained by the tail-cuff method were significantly elevated in DS rats compared to DR rats. However, blood pressure was significantly lower in DS rats that were given taurine-containing water than in DS rats given tap water. Systolic blood pressure was higher in taurine-treated DS rats than in DR rats (Fig. 1). In DR rats, high salt diets did not elevate the blood pressure and taurine did not change the blood pressure. Changes in body weight were significantly depressed in taurine-treated rats and averaged (g ± SEM): DR: 322 ± 10; DR given taurine: 275 ± 11; DS: 291 ± 16; and DS given taurine: 275 ± 8, respectively, by the 12th week (F ratio of 8.82, significant at 0.2%). Heart rate did not show any difference among the groups at the 12th week (Fig. 1).

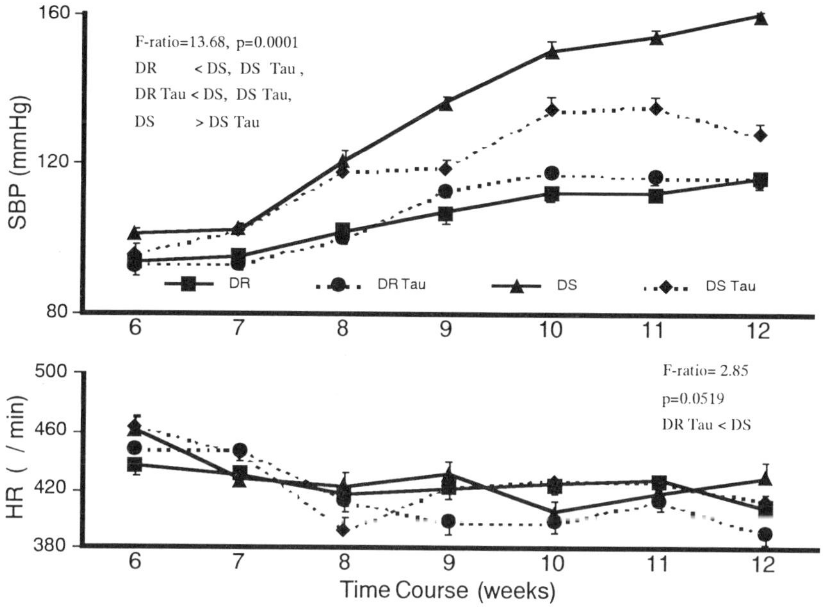

Figure 1. Time course of systolic blood pressure and heart rate in Dahl salt-resistant (DR) and Dahl salt-sensitive (DS) rats after salt loading. DR: DR rats fed with tap water (n=9); DR Tau: DR rats fed with 2% taurine solution (n=9); DS: DS rats fed with tap water (n=8); DS Tau: DS rats fed with 2% taurine solution (n=8).

Urine Volume and Electrolyte Excretion

After systolic blood pressure had been measured, rats were moved individually into metabolic cages at the 12th week, and water intake and urine volumes were measured. Water consumption and urine excretion did not differ among the four groups (Fig. 2). Sodium and K^+ excretion were also unchanged by taurine treatment (Fig. 3).

Pressor Response to Intravenous Injection of Norepinephrine

To determine whether pressor responsiveness to injected norepinephrine is altered by taurine treatment, blood pressure was recorded during intravenous injection in urethane anesthetized rats. Group differences of blood pressure and heart rate at the 12th week recorded directly from femoral catheters were the same as those detected indirectly with the tail-cuff method while the rats were awake. Mean blood pressure was significantly higher in DS rats than in DR rats, while mean blood pressure in taurine-treated DS rats was significantly lower than in DS rats, but significantly higher than in DR rats. Taurine treatment did not change blood pressure in DR rats. The four groups did not differ in either the heart rate or the pressor responses to norepinephrine (Table 1). Reflex bradycardia during pressor responses was augmented only in DR rats given taurine (Fig. 4).

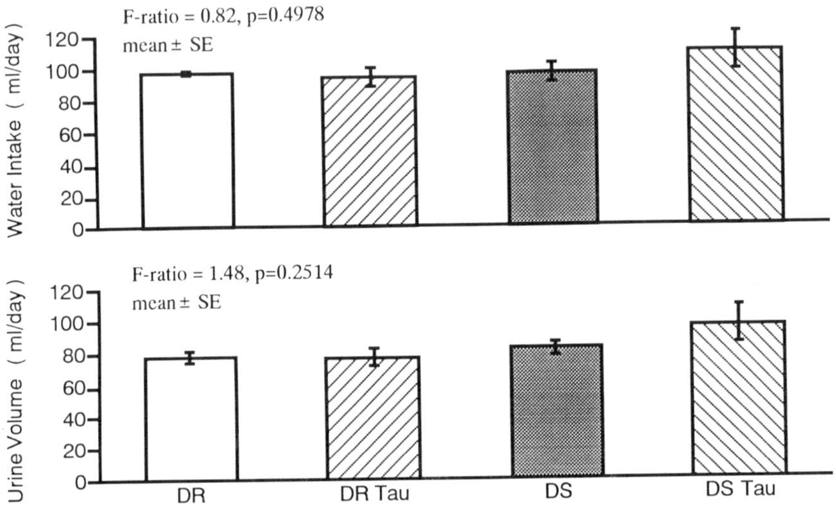

Figure 2. Water intake and urine volume. Abbreviations as for Figure 1.

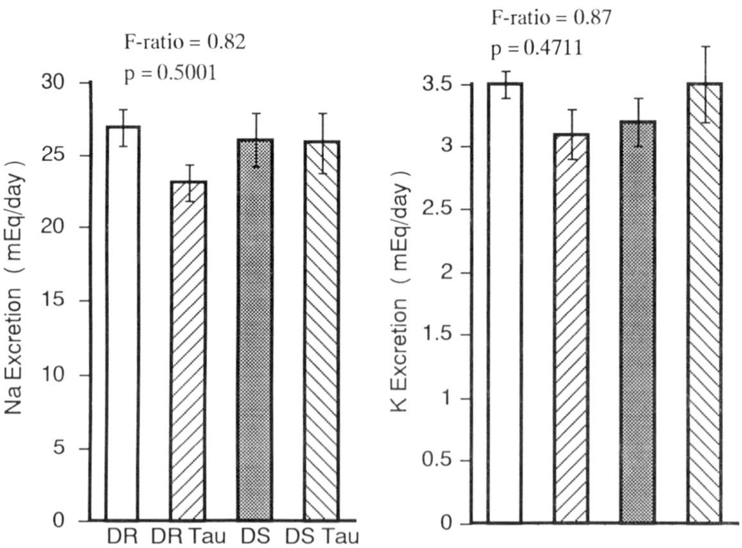

Figure 3. Urinary electrolyte levels. Data are means ± SEM for 6 animals per group. Abbreviations as for Fig. 1.

Table 1. Mean blood pressure and heart rate in urethan-anesthetized rats

	Dahl Salt-resistant		Dahl Salt-sensitive		
	Control	**Taurine**	**Control**	**Taurine**	**F value**
Mean blood pressure (mm Hg)	95 ± 3	98 ± 3	156 ± 6*	129 ± 10**	18.31
Heart rate (beat/min)	421 ± 14	344 ± 14	408 ± 35	393 ± 25	2.03

Data are means ± SEM. *p < .05 compared to all other groups; **p < .05 compared to DR group

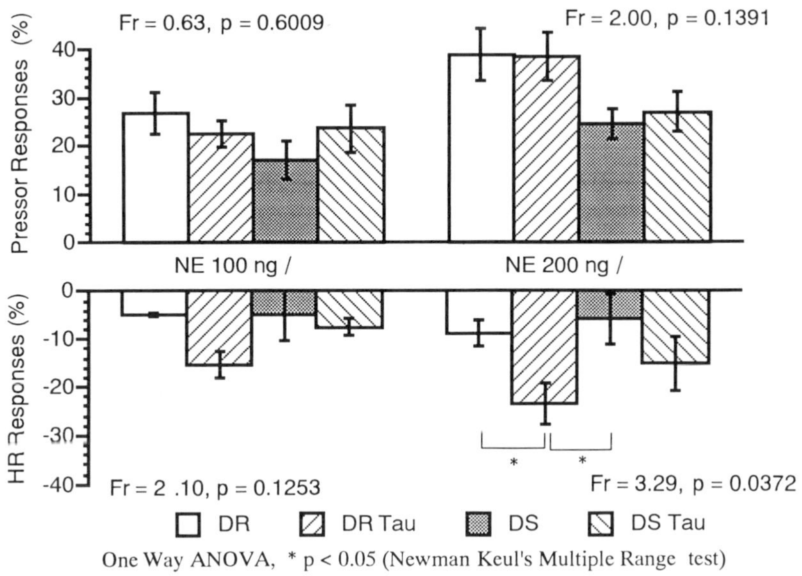

Figure 4. Pressor and bradycardic responses to intravenous injection of norepinephrine, shown as percent magnitude of increases relative to basal. Data are means ± SEM for 8-9 animals per point.

Depressor Response to Intravenous Hexamethonium

Upon the induction of ganglionic blockade by i.v. injection of hexamethonium (40 mg/kg), the blood pressure fall in mm Hg was largest in DS rats without taurine treatment. This response was reduced in taurine-treated DS rats. Taurine treatment was without effect on the depressor responses of DR rats. The bradycardic response of each groups was same as the depressor response (Fig. 5).

Pressor Response to Intracisternal Injection of Hypertonic Saline

To examine whether pressor responsiveness to central administration of hypertonic saline is altered by taurine, 10% sodium solution was injected intracisternally (i.c.) in all rats. In all rats, blood pressure was elevated one min after injection and peaked at 5 to 10 min. In DS rats treated with taurine, pressor responses were significantly smaller than in other rats (Fig. 6).

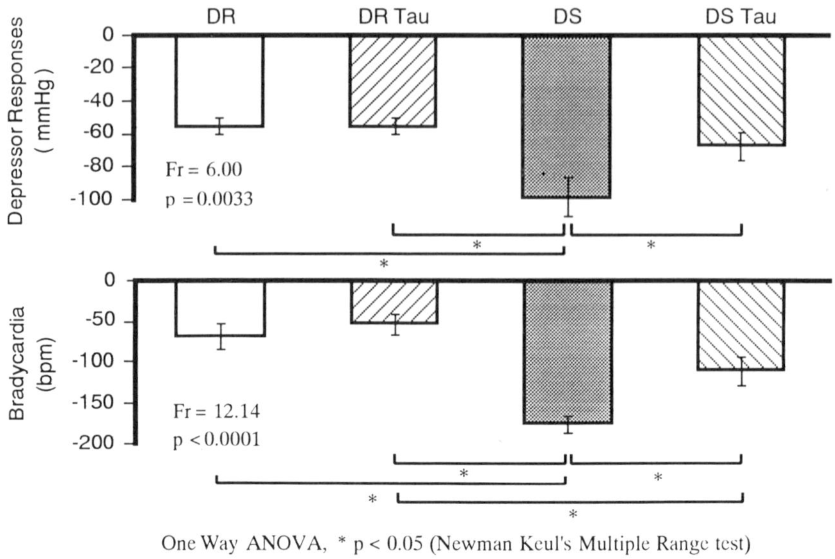

Figure 5. Depressor responses to intravenous injection of hexamthonium.

Taurine Concentration in the Brain

Taurine content was higher in the hypothalamus than in the medulla in both DR and DS rats. The two strains did not differ in their taurine content. Taurine supplementation significantly increased the taurine content of the medulla and hypothalamus in both DR ($P<0.05$) and DS rats ($P<0.01$) (Table 2).

Table 2. Taurine concentration of brain

	Dahl Resistant		Dahl Sensitive	
	Control	**Taurine**	**Control**	**Taurine**
Medulla	2.72 ± 0.10	4.83 ± 0.25*	2.37 ± 0.08	4.85 ± 0.34*
Hypothalamus	3.67 ± 0.10	5.85 ± 0.24*	3.40 ± 0.14	6.04 ± 0.34*

*p < .05 compared to control

Blood Pressure Response to i.c.v. Injection of Taurine

To determine whether the central depressor effect of taurine is different in DS rats from DR rats with salt loading, taurine was injected intracerebroventricularly in DS and DR rats fed either normal or high salt diet. After 6 weeks on the diet, systolic blood pressure was significantly higher in DS rats (184 ± 12 versus 136 ± 8 mm Hg, $P<0.01$). Mean blood pressure was also higher in DS rats given a high salt diet than in DS rats given a normal diet (136 ± 6 versus 94 ± 6 mm Hg, $p < 0.01$). In DR rats, the high salt diet

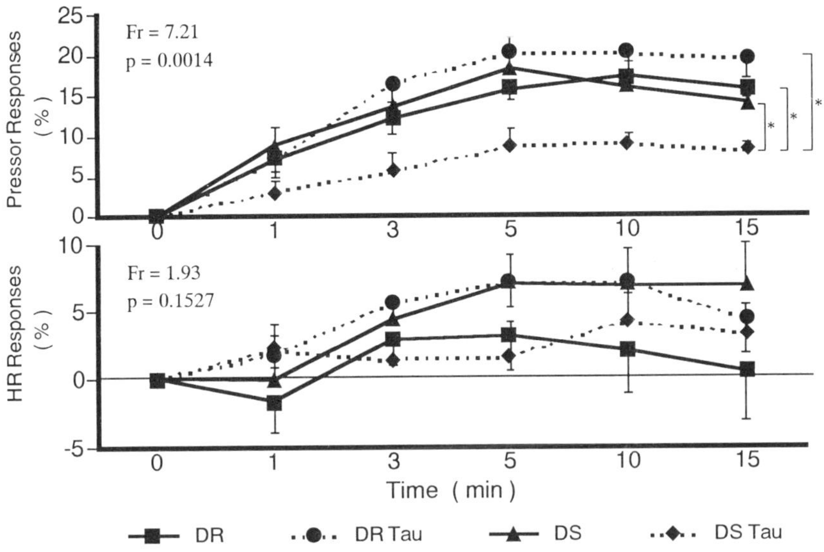

Repeated Measure ANOVA, * p < 0.05 (Newman Keul's Multiple Range test)

Figure 6. Pressor responses to intracranial injection of 10% NaCl Solution. Time (min): after injections. Abbreviations as for Fig. 1. *Indicated groups are significantly different (p < .05).

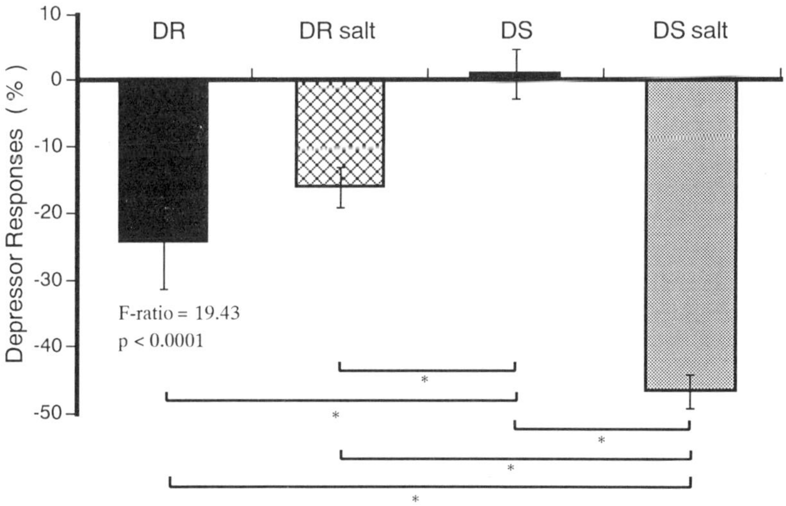

One Way ANOVA, * p < 0.05 (Newman Keul's Multiple Range test)

Figure 7. Depressor response to intracerebroventricular injection of taurine (150 μg). Responses are shown 15 min after injections. *Indicated groups are significantly different (p < .05).

increased mean blood pressure (91 $\pm$ 6 versus 83 $\pm$ 6 mm Hg, $p < 0.05$). Depressor responses to i.c.v.injection of taurine (150 μg) were obtained in DR rats fed a normal diet. However, no change of blood pressure was observed in DS rats. In DS rats fed with high salt diet, the depressor response to i.c.v. injection of taurine was the largest of the four groups. (Fig. 7).

DISCUSSION

In the present study, 2% taurine supplementation in Dahl salt-sensitive rats prevented the development of hypertension after salt loading. Concomitantly with the antihypertensive effect of taurine, its administration restored the augmented depressor responses to ganglionic blockade in DS rats to the level of DR rats. In neither strain was urine volume or sodium excretion altered by taurine supplementation. Moreover, taurine did not change the pressor responsiveness to norepinephrine. Oral administration of taurine prevents the development of DOCA/salt hypertensive rats (6,7,18) and spontaneously hypertensive rats (4,9). As for the mechanism of antihypertensive effect of taurine, the lack of change in urine volume and Na^+ excretion makes natriuresis by taurine. Furthermore, fluid retention has been demonstrated even after an antihypertensive effect obtained by taurine treatment in DOCA/salt hypertensive rats (6). Sato et al. showed that cardiac and splenic norepinephrine turnover was augmented during either normal conditions or cold exposure in DOCA/salt hypertensive rats. Oral supplementation of taurine inhibited these enhanced turnovers (18). These findings suggested that taurine inhibited the sympathetic hyperactivity in DOCA/salt hypertensive rats, as was demonstrated by previous studies (4,19). The fact that the depressor effect of i.v. taurine disappears after sympathetic denervation suggests that the antihypertensive effect of taurine may be due to sympatho-inhibitory action (12). Oral supplementation with taurine inhibits pressor responses to shaker stress in spontaneously hypertensive rats (24). These findings also suggest that taurine inhibits the activation of the sympathetic nervous system.

A contribution of the sympathetic nervous system to the hypertension of Dahl salt-sensitive rats has been demonstrated (13,16,22). This sympathetic hyperactivity may be delivered from the brain, because an enhanced sympathetic response to hypothalamic stimulation is observed with exaggerated pressor response in DS rats (3), while PVN or AV3V lesions prevent the development of hypertension in DS rats (2,8). In the present study, the depressor responses to ganglionic blockade was significantly larger in DS rat than in DR rats. This suggests that the salt-induced hypertension in DS rats is dependent on sympathetic nerve activation. Taurine treatment corrected this augmented response. These findings suggest that the sympathetic inhibition by taurine contributes to the antihypertensive effect of taurine in DS rats with salt loading. Moreover, this possibility is supported by the fact that taurine treatment did not alter the pressor response to norepinephrine in either DR or DS rats.

We previously reported that i.c.v. administration of taurine induced vasodepressor responses with sympathetic nerve inhibition and this response was augmented in urethane-anesthetized DOCA/salt hypertensive rats (11,14). In renovascular hypertensive rats, central administration of taurine induced vasodepressor responses (21). These findings suggest the possibility that orally administered taurine acts centrally to inhibit sympathetic nervous system, resulting in an antihypertensive effect. Actually, in the present study, i.c.v. injections of taurine elicited augmented depressor responses in DS rat fed a high salt diet. Pressor response to intracisternal injection of hypertonic saline has been known to be derived from sympathetic activation and by vasopressin release (17). Taurine treatment inhibited the pressor responses to i.c. injection of hypertonic saline only in DS rat in this experiment. These findings suggest that taurine acts centrally only in DS rats. The increased taurine concentration in hypothalamus and medulla in the rats given oral supplementation with taurine indicates that taurine possibly acts in the brain.

The site in the brain at which taurine may act has not been defined in the present study. However, the hypothalamus may be a candidate, since oral treatment with taurine restored the reduced norepinephrine turnover of hypothalamus concomitantly with inhibition of increased norepinephrine turnover of heart in DOCA/salt hypertensive rats (7). Moreover, taurine may act on the preoptic area, because the responses of water intake on the renin and angiotensin injections into preoptic area are reported to be inhibited by oral administration of taurine in SHR with the prevention of the development of hypertension (1).

REFERENCES

1. Abe, M., Shibata, K., Matsuda, T., and Furukawa, T. 1988, Inhibition of hypertension and salt intake by oral taurine treatment in hypertensive rats, *Hypertension*, 10:383-389.
2. Azar, S., Ernsberger, P., Livingston, S., Azar, P., and Iwai, J. 1981, Paraventricular-suprachiasmatic lesion prevent salt-induced hypertension in Dahl rats, *Clin.Sci.*, 61:49s.
3. Bunag, R.D., Butterfield, J., and Sasaki, S. 1983, Hypothalamic pressor responses and salt-induced hypertension in Dahl rats, *Hypertension*, 5:460.
4. De Champlain, J., Krakoff, L.R., and Axerol, J. 1968, Relationship between sodium and norepinephrine storage during the development of experimental hypertension, *Circ.Res.*, 23:479.
5. Fujita, T., Ando, K., Noda, H., Ito, Y., and Sato, Y. 1987, Effects of increased adrenomedullary activity and taurine in young patients with borderline hypertension, *Circulation*, 75:525-532.
6. Fujita, T. and Sato, Y. 1986, Changes in blood pressure and extracellular fluid with taurine in DOCA-salt rats, *Am.J.Physiol.*, 19:R1014-R1020.
7. Fujita, T., Sato, Y., and Ando, K. 1986, Changes in cardiac and hypothalamic noradrenergic activity with taurine in DOCA-salt rats, *Am.J.Physiol.*, 251:H926-H933.
8. Goto, A., Ganguli, M., Tobian, L., Johnson, M.A., and Iwai, J. 1982, Effect of an anteroventral third ventricle lesion on NaCl hypertension in Dahl salt-sensitive rats, *Am.J.Physiol.*, 243:H614.
9. Horie, R., Yamori, Y., Nara, Y., Sawamura, M., and Mano, M. 1987, Effects of sulphur amino acids on the development of hypertension and atherosclerosis in stroke-prone spontaneously hypertensive rats, *Hypertension*, 5 (Suppl 5):S223.
10. Inoue, A., Takahashi, H., Lee, L., Iyoda, I., Sasaki, S., Okajima, H., Takeda, K., Yoshimura, M., Nakagawa, M., and Ijichi, H. 1985, Centrally induced vasodepressor and sympathetic nerve responses to taurine, *Jpn.Circ.J.*, 49:1180-1184.
11. Inoue, A., Takahashi, H., Lee, L.-C., Sasaki, S., Takeda, K., Yoshimura, M., Nakagawa, M., and Ijichi, H. 1986, Hypotensive responses to centrally administered taurine in DOCA-salt hypertensive and spontaneously hypertensive rats, *Jpn.Circ.J.*, 50:1215-1223.
12. Ito, R., Uchiyama, T., Yoda, S., Homma, N., and Furukawa, K. 1981, Cardiovascular actions of taurine, *gamma-aminobutyric acid (GABA) and *gamma-amino-β-hydroxybutyric acid (GABOA) after chemical denervation, in: "The Effects of Taurine on Excitable Tissues", Schaffer, S.W., Baskin, S.I. and Kocsis, J.J., eds., Spectrum Publications, New York, pp. 313-327.
13. Mark, A.L. 1991, Sympathetic neural contribution to salt-induced hypertension and in Dahl rats, *Hypertension*, 17 (Suppl I):I-86.
14. Nakagawa, M., Inoue, A., Toyoda, T., Sawada, S., Uno, M., and Takeda, K. 1990, Central antihypertensive effect of taurine, in: "Taurine: Functional Neurochemistry, Physiology, and Cardiology", Pasantes-Morales, H., Martin, D.L., Shain, W. and del Río, R.M., eds., Wiley-Liss, New York, pp. 463-470.
15. Nara, Y., Yamori, Y., and Lovenberg, W. 1978, Effects of dietary taurine on blood pressure in spontaneously hypertensive rats, *Biochem.Pharmacol.*, 27:2689-2692.
16. Rapp, J.P. 1982, Dahl salt-susceptible and salt-resistant rats, *Hypertension*, 4:753.
17. Sasaki, S., Takeda, K., Okajima, H., Takahashi, H., Yoshimura, M., Nakagawa, M., and Ijichi, H. 1984, Pressor responses to intracisternal injection of hypertonic NaCl in rats, *J.Cardiovasc.Pharmacol.*, 6:349.
18. Sato, Y., Ando, K., and Fujita, T. 1987, Role of sympathetic nervous system in hypotensive action of taurine in DOCA-salt rats, *Hypertension*, 9:81-87.
19. Takeda, K. and Bunag, R.D. 1978, Sympathetic hyperactivity during hypothalamic stimulation spontaneously hypertensive rats, *J.Clin.Invest.*, 62:642.
20. Takeda, K. and Bunag, R.D. 1980, Augmented sympathetic nerve activity and pressor responsiveness in DOCA hypertensive rats, *Hypertension*, 2:97.
21. Takemoto, Y. 1991, Central depressor effects of amino acids in conscious normotensive and two-kidney, one-clip renovascular hypertensive rats, *Jpn.J.Physiol.*, 41:717.
22. Takeshita, A., Mark, A.L., and Brody, M.J. 1979, Prevention of salt-induced hypertension in the Dahl strain by 6-hydroxydopamine, *Am.J.Physiol.*, 236:H48.

23. Urquhart, N., Perry, T.L., Hansen, S., and Kennedy, J. 1974, Passage of taurine into adult mammalian brain, *J.Neurochem.*, 22:871-872.
24. Yamamoto, J., Akabane, S., Yoshimi, H., Nakai, M., and Ikeda, M. 1985, Effects of taurine on stress-evoked hemodynamic and plasma catecholamine changes in spontaneous hypertensive rats, *Hypertension*, 7:913-922.

THE EFFECT OF TAURINE ON BLOOD PRESSURE, AND URINARY SODIUM, POTASSIUM AND CALCIUM EXCRETION

Michael J. Meldrum, Raymond Tu, Tucker Patterson, Ralph Dawson Jr. and Trish Petty

Department of Pharmacodynamics, College of Pharmacy
University of Florida, JHMHC BOX 100487
Gainesville, Florida 32610

INTRODUCTION

Taurine is a sulfur-containing amino acid found in large quantities in muscle, brain, heart and blood. Its numerous physiological and biological properties are still not fully understood. Studies have examined the physiological effects of taurine in several different areas. Proposed functions include: major inhibitory neurotransmitter; inhibitory neuromodulator; regulator of Ca^{2+} movements in muscle (cardiac, smooth and skeletal); transmitter within the retina; growth factor in brain development; and major osmoregulater within cells in the periphery and the central nervous system (CNS) (2,6-8,10,14,15).

The acute central and chronic oral administration of taurine have been shown to lower blood pressure in DOCA-salt, spontaneously hypertensive and two-kidney-one-clip models of hypertension (4,5,11,13,16). The exact mechanism by which taurine lowers blood pressure is not completely understood.

The purpose of the present study was to investigate the effects of oral taurine ingestion on Na^+, K^+ and Ca^{2+} excretion in hypertension. This was in an effort to determine if taurine may be affecting blood pressure through renal mechanisms.

METHODS

Nine-week-old male spontaneously hypertensive rats (Taconic Farms, Germantown NY) were placed in metabolic cages and divided into two groups. One group received tap water while the other group received taurine in the drinking water. Two separate studies were done using two different concentrations of taurine. In the first study 1% taurine was given in the drinking water and in the second 3% taurine was given. In the treated animals the taurine solution was the only drinking source available. The following metabolic parameters were measured: fluid intake; urinary volume; food intake; body weight; and

Taurine in Health and Disease, Edited by R. Huxtable and D.V. Michalk, Plenum Press, New York, 1994

urinary excretion of Na^+, K^+ and Ca^{2+}. The urine was collected in sealed beakers containing 500 μl of 10 N HCl to acidify the urine. Urinary sodium and potassium levels were measured immediately after collection using ion selective electrodes (NOVA 1 Na/K analyzer) and expressed as mEq excreted per 24 h. Samples were also taken twice weekly (Wednesday and Friday) for the determination of total Ca^{2+} using a spectrophotometric assay (Stanbio Total Calcium Kit). The two measurements were averaged to give a weekly average 24 h excretion of Ca^{2+}. Blood pressures were measured weekly (Friday 10:00 am) using the tail-cuff method.

After two weeks of basal measurements, half of the animals received taurine (Sigma Chemical Co) in the drinking water. In the first study the treated animals received 1% taurine and in the second study they received 3% taurine. The metabolic parameters were then measured for 4 weeks. At the end of the treatment the 3% animals were sacrificed and

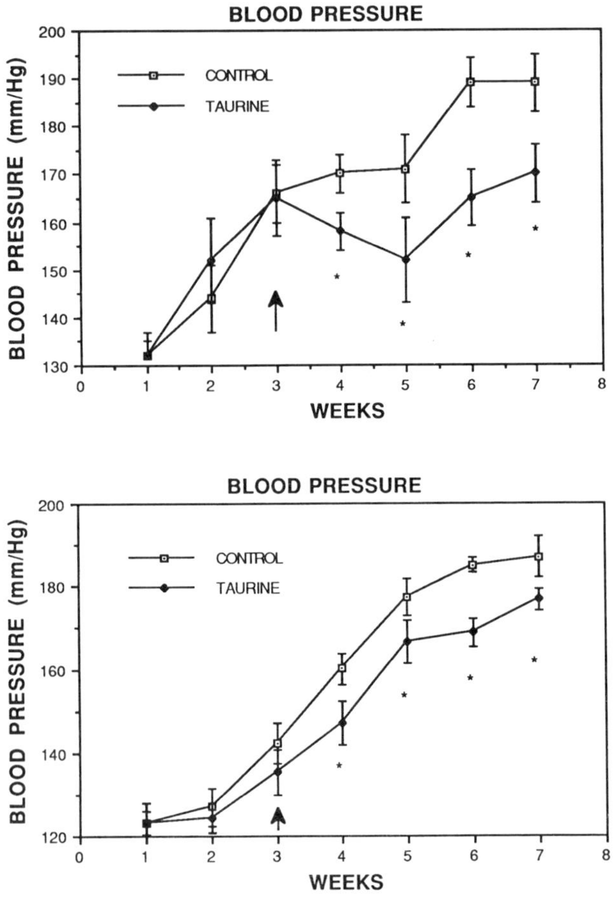

Figure 1. Effects of 1 and 3% taurine in the drinking water on blood pressure. The top panel shows responses to 1% and the bottom panel shows responses to 3% taurine. The arrow shows the start of taurine in the drinking water. * Significantly different from control spontaneously hypertensive rats ($p < 0.05$).

trunk blood was collected and allowed to clot. The blood was then centrifuged at 2500 rpm for 10 min and the serum was collected. Serum taurine levels were measured using HPLC with electrochemical detection in both treated and control animals. Tissues were also collected and frozen on dry ice. The tissues collected were: atrium, ventricles, kidney,

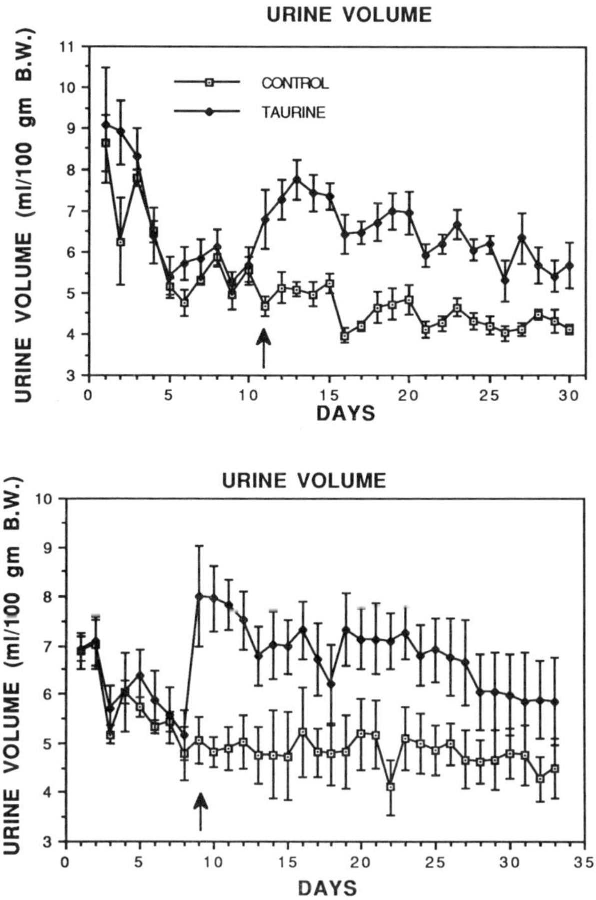

Figure 2. Effects of taurine treatment on urine volume. Top panel: Responses to 1% taurine in the drinking water; Bottom panel 3% taurine. The taurine was started at the arrow in each panel. Values are expressed as mls excreted per 100 g of body weight. Urine volume increased on the first day of treatment.

caudal artery, leg muscle, cortex, hippocampus and hypothalamus. The tissues were homogenized, centrifuged and taurine content determined using HPLC with electrochemical detection.

In a separate study serum and brain levels of taurine were measured in 10 week old spontaneously hypertensive rats and age matched Wistar-Kyoto animals (Taconic Farms).

Serum was collected as described above. Animals were sacrificed and cortex and hypothalamus were quickly removed and frozen on dry ice and kept frozen at -20°C until assayed by HPLC.

RESULTS

Taurine treatment did not significantly affect body weight at either of the concentrations used in this study. There were no major differences between the 1% and 3% treatment groups, except that the 3% treatment group showed an increased variability in the individual measurements. The animals appeared healthy and showed no adverse signs from the 4 week treatment.

The effects of taurine treatment on blood pressure are shown in Figure 1. The control spontaneously hypertensive animals demonstrated the normal increase in blood pressure over the time course of the study, increasing from approximately 130 mm Hg up to 190 mm Hg. Both concentrations of taurine blunted the increase in blood pressure compared to controls. A significant difference in blood pressure was seen after the first seven days of treatment. The blood pressure increased over the 4 week period, but not to the same extent as seen in the control animals. The effect was significant over the duration of the treatment.

Food intake and body weight were affected by neither concentration of taurine over the 4 week treatment period (data not shown). Urine volume, however, significantly increased on the first day of taurine treatment (Figure 2) and remained elevated throughout the treatment period. Fluid intake also increased significantly in both treatment groups. However the increase in fluid intake was not seen until the second day of treatment (Figure 3). Fluid intake also remained elevated for the duration of the treatment.

Urinary Na^+ excretion showed no difference between treated and control animals when expressed as mEq excreted per 24 h (data not shown). Urinary K^+ excretion showed a significant elevation starting with the first day of taurine treatment (Figure 4). The difference in effects between Na^+ and K^+ suggests the differences were not due merely to increases in urinary volume, but to differential changes in the handling of the two ions. The effects of taurine treatment on urinary Ca^{2+} excretion were interesting (Figure 5). In the 1% treatment group, the taurine was started during the second week of study. However the increase in urinary Ca^{2+} excretion showed a one week delay. During the first week of taurine treatment the Ca^{2+} excretion was equal to the control animals. Once elevated, the Ca^{2+} excretion remained significantly increased for the remainder of the study. In the 3% animals, the taurine was also started during the second week but the increased excretion only reached significance during the last two weeks of treatment. There was, however, a nonsignificant elevation in week five.

Serum and tissue samples were taken from the 3% taurine and their control group for the determination of taurine content. The control serum taurine content was 364 $\pm$ 34 nmol/ml while the taurine treated group had levels of 1545 $\pm$ 728 nmol/ml. A group of animals given 3% taurine for 4 weeks and then tap water for 7 days had serum taurine levels of 374 $\pm$ 28 nmol/ml. Tissue taurine levels measured in cortex, hippocampus and hypothalamus from treated animals were not different from the control values. However, in kidney and skeletal muscle the taurine treated animals had significantly higher levels than control animals (kidney control 0.71 $\pm$ 0.03, kidney treated 2.29 $\pm$ 0.55 nmol/mg protein; skeletal muscle control 0.114 $\pm$ 0.03, skeletal muscle treated 0.283 $\pm$ 0.08 nmol/mg protein $p < 0.05$).

Serum and brain levels of taurine were also measured in a separate group of 10 week old control spontaneously hypertensive rats and compared with age-matched Wistar-Kyoto animals. The levels of taurine are shown in Table 1.

DISCUSSION

Taurine treatment produced a small but significant decrease in blood pressure in spontaneously hypertensive rats compared to untreated spontaneously hypertensive rats after one week of treatment. This decrease was maintained over the 4 week treatment period. The

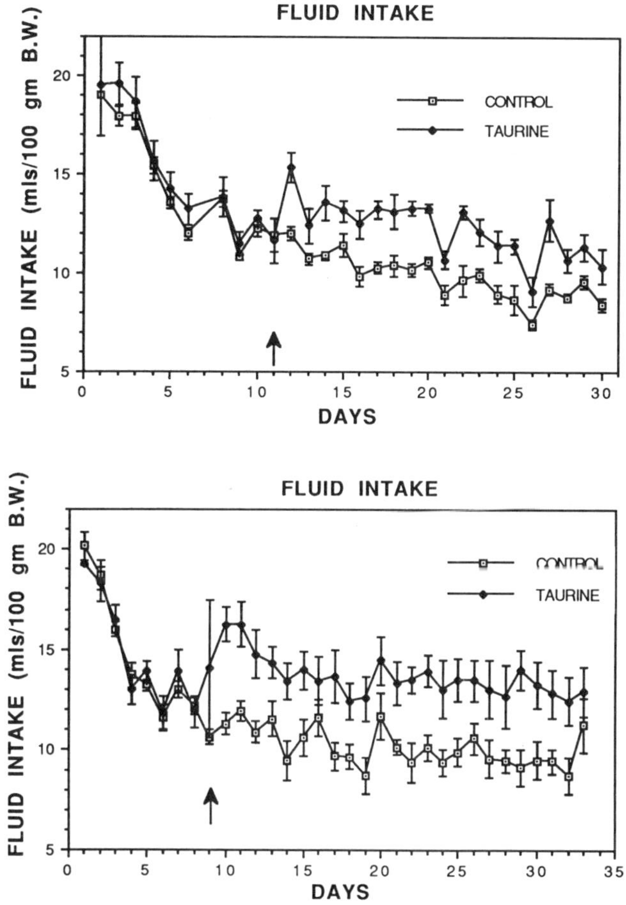

Figure 3. Effects of taurine treatment on fluid intake. Top panel: 1% taurine in drinking water; Bottom panel 3% taurine. Taurine was started at the arrow. Increases in fluid intake are seen on the second day of treatment.

decrease averaged 10 -20 mm/Hg. Taurine has previously been reported to decrease blood pressure slightly in spontaneously hypertensive rats, moderately in stroke-prone spontaneously hypertensive rats, significantly in DOCA-salt hypertensive rats, but had no effect on Wistar-Kyoto rats (1,3,12). The decreases in the present study were of the same magnitude as reported previously.

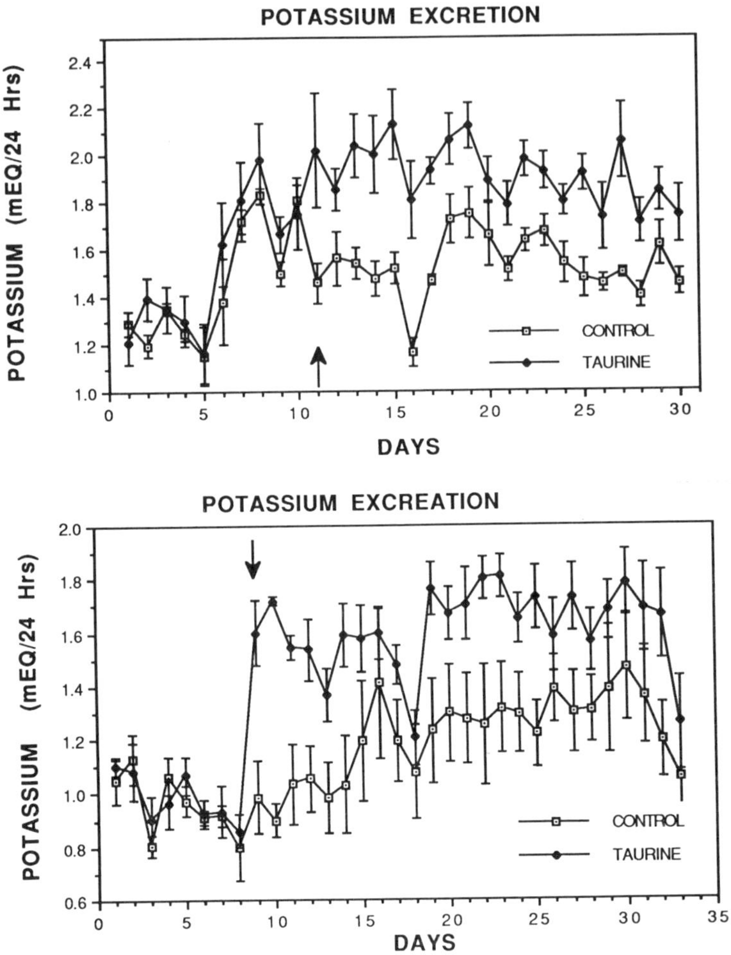

Figure 4. Effects of taurine treatment on urinary excretion of K^+. Top panel 1% taurine; Bottom panel 3% taurine. Taurine was started at arrow. Potassium increased on the first day of treatment and remained elevated. Values are expressed as mEq excreted per 24 h.

In addition to blood pressure measurements, metabolic measurements were done to elucidate further the mechanism of action of taurine. Taurine treatment did not affect the animals observable behavior and did not affect food intake or body weight gain at either of the concentrations used. The treatment did however increase urinary volume in both treatment groups during the first 24 h. The increase averaged 2-3 ml/100 g body weight/day and remained elevated throughout the treatment period. Fluid intake also increased but not until the second day of treatment. The increase in intake was about the same as the increase in urinary volume and remained throughout the treatment period. Urinary Na^+ excretion (mEq/24 h) did not change with the treatment. Urinary K^+, however, was significantly

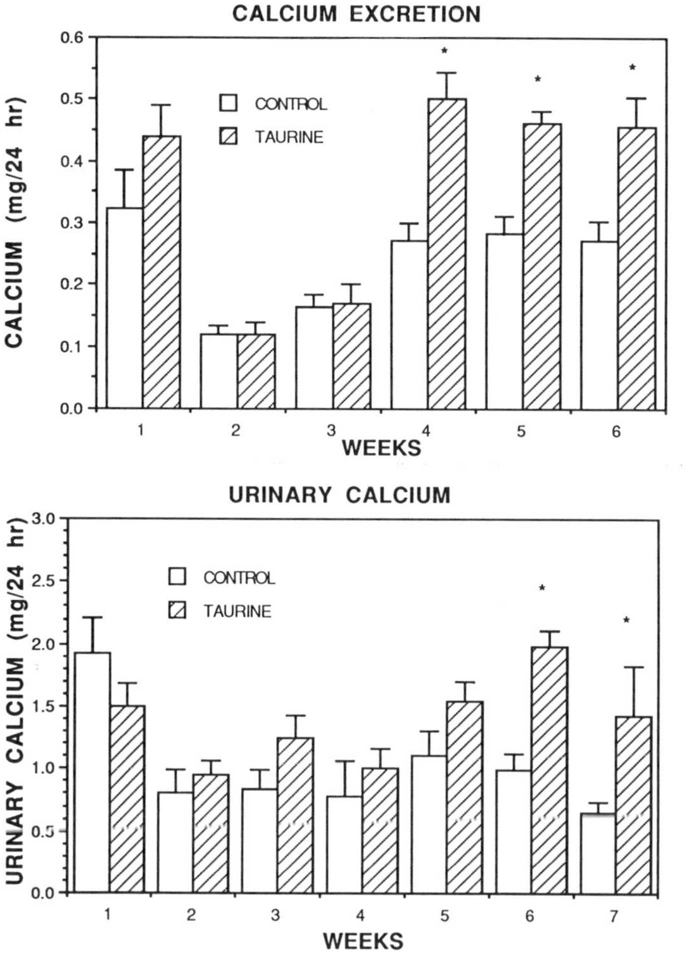

Figure 5. Effects of taurine treatment on the excretion of urinary Ca^{2+}. Top panel; 1% taurine; Bottom panel 3% taurine. Values are the means of two determinations per week of urinary Ca^{2+} excretion. Samples were taken twice weekly and the values averaged to give the weekly mean. The weekly means for each animal were then averaged for the treatment groups. * ($p < .05$).

elevated starting with the first day of treatment. Urinary Ca^{2+} excretion was not measured daily but there was a significant increase in the weekly urinary Ca^{2+} output on 1% taurine starting with a one week delay after the initiation of treatment. The 3% group showed a smaller increase that was not evident until the second week of treatment. The reason for this is not completely understood. However, the 3% treatment produced higher levels of variation in all measurements versus the 1% treatment. This may suggest there is a maximally effective concentration above which the effects are diminished. The taurine-treated spontaneously hypertensive rats showed significant elevations in serum taurine.

Table 1. Taurine levels in serum and brain from control spontaneously hypertensive and Wistar-Kyoto rats.

	Amount (nmol/ml or nmol/g)
Serum	
Spontaneously hypertensive	387 ± 19*
Wistar-Kyoto	317 ± 19
Cortex	
Spontaneously hypertensive	4.30 ± 0.05*
Wistar-Kyoto	4.67 ± 0.09
Hypothalamus	
Spontaneously hypertensive	1.97 ± 0.05
Wistar-Kyoto	1.78 ± 0.26

*Significantly different from Wistar-Kyoto control ($p < .05$)

Increases were also seen in kidney and skeletal muscle taurine content during treatment. The elevations in taurine were not just across the board as there were no increases in brain tissues. The elevations in kidney taurine and the changes in the urinary excretion of ions were interesting. These data suggest that the increase in K^+ and urinary volume may be associated with a diuretic effect of taurine associated with the increased intrarenal levels The lack of effects on Na^+ excretion suggest it is not just a nonspecific diminished reabsorption in the kidneys but may involve specific actions on the kidney. The lack of effect on Na^+ was in contrast to that reported by Kohashi et al. (9) who reported a slight increase in urinary Na^+ excretion in rats given 3% taurine to drink during the second week of treatment. This increase was maintained for four weeks of treatment. The reasons for this discrepancy are not clear. The increased serum levels and decreased cortical levels of taurine in control spontaneously hypertensive rats compared to control WKY animals were also interesting. It is at present not clear if these differences play a role in hypertension in the spontaneously hypertensive rat. Further work will be required to determine the exact mechanism of action of taurine.

ACKNOWLEDGEMENTS

These studies were partially supported by a grant from the American Heart Association - Florida Affiliate. We thank Linda Panchou for assistance in preparing this manuscript.

REFERENCES

1. Abe, M., Shibata, K., Matsuda, T., and Furukawa, T. 1988, Inhibition of hypertension and salt intake by oral taurine treatment in hypertensive rats, *Hypertension*, 10:383-389.

2. Barbeau, A., Inoue, A., Tsukada, Y., and Butterworth, R.F. 1975, The neuropharmacology of taurine, *Life Sci.*, 17:669-677.
3. Fujita, T. and Sato, Y. 1984, The antihypertensive effect of taurine in DOCA-salt rats, *J.Hypertension*, 2 (Suppl. 3):563-565.
4. Fujita, T. and Sato, Y. 1986, Changes in blood pressure and extracellular fluid with taurine in DOCA-salt rats, *Am.J.Physiol.*, 19:R1014-R1020.
5. Fujita, T. and Sato, Y. 1988, Hypotensive effect of taurine. Possible involvement of the sympathetic nervous system and endogenous opiates, *Journal of Clinical Investigation*, 82:993-997.
6. Hanretta, A.T. and Lombardini, J.B. 1987, Is taurine a hypothalamic neurotransmitter?: a model of the differential uptake and compartmentalization of taurine by neuronal and glial cell particles from the rat hypothalamus, *Brain Res.Rev.*, 12:167-201.
7. Huxtable, R.J. 1989, Taurine in the central nervous system and the mammalian actions of taurine, *Prog.Neurobiol.*, 32:471-533.
8. Huxtable, R.J. and Sebring, L.A. 1986, Towards a unifying theory for the action of taurine, *Trends Pharmacol.Sci.*, 7:481-485.
9. Kohashi, N., Okabayashi, T., Horiuchi, M., Nishiyama, H., Takenaka, T., and Katori, R. 1990, Mechanism of taurine natriuresis in rats, *Adv.Exp.Med.Biol.*, 247A:635-640.
10. Kuriyama, K. 1980, Taurine as a neuromodulator, *Federation Proceedings*, 39:2680-2684.
11. Nakagawa, M., Inoue, A., Toyoda, T., Sawada, S., Uno, M., and Takeda, K. 1990, Central antihypertensive effect of taurine, in: "Taurine: Functional Neurochemistry, Physiology, and Cardiology", Pasantes-Morales, H., Martin, D.L., Shain, W. and del Río, R.M., eds., Wiley-Liss, New York, pp. 463-470.
12. Nara, Y., Yamori, Y., and Lovenberg, W. 1978, Effects of dietary taurine on blood pressure in spontaneously hypertensive rats, *Biochem.Pharmacol.*, 27:2689-2692.
13. Paakkari, P., Paakkari, I., Karppanen, H., and Paasonen, M.K. 1983, Mechanisms of the inhibitory cardiovascular effects of taurine and homotaurine, *Acta Med.Scand.*, S677:134-137.
14. Saransaari, P. and Oja, S.S. 1992, Release of GABA and taurine from brain slices, *Prog.Neurobiol.*, 38:455-482.
15. Schaffer, S.W. and Azuma, J. 1992, Review: Myocardial physiological effects of taurine and their significance, *Adv.Exp.Med.Biol.*, 315:105-120.
16. Takemoto, Y. 1991, Central depressor effects of amino acids in conscious normotensive and one-kidney, one-clip renovascular hypertensive rats, *Japan.J.Physiol.*, 41:717-724.

DUAL EFFECTS OF TAURINE ON MEMBRANE IONIC CONDUCTANCES OF RAT SKELETAL MUSCLE FIBERS

Sabata Pierno[1], Annamaria De Luca[1],
Ryan J. Huxtable[2] and Diana Conte Camerino[1]

[1]Unit di Farmacologia, Dipartimento Farmacobiologico
Facoltà di Farmacia, Università di Bari, Italy
[2]Department of Pharmacology, College of Medicine
University of Arizona, Tucson, USA

INTRODUCTION

Taurine (2-aminoethane sulfonic acid) levels are high in excitable tissues of mammals, including skeletal muscle. However, its physiological role at the molecular level remain unclear. Results from our laboratory indicate that taurine is a membrane stabilizer. Taurine, applied in vitro, reduces membrane excitability of rat skeletal muscle by specifically increasing membrane chloride conductance (G_{Cl}) (3). In mammalian skeletal muscle, the electrical stability of sarcolemma is controlled by a large G_{Cl}. An abnormally low G_{Cl} is responsible for the hyperexcitability characteristic of myotonic muscles (1,21). It has been proposed that many pharmacological actions of taurine are due to interaction with a low affinity binding site on membrane phospholipids (14). In rat skeletal muscle, this amino acid might modify G_{Cl} by a direct action on the low affinity site, as suggested by the findings that taurine restores G_{Cl} lowered by 20,25-diazacholesterol, which indirectly affects chloride channels by altering the lipid environment of the sarcolemma (5). In order to evaluate better the interaction between taurine and its low affinity binding site, the effects of a series of newly synthetized analogues of taurine (Figure 1) have been tested in vitro, on rat skeletal muscle and compared with those of taurine and other well known analogues such as 3-aminopropane sulfonic acid (homotaurine) (10) and guanidinoethane sulfonate (14).

Also, taurine may exert a long-term regulatory role on ionic conductances of rat skeletal muscle, since our previous results showed that taurine depletion, achieved by in vivo chronic administration of guanidinoethane sulfonate, a specific inhibitor of taurine transport in several tissues (12), produces a decrease of G_{Cl} and an increase of potassium conductance (G_K) (7). In this regard it is known that taurine, by increasing Ca^{2+} binding to neutral phospholipids and promoting Ca^{2+} uptake by sarcoplasmic reticulum (13), is able to regulate the amount of intracellular calcium. Recently it has been proposed that taurine may inhibit specific protein phosphorylation in rat brain by reducing the optimal Ca^{2+} levels (19). This may have important consequences in skeletal muscle, since the activation of Ca^{2+} /phospholipid-dependent protein kinase C reduces G_{Cl} (2,22). Thus the present study was

A-D

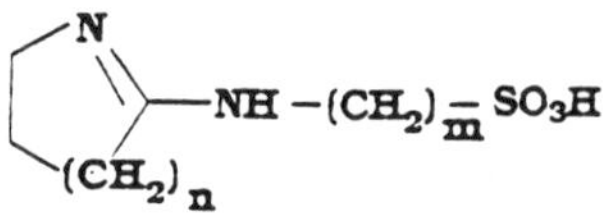

	n	m
A	3	2
B	1	2
C	3	3
D	1	3

Figure 1. Structures of taurine analogues synthetized and tested on rat extensor digitorum longus muscle.

also aimed at investigating in more details the possible long-term regulatory role, via Ca^{2+}, exerted in vivo by taurine, on chloride and potassium conductance of rat skeletal muscle. For this purpose, we evaluated the effects of an in vivo taurine loading in comparison with those observed in taurine-depleted muscles. In addition, the increase of intracellular Ca^{2+} levels by the application of Ca^{2+} ionophore, A23187, was also evaluated.

METHODS

Adult male Wistar rats were used for all the experiments. The in vivo treatment was performed on rats receiving daily 1 g/kg taurine i.p. (O-Due, Nativelle, Italy) for 23 days. The electrophysiological experiments were done in vitro on extensor digitorum longus muscles removed under urethane anaesthesia from untreated and taurine-treated rats and placed in a 20 ml muscle bath maintained at 30°C and perfused with physiological solutions (7).

Taurine and homotaurine were obtained respectively from Merck (Darmstadt, Germany) and Aldrich (USA). The taurine analogues (Fig. 1) were synthetized in the laboratories of Professor G. Casini (University of Bari, Italy) by reacting the corresponding aminosulfonic acids with lactim ethers (O-methylcaprolactim and O-methylbutyrolactim) in equimolar ratio: details of their preparation and subsequent cyclization to sultams will be reported elsewhere. Guanidinoethane sulfonate was synthesized as previously described (12). A23187 was obtained from Sigma (St. Louis, USA).

The electrophysiological intracellular recordings were made before and after the addition to muscle preparations of different concentrations (450 μM-30 mM) of all the examined compounds dissolved in chloride-containing and chloride-free solutions. The normal physiological solution had the following composition (mM): NaCl 148; KCl 4.5; $CaCl_2$ 2.0; $MgCl_2$ 1.0; $NaHCO_3$ 12.0; NaH_2PO_4 0.44 and glucose 5.55. The chloride-free solution was prepared by equimolar substitution of methylsulfate salts for NaCl and KCl and nitrate salts for $CaCl_2$ and $MgCl_2$. All the solutions were continuously gassed with 95% O2 and 5% CO_2 and the pH was carefully maintained between 7.2-7.3 during each experiment, since it has been demonstrated that the effects of taurine are evident in this pH range, but disappear at pH values higher than 7.5 (3). To suppress spontaneous contractions of muscle preparations 1 μM tetrodotoxin (TTX, Sigma, St. Louis, USA) was added to chloride-free solutions.

A pair of microelectrodes, one for recording potentials and the other for passing current, was used for making intracellular measurements of membrane potentials, cable properties and membrane excitability of single muscle fibers (7). The current pulse generation, the acquisition of the voltage records and the calculation of the fiber constants were done in real time under computer control as described in detail elsewhere (2). The electrotonic potentials elicited by hyperpolarizing square wave constant current pulses in chloride-containing and in chloride-free solutions, before and after the addition of the tested compounds, were used for the estimates of Rm by standard cable analysis (7). The total membrane conductance, Gm, was 1/Rm in the normal physiological solution. The potassium conductance, G_K, was 1/Rm in chloride-free solution. The mean chloride conductance, G_{Cl}, was calculated as the mean Gm minus the mean G_K.

The excitability characteristics of the sampled fibers were determined by recording the intracellular membrane potentials response to square-wave constant current pulses (100 ms duration). In each fiber the membrane potential was set by a steady holding current to -80 mV, before passing the depolarizing pulses (4,7).

The data are expressed as mean $\pm$ standard error (S.E.). The estimates for S.E. and number of fibers (N) of G_{Cl} were obtained from the variance of Gm and G_K, assuming no covariance and the number of fibers sampled for Gm and G_K, respectively (9). The estimates of S.E. and N of normalized G_{Cl} values have been obtained as described by Green and Margerison (11). Statistically significance of difference between mean values has been calculated by unpaired Student's t-test.

RESULTS

In Vitro Application of Taurine and Analogues on Ionic Conductances and Excitability Parameters of Rat Skeletal Muscle Fibers

In agreement with previous studies (3), the in vitro application of taurine (450 μM-30 mM) to extensor digitorum longus muscles caused a significant and concentration-dependent increase of membrane G_{Cl} (Table 1). A nonsignificant increase of G_K was observed after application of 18 mM and 30 mM taurine. Moreover, as previously described (3), muscle fibers exposed to 30 mM taurine showed a decrease in membrane excitability and, in particular, an increase of the threshold current necessary to elicit the first action potential (I_{th}), a decrease of the delay between onset of the current pulse and initiation of action potential (latency) and a decrease of the firing capability.

All the analogues tested produced a concentration-dependent increase of G_{Cl}, although with a lower potency with respect to taurine.

Homotaurine, applied in vitro to extensor digitorum longus muscles in the concentration range of 450 μM - 30 mM caused an increase of G_{Cl} that was significant only at 18 mM and 30 mM (Table 1), whereas G_K was unchanged. Also, the excitability parameters were less affected than they were by taurine. Guanidinoethane sulfonate (450 μM-30 mM), showed much lower activity in increasing G_{Cl} in comparison to both taurine and homotaurine. Indeed, this increase was significant only at a concentration of 30 mM (Table 1). Accordingly, a slight effect was observed on G_K and membrane excitability.

The newly synthesized taurine analogues A and B tested in vitro on extensor digitorum longus muscles (450 μM-30 mM) produced a significant increase of G_{Cl} only at the concentration of 18 mM and 30 mM (Table 2). The analogues C and D (4.5 mM-30 mM) were even less effective in increasing G_{Cl} (Table 2). None of these compounds affected G_K, but all produced a significant and reversible membrane depolarization that altered the excitability parameters; in particular decreasing the firing capability.

Table 1. Increase of resting chloride conductance (G_{Cl}) produced in rat skeletal muscle fibers by in vitro application of taurine, homotaurine and guanidinoethane sulfonate guanidinoethane sulfonate

Experimental conditions	Dose (mM)	N	G_{Cl} ($\mu S/cm^2$)
Control		64	2667 ± 95
Taurine	0.45	34	3085 ± 142*
	4.5	26	3340 ± 140*
	18	17	3678 ± 245*
	30	31	3705 ± 257*
Control		18	2633 ± 182
Homotaurine	0.45	13	2983 ± 307
	4.5	14	3094 ± 233
	18	12	3422 ± 200*
	30	16	3552 ± 311*
Control		10	2924 ± 207
Guanidinoethane sulfonate	0.45	7	3030 ± 387
	4.5	10	3277 ± 177
	18	13	3481 ± 240
	30	13	3552 ± 201*

The columns from left to right are as follows: Experimental conditions in the absence (control) and in the presence of the examined compounds; doses tested; N: number of fibers; G_{Cl}: component chloride conductance. Each value is expressed as mean ± S.E. *Significantly different from control values ($p < 0.05$).

Table 2. Normalized percent increase of G_{Cl} produced by in vitro application of taurine and its analogues on rat extensor digitorum longus muscle fibers

Compound tested	18 mM	30 mM
Taurine	37.9 ± 7*	38.9 ± 8*
Homotaurine	29.9 ± 8*	34.9 ± 10*
Compound A	29.6 ± 6*	31.6 ± 6*
Compound B	27.0 ± 6*	32.0 ± 9*
Guanidinoethane sulfonate	19.0 ± 8	21.5 ± 8*
Compound C	18.0 ± 4*	17.0 ± 5*
Compound D	11.0 ± 5	22.0 ± 8*

The columns from left to right show the compounds and the concentrations tested. Values represent normalized percent increase ± S.E. (11), calculated from mean values of G_{Cl} (7-48 fibers) with respect to mean control values (10-64 fibers). Test of significance was performed by comparing mean absolute values of G_{Cl} in the absence (control) and in the presence of the tested compounds. Structures of compounds A-D are shown on Fig. 1. *Significantly different from control ($p < 0.05$).

In Vivo Taurine Loading and Depletion on Ionic Conductances and Excitability Parameters of Rat Skeletal Muscle Fibers

Previously, in rat skeletal muscle taurine-depleted by chronic treatment with guanidinoethane sulfonate, we have found a 32% decrease of G_{Cl} and a significant increase of G_K (Figure 2). The membrane excitability parameters changed accordingly to the alteration of resting ionic conductances (Table 3). In agreement with these findings, after 23 days of taurine treatment, we observed a 20% increase of G_{Cl} with respect to control value (Figure 2). In particular G_{Cl} increased significantly from a mean control value of 2917 $\pm$ 37 $\mu S/cm^2$ (n=15) to 3517 $\pm$ 223 $\mu S/cm^2$ (n=15) in taurine-treated muscles. However G_K did not change significantly with respect to control value (Figure 2).

As expected by the observed increase of G_{Cl}, we recorded a decrease of membrane excitability parameters in taurine-loaded rats. The I_{th} was increased and the ratio between I_{th} and the current necessary to elicit the second action potential (I_{th}/I_{2nd}) was decreased by taurine treatment. Accordingly, the firing capability of taurine-treated fibers showed a tendency to decrease. The latency as well as the duration of the action potential (Dur) did not change in taurine treated-fibers (Table 3).

To test whether the long-term in vivo effect of taurine may be via Ca^{2+} availability, we compared the effects observed in both taurine-depleted and taurine-loaded muscle with those produced by in vitro application of the calcium ionophore, A23187. The increase of the cytosolic calcium levels in extensor digitorum longus muscle fibers produced by A23187 caused a significant reduction of G_{Cl} with respect to the control value. This decrease amounted to 28% and 43% at concentrations of 1 μM and 3 μM, respectively (Figure 2). A nonsignificant increase of G_K, of 24% and 27%, respectively, was observed at the same concentrations (Figure 2).

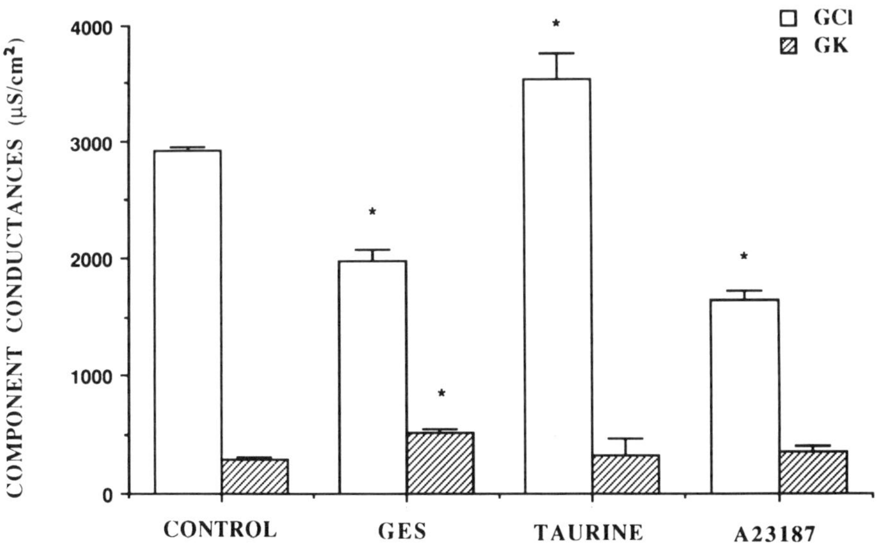

Figure 2. Effects of taurine depletion (by guanidinoethane sulfonate administration in drinking water at the dose of 1% for 4 weeks) and taurine loading (by i.p. daily injection at the dose of 1 g/kg for 23 days) in comparison with the effects of in vitro application of calcium ionophore, A23187, (3 μM) on component ionic conductances of rat skeletal muscle fibers. Each bar represents the mean value $\pm$ S.E. of 15-38 fibers. *Significantly different with respect to the control value ($p < 0.05$).

Table 3. Effects of taurine depletion by chronic treatment with guanidinoethane sulfonate and taurine loading with 1 g/kg taurine on the excitability parameters of rat skeletal muscle fibers

Treatment	N	RP (-mV)	AP (mV)	I_{th} (nA)	I_{th}/I2nd	Lat (ms)	Dur (ms)	N spikes
Control	7	67±2	102±7	129±6	5.3±0.1	9.4±0.4	9.4±0.4	3.4±0.5
GES	7	70±2	98±2	131±11	0.84±0.03	7.7±0.5*	13.2±0.7*	5.2±0.8
Taurine	6	67±3	118±4	138±14	0.66±0.003	5.0±0.6	9.0±0.7	2.0±1.1

The columns from left to right are as follows: guanidinoethane sulfonate (GES) was administered in drinking water at a dose of 1% for 4 weeks; taurine was administered by i.p.injection at the dose of 1 g/kg for 23 days; N: number of fibers; RP: resting membrane potential; AP: amplitude of the action potential; I_{th}: threshold current; I_{th}/I2nd: ratio between threshold current and the current necessary to elicit the second action potential; Lat (latency): time interval between the onset of the current pulse and the beginning of an action potential; Dur: duration of the action potential; N spikes: maximum number of action potential elicited. Each value is expressed as mean ± S.E. *Significantly different with respect to control value ($p < 0.01$).

DISCUSSION

Our results suggest that the increase of muscle G_{Cl} observed after in vitro application of taurine is mediated by the interaction of the amino acid with a low affinity site present on skeletal muscle membrane and already described on membrane phospholipids of various tissues (14). In fact, slight modification of the chemical structure of taurine, as with known taurine analogues or our newly synthesized compounds, led to a marked decrease of the taurine-like activity on G_{Cl} that could likely be due to a weaker interaction with a specific site near chloride channels. The less potent activity in increasing G_{Cl} shown by homotaurine, which differs from taurine by having an additional methylene group, suggests that the distance between charged groups plays an important role in maintaining the taurine-like activity. It has been reported that guanidinoethane sulfonate displaces taurine from the low affinity site on sarcolemma by only 7% (14). This compound at physiological pH range is mainly present as a zwitterion having a positive charge delocalized on the guanidinium head, which may cause a further reduction in the ability to increase the G_{Cl} of rat extensor digitorum longus muscle. Thus, charge density might be a even more important factor than the distance between charged groups in determining taurine-like binding on the low affinity site. On the other hand, the taurine analogues A and B (Fig. 1), in which the positive charge is delocalized on a lesser number of atoms than in guanidinoethane sulfonate, show only a slightly reduced taurine-like activity. Moreover, the much lower activity showed by compounds C and D with respect to compounds A and B and homotaurine could be related to either lower charge density or increased distance between the two functional groups.

However, our findings show that taurine, apart from the increase of muscle chloride conductance produced after in vitro application, by a direct action on chloride channel also exerts an indirect long-term modulation of the same parameter. Indeed, a 50% reduction of taurine content by chronic treatment with guanidinoethane sulfonate produces a significant 32% reduction of membrane G_{Cl} and a significant 80% increase of G_K (7). In the present study we found that chronic treatment with taurine causes an increase of G_{Cl} (20%) in adult extensor digitorum longus muscle fibers, whereas G_K was almost unaffected. The observed effects on G_{Cl} and G_K, in consequence of a depletion of taurine content as well as of taurine loading, might be due to a Ca^{2+} mediated mechanism, rather than to an interaction with the low affinity site responsible of the acute action of this amino acid on skeletal muscle.

It has been proposed that taurine, by modulating Ca^{2+} binding to membrane phospholipids, decreases the concentration of cytosolic Ca^{2+} (15). The decrease of G_{Cl} observed in taurine-depleted muscles might be due to the increased intracellular Ca^{2+} concentration, since we have demonstrated that application of Ca^{2+} ionophore, A23187, produces similar effects. In support of this hypothesis the mechanical threshold of taurine-depleted muscle fibers is significantly changed. Indeed, the guanidinoethane sulfonate-treated fibers contract at more negative potentials than normal controls (7), similarly to what is observed in situations characterized by an increase of cytosolic Ca^{2+} either during ageing (16,6) or in vitro application of the ionophore A23187 (20). The mechanism through which Ca^{2+} ions controls ionic conductances remains to be clarified. It is known that activation of Ca^{2+}-dependent protein kinase C decreases G_{Cl} (2,22). Furthermore, it is tempting to speculate that the increase of G_{Cl} observed in taurine-treated fibers might be due to decreased intracellular calcium levels and therefore to a modulation of protein kinase C activity. In this regard, Lombardini (19), reports that taurine, by reducing the optimal Ca^{2+} levels, might inhibit phosphorylating processes in rat brain catalyzed by protein kinase C. This mechanism, however, remain to be better investigated in skeletal muscle.

Interestingly, G_K appeared to be sensitive to long-term depletion of intracellular muscle taurine content in contrast with the nonsignificant effect observed on this parameter by acute in vitro application of the amino acid (3). Again the long-term action of taurine on G_K might occur via Ca^{2+}. Indeed, the taurine-related change in Ca^{2+} homeostasis (18) would in turn modulate the open/shut probability of Ca^{2+}-dependent potassium channels (17).

Thus, appropriate levels of intracellular taurine may play a crucial role in muscle performance. It is important to note that taurine-depleted muscles are much more sensitive to acute in vitro application of the amino acid (7), suggesting that intracellular taurine levels may also modulate the actions of taurine on the low affinity binding site and supporting the use of this latter pharmacological effect in pathological situations, such as some dystrophic syndromes (8), in which alteration of taurine levels may occur.

ACKNOWLEDGEMENTS

The financial support of the Italian CNR 93-227, 93-263 and of Telethon, Italy for the project, "Abnormal ion channel function of skeletal muscle membrane in genetically occurring diseases: electrophysiology and pharmacology" are gratefully acknowledged.

REFERENCES

1. Adrian, R.H., and Bryant, S.H., 1974, On the repetitive discharge in myotonic muscle fibres, *J. Physiol.* 240:505-510.
2. Bryant, S.H., and Conte Camerino, D., 1991, Chloride channel regulation in the skeletal muscle of normal and myotonic goats, *Pflügers Arch.* 417:605-610.
3. Conte Camerino, D., Franconi, F., Mambrini, M., Bennardini, F., Failli, P., Bryant, S.H., and Giotti, A., 1987, The action of taurine on chloride conductance and excitability characteristics of rat striated muscle fibers, Pharmacol. *Res. Commun.* 19:685-701.
4. Conte Camerino, D., Mambrini, M., De Luca, A., Tricarico, D., Bryant, S.H., Tortorella, V., and Bettoni, G., 1988, Enantiomers of clofibric acid analogs have opposite actions on rat skeletal muscle chloride channels, *Pflügers Arch.* 413:105-107.
5. Conte Camerino, D., De Luca, A., Mambrini, M., Ferrannini, E., Franconi, F., Giotti, A., and Bryant, S.H., 1989, The effects of taurine on pharmacologically induced myotonia, *Muscle and Nerve* 12:898-904.
6. De Luca, A., and Conte Camerino, D., 1992, Effect of aging on the mechanical threshold of rat skeletal muscle fibers, *Pflügers Arch.* 420:407-409.

7. De Luca, A., Pierno, S., Huxtable, R.J., Failli, P., Franconi, F., Giotti, A., and Conte Camerino, D., 1992, Effects of taurine depletion on membrane electrical properties of rat skeletal muscle, *in:* "Taurine Nutritional Value and Mechanisms of Action", Lombardini, J.B. Schaffer, S.W., and Azuma, J., eds., Plenum Press, New York, pp. 199-205.
8. Durelli, L., Mutani, R., and Fassio, F., 1983, The treatment of myotonia: evaluation of chronic oral taurine therapy, *Neurol.* 33:559-603.
9. Eisenberg, R.S., and Gage, P.W., 1969, Ionic conductance of the surface and trasverse tubular membranes of frog sartorius fibers, *J. Gen. Physiol.* 53:279-297.
10. Franconi, F., Martini, F., Manghi, N., Galli, A., Bennardini, F., and Giotti, A., 1981, Uptake of [3H] taurine into myocardial membranes, *Biochem. Pharmacol.* 30:77-80.
11. Green, J.R., and Margerison, D., 1978, "Statistical treatment of experimental data", New York, Elsevier pp 86-88.
12. Huxtable, R.J., Laird, H.E., and Lippincott, S.E., 1979, The transport of taurine in the heart and the rapid depletion of tissue taurine content by guanidinoethyl sulfonate, *J. Pharmacol. Exp. Ther.* 211:465-471.
13. Huxtable, R.J., and Sebring, L.A., 1986, Towards a unifying theory for the action of taurine, *TIPS* 7:481-485.
14. Huxtable, R.J., 1992, The physiological actions of taurine, *Physiol. Rev.* 72:101-163.
15. Kuriyama, K., 1980, Taurine as a neuromodulator, *Federation Proc.* 39:2680-2684.
16. Larsson, L., and Salviati, G., 1989, Effects of age on calcium transport activity of sarcoplasmic reticulum in fast- and slow-twitch rat muscle fibres, *J. Physiol.* 419:253-264.
17. Latorre, R., Oberhanser, A., Labarca, D., and Alvarez, O., 1989, Varieties of calcium-activated potassium channels, *Ann. Rev. Physiol.* 51:385-399.
18. Lleu, P-L., Huxtable, R.J., 1992, Phospholipid methylation and taurine content of synaptosomes from cerebral cortex of developing rat, *Neurochem. Int.* 21 1:109-118.
19. Lombardini, J.B., 1992, Effects of taurine on protein phosphorylation in mammalian tissues, *in:* "Taurine Nutritional Value and Mechanisms of Action", Lombardini, J.B. Schaffer, S.W. Azuma, J., eds., Plenum Press, New York, pp. 309-318.
20. Morgan, K.G., and Bryant, S.H., 1977, The mechanism of action of dantrolene sodium, *J. Pharmacol. Exp. Ther.* 201:138-147.
21. Rüdel, R., and Lehmann-Horn, F., 1985, Membrane change in cells from myotonia patients, *Physiol. Rev.* 65:310-356.
22. Tricarico, D., Conte Camerino, D., Govoni, S., and Bryant, S.H., 1991, Modulation of rat skeletal muscle chloride channels by activators and inhibitors of protein kinase C, *Pflügers Arch.* 418:500-503.

NEW DATA ON THE REGULATION OF TAURINE UPTAKE IN CULTURED NERVOUS CELLS

Gérard Rebel[1], Valérie Petegnief[1],
Pierre-Louis Lleu[1], Ramesh C.Gupta[22],
Paulette Guérin[1] and Jacques Bourguignon[1]

[1]Centre de Neurochimie du CNRS
5, rue Blaise Pascal
67084-Strasbourg-Cédex, France
[2]North Eastern hill University
Sasid Medziphena
Nagaland 797106, India

INTRODUCTION

Uptake of taurine by different tissues or cell types has been extensively studied these last decades (8,10,20). The high affinity taurine uptake displayed by the majority of the cells studied is a Na^+, Cl^- dependent system. Various hormones, drugs, and amino acids have been shown to affect this uptake. These compounds have been very often tested in vitro on cells in culture. However, they were usually tested by addition to the incubation media used for testing the taurine uptake. Some of these incubation media contain HEPES (N,N'hydroxyethylpiperazineethane sulfonate), which shows structural similarities with taurine (2-aminoethane sulfonate). We have previously shown that HEPES inhibits the taurine uptake of rat glial cells in primary cultures. This occurs by two mechanisms (11,12). A fast inhibition (within a few minutes) is observed when a Krebs-Ringer buffer supplemented with HEPES is used for the incubation of cells during taurine uptake determinations. On the contrary, when the glial cells are grown for a few days in a culture medium buffered with HEPES, instead of the more normal Na^+ bicarbonate system, an inhibition of the taurine uptake is observed when cells are tested in the Krebs-Ringer incubation medium (11). This inhibition increases slowly with the culture length in the presence of HEPES (11). These results suggest the existence of two different mechanisms; a fast one (F) and a slow one (S), which may depend on the exposure time and conditions, and by which taurine uptake could be modulated. Different taurine analogs such as guanidinoethane sulfonate β-alanine, or hypotaurine are generally used to characterize the uptake of taurine by cultured cells by addition to the incubation media used for taurine uptake measurements. We have compared the effects obtained for these compounds on the

Taurine in Health and Disease, Edited by R. Huxtable
and D.V. Michalk, Plenum Press, New York, 1994

taurine uptake upon addition to either the culture media (S condition) or the incubation media (F condition).

MATERIAL AND METHODS

Cell Culture

Primary cultures of glial cells were established by mechanical dispersion of new born rat hemispheres (2). Cells were plated in Petri dishes (Costar, 30 mm diameter) and grown in low glucose Dulbecco's Modified Eagle's Medium (DMEM) (Gibco 74-1600) buffered with 2 g/l $NaHCO_3$ supple-mented with decreasing amounts of selected heat-inactivated fetal calf serum (20% during the 4 first days of culture; down to 5% after 10 days in culture). They were maintained in an air-5% CO_2 atmosphere at 37°C.

The C6 rat glioma cell line (1) was routinely maintained in DMEM+5% fetal calf serum, in an air-5% CO_2 atmosphere at 37°C. For the experiments, C6 cells were detached with trypsin-EDTA and the cell suspension was plated on Petri dishes (Costar) and further grown in the same medium.

Taurine Uptake

Fifteen day-old primary glial cells or confluent C6 cultures were used for taurine uptake measurement. The culture medium was withdrawn and the cells were washed three times on the plates with oxygenated Krebs-Ringer buffer, pH 7.4. Cells were preincubated 5 min with 3 ml of the same buffer at 37°C in a rotary water bath. 100 μl of 40 μM cold taurine containing 0.4 μCi [^{3}H]taurine (NEN, specific activity 25.6Ci/mmol) were added to the cells. After 10 min of incubation with the radiolabeled taurine, the cells were washed 3 times with NaCl (0.9%). Saline was withdrawn and cells on the plates were immediately frozen. Later, 1 ml/dish of NaOH 0.1N was added and the dishes were subjected to an incubation at 37°C for 1 h, followed by a homogenization of the cells through repetitive pipetting. Samples were withdrawn for radioctivity counting and for protein determination (13).

The different compounds used for this study were added either into the culture medium of the cells 48 h before the uptake measurements or to the Krebs-Ringer buffer. At the molarity used for this study, none of the compounds have an effect on the osmotic pressure of the culture medium or on the incubation buffer. This was checked using a Rœbling osmometer.

Tested Compounds

Guanidinoethane sulfonate and acetyltaurine were synthesized as previously described (4,9). 2-Aminoethane phosphonic acid was obtained from Aldrich. All other compounds were from Sigma. When required, the pH of the media supplemented with the analogs was adjusted to pH 7.4 with NaOH. We have verified that the addition of the small amount of sodium required for buffering did not affect the taurine uptake by the cells.

RESULTS

Effects of Taurine Analogs on the Taurine Uptake

Effects on Primary Glial Cells. A strong inhibition of the taurine uptake was observed when hypotaurine, β-alanine, guanidinoethane sulfonate or acetyltaurine were individually added to the Krebs-Ringer buffer used for the taurine uptake measurements. GABA or N-methyltaurine are weaker inhibitors. Cysteic acid and cysteine sulfinic acid were poor inhibitors whereas isethionate (2-hydroxyethane sulfonate) or 2-aminoethane phosphonate are without effect (Table 1). A notably different result was obtained when the same compounds were added for two days to the

Table 1. Effect of some taurine analogs on the taurine uptake by primary glial cell cultures

Analog	Taurine uptake in % of the control culture when the analogs are added to:	
	culture medium	Krebs Ringer buffer
Hypotaurine	45	2
β-Alanine	No inhibition	6
Guanidinoethane sulfonate	56	14
Acetyltaurine	42	14
GABA	No inhibition	34
N-Methyltaurine	74	36
Cysteine sulfinic acid	Not tested	73
Cysteic acid	Not tested	75
Isethionic acid	62	100
2-Aminoethane phosphonic acid	93	100

Confluent cultures were used. Compounds were tested at a concentration of 1 mM. They were added either to the cell culture medium for two days or to the Krebs Ringer medium. Taurine uptake was measured as described in "Material and Methods". Each result is the mean of four determinations. Variations between the extreme values were less than 5% of the mean value. A similar variation was also observed for all the other results shown

culture medium. The main change concerns β-alanine and GABA, which lost completely their inhibitory effect. The other compounds were less potent inhibitors when they were added in the culture medium than when present in the Krebs-Ringer buffer. Cysteic acid and cysteine sulfinic acid were tested only at 5 mM concentration. At this concentration, they inhibited the taurine uptake by 36% and 42%, respectively. Under the same conditions, N-methyltaurine (5 mM) inhibited the uptake by 48%. We can therefore expect that at 1 mM the two α-amino acids will behave as very weak inhibitors.

The effect of isethionic acid on the taurine uptake was opposite to the results obtained with the other tested compounds. Indeed, the inhibitory effect of the isethionic acid showed a notable increase when this compound was present in the culture medium for two days.

Effects on C6 Glioma Cells. As shown in Table 2, guanidinoethane sulfonate, β-alanine and GABA, when present in the incubation medium, inhibited similarly the taurine uptake in the C6 glioma cells and in primary glial cells. Addition of either guanidinoethane sulfonate, or GABA or β-alanine to the culture medium results in a decrease of the inhibitory efficiency of guanidinoethane sulfonate and complete loss of the effects of β-alanine and GABA on the uptake.

Addition of β-alanine to the culture medium of glial or glioma cells (for two days) induces the loss of its inhibitory effect on taurine uptake. The same results were obtained with β-alanine when it was present in the culture medium for 4 days (Fig. 1).

Table 2. Effect of guanidinoethane sulfonate, β-alanine and GABA on taurine uptake by C6 glioma cells

Analog	Taurine uptake (pmoles/min/mg protein) when the analogs were added to:	
	culture medium	Krebs Ringer buffer
None	13.24 ± 0.65 (100)	21.86 ± 0.40 (100)
Guanidinoethane sulfonate	7.66 ± 0.40 (58)	1.49 ± 0.05 (7)
β-Alanine	12.93 ± 0.53 (98)	1.25 ± 0.03 (6)
GABA	10.68 ± 0.37 (81)	6.40 ± 0.27 (29)

Confluent cultures were used. Compounds were tested at a concentration of 1 mM. They were added either to the cell culture medium for two days or to the Krebs Ringer medium. Taurine uptake was measured as described in "Material and Methods". Each result is the mean of four determinations. Values in parentheses indicated the taurine uptake as a percent of the control cells.

However, a strong inhibition was immediately observed when β-alanine was added to the culture medium and the Krebs Ringer buffer (Fig. 1).

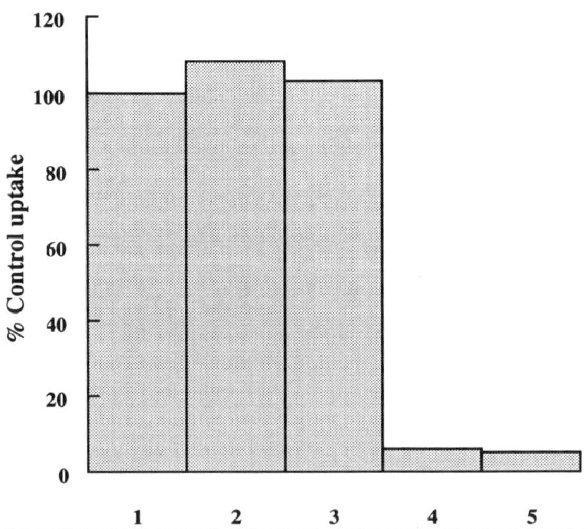

	1	2	3	4	5
ß-alanine in culture medium	**none**	**2 days**	**4 days**	**none**	**2 days**
ß-alanine in incubation medium	**none**	**none**	**none**	**+**	**+**

Figure 1. Confluent cultures were grown for four days (Bar 3) or two days without (Bars 1,4) or with (Bars 2,3,5) 1 mM β-alanine. For taurine uptake measurements, cultures were incubated in Krebs Ringer buffer supplemented (Bars 4,5) or not (Bars 1,2,3) with 1 mM β-alanine. Results are expressed in % of taurine uptake in control cells (Bar 1). Each result is the mean of four values.

Metabolism of β-Alanine Added in the Medium of Cells in Culture. Our results show that β-alanine present in the culture medium did not affect the taurine uptake. When [^{3}H]β-alanine was added to the culture medium of rat primary glial cells, nearly all the added radioactivity was recovered from the medium after two days (Table 3). β-Alanine is catabolized in mammalian cells to acetyl CoA and then to CO_2 (5), resulting in a decrease in radioactivity in the culture media. It has been shown that β-alanine transaminase, the first enzyme involved in β-alanine catabolism, is not secreted in the culture medium of kidney cells, which catabolize this amino acid very actively (16). Furthermore, it is well known that transaminases are usually released from cells only in pathological situations or after cell

Table 3. Proportion of [^{3}H]β-alanine remaining in the culture medium after 2 days

Addition to the culture medium (day 13)	Activity (cpm/100 μl) in culture medium measured on	
	Day 13	Day 15
5 mM β-alanine	6186 (100%)	5544 ± 54 (90%)

13 day-old confluent cultures of primary glial cells were used. The culture medium was removed at day 13 and replaced by fresh medium: 2 ml per dish of DEMEM + 5% fetal calf serum + 5 mM β-alanine containing or not 100,000 cpm [^{3}H]β-alanine. Radioactivity in the culture medium was determined at the start of the experiment and 2 days after. Each result is the mean of 3 values.

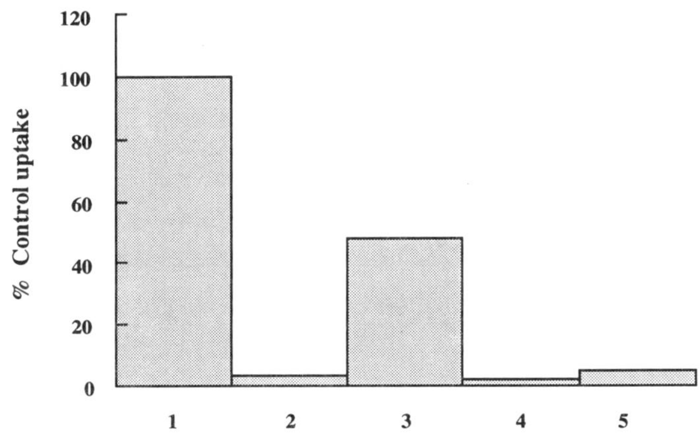

Incubation	KR	KR	DMEM	DMEM	DMEM
+ GES	-	+	-	+	-
+ ß-alanine	-	-	-	-	+

Figure 2. Confluent cultures were incubated for taurine uptake measurement in Krebs Ringer buffer (Bars 1,2) or DMEM (Bars 3,4,5) containing 1 mM GES (Bars 2,4) or β-alanine (Bar 5). Results are in percent of uptake by cells incubated in Krebs Ringer buffer. Each result is the mean of four values.

death (15). We can conclude from the amount of [^{3}H]β-alanine remaining in the culture medium after two days, that β-alanine is not catabolized under our experimental conditions. Thus, the absence of inhibition of taurine uptake by β-alanine when present in the culture medium is due to the inability of the transport system to transport β-alanine.

Effects of Incubation Medium Composition on Guanidinoethane Sulfonate and on β-Alanine Inhibitory Activity

Cells are grown in a complex formulated medium containing carbohydrates, amino acids, vitamins, ions. For the taurine uptake determinations, cells were incubated in a comparatively simple buffer containing only ions and glucose. When the taurine uptake was measured in cells incubated in DMEM instead of Krebs-Ringer buffer, a dramatic decrease of the taurine uptake was observed (Fig. 2). guanidinoethane sulfonate and β-alanine remain strong inhibitors when added to the DMEM during the uptake measurement. For the experiments depicted in Tables 1 and 2, fetal calf serum (5%) was present in the culture medium but not in the Krebs-Ringer buffer. Supplementation of the Krebs-Ringer buffer or the DMEM used for incubation of the cells during the taurine uptake with fetal calf serum (5%) induces a decrease in the basal taurine uptake (Fig. 3). Under the same conditions (presence of fetal calf serum) the guanidinoethane sulfonate inhibitory potency was not affected.

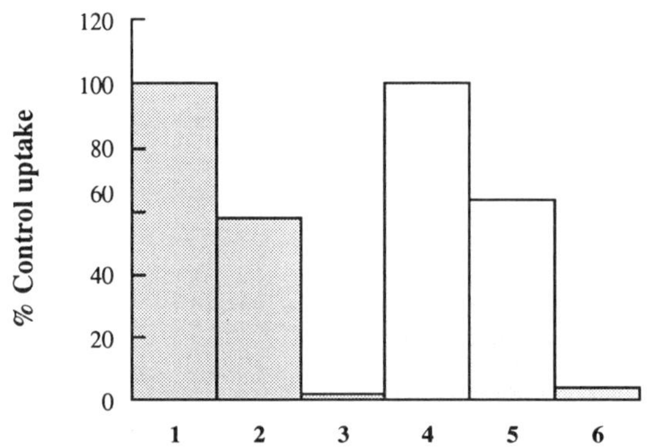

Incubation medium	KR	KR	KR	DMEM	DMEM	DMEM
FCS	-	+	+	-	+	+
GES	-	-	+	-	-	+

Figure 3. Confluent cultures were incubated for taurine uptake measurements either in Krebs Ringer medium (Bars 1,2,3) or in DMEM (Bars 4-6) with or without fetal calf serum or guanidinoethane sulfonate as indicated under the figure. Incubation in the presence of DMEM was done in an incubator with an atmosphere of air-5% CO_2. Results are in percent of the corresponding control cultures (Bars 1 and 4 respectively). Each result is the mean of four values

α-Amino acids and β-amino acids are taken up by mammalian cells by two uptake systems. When tested alone, α-amino acids such as leucine have no inhibitory effects on the taurine uptake. Tables 1 and 2 show that cysteic acid and cysteine sulfinic acid, α-amino acids structurally related to taurine, were poor inhibitors of taurine uptake by primary glial cells. However, we cannot exclude that the complex mixture of amino acids contained in

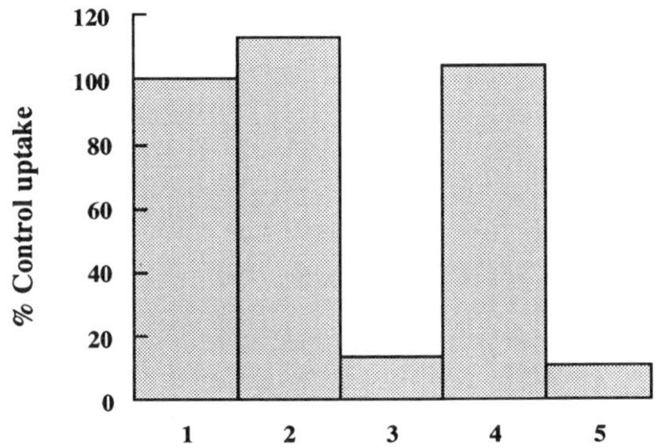

	1	2	3	4	5
Addition to the culture medium	-	ß-alanine	-	-	-
Addition to the Krebs Ringer medium	-	-	ß-alanine	AA	AA + ß-alanine

Figure 4. Confluent glial cells were grown for two days in (DMEM + 5% fetal calf serum) containing or not 1 mM β-alanine. Cell cultures were incubated for taurine uptake measurements in Krebs Ringer buffer containing the additives as indicated. Results are in percent of uptake in control cells (Bar 1). Each result is the mean of four values. AA = MEM amino acid solution from Gibco

a culture medium could affect the uptake of taurine. Fig. 4 shows that when the MEM amino acid solution of Gibco (◇ 04301130, diluted 50X) was added to the Krebs-Ringer buffer (used for the incubation of the cells for taurine uptake) it did not change the basal level of the taurine uptake and did not affect the inhibitory capacity of β-alanine.

DISCUSSION

Our observations concerning the effects of the different taurine analogs, when added to the incubation medium used for taurine uptake measurements in primary rat glial cells, are in agreement with data reported by Holopainen and Kontro for the same type of cells (7). Comparison of the present results with those obtained for other types of glial or glioma cells show that these compounds were generally less efficient for inhibiting taurine uptake by the fast mechanism (3,6,14,17-19). However, guanidinoethane sulfonate, β-alanine and hypotaurine were always the most potent inhibitors, GABA being less active.

Addition of the same analogs to the culture medium of primary glial cell from rat causes a decrease of their inhibitory capacity. Surprising was the complete loss of effect of β-alanine and of GABA. This is not specific for the primary glial cells as the same result was obtained with the C6 rat glioma cell line. This result could be related to the fact that DMEM and Krebs Ringer buffer have a very different composition. However, substitution of the Krebs Ringer buffer by DMEM for the taurine uptake determination did not change the inhibitory potency of β-alanine by the fast mechanism.

When the taurine uptake was made in DMEM, instead of Krebs Ringer, a notable decrease of the basal uptake was observed in primary glial cells. The respective amounts of Na^+, K^+ and Cl^- in DMEM and in Krebs Ringer buffer are in the same range and cannot account for the taurine uptake decrease observed. The lower taurine uptake activity measured in cells incubated in DMEM is not linked to the presence in DMEM of amino acids, since addition of an amino acid mixture to the Krebs-Ringer buffer did not alter the uptake. Moreover, the basal taurine uptake activity was the same whether measured in cells incubated with DMEM containing phenol red or in cells incubated with DMEM devoid of the pH indicator (not shown). Another difference between cells grown in DMEM and cells incubated in Krebs Ringer was the presence of serum in the culture medium. Addition of fetal calf serum to the incubation medium during the taurine uptake determinations induces an inhibition of this uptake whatever the incubation medium used (Krebs Ringer or DMEM) but did not change the inhibitory potency of guanidinoethane sulfonate. This effect of serum on the level of taurine uptake is probably linked to the presence of a fast acting inhibitory component of the serum. This result confirms our previous observation showing that the taurine uptake is notably increased in rat primary glial cells grown for two days in a hormone supplemented, serum-free medium (12). All together, these results demonstrate that the taurine uptake system of primary glial cells is notably influenced by the cell environment. Despite this variability in the level of basal taurine uptake, the degree of inhibition by compounds such as guanidinoethane sulfonate or β-alanine seems to be unaffected in the same cells.

REFERENCES

1. Benda, P., Lightbody, J., Sato, G., Levine, L., and Sweet, W. 1968, Differentiated rat glial cell strain in tissue culture, *Science*, 161:370-371.
2. Booher, J. and Sensenbrenner, M. 1972, Growth and cultivation of dissociated neurons and glial cells from embryonic chick, rat and human brain in flask cultures, *Neurobiology*, 2:97-105.
3. Borg, J., Balcar, V.J., Mark, J., and Mandel, P. 1979, Characterization of taurine uptake by neuronal and glial cells in culture, *J.Neurochem.*, 32:1801-1805.
4. Field, L., Owen, T.C., Crenshaw, R.R., and Bryan, A.W. 1961, Organic disulfids and related substances. Thiosulfonates and disulfides containing 2-aminoethyl moieties, *J.Amer.Chem.Soc.*, 83:4414-4417.
5. Gray, R.G., Pollit, R.J., and Webley, J. 1987, Methylmalonic semialdehyde dehydrogenase deficiency: demonstration of defective valine and β-alanine metabolism and reduced malonic semialdehyde dehydrogenase activity in cultured fibroblasts, *Biochem.Med.Metab.Biol.*, 38:121-124.
6. Holopainen, I., Malminen, O., and Kontro, P. 1988, Sodium-dependent high-affinity uptake of taurine in cultured cerebellar granule cells and astrocytes, *J Neurosci Res*, 18:479-483.
7. Holopainen, L. and Kontro, P. 1986, High-affinity uptake of taurine and beta-alanine in primary cultures of rat astrocytes, *Neurochem.Res.*, 11:207-215.
8. Huxtable, R.J. 1989, Taurine in the central nervous system and the mammalian actions of taurine, *Prog.Neurobiol.*, 32:471-533.
9. Huxtable, R.J., Laird, H.E., and Lippincott, S.E. 1979, The transport of taurine in the heart and the rapid depletion of tissue taurine content by guanidinoethyl sulfonate, *Journal of Pharmacology and Experimental Therapeutics*, 211:465-471.
10. Jacobsen, J.G. and Smith, L.H. 1968, Biochemistry and physiology of taurine and taurine derivatives, *Physiol.Rev.*, 48:424-511.

11. Lleu, P.L. and Rebel, G. 1989, Effect of HEPES on the taurine uptake by cultured glial cells, *J.Neurosci.Res.*, 23:78-86.
12. Lleu, P.L. and Rebel, G. 1990, Effect of HEPES on the Na^+,Cl^--dependent uptake of taurine and β-alanine by cultured glial cells. Modulation by composition and osmolarity of medium, *Neuropharmacology*, 29:719-725.
13. Lowry, O.J., Rosebrough, N.J., Farr, A.L., and Randall, R.J. 1951, Protein measurement with folin phenol reagent, *J.Biol.Chem.*, 193:265-275.
14. Martin, D.L. and Shain, W. 1979, High affinity transport of taurine and β-alanine and low affinity transport of τ-aminobutyric acid by a single transport system in cultured glioma cells, *J.Biol.Chem.*, 254:7076-7084.
15. Mitchell, D.B., Santone, K.S., and Acosta, D. 1980, Evaluation of cytotoxicity in cultured cells by enzyme leakage, *J.Tissue Cult.Meth.*, 6:113-116.
16. Nutzenadel, W. and Scriver, C.R. 1976, Uptake and metabolism of β-alanine and L-carnosine by rat tissues in vitro: role of nutrition, *Am.J.Physiol.*, 230:643-651.
17. Schousboe, A., Fosmark, H., and Svenneby, G. 1976, Taurine uptake in astrocytes cultured from dissociated mouse brain hemispheres, *Brain Res.*, 116:158-164.
18. Schrier, B.K. and Thompson, E.J. 1974, Uptake, excretion and metabolism of putative neurotransmitters by cultured glial tumor cells, *J. Biol. Chem.*, 249:1769-1780.
19. Tasaka, J., Sakai, S., Tosaka, T., and Yoshihama, I. 1989, Glial uptake system of GABA distinct from that of taurine in the bullfrog sympathetic ganglia, *Neurochem.Res.*, 14:271-277.
20. Wright, C.E., Tallan, H.H., Lin, Y.Y., and Gaull, G.E. 1986, Taurine: biological update, *Ann.Rev.Biochem.*, 55:427-453.

CHANGES IN THE LOCALIZATION OF TAURINE-LIKE IMMUNOREACTIVITY DURING DEVELOPMENT AND REGENERATION IN THE RAT BRAIN

Kathy R. Magnusson

Department of Anatomy and Neurobiology
Colorado State University
Fort Collins, CO 80523

INTRODUCTION

Taurine is essential to the normal development of the mammalian brain (31). A deficiency of taurine is associated with arrested cell migration in the developing cerebellum in kittens (32, 33) and weaver mice (24, 34) and with morphological abnormalities in the visual cortex of cats during development (35). The role of taurine in development of many other brain structures, however, remains to be elucidated. Taurine concentrations in the brain are four- to six-fold higher in the neonate than in the adult and decline to adult concentrations by 40-50 days post-natal (1, 19, 20, 21, 30, 31). It is unclear, in many brain regions, whether this decline is due to a decrease in concentration within the same cells that contain taurine during adulthood or whether there is a change in localization during development. A localization of taurine in the neonate that differed from the localization in the adult brain would suggest that taurine plays different roles during development than during adulthood.

Changes in subcellular distribution of taurine have been studied in rat brain homogenates across development (23, 32, 33). However, visualization of changes in taurine localization during development within cell types and subcellular structures in specific brain regions has only recently been reported. Changes in localization of taurine immunoreactivity during development have been reported in the rat optic nerve (8) and retina (9). Taurine-like immunoreactivity (Tau-LI) also has been shown to change localization in the cerebellum of kittens between birth and 8 weeks of age and the localization at 8 weeks and the intensity of Tau-LI is influenced by taurine deficiency in these kittens (13). Cats are dependent on dietary sources of taurine, while rats are able to synthesize it in vivo. Because of the importance of taurine to the development of the cerebellum in both cats and rodents and the difference in their dietary dependence, we were interested in determining if the localization of taurine also changes in the rat cerebellum during development.

Taurine in Health and Disease, Edited by R. Huxtable and D.V. Michalk, Plenum Press, New York, 1994

One would also expect that because taurine is important in developmental processes in the nervous system, it might also be a necessary component of regenerative processes within the adult nervous system. If this is true, taurine should increase in concentration in the regenerative region and reestablish a developmental localization, if such a differential localization exists between the early postnatal period and adulthood. A model of regeneration in the adult brain has been developed and characterized and involves lesioning the entorhinal cortex (6, 14, 17). This removal of synaptic inputs induces axonal sprouting in the other, nonlesioned pathways that normally synapse in the molecular layer of the dentate gyrus. This model was used in the present study to determine whether taurine localization changes in the presence of regenerative processes. In addition, Tau-LI was examined during development in the hippocampus and dentate gyrus in order to be able to compare the localization during regeneration with the localization during development in this region. The information gained by these studies should give us insights into the potential functions of taurine during periods of intensive neuronal growth and how they may differ from or be similar to the functions in the normal adult brain.

METHODS

Development Study

Male and female Sprague-Dawley rats at 7, 14, 28 or 56-60 days of age were anesthetized with halothane and sacrificed by transcardial perfusion of calcium-free Tyrode's buffer solution, followed by 4% paraformaldehyde with 2% glutaraldehyde. Brains were sectioned (50 μm) in the coronal plane through the hippocampus and cerebellum with the use of a vibratome. Free-floating sections were processed with the Vectastain avidin-biotin-peroxidase method (Vector Labs, Burlingame, CA) as follows, with three rinses in phosphate-buffered saline (pH 7.2) between each incubation step. Tissue sections were incubated overnight with Tau1 culture supernate (1:100) (16) and normal horse serum (1:66), diluted in phosphate-buffered saline + 0.3% Triton. Sections were incubated for 30 min in secondary biotinylated antibody (1:1600 in phosphate-buffered saline) and 30 min in avidin-biotin-peroxidase complex (1:400 in phosphate-buffered saline), followed by reaction with diaminobenzidine and hydrogen peroxide to produce a homogeneous brown reaction product. Absorption controls were performed by incubating the Tau1 antibody with taurine conjugated to bovine serum albumin with glutaraldehyde.

Regeneration Study

Male Sprague-Dawley rats (2-3 months old) received electrolytic lesions in the right entorhinal cortex or were sham-operated and were allowed to survive for 7 or 56 days post-lesioning. Animals were anesthetized with intramuscular injections of ketamine (85 mg/kg), xylazine (12.5 mg/kg) and atropine (0.2 ml). Entorhinal lesions were performed as described by Scheff and coworkers (26). Briefly, a small hole was drilled in the calvaria and an electrode lowered into the entorhinal cortex with the use of a stereotaxic apparatus. The right entorhinal cortex was lesioned at nine separate sites (1 mA; 45 s @ site). Sham-operated animals received identical pre-operative and postoperative treatment except for passage of the electrode and current into the entorhinal cortex. Following the appropriate survival periods, each animal was anesthetized with ketamine and xylazine (see above) and transcardially perfused with calcium-free Tyrode's solution and 4% paraformaldehyde with 0.5% glutaraldehyde. Immunohistochemical staining of the hippocampus and dentate gyrus was performed as described above in the development study. Acetylcholinesterase staining

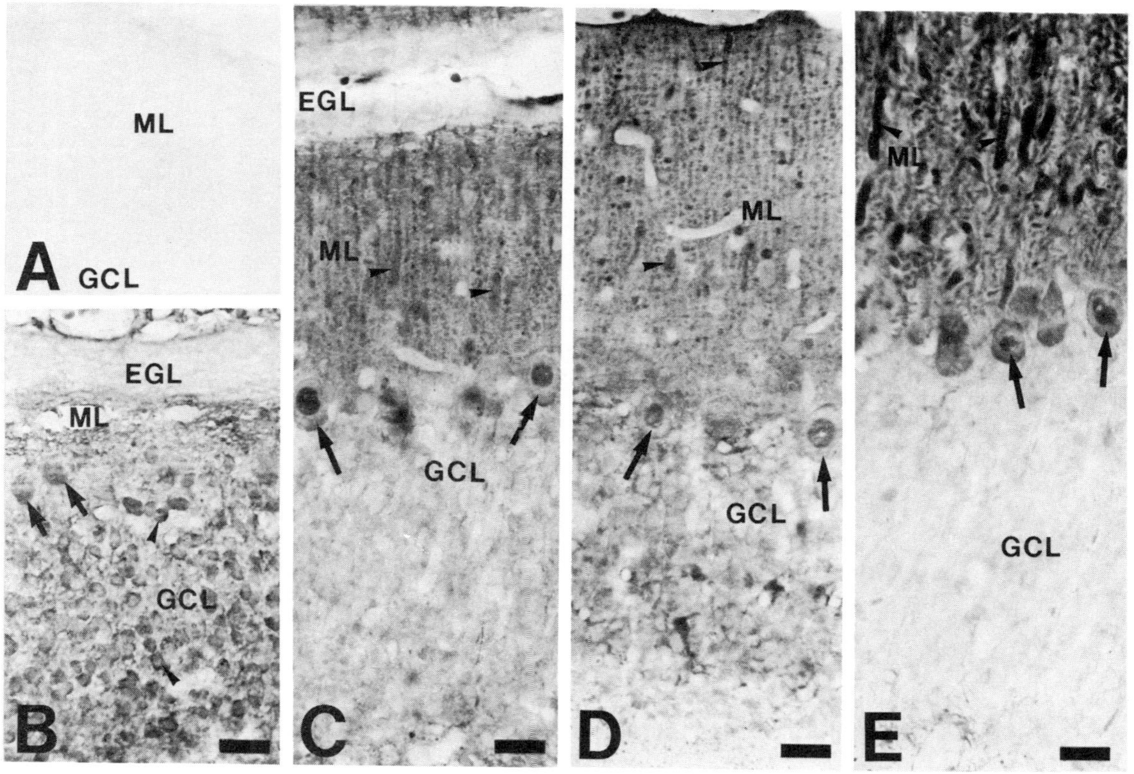

Figure 1. Localization of Tau-LI in the cerebellum of the rat during development. A) Absorption control for taurine-like immunostaining in the seven day old cerebellum. B-E) Tau-LI in the cerebellum of rats at postnatal day 7 (B), day 14 (C), day 28 (D), and day 60 (adult; E). EGL, external granule cell layer; ML, molecular layer; GCL, granule cell layer. Bars = 25 μm.

(18) was performed to verify the effectiveness of the lesion in promoting axonal sprouting (26).

RESULTS AND DISCUSSION

Localization of Taurine in the Developing Cerebellum

Tau-LI was present in Purkinje cell bodies (large arrows; Fig. 1B) and in the granule cells (arrowheads; Fig. 1B) of the inner granule cell layer at post-natal day 7 in the rat. At this same age, there was a homogeneous band of Tau-LI above the Purkinje cell bodies covering the inner half of the molecular layer. The external granule layer was relatively unstained at post-natal day 7 (Fig. 1B). By post-natal day 14 (Fig. 1C), the molecular layer was more intensely labeled with Tau-LI than the inner granule cell layer and the granule cells themselves appeared to be relatively lightly stained. The Purkinje cell bodies (large arrows; Fig. 1C) contained Tau-LI and presumed Purkinje cell dendrites (arrowheads; Fig. 1C) were immunostained more intensely than the neuropil of the molecular layer. A larger proportion of the molecular layer was labeled with Tau-LI on post-natal day 14 than on post-natal day 7. The external granule layer was still relatively unstained on post-natal day 14. The entire molecular layer, extending from the Purkinje cell layer to the surface of the cerebellum, was immunostained with Tau-LI by post-natal day 28 (Fig. 1D) and post-natal day 60 (Fig. 1E) with presumed Purkinje cell dendrites (arrowheads; Fig. 1D,E) labeled more intensely than the surrounding neuropil. Purkinje cell bodies (large arrows; Fig. 1D,E) were still labeled with Tau-LI, while the granule cells showed less staining than the surrounding neuropil. All of the Tau-LI was absorbed out by incubation of the primary antibody with a taurine-carrier protein antigen in all age groups (Fig. 1A (post-natal day 7 shown)).

The band of Tau-LI widened in the molecular layer over the same time course as the Purkinje cell dendrites were expanding out from the Purkinje cell layer to the surface of the molecular layer (3). Thus it appeared that taurine may be involved in the development of the dendritic tree of the Purkinje cells. A similar band of Tau-LI is present in kittens during a comparable period of Purkinje cell development (13). Taurine appeared to be present within the Purkinje cell dendrites by post-natal day 14 and was in the Purkinje cell bodies from post-natal day 7 to post-natal day 60, suggesting that fairly early on in post-natal development, taurine has obtained a localization resembling the adult, at least with respect to the Purkinje cells. Excitatory parallel fiber synapses are being formed on these Purkinje cell dendrites as early as post-natal day 7 (3). If taurine plays a protective role at glutamatergic synapses in the adult (7, 10, 11, 12, 15, 34), then perhaps it needs to be present from the initiation of these synaptic contacts during development as well. Release of taurine stimulated by excitatory amino acids is greater in the developing mouse brain than in the adult (25), providing further support for a protective role of taurine during development.

The Tau-LI in the granule cell layer was strongest on post-natal day 7 which is the period when climbing and mossy fibers are migrating into the inner granule cell layer (2), but there appeared to be less need for high concentrations of taurine during the period of maximum migration and maturation of granule cells and synapse formation in this region (4). Taurine is localized to axons in the optic nerve during the establishment of retino-geniculate connections and then changes to a predominantly glial localization (8). Some of the Tau-LI in the granule cell layer of the cerebellum at post-natal day 7 in the rat also may be localized to migrating axons.

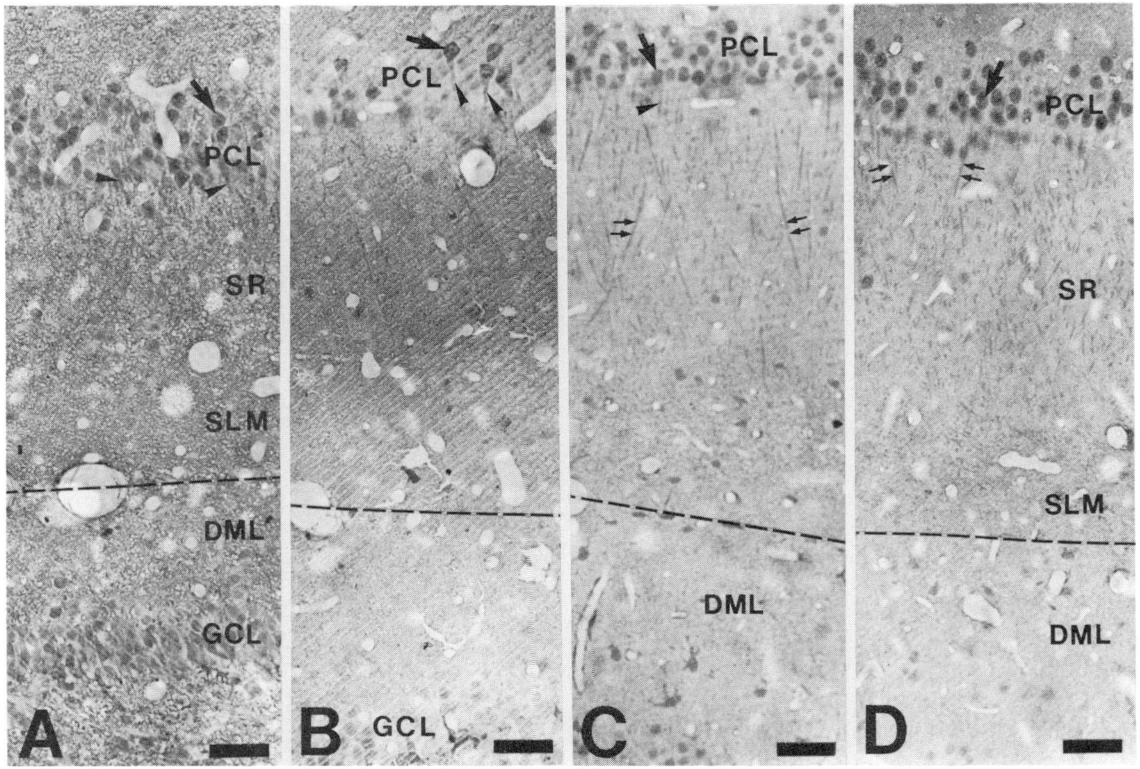

Figure 2. Localization of Tau-LI in the hippocampus of the rat during development at postnatal day 7 (A), day 14 (B), day 28 (C) and day 60 (adult; D). PCL, CA1 pyramidal cell layer; SR, CA1 stratum radiatum; SLM, CA1 stratum lacunosum/moleculare; DML, dentate gyrus molecular layer; GCL, dentate gyrus granule cell layer. Bars = 50 μm.

Localization of Taurine in the Developing Hippocampus

Tau-LI was fairly homogeneous in intensity of staining throughout the CA1 region and dentate gyrus on post-natal day 7 (Fig. 2A). By post-natal day 14, the staining in the dentate gyrus and stratum laçunosum/moleculare was lighter than the pyramidal cell layer and stratum radiatum of the CA1 region (Fig. 2B). At post-natal day 28 (Fig. 2C) and in the adult (Fig. 2D), the neuropil staining throughout the stratum radiatum and stratum lacunosum/moleculare were similar to the dentate gyrus molecular layer. The pyramidal cell bodies (large arrows) and proximal dendrites (arrowheads) were labeled with Tau-LI throughout the ages examined (Fig. 2A-D). Labeled pyramidal cell dendrites (small double arrows) became apparent above the neuropil staining in the stratum radiatum by post-natal day 28. All of the Tau-LI was absorbed out by incubation of the primary antibody with a taurine-carrier protein antigen in all age groups (not shown; see Fig. 3F)

Taurine appeared to be present in pyramidal cell bodies and at least the proximal dendrites by post-natal day 7. The more distal dendrites also may have been labeled at that age, but may not have been distinguishable from the other neuropil staining. This again suggests that taurine obtained an adult localization within these cells early in development and may already serve some of the same functions as in the adult.

The primary dendrites of the pyramidal cells are fully developed lengthwise at post-natal day 7 (22), the cells having been formed during gestation (5). Outgrowth of branches along the primary dendrite starts before post-natal day 7 and continues until post-natal day 48, but the majority of branches are formed by post-natal day 24 (22). The homogeneous staining of Tau-LI in the molecular layer of CA1 on post-natal days 7 and 14 suggests that taurine may be localized to other structures besides the dendrites and if so, the timing suggests that taurine may be involved in the processes of dendritic branching and/or synapse formation in this region.

Localization of Taurine in the Regenerating Dentate Gyrus

Following unilateral entorhinal lesions, the pattern of acetylcholinesterase staining in the molecular layer of the dentate gyrus changed on the lesioned side (Fig. 3A) as compared to the non-lesioned side (Fig. 3B) in both survival periods. Acetylcholinesterase staining intensified in a band in the outer molecular layer on the lesioned side, which indicated that sprouting of cholinergic axons had occurred (6, 27). The change in the pattern of acetylcholinesterase histochemistry verified that the lesions were successful (6, 27).

The pattern of Tau-LI in the molecular layer of the dentate gyrus also changed on the lesion side (Fig. 3C,E), as compared to the non-lesioned side (Fig. 3D). The intensity of staining in the outer dentate gyrus molecular layer was increased over the staining in the inner dentate gyrus molecular layer on the lesioned side. There was also an increase in the intensity of staining in the stratum lacunosum/moleculare of CA1 on the lesioned side. This change in pattern of Tau-LI was present by 7 days post-lesioning (Fig. 3E) and persisted for at least 56 days after surgery (Fig. 3C). All the Tau-LI could be absorbed by incubation with a taurine-carrier protein antigen (Fig. 3F). The pattern of Tau-LI on the "lesion side" of sham-operated controls (not shown) resembled the staining on the non-lesioned side of lesioned animals (Fig. 3D).

Taurine appeared to increase in concentration in the regions that were undergoing regenerative changes following lesioning of the entorhinal cortex, at least relative to surrounding brain regions. No obvious change in intensity of Tau-LI was seen in non-regenerative regions of the brain on the lesioned side when compared to the non-lesioned side. This would favor the hypothesis that there was an increase in taurine concentration in the regenerative region and not just a redistribution, although redistribution cannot be entirely ruled out at this point. Although the dentate gyrus receives the majority of entorhinal inputs to the hippocampus, the stratum lacunosum/moleculare also receives inputs from the entorhinal cortex and undergoes some changes following removal of the perforant pathway (28). Tau-LI also increased in this region following lesioning.

These studies do not prove unequivocally that the change in taurine localization was due to a role in axonal sprouting rather than a role in fiber degeneration. The majority of degenerating synapses have been shown to rapidly disappear by 10 days post-lesioning, followed by a slower phase of removal, leaving only 23% remaining by 60 days post-

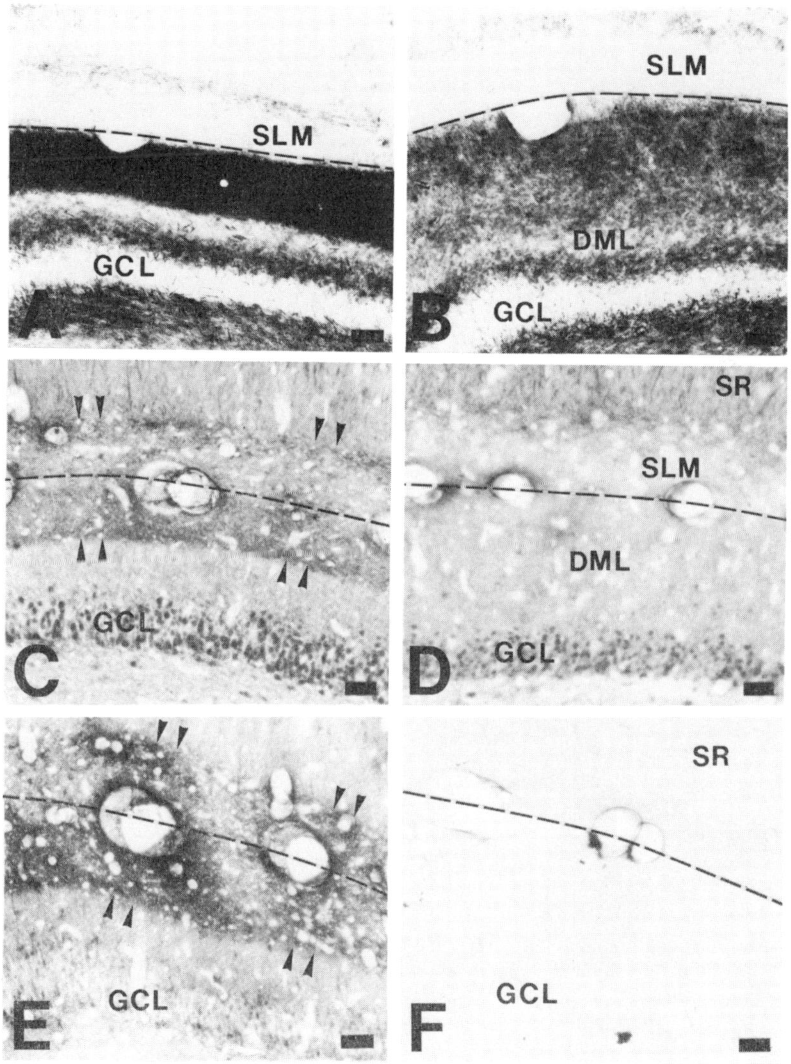

Figure 3. Changes in localization of Tau-LI during regeneration in the dentate gyrus following unilateral lesioning of the entorhinal cortex. A,B) Acetylcholinesterase staining in the dentate gyrus on the lesioned side (A) and on the control side (B) of a rat that survived for 56 days post-lesioning. C-E) Tau-LI in the dentate gyrus and CA1 stratum lacunosum/moleculare on the lesioned side (C,E) and the control side (D) of rats that survived for 56 days (C,D) and 7 days (E) post-lesioning. F) Absorption control for Tau-LI in the dentate gyrus on the lesioned side of a rat that survived for 56 days post-lesioning. SR, CA1 stratum radiatum; SLM, CA1 stratum lacunosum/moleculare; DML, dentate gyrus molecular layer; GCL, dentate gyrus granule cell layer. Dashed line indicates hippocampal fissure. Bars = 50 μm.

surgery (17). Degenerating fibers are mostly cleared by 10 days and only rarely seen by 30 days post-lesioning (17). The addition of new synapses, however, appears to continue for at least 7 months post-surgery (29). The change in taurine labeling persisted for at least 56 days after surgery. Additional support for a role in regeneration is given by the homogeneous pattern of Tau-LI in the region of axonal sprouting, which was similar to the pattern of taurine labeling in the developing hippocampus during a developmental period of synapse formation.

SUMMARY

In summary, taurine appeared to be present in certain cell types, such as cerebellar Purkinje cells and hippocampal pyramidal cells, throughout development to adulthood and a differential function for taurine between these periods would be difficult to hypothesize simply based on localization. However, in both the cerebellum and hippocampus, there was a period including post-natal day 7 in the cerebellum and including both post-natal days 7 and 14 in the hippocampus in which taurine appeared not to be confined only to the dendrites of the aforementioned cells, but seemed ubiquitously present in the molecular layers of these two brain regions. This suggests that the taurine may be present in significantly higher concentrations in certain cell types or subcellular structures during development than in the adult rat brain. The elucidation of these taurine-containing structures with the use of electron microscopy may provide some insight into the functions of taurine during these critical periods in development. Finally, taurine appeared to reverse its developmental decline in concentration in the presence of regeneration, suggesting that it may play a role in axonal sprouting and/or synapse formation.

REFERENCES

1. Agrawal, H.C., Davison, A.N. and Kaczmarek, L.K., 1971, Subcellular distribution of taurine and cysteinesulphinate decarboxylase in developing rat brain, *Biochem. J.*, 122: 759-763.
2. Alcantara, A.A., Pfenninger, K.H. and Greenough, W.T., 1992, 5B4-CAM expression parallels neurite outgrowth and synaptogenesis in the developing rat brain, *J. Comp. Neurol.*, 319: 337-348.
3. Altman, J., 1972, Postnatal development of the cerebellar cortex in the rat. II. Phases in the maturation of Purkinje cells and of the molecular layer, *J. Comp. Neurol.*, 145: 399-464.
4. Altman, J., 1972, Postnatal development of the cerebellar cortex in the rat. III. Maturation of the components of the granular layer, *J. Comp. Neurol.*, 145: 465-514.
5. Bayer, S.A., 1980, Development of the hippocampal region in the rat I. Neurogenesis examined with 3H-thymidine autoradiology, *J. Comp. Neurol.*, 190: 87-114.
6. Cotman, C.W., Matthews, D.A., Taylor, D. and Lynch, G., 1973, Synaptic rearrangement in the dentate gyrus: Histochemical evidence of adjustments after lesions in immature and adult rats, *Proc. Natl. Acad. Sci. USA,* 70: 3473-3477.
7. French, E.D., Vezzani, A., Whetsell Jr., W.O. and Schwarcz, R., 1986, Antiexcitotoxic actions of taurine in the rat hippocampus studied in vivo and in vitro, *Adv. Exp. Med. Biol.*, 203: 349-362.
8. Lake, N., 1992, Taurine, GABA and GFAP immunoreactivity in the developing and adult rat optic nerve, *Brain Res.*, 596: 124-132.
9. Lake, N., De Marte, L. and Verdone-Smith, C., 1989, Immunocytochemical localization of taurine in the developing rat retina, *Invest. Opthalmol. Vis. Sci.*, 30: ARVO Suppl. 122.
10. Lehmann, A., Hagberg, H. and Hamberger, A., 1984, A role for taurine in the maintenance of homeostasis in the central nervous system during hyperexcitation, *Neuroscience Lett,* 52: 341-346.
11. Lehmann, A., Isacsson, H. and Hamberger, A., 1983, Effects of in vivo administration of kainic acid on the extracellular amino acid pool in the rabbit hippocampus, *J. Neurochem.*, 40: 1314-1320.
12. Lehmann, A., Lazarewicz, J.W. and Zeise, M., 1985, N-Methylaspartate-evoked liberation of taurine and phosphonoethanolamine in vivo: site of release, *J. Neurochem.*, 45: 1172-1177.

13. Lu, P., Schuller-Levis, G. and Sturman, J.A., 1991, Distribution of taurine-like immunoreactivity in cerebellum of kittens from taurine-supplemented and taurine-deficient mothers, *Int. J. Dev. Neurosci.*, 9: 621-629.

14. Lynch, G., Deadwyler, S. and Cotman, C., 1973, Postlesion axonal growth produces permanent functional connections., 180: 1364-1366.

15. Magnusson, K.R., Koerner, J.F., Larson, A.A., Smullin, D.H., Skilling, S.R. and Beitz, A.J., 1991, NMDA-, kainate- and quisqualate-stimulated release of taurine from electrophysiologically-monitored rat hippocampal slices, *Brain Res.*, 549: 1-8.

16. Magnusson, K.R., Madl, J.E., Clements, J.R., Wu, J., Larson, A.A. and Beitz, A.J., 1988, Colocalization of Taurine- and Cysteine Sulfinic Acid Decarboxylase-like Immunoreactivity in the Cerebellum of the Rat with Monoclonal Antibodies Against Taurine, *J. Neurosci.*, 8: 4551-4564.

17. Matthews, D.A., Cotman, C. and Lynch, G., 1976, An electron microscopic study of lesion-induced synaptogenesis in the dentate gyrus of the adult rat. 1. Magnitude and time course of degeneration, *Brain Res.*, 115: 1-21.

18. Naik, N.T., 1963, Technical variations in Koelle's histochemical method for demonstrating cholinesterase activity, *Quart. J. Microsc. Sci.*, 104: 89-100.

19. Oja, S.S. and Kontro, P., 1989, Release of endogenous taurine and gamma-aminobutyric acid from brain slices from the adult and developing mouse, *J Neurochem.*, 52: 1018-1024.

20. Oja, S.S., Uusitalo, A.J., Vahvelainen, M. and Piha, R.S., 1968, Changes in cerebral and hepatic amino acids in the rat and guinea pig during development, *Brain Res.*, 11: 655-661.

21. Palkovits, M., Elekes, I., Lang, T. and Patthy, A., 1986, Taurine levels in discrete brain nuclei of rats, *J. Neurochem.*, 47: 1333-1335.

22. Pokorny, J. and Yamamoto, T., 1981, Postnatal ontogenesis of hippocampal CA1 are in rats. I. Development of dendritic arborisation in pyramidal neurons, *Brain Res. Bull.*, 7: 113-120.

23. Rassin, D.K., Sturman, J.A. and Gaull, G.E., 1977, Taurine in developing rat brain: subcellular distribution and association with synaptic vesicles of [35S]taurine in maternal, fetal and neonatal rat brain, *J. Neurochem.*, 28: 41-50.

24. Roffler-Tarlov, S. and Turey, M., 1982, The content of amino acids in the developing cerebellar cortex and deep cerebellar nuclei of granule cell deficient mutant mice., 247: 65-73.

25. Saransaari, P. and Oja, S.S., 1991, Excitatory amino acids evoke taurine release from cerebral cortex slices from adult and developing mice, *Neuroscience*, 45: 451-459.

26. Scheff, S.W., Benardo, L.S. and Cotman, C.W., 1980, Decline in reactive fiber growth in the dentate gyrus of aged rats compared to young adult rats following entorhinal cortex removal, *Brain Res*, 199: 21-38.

27. Stanfield, B.B. and Cowan, W.M., 1982, The sprouting of septal afferents to the dentate gyrus after lesions of the entorhinal cortex in adult rats, *Brain Res.*, 232: 162-170.

28. Steward, O., 1977, Topographic organization of the projections from the entorhinal area to the hippocampal formation of the rat, *J. Comp. Neurol.*, 167: 285-314.

29. Steward, O. and Vinsant, S.L., 1983, The process of reinnervation in the dentate gyrus of the adult rat: A quantitative electron microscopic analysis of terminal proliferation and reactive synaptogenesis, *J. Comp. Neurol.*, 214: 370-386.

30. Sturman, J.A., Nutritional taurine and central nervous system development, *Annals New York Academy of Sciences*, 196-213.

31. Sturman, J.A., 1993, Taurine in development, *Physiol. Rev.*, 73: 119-147.

32. Sturman, J.A., Moretz, R.C., French, J.H. and Wisniewski, H.M., 1985, Postnatal taurine deficiency in the kitten results in a persistence of the cerebellar external granule cell layer: correlation by taurine feeding, *J. Neurosci. Res.*, 13: 521-528.

33. Sturman, J.A., Moretz, R.C., French, J.H. and Wisniewski, H.M., 1985, Taurine deficiency in the developing cat: persistence of the cerebellar external granule cell layer, *J. Neurosci. Res.*, 13: 405-416.

34. Trenkner, E., 1990, The role of taurine and glutamate during early postnatal cerebellar development of normal and weaver mutant mice, *Adv. Exp. Med. Biol.*, 268: 239-244.

35. Xu, Y., Lu, P., Imaki, H. and Sturman, J.A., 1992, Feline maternal taurine deficiency: a quantitative morphometric and immunohistochemical study of newborn visual cortex, *Dendron*, 1: 129-145.

TRANSMITTER AND ELECTRICAL STIMULATION OF [^{3}H]TAURINE RELEASE FROM RAT SYMPATHETIC GANGLIA

Robert A. Waniewski[1,2], Jonathan S. Carp[1] and David L. Martin[1,2]

[1]Wadsworth Center for Laboratories and Research
New York State Department of Health
[2]Department of Environmental Health and Toxicology
Department of Biomedical Sciences
School of Public Health, State University of New York at Albany
Albany, NY, 12201-0509

INTRODUCTION

Astroglial cells release taurine in response to stimulation with neurotransmitters. This process has been studied most extensively with primary cultures of astrocytes and LRM55 glial cells. These studies have demonstrated that several transmitters can elicit release. The second messenger systems involved in activating release have been characterized (10, 15, 19). An important issue concerning all studies of this type is the applicability of results obtained with glia in culture to glia in vivo. We have chosen the rat superior cervical ganglion as a nervous tissue having the potential for exploring taurine release from glial cells in situ. The major neuronal composition of the ganglion consists of preganglionic nerve terminals providing cholinergic input and principal neurons providing noradrenergic output. The superior cervical ganglion also contains a very small population of dopamine-containing intrinsic neurons known as SIF cells (3). The glial population of the superior cervical ganglion is composed of Schwann cells responsible for myelination and satellite glia, immunoreactive to glial fibrillary acidic protein, that surround the cell bodies of the principal neurons (1, 7). Currently available data suggest that taurine is selectively taken up by the satellite glia. Autoradiographic studies have demonstrated that the ω-amino acids GABA and β-alanine are selectively accumulated by satellite glial cells in the superior cervical ganglion (5, 20), while transport studies have shown that taurine inhibits [^{3}H]GABA uptake and that GABA inhibits [^{14}C]taurine uptake by the superior cervical ganglion (5). The demonstration that potassium-stimulated efflux of [^{3}H]GABA from the superior cervical ganglion is not reduced by preganglionic denervation also supports the glial localization of ω-amino acid transporters in this tissue (4).

We have found that taurine is one of the most abundant free amino acids in the superior cervical ganglion, having a concentration of 44 nmol/mg protein, and that

Taurine in Health and Disease, Edited by R. Huxtable
and D.V. Michalk, Plenum Press, New York, 1994

[^{3}H]taurine is taken up by the superior cervical ganglion against a large concentration gradient of endogenous taurine. Accumulated labelled taurine is also strongly retained by the superior cervical ganglion. After an initial wash-out period, the perfused superior cervical ganglion releases labelled taurine at a very slow rate, retaining most of it even over a period of several hours (18). The purpose of the present study was to determine whether neurotransmitters or preganglionic nerve stimulation could increase the release of [^{3}H]taurine from the superior cervical ganglion.

METHODS

Superior Cervical Ganglion Dissection

Adult male Wistar rats were anesthetized by inhalation of carbon dioxide and decapitated. This procedure was reviewed and approved in advance by the Institutional Animal Care and Use Committee. The superior cervical ganglia were bilaterally dissected and placed in ice-cold HEPES-buffered Hanks' balanced salt solution. The blood vessels and surrounding connective tissue were dissected and the ganglia were desheathed. The nerve branches entering and exiting the ganglion were cut as close to the body of the ganglion as possible except for electrophysiological studies, when the preganglionic nerve and the largest postganglionic nerve were cut with at least 1 mm of the ends remaining so that they could be drawn into suction electrodes. The preganglionic nerve was cut to approximately twice the length of the postganglionic nerve to permit proper orientation of the superior cervical ganglion with respect to the stimulating and recording electrodes.

[^{3}H]Taurine Release

Four superior cervical ganglia were allowed to accumulate [^{3}H]taurine by incubation at 37°C for 2 ($\pm$ 0.25) h in 400 μl buffer containing 3.9 μM [^{3}H]taurine. For neurotransmitter stimulation experiments, superior cervical ganglia were rinsed in fresh buffer and placed together in a sealed perfusion chamber with an internal volume of approximately 100 μl. Superior cervical ganglia were perfused at 37°C with Hanks' balanced salt solution pumped at a rate of 0.5 ml/min. The perfusate was mixed with a scintillation cocktail pumped at a rate of 2.5 ml/min (Flo-scint V, Radiomatic Instruments, Meriden, CT) and directed through a flow-through, scintillation detector (Flo-one\ßeta, Radiomatic). The tritium released from the superior cervical ganglion was monitored over the entire perfusion period (usually 120 min). Solutions containing neurotransmitter agonists or antagonists were applied for 3 min intervals by computer-controlled valves which directed the perfusate flow. A 17 min recovery period followed each drug application. After 120 min, the superior cervical ganglia were removed from the chamber, digested overnight in 1 M NaOH, homogenized by sonication and analyzed for ^{3}H remaining in the tissue.

[^{3}H]Taurine Release Combined with Electrophysiological Measurements

After accumulating [^{3}H]taurine, individual superior cervical ganglia were placed in a plexiglass recording chamber filled with buffer (100 μl internal volume) and superfused with fresh Hanks' buffer at room temperature. Throughout the electrophysiological protocol described below, the superfusate was drawn out of the recording chamber by a paper wick and carried away by a peristaltic pump (Rabbit, Rainin, Woburn, MA).

The preganglionic and postganglionic nerves of the superior cervical ganglion were gently but securely mounted in glass suction electrodes constructed from 1.2 mm O.D. micropipette blanks and used for stimulation and recording, respectively. Ganglionic responses were elicited by 0.5 ms duration, square pulse stimuli (S88 stimulator and SIU5 isolation unit; Grass Instrument Co., Quincy, MA) at twice the minimum intensity required to elicit a maximal response (6 - 10 volts) (6). These responses were bandpass-filtered and amplified (0.1 - 3000 Hz, gain = 5000; P511 amplifier, Grass) to permit analog-to-digital conversion by computer at 10,000 samples per second for subsequent analysis of the amplitude of the short-latency depolarizing component of the recorded potential. One min fractions of superfusion buffer were collected in an automated fraction collector, mixed with 5 ml scintillation cocktail and processed overnight for liquid scintillation counting (LS 1801, Beckman, Fullerton, CA).

Postganglionic nerve recordings were made throughout the experiment. Test pulses of 0.2 Hz at supramaximal intensity were applied every 5 min when the 10 Hz stimulus train was off. The postganglionic responses were monitored and recorded to insure that the superior cervical ganglion was properly positioned in the electrodes, that it was receiving the test stimuli, and that it was responding appropriately to those stimuli. Collection of the superfusate did not begin until stable recordings were established and after a 15 min period had passed to allow loosely associated taurine to wash out of the superior cervical ganglion as in the experiments described above.

A 10 Hz stimulus was applied for 5 min at supramaximal intensity 10 and 30 min after the start of the superfusate collection period. After the two 10 Hz stimulus intervals, the superfusing solution was manually switched to a carbachol-containing buffer (100 μM), which was applied for a 5 min period.

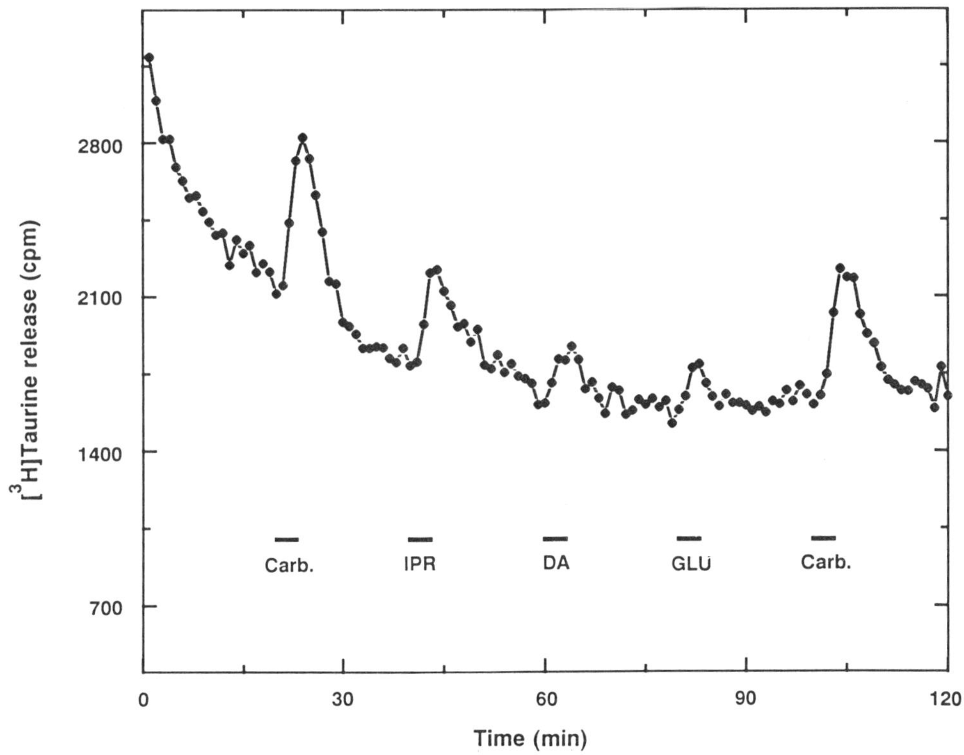

Figure 1. Neurotransmitter agonists stimulate [^{3}H]taurine release from superior cervical ganglion. Carb: carbachol; IPR: isoproterenol; DA: dopamine; GLU: glutamate.

Data Analysis

The output from the flow-through radioactivity detector was collected and stored as counts accumulated in 6 s intervals. These data were then summed for 1 min intervals and expressed as counts per min in the perfusion medium. To calculate the evoked release the cpm in the perfusion medium were summed during the peak of release following the application of drug-containing solutions or following the start of the 10 Hz stimulus intervals and then baseline release was subtracted. Baseline release was obtained by interpolation from the beginning to the end of the peak.

RESULTS

Carbachol, isoproterenol, dopamine and glutamate were all found to stimulate taurine release from the superior cervical ganglion (Fig. 1). Carbachol was the most effective and glutamate the least effective of these agonists when each was applied at 100 μM. Carbachol is an analogue of acetylcholine, the major presynaptic neurotransmitter in sympathetic ganglia.

Since carbachol was the most effective of the neurotransmitter agonists tested in releasing [^{3}H]taurine from the superior cervical ganglion, the properties of carbachol-stimulated release were examined in more detail. First, it was important to determine how long the superior cervical ganglion would continue to respond to carbachol, and second, how consistent were the responses to carbachol. Carbachol was repeatedly applied for 3 min periods. The superior cervical ganglion continued to respond by releasing [^{3}H]taurine in a highly reproducible manner from the second to the fifth applications: thereafter the response

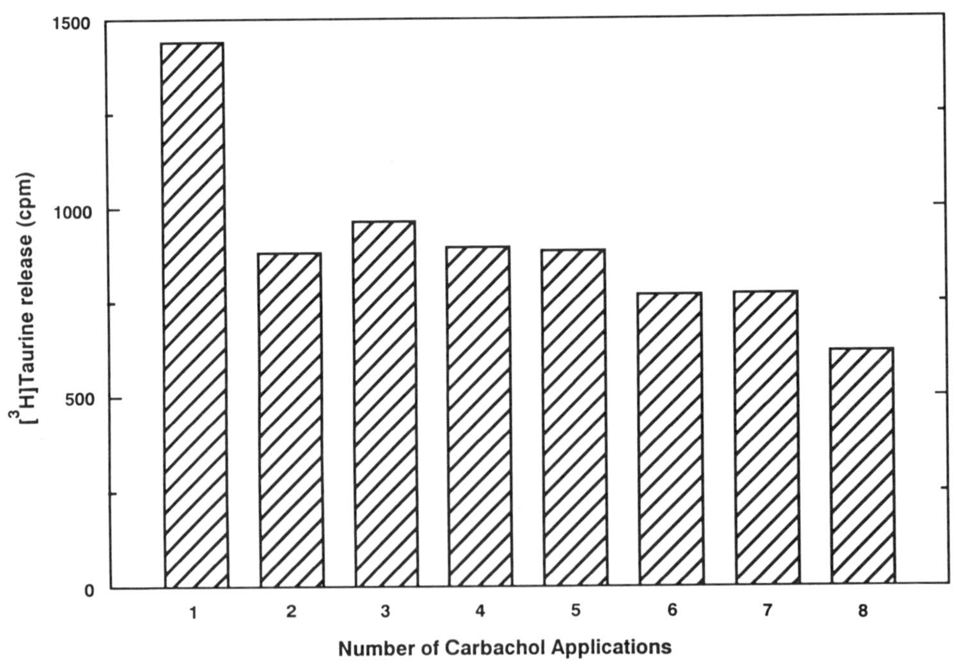

Figure 2. Carbachol continues to stimulate release of [^{3}H]taurine for at least eight applications.

gradually decayed (Fig. 2). In the two experiments summarized in Figure 2, the first application of carbachol produced a much larger release of [^{3}H]taurine than any subsequent application. However, this pattern was not regularly observed in other experiments. The number of drug applications was limited to five in all subsequent experiments. The same concentration of carbachol was usually applied at the beginning and the end of each experiment and the responses to the two applications were averaged.

The results of a typical carbachol concentration-response experiment are presented in Figure 3. Three concentrations of carbachol were tested in between two standard applications of 100 μM carbachol. As can be seen, the stimulated release of taurine increased with increasing concentrations of carbachol. By repeating this type of experiment with several sets of ganglia using many different concentrations of carbachol, it was possible to construct a concentration-response curve for carbachol-stimulated release of taurine. This allowed us to determine an EC_{50} value of 26 μM for carbachol.

Since acetylcholine depolarizes the post-ganglionic neuron by activation of nicotinic receptors, we tested whether nicotinic receptor blockade would prevent the stimulation of taurine release produced by carbachol. Hexamethonium, a selective nicotinic antagonist, was found to be very effective in blocking the stimulated release of [^{3}H]taurine. A profile of taurine release in a typical hexamethonium blockade experiment is shown in Figure 4. The response to 30 μM carbachol alone (20-23 min) was almost completely eliminated by simultaneous application of carbachol and 1 mM hexamethonium (40-43 min). Reapplication of carbachol alone (60-63 min and 100-103 min) producednearly complete recovery of the response. Hexamethonium alone (80-83 min) did not alter basal release of taurine. This indicates that a major proportion of carbachol-stimulated release is sensitive to nicotinic blockade.

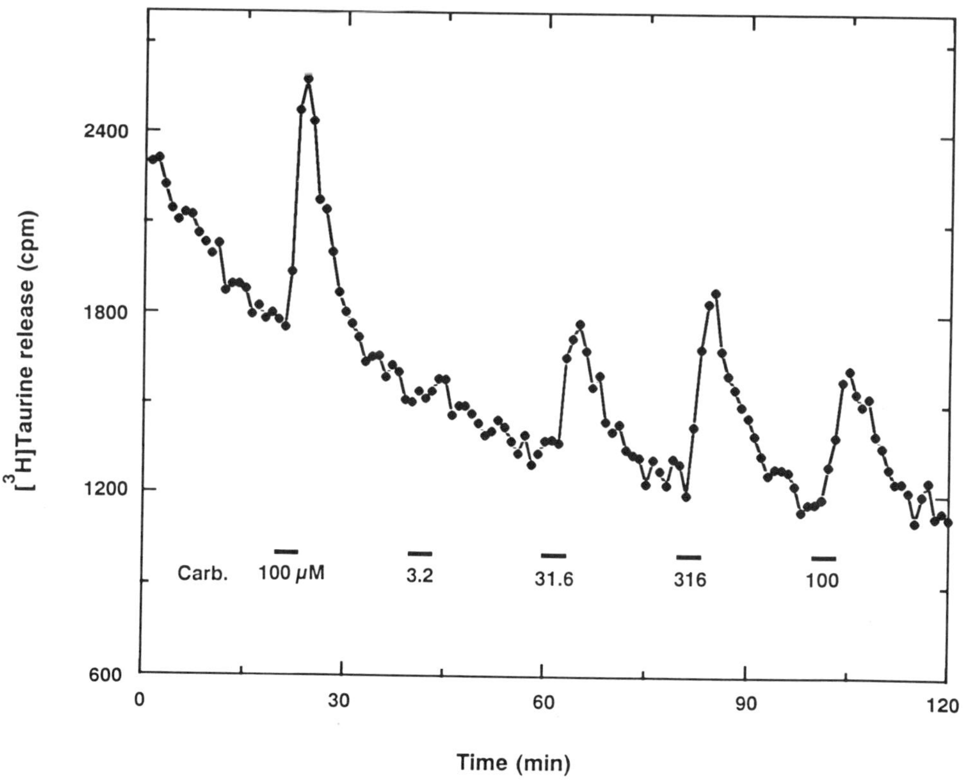

Figure 3. Carbachol (Carb.) stimulates [^{3}H]taurine release in a concentration-dependent manner.

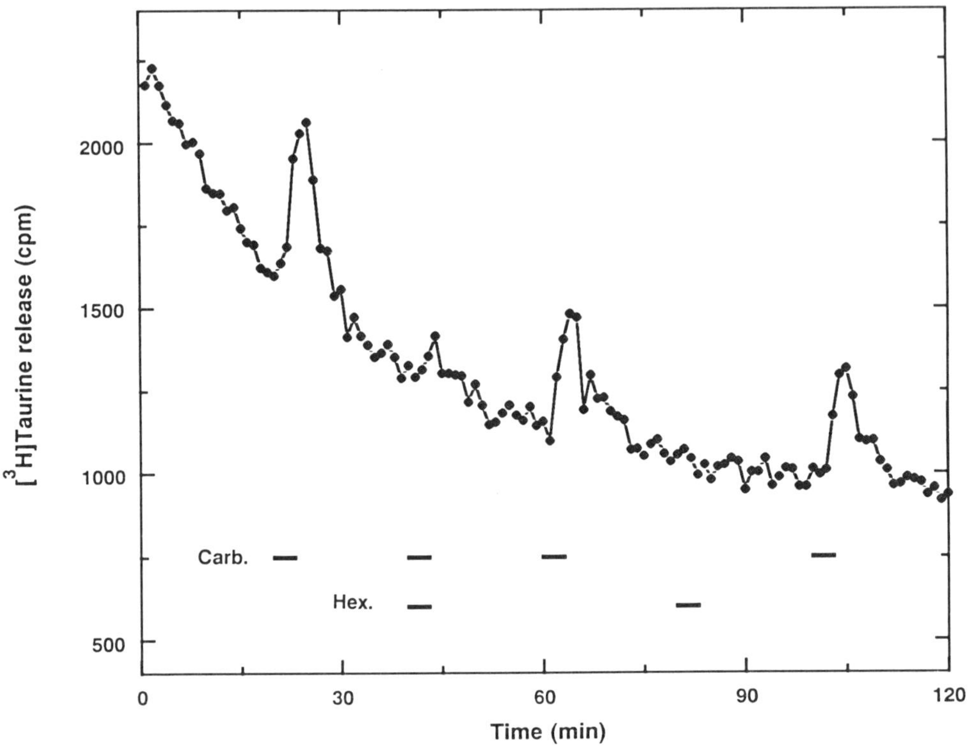

Figure 4. Hexamethonium (Hex.) inhibits carbachol-stimulated [^{3}H]taurine release. Carb: carbachol.

The application of the nicotinic agonist 1,1-dimethyl-4-phenylpiperazinium (DMPP) also stimulated the release of taurine in a concentration-dependent fashion (Fig. 5). The responses to DMPP were very similar to those produced by carbachol and the same EC_{50} was obtained. These results demonstrate that carbachol stimulates taurine release through nicotinic receptor activation.

(±)Muscarine was applied to superior cervical ganglion to determine whether there was a muscarinic component to the release stimulated by carbachol. The results of a typical (±)muscarine experiment are shown in Figure 6. (±)Muscarine stimulated [^{3}H]taurine release in a concentration-dependent manner but much less effectively than carbachol or DMPP. The release stimulated by 100 μM muscarine was the maximum achievable (data for higher concentrations not shown). Other studies with the muscarinic antagonist atropine, which blocked about 30% of the carbachol-stimulated release, are consistent with this finding.

The results of these experiments demonstrate that several transmitters are able to stimulate [^{3}H]taurine release from the superior cervical ganglion. The acetylcholine analogue is the most effective of those transmitters tested. The cholinergic stimulation of release is mediated by both nicotinic and muscarinic receptors. Ganglionic transmission in the superior cervical ganglion is mediated by endogenous acetylcholine and results in postganglionic changes in membrane potential that are produced by both nicotinic and muscarinic receptors. Since we had demonstrated that analogues of acetylcholine could evoke taurine release by both nicotinic and muscarinic receptors, we next wanted to determine whether direct electrical stimulation of the preganglionic nerve would evoke taurine release from the superior cervical ganglion.

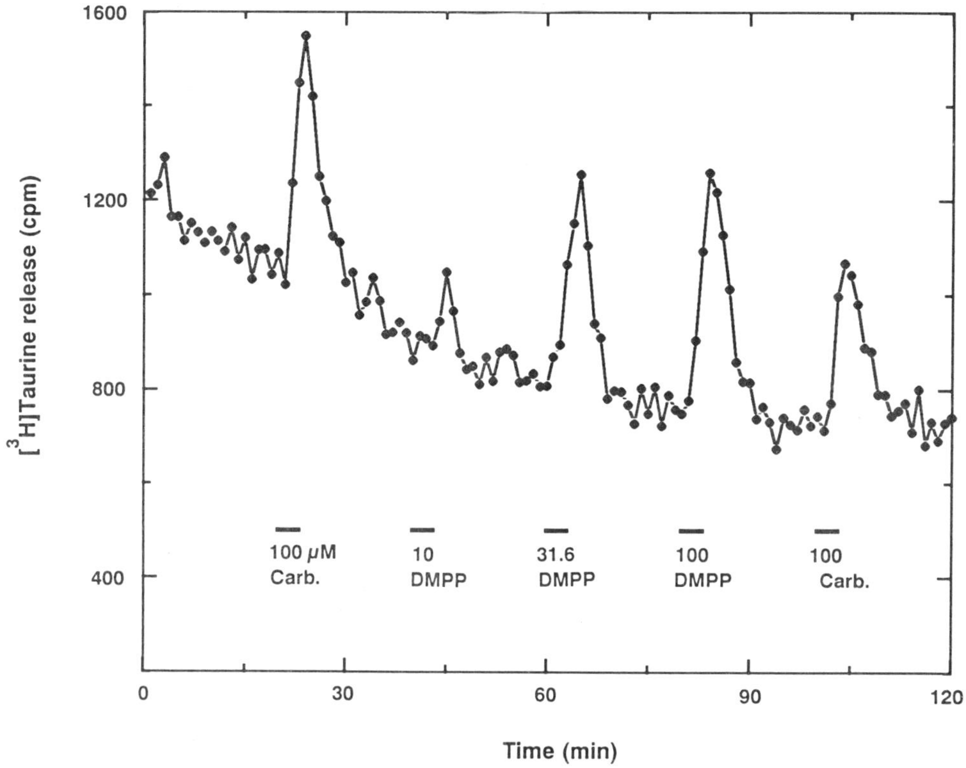

Figure 5. The nicotinic agonist DMPP stimulates taurine release in a concentration-dependent manner. Carb: carbachol; DMPP: 1,1-dimethyl-4-phenylpiperazinium.

Application of low-frequency (0.2 Hz), supramaximal electrical test stimuli to the preganglionic nerve before application of the 10 Hz stimulus train elicited a rapid, multiphasic depolarizing response in the postganglionic nerve (Fig. 7, profile labelled "Before"). This profile represents the average response from eight successive low-frequency stimuli. The reponse to a single 10 Hz stimulus was recorded near the end of the second, 5 min interval of high-frequency stimulation (Fig. 7, profile labelled "Stimulus"). The peak amplitude was decreased and delayed compared with the response elicited by the 0.2 Hz stimuli. This could be due either to decreased release of preganglionic transmitter or to desensitization of the postganglionic receptors with prolonged high-frequency stimulation. The response to a second application of low-frequency test stimuli (0.2 Hz) applied after the 5 min interval of 10 Hz stimulation showed almost complete recovery within 1 min (Fig. 7, profile labelled "After"; average of eight sweeps).

The profile of [^{3}H]taurine release into the superfusion medium collected during two 5 min intervals of high frequency (10 Hz) electrical stimulation of the preganglionic nerve followed by a 5 min application of 100 μM carbachol is shown in Figure 8. Electrical stimulation elicited an increase in [^{3}H]taurine release from the superior cervical ganglion. The second high-frequency stimulus interval appeared to evoke a diminished response compared with the first interval. The release evoked by carbachol was very similar to that evoked by the first 10 Hz stimulus interval with respect to both the time course and peak height. Nearly identical results were obtained in three separate experiments. Preganglionic electrical stimulation of the superior cervical ganglion at supramaximal intensities appears

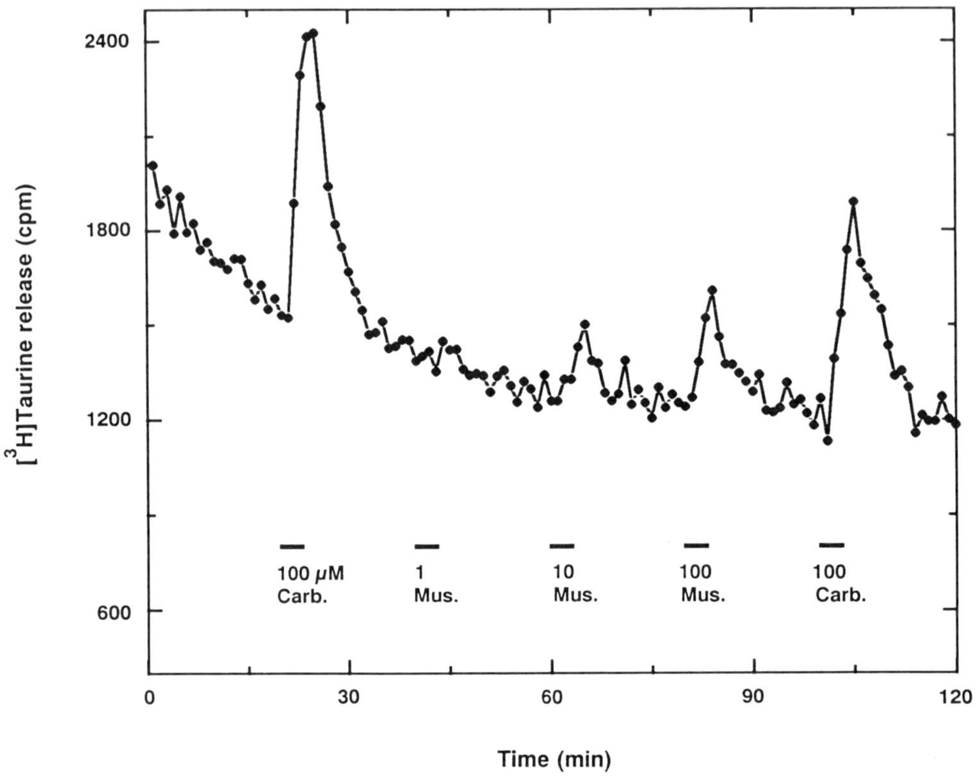

Figure 6. (±)Muscarine (Mus.) stimulates taurine release but much less effectively than carbachol (Carb.).

to produce the same effect on [^{3}H]taurine release as does application of a supramaximal concentration of carbachol.

DISCUSSION

At least four neurotransmitter agonists stimulate [^{3}H]taurine release from the superior cervical ganglion. Carbachol was the most effective of the agonists tested and appears to act through both nicotinic and muscarinic receptors. Electrophysiological stimulation of the preganglionic nerve also increases the release of [^{3}H]taurine, with a response profile similar to that evoked by 100 μM carbachol. It is possible that electrophysiological stimulation is doing the same thing as carbachol stimulation. That is, stimulation of the preganglionic nerve releases acetylcholine from the terminals, which then activates nicotinic and muscarinic receptors, which in turn increase the release of [^{3}H]taurine. To test this hypothesis it will be important to examine the effects of cholinergic antagonists on electrically stimulated taurine release.

At this point we do not know which cells are responsible for releasing taurine. Previous work with GABA and β-alanine indicates that satellite glial cells are the site of ω-amino acid accumulation and release in the superior cervical ganglion (4, 5, 20). Our preliminary autoradiographic findings indicate that [^{3}H]taurine is accumulated by satellite glial cells and not by the cell bodies of the principal neurons in the superior cervical ganglion (unpublished results). However, we cannot eliminate the possibility that preganglionic nerve terminals are also accumulating [^{3}H]taurine and that release from these terminals is selectively increased by preganglionic nerve stimulation. Electrical stimulation of rat brain slices has recently been shown to increase the release of endogenous taurine. The increase was calcium-dependent for slices prepared from the cortex, cerebellum, and hippocampus, but not for the medulla oblongata (8). It will be important to determine whether the release of taurine from the superior cervical ganglion is calcium-dependent. This may provide some clues to the origin of the [^{3}H]taurine since release from glial cells in culture has been shown to be calcium-independent (11, 16).

We have not yet explored the possibility that the stimulated release of [^{3}H]taurine from the superior cervical ganglion is related to the osmoregulatory function ascribed to taurine (14, 17). The neurotransmitter agonists are not likely to be producing a direct osmotic effect since the concentrations used were only 0.1 to 0.2 mOsmolar. Furthermore, electrical stimulation was applied under isosmotic conditions. However, it is possible that both of these stimulus modalities could have produced a localized change in the extracellular ion concentration that would be large enough to have an osmotic effect. Electrical and neurotransmitter stimuli will be applied to superior cervical ganglion under hyper- and hypo-osmotic conditions as employed previously with glial cultures (12) in order to test this possibility.

The stimulus intensity and frequency used in the present study are the same as those previously used to evoke the increased synthesis of cyclic AMP and cyclic GMP in the superior cervical ganglion (6). These changes were also mimicked by carbachol treatment (6). The increase in cyclic AMP occurred in the satellite glia of the superior cervical ganglion while the increase in cyclic GMP was neuronal (2). The isoproterenol-stimulated release of [^{3}H]taurine from LRM55 glial cells is mediated by an increase in cyclic AMP

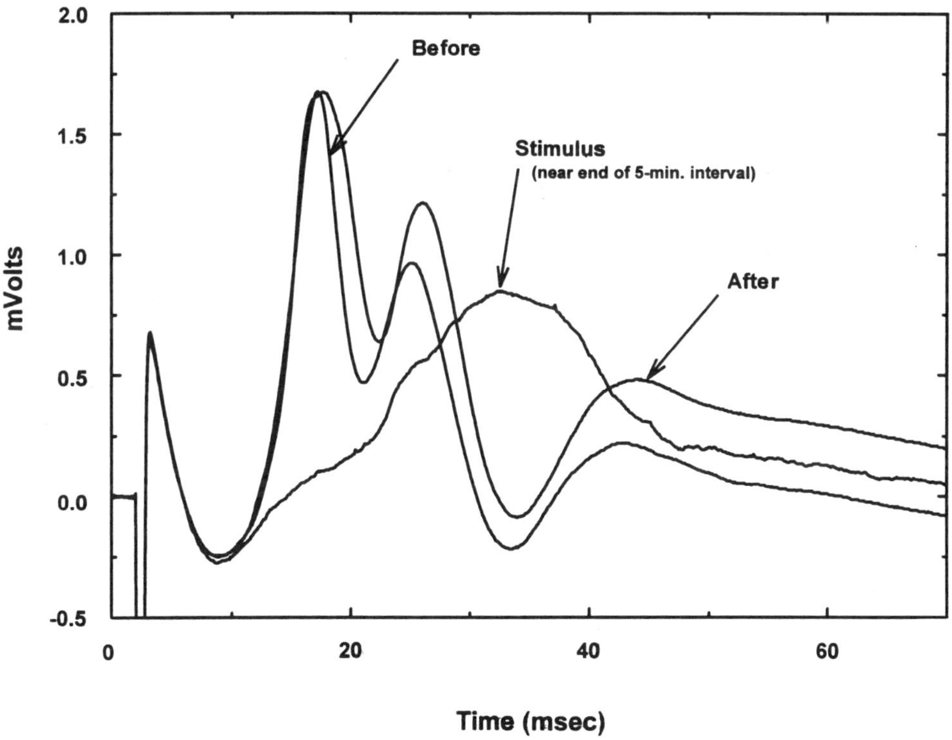

Figure 7. Electrophysiological responses recorded from the postganglionic nerve following stimulation.

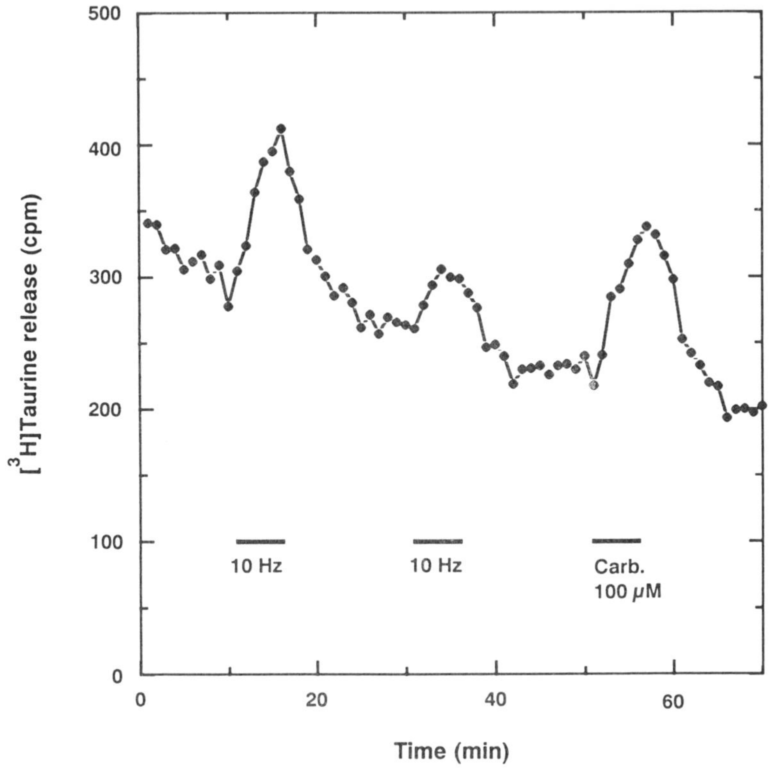

Figure 8. Electrical stimulation of the preganglionic nerve evokes an increased release of [^{3}H]taurine. Carb: carbachol.

synthesis (9). It is difficult to predict how the increases in cyclic nucleotides are related to the release of taurine in the present experiments since the electrical stimulus was applied for 5 min in the present study and only for 30 or 60 s in the previously described study (6). In that study, differences in the ability of antagonists to block the increase in cyclic nucleotide synthesis were observed between these two shorter stimulus intervals.

Electrophysiologically stimulated taurine release may be functionally significant. Muramatsu et al. (13) have demonstrated that taurine inhibits the release of [^{14}C]acetylcholine from the rat superior cervical ganglion. It is possible that taurine release is increased during periods of high neuronal activity. The elevated extracellular taurine may then act upon the preganglionic nerve terminals in the superior cervical ganglion in a short feedback loop to inhibit excessive release of acetylcholine.

REFERENCES

1. Alvarez, M. P., Solas, M. T., Suarez, I., and Fernandez, B., 1989, Glial fibrillary acidic protein-like immunoreactivity in cat satellite cells of sympathetic ganglia, *Acta Anat.*, 136: 9-11.
2. Ariano, M. A., Briggs, C. A., and McAfee, D. A., 1982, Cellular localization of cyclic nucleotide changes in rat superior cervical ganglion, *Cell. Molec. Neurobiol.*, 2: 143-156.
3. Borghini, N., Dalmaz, Y., and Peyrin, L., 1991, Effect of guanethidine on dopamine in small intensely fluorescent cells of the superior cervical ganglion of the rat, *J. Autonom. Nerv. Syst.*, 32: 13-20.
4. Bowery, N. G., and Brown, D. A., 1972, γ-Aminobutyric acid uptake by sympathetic ganglia, *Nature New Biol.*, 238: 89-91.

5. Bowery, N. G., Brown, D. A., White, R. D., and Yamini, G., 1979, [^{3}H]γ-Aminobutyric acid uptake into neuroglial cells of rat superior cervical sympathetic ganglia, *J. Physiol.*, 293: 51-74.
6. Briggs, C. A., Whiting, G. J., Ariano, M. A., and McAfee, D. A., 1982, Cyclic nucleotide metabolism in the sympathetic ganglion, *Cell. Molec. Neurobiol.*, 2: 129-141.
7. Elfvin, L. G., Bjorklund, H., Dahl, D., and Seiger, A., 1987, Neurofilament-like and glial fibrillary acidic protein-like immunoreactivities in rat and guinea-pig sympathetic ganglia in situ and after perturbation, *Cell Tissue Res.*, 250: 79-86.
8. Kubo, T., Takano, A. Tokushige, N., Miyata, N., Sato, M., and Hatakeyama, S., 1992, Electrical stimulation-evoked release of endogenous taurine from slices of the hippocampus, cerebral cortex and cerebellum of the rat, *J. Pharmacobio-Dyn.*, 15: 519-525.
9. Madelian, V., Martin, D. L., Lepore, R., Perrone, M., and Shain, W., 1985, ß-receptor-stimulated and cyclic adenosine 3',5'-monophosphate-mediated taurine release from LRM55 glial cells, *J. Neurosci.*, 5: 3154-3160.
10. Martin, D. L., Shain, W., and Madelian, V., 1988, Receptor-mediated release of taurine from glial cells and signaling between neurons and glia, *in*: "Glial Cell Receptors", H. K. Kimelberg, ed., Raven Press, Ltd., New York, pp. 183-195.
11. Martin, D. L., Madelian, V., and Shain, W., 1989, Spontaneous- and beta-adrenergic receptor-mediated taurine release from astroglial cells do not require extracellular calcium, *J. Neurosci. Res.*, 23: 191-197.
12. Martin, D. L., Madelian, V., Seligmann, B., and Shain, W., 1990, The role of osmotic pressure and membrane potential in K^+-stimulated taurine release from astrocytes and LRM55 cells, *J. Neurosci.*, 10: 571-577.
13. Muramatsu, M., Kakita, K., Nakagawa, K., and Kuriyama, K., 1978, A modulating role of taurine on release of acetylcholine and norepinephrine from neuronal tissues, *Japan J. Pharmacol.*, 28: 259-268.
14. Pasantes-Morales, H., and Schousboe, A., 1988, Volume regulation in astrocytes: A role for taurine as an osmoeffector, *J. Neurosci. Res.*, 20: 505-509.
15. Shain, W. G., and Martin, D. L., 1984, Activation of β-adrenergic receptors stimulates taurine release from glial cells, *Cell. Mol. Neurobiol.*, 4: 191-196.
16. Shain, W., Connor, J. A., Madelian, V., and Martin, D. L., 1989, Spontaneous and beta-adrenergic receptor-mediated taurine release from astroglial cells are independent of manipulations of intracellular calcium, *J. Neurosci.*, 9: 2306-2312.
17. Thurston, J. H., Hauhart, R. E., and Dirgo, J. A., 1980, Taurine: a role in osmotic regulation of mammalian brain and possible clinical significance, *Life Sci.*, 26: 1561-1568.
18. Waniewski, R. A., and Martin, D. L., 1992, Cholinergic stimulation of taurine release from sympathetic ganglia, *Trans. Am. Soc. Neurochem.*, 23: 221.
19. Waniewski, R. A., Martin, D. L., and Shain, W., 1991, Isoproterenol selectively releases endogenous and [^{14}C]-labelled taurine from a single cytosolic compartment in astroglial cells, *Glia*, 4: 83-90.
20. Young, J. A. C., Brown, D. A., Kelly, J. S., and Schon, F., 1973, Autoradiographic localization of sites of [^{3}H]γ-aminobutyric acid accumulation in peripheral autonomic ganglia, *Brain Res.*, 63: 479-486.

CYSTEINE SULFINATE DECARBOXYLASE IN BRAIN: IDENTIFICATION, CHARACTERIZATION AND IMMUNOCYTOCHEMICAL LOCATION IN ASTROCYTES

Marcel Tappaz,[1] Khalid Almarghini,[1] and Kim Do[2]

[1] INSERM U 171. Centre Hospitalier Lyon-Sud. F 69310 Pierre Bénite
[2] Brain Research Institute. University of Zürich. CH Zürich

INTRODUCTION

Evidence that taurine synthesis may occur in the brain was provided by the finding that brain cortex slices produced labelled taurine after incubation with [^{35}S]methionine or [^{35}S]cysteine (39). Radioactive taurine was also found in the brain after intracisternal injection of radiolabelled cysteine (11). Cysteine sulfinate was found as an endogenous compound (3,4) or as a labelled precursor of taurine (36) in brain tissue while cysteamine or cystamine could not be detected (23). Accordingly, biosynthesis of taurine in the brain is thought to occur from cysteine through the so-called cysteine sulfinate pathway; i.e. oxidation of the thiol group of cysteine to form cysteine sulfinic acid, followed by decarboxylation to hypotaurine and subsequent oxidation to taurine. Cysteine sulfinate might be also first oxidized to cysteic acid followed by decarboxylation directly to taurine. In the liver, where this pathway was first demonstrated (22), decarboxylation of both cysteine sulfinate and cysteate was shown to be performed by the same enzyme, called cysteine sulfinate decarboxylase (E.C.4.1.1.29) rather than cysteate decarboxylase since it has a better affinity for cysteine sulfinate than for cysteate (21). In vivo, cysteine sulfinate decarboxylase was suggested to be the rate-limiting enzyme for taurine biosynthesis in liver (14).

BRAIN CYSTEINE SULFINATE DECARBOXYLASE : IDENTIFICATION

In the mid-fifties, brain homogenates were shown to be able to decarboxylate cysteine sulfinate as well as cysteate (13). However, characterization of the cysteine sulfinate decarboxylase activity detected in brain tissue remained elusive for years until two cysteine sulfinate decarboxylase activities were separated from rat brain extracts through ion exchange chromatography. These were named cysteine sulfinate decarboxylases I and II,

Taurine in Health and Disease, Edited by R. Huxtable
and D.V. Michalk, Plenum Press, New York, 1994

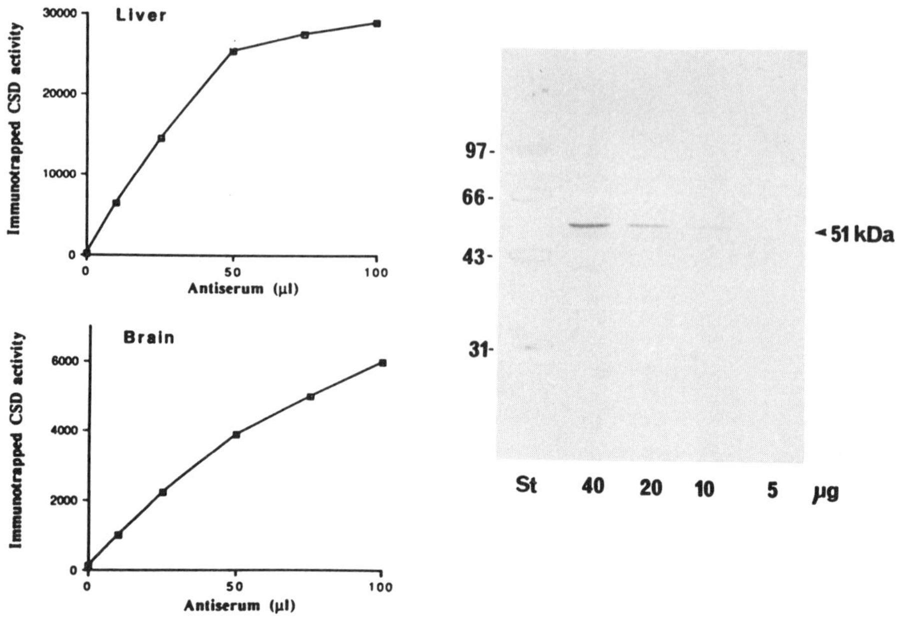

Figure 1. Characterization of the cysteine sulfinate decarboxylase antiserum: Left: Immunotrapping curves of cysteine sulfinate decarboxylase activity: Immunotrapping was carried out by incubating increasing amounts of antiserum made up to 100 µl with preimmune serum with liver or brain homogenate and protein-sepharose. The enzymatic activity of the pellets was determined by measuring carbon dioxide formed from radiolabelled cysteine sulfinate and expressed in arbitrary units (cpm/h). Blank values obtained following incubation with preimmune serum only were around 100 cpm/h. Right: Immunoblots of an intermediate fraction : In this fraction cysteine sulfinate decarboxylase was estimated to represent about 1% of the total protein. Dilution of the antiserum was 1/5000 and the amount of protein in µg for each lane as indicated. Standards (St) on the left were stained reversibly with Ponceau red before immunolabelling. The antiserum labelled a major band at 51 kDa. CSD: Cysteine sulfinate decarboxylase.

according to their elution order (32). Similar results were reported in calf brain extract (47). Further work established the identity of each of them (45). The so-called cysteine sulfinate decarboxylase II, which represents the major contribution (around 80%) to total cysteine sulfinate decarboxylase activity in rat brain extract, was found to be a side activity, expressed in vitro, in the presence of relatively high concentration of cysteine sulfinate, of glutamate decarboxylase, the biosynthetic enzyme for GABA (24,31,32,47). We proposed to call this GAD-associated cysteine sulfinate decarboxylase activity (GAD/CSD). Chronic reduction of the activity of this enzyme in vivo, did not produce a parallel reduction of endogenous taurine content (25). Thus cysteine sulfinate decarboxylase II, i.e. GAD/CSD, does not seem to play a functional role in the biosynthesis of taurine in vivo.

The other enzyme activity, previously called cysteine sulfinate decarboxylase I, which represents about 20% of the total cysteine sulfinate decarboxylase activity in a rat brain extract, had the same kinetic characteristics (Km, substrate specificity, pH sensitivity) as the cysteine sulfinate decarboxylase enzyme activity found in liver (32). Both enzymes showed the same elution pattern on various chromatography columns (32,40). In addition a specific antiserum against liver cysteine sulfinate decarboxylase immunoprecipitated both activities with strictly parallel immunoprecipitation curves and immunolabelled the same protein subunit (MW 51 kDa) on immunoblots of liver homogenate and brain cysteine sulfinate decarboxylase-enriched fraction (40).

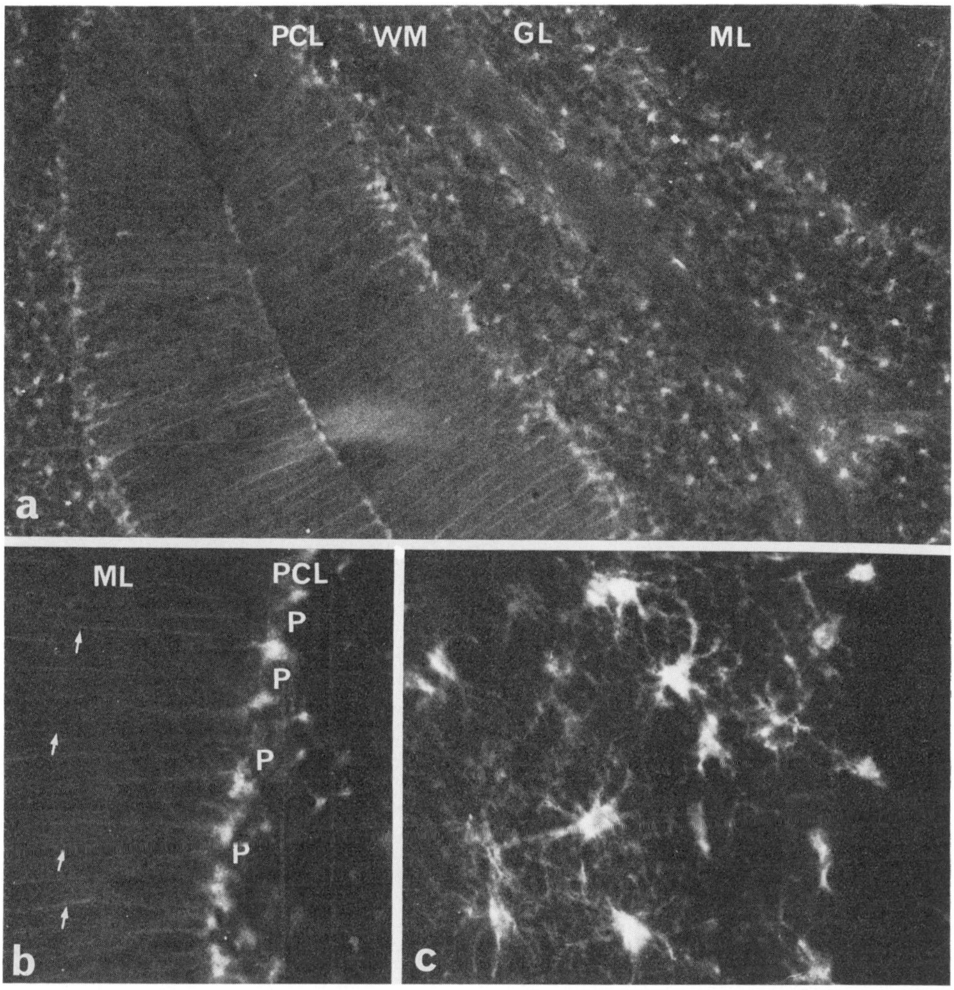

Figure 2. Cysteine sulfinate decarboxylase immunofluorescence in the cerebellar cortex: a) General view of the cerebellar cortex : numerous small cells are immunolabelled in the white matter (WM), in the granular layer (GL), as well as in the Purkinje cell layer (PCL). Faintly immunolabelled radial fibers are visible in the the molecular layer (ML). b) Enlarged view of the Purkinje cell layer. Small immunolabelled cells are satellites of the Purkinje cells (P). They send faintly immunolabelled radial fibers (arrows) to the molecular layer. The Purkinje cells are devoid of any labelling. In the molecular layer no immunolabelled cells were ever observed. c) Enlarged view of the white matter showing numerous star-shaped immunolabelled cells.

Accordingly, cysteine sulfinate decarboxylase I from brain is considered to be identical to cysteine sulfinate decarboxylase from liver, and we proposed to call it simply cysteine sulfinate decarboxylase (45). Of the two cysteine sulfinate decarboxylase activities found in the brain this latter one was shown to have the higher affinity and selectivity for the substrate cysteine sulfinate (32,47). It is thus considered to be the enzyme involved in the biosynthesis of taurine in vivo and consequently represents a putative marker of taurine-producing cells. However, it must be noted that direct demonstration is still missing since no specific inhibitor of cysteine sulfinate decarboxylase is available.

ANTISERA TO CYSTEINE SULFINATE DECARBOXYLASE

A first attempt to produce an antiserum against cysteine sulfinate decarboxylase was reported by Wu et al. (47), using as the antigen a cysteine sulfinate decarboxylase fraction from calf brain. It is not possible to ascertain the quality of this antiserum since it was not characterized in this or subsequent reports by any of the usual tests that are necessary to assess the specificity of an antiserum : sodium dodecyl sulfate (SDS)-polyacrylamide gel electrophoresis (PAGE) of the fraction used for immunization; immunoprecipitation of enzymatic activity by the immunoglobulins fixed on a solid phase; or immunoblots of partially purified fractions. Nonetheless, as it remained the only available antiserum for years, it has been widely used for immunocytochemical studies (8,9,27,28,44) in spite of a striking lack of careful characterization. Taking advantage of the suspected identity of brain and liver enzymes suggested from our early kinetics data (32), we produced an antiserum against cysteine sulfinate decarboxylase using as the antigen an homogeneous liver fraction about enriched about 2000-fold in cysteine sulfinate decarboxylase activity (40). The final fraction was ultimately isolated in the final step through preparative SDS-PAGE; i.e. in its denatured form. This antiserum immunoprecipitated a cysteine sulfinate decarboxylase activity from brain extract and immunolabelled only one band on immunoblots of brain fractions. The antiserum was used to establish the identity of brain cysteine sulfinate decarboxylase I and liver cysteine sulfinate decarboxylase on the basis of immunochemical criteria as recalled above (40).

This antiserum was also used to develop cysteine sulfinate decarboxylase immunocytochemistry in the cerebellum, where numerous small cells were found to be immunolabelled (1). In contrast to the earlier results reported by Chan-Palay et al. (18,9) and Magnusson et al. (27) who both used Wu's antiserum (47), our data provided no evidence of immunolabelling of nerve cell bodies (for example Purkinje cells or stellate cells) or corresponding nerve endings, but showed that the small cell satellites of the Purkinje cells, i.e. the Golgi epithelial cells as well as their Bergmann radial fibers in the molecular layer, were immunolabelled. These findings provided the first evidence for a glial localization of cysteine sulfinate decarboxylase in the brain (1). While the specificity of this antiserum was satisfactory, its use in immunocytochemisty proved difficult since this antiserum recognized primarily the denatured form of cysteine sulfinate decarboxylase on fixed tissue because of the way it had been raised, As a result satisfactory immunolabelling could be obtained only on tissue sections which had been treated by the denaturing agent SDS prior to incubation. Our recent effort in this field has been devoted to the production of a specific anti-cysteine sulfinate decarboxylase antiserum (2) which provides a satisfactory labelling on regularly fixed tissue and permits double immunlabelling in combination with other antisera for identification of the immunolabelled cells.

PRODUCTION AND CHARACTERIZATION OF A NEW ANTI- CYSTEINE SULFINATE DECARBOXYLASE ANTISERUM

Cysteine sulfinate decarboxylase was purified in its native form from rat liver. The purification was performed in eight steps which included conventional chromatography (diethylaminoethylcellulose, hydroxylapatite) followed by high performance liquid chromatography (hydrophobic, adsorption and ion-exchange HPLC). The purification factor was 11 000-fold and the final yield was around 2%. The procedure led to the enrichment of a protein, the molecular weight of which was 51 kDa as determined by SDS-polyacrylamide gel electrophoresis. The final fraction was more than 90%

homogeneous. With this fraction as the antigen, an antiserum was raised in rabbit which quantitatively immunoprecipitated cysteine sulfinate decarboxylase activity from liver and brain extract. This activity was determined by the formation of labelled carbon dioxide from [1-^{14}C]cysteine sulfinate (Fig. 1) as well as by the formation of hypotaurine identified through HPLC analysis from cysteine sulfinate (Fig. 5). In addition, the enzyme immunotrapped from brain and liver extracts formed taurine when incubated with cysteate (Fig. 5). On the other hand, this antiserum immunolabelled specifically one protein (MW 51 kDa) on immunoblots of partially cysteine sulfinate decarboxylase-enriched fractions from liver (Fig. 1). That the 51 kDa protein is constitutive subunit of cysteine sulfinate decarboxylase relies on a body of indirect but consistent evidence (2) which is as follows: a) cysteine sulfinate decarboxylase activity was measured in the final fraction after around an 11,000-fold purification where the 51 kDa protein represented the major constituent on SDS-PAGE; b) During the purification of cysteine sulfinate decarboxylase, the 51 kDa was progressively enriched; c) for a given step of purification, namely hydroxylapatite HPLC, the enrichment of the 51 kDa protein labelled by the antiserum and the enrichment of cysteine sulfinate decarboxylase-specific activity were found to be parallel; d) Finally, as previously reported, we produced an antiserum that was able to immunoprecipitate a cysteine sulfinate decarboxylase activity by using as immunogen the 51 kDa protein isolated through preparative SDS electrophoresis (40). From this set of evidence we conclude that the 51 kDa protein specifically immunolabelled by the antiserum is cysteine sulfinate decarboxylase.

CYSTEINE SULFINATE DECARBOXYLASE IMMUNOCYTOCHEMISTRY

Numerous small cells were immunolabelled by the antiseum throughout the brain on conventionally fixed tissues. We will illustrate the results we have obtained in the cerebellum and hippocampus.

Cerebellum

Small immunolabelled cells were found in the white matter, in the granular layer and in the Purkinje cell layer (Fig. 2). At this latter level they were located close to the Purkinje cells and sent faintly immunolabelled radial fibers through the molecular layer. These cells were identified as Golgi epithelial cells and their Bergmann fibers by their location and shape. Our observations did not provide any evidence of immunolabelling of the known nerve cells or nerve endings in the cerebellar cortex. The Purkinje cell bodies were devoid of any labelling. No staining was observed that could correspond to the stellate cell bodies in the molecular layer. Moreover, we never observed any immunoreactive puncta typical of the nerve endings of local interneurons in the cortical layers; e.g. nerve endings of stellate cells in the molecular layer or those of Golgi cells in the cerebellar glomeruli of the granular layer. No immunoreactive puncta were visible in the deep cerebellar nuclei where the Purkinje nerve endings are located. Qualitatively the immunolabelling was similar to the labelling which we observed previously (1) with an anti-cysteine sulfinate decarboxylase antiserum that had been raised using the 51 kDa protein as the immunogen (40). In particular, both antisera similarly immunolabelled the Golgi epithelial cells and their Bergmann fibers and did not reveal any detectable labelling over identified nerve cells or endings in the cerebellar cortex. In the white matter however, we did not confirm the light immunolabelling we observed previously (1) on the oligodendrocytes in the white matter.

Hippocampus

Numerous small cells were immunolabelled in most layers of the gyrus dentate and of the Ammon's horn (Fig. 3). No cell was found immunolabelled in the granular cell layer and virtually none in the pyramidal cell layers. In the hilus of the dentate gyrus, the immunolabelled cells showed a strong staining and a typical star-like shape. At the border of the hilus they send processes into the granular layer. In the molecular layer the immunolabelled cells as well as their processes were oriented perpandicular to the layer main axis. Numerous immunolabelled cells were found in the stratum lacunosum of the Ammons' horn. Their general distribution tended to be parallel to the main axis of the layer. Immunolabelled processes abuting on the numerous blood vessels of this layer were commonly observed at this level. Small immunolabelled cells were scattered in the stratum radiatum and in the stratum oriens. In the alveus the labelled cells showed a strong immunolabelling. In the subiculum a high density of strongly immunolabelled cells was also observed. Finally small immunolabelled cells with short and fine processes were also scattered in the fimbria and in the corpus callosum. Immunolabelled puncta were never seen in the hippocampus.

DOUBLE IMMUNOLABELLING FOR CYSTEINE SULFINATE DECARBOXYLASE AND GLIAL FIBRILLARY ACIDIC PROTEIN

The small size and star-like shape of the immunolabelled cells, their location within the white matter, the typical orientation of their processes, which in many cases abut directly on blood vessels, suggested that they were astrocytes. To ascertain this interpretation we have carried out a double immunofluorescence study (Fig. 4) with glial fibrillary acidic protein, an established marker of astrocytes (5,15). Quantitative analysis revealed that the overwhelming majority (more than 97%) of cysteine sulfinate decarboxylase-positive cells in the granular layer and in the white matter of the cerebellum were also positive for glial fibrillary acidic protein. Similarly between 85 and 95% of cysteine sulfinate decarboxylase-positive cells, depending upon the layer, were also immunolabelled for glial fibrillary acidic protein in the hippocampus. These results demonstrated that the cellular location of cysteine sulfinate decarboxylase is restricted to astrocytes in those regions.

SPECIFICITY OF CYSTEINE SULFINATE DECARBOXYLASE FOR SULFUR-CONTAINING AMINO ACIDS

The cellular location of cysteine sulfinate decarboxylase in astrocytes in the cerebellum substantially differs from the location of taurine (see paragraph below) but its location in astrocytes, especially in Golgi epithelial cells and Bergmann radial fibers, closely resembled that of another sulfur-containing amino acid, homocysteic acid (17,49). These observations prompted us to investigate whether or not cysteine sulfinate decarboxylase could be involved in the decarboxylation of homocysteate. We have studied in vitro the specificity of cysteine sulfinate decarboxylase for cysteine sulfinate, and homocysteate, as well as for the related analogs, homocysteine sulfinate and cysteate. Cysteine sulfinate decarboxylase was first immunotrapped from brain and liver tissue extracts, using the specific anti-cysteine sulfinate decarboxylase antiserum, and then incubated with the various sulfur-containing amino acids. Reaction products were identified and quantified by HPLC with precolumn o-phthalaldehyde derivatization. Cysteate (10 mM) as well homocysteate inhibited the formation of hypotaurine from

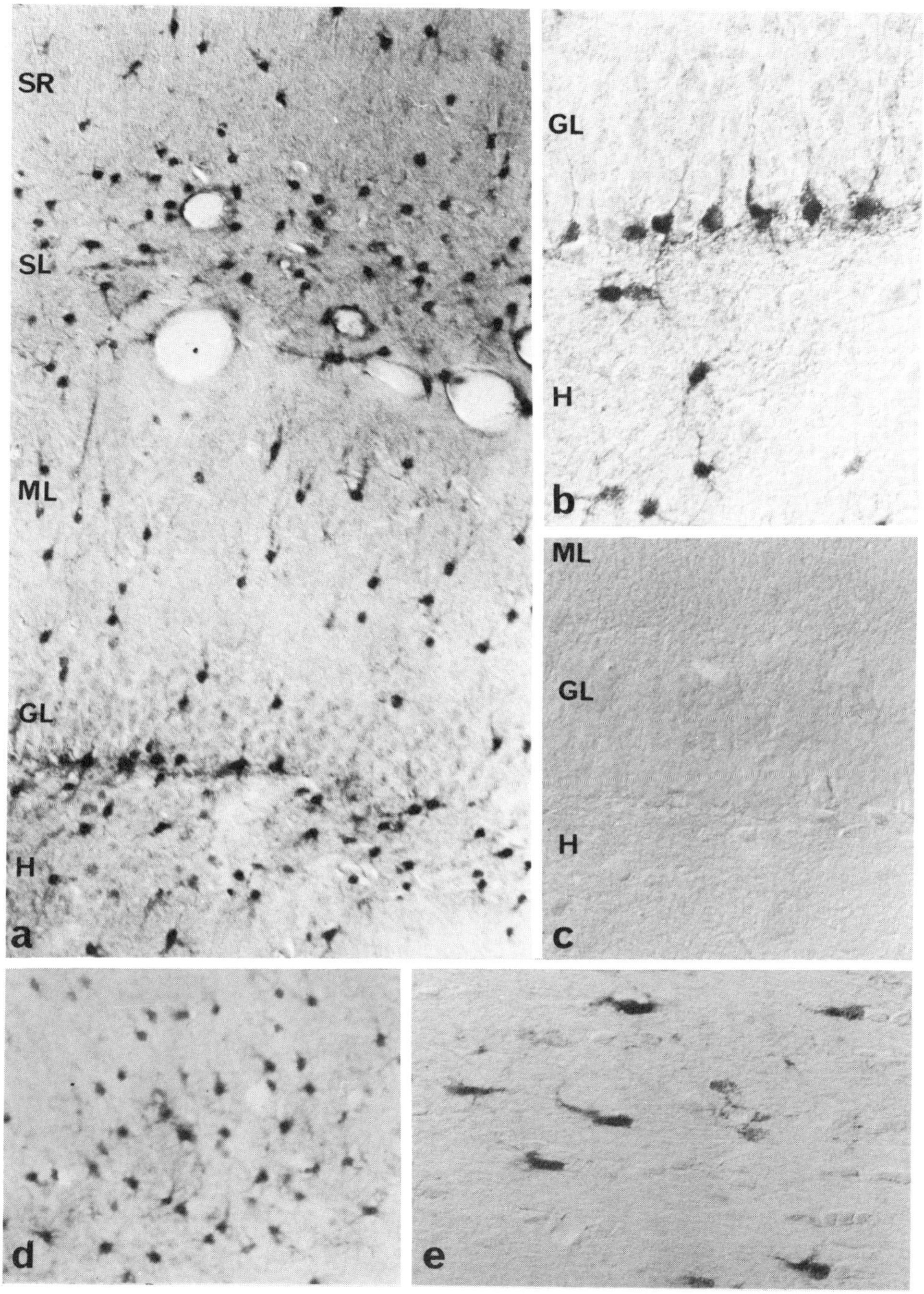

Figure 3. Cysteine sulfinate decarboxylase immunocytochemistry in the hippocampus: a) General view showing numerous immunolabelled cells in the various hippocampal layers, hilus (H), granular layer (GL), and molecular layer (ML) of the gyrus dentate as well as stratum lacunosum (SL) and stratum radiatum (SR) of the Ammon's horn. b) Enlarged view showing immunolabelled cells at the border of the hilus that send processes within the granular layer. The granule cells are devoid of immunoreactivity. c) Control with preimmune serum. d) Numerous small immunolabelled cells are visible in the subiculum. e) Elongated isolated cells are immunolabelled in the corpus callosum.

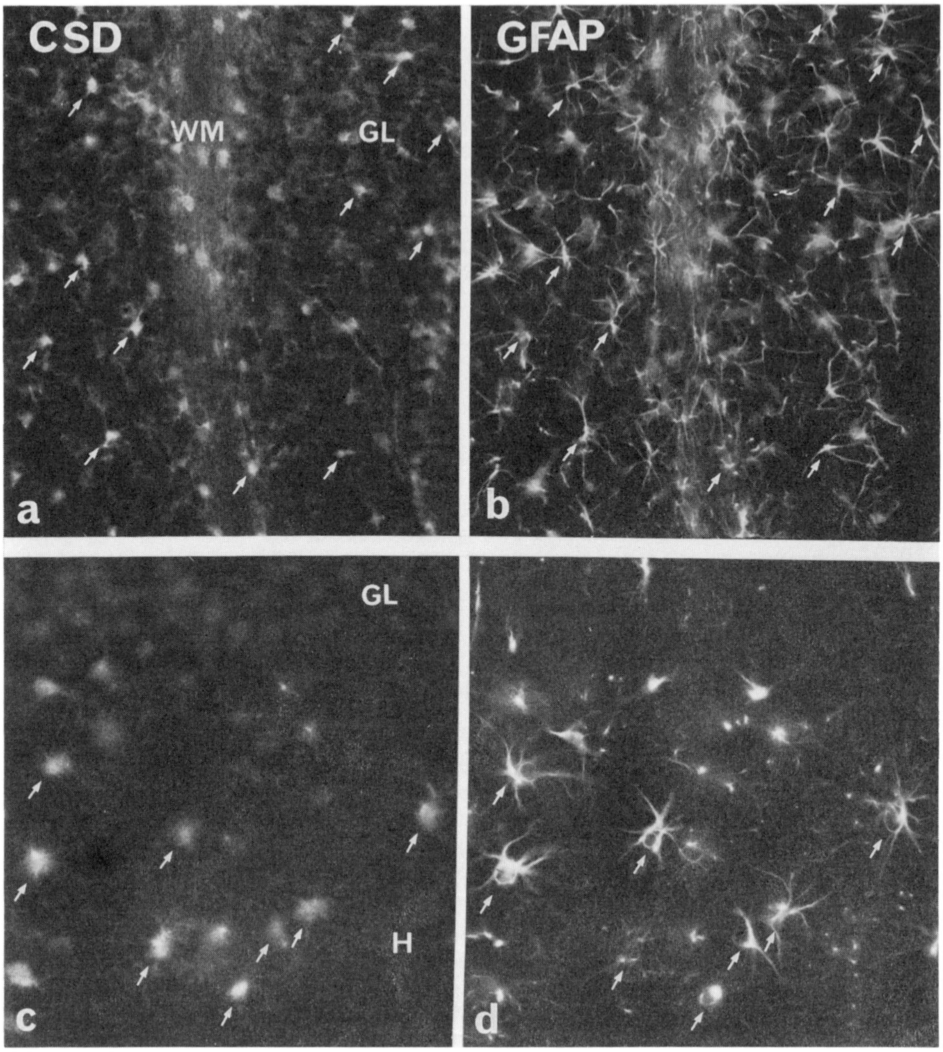

Figure 4. Double immunofluorescence for cysteine sulfinate decarboxylase and glial fibrillary acidic protein in the cerebellum (a,b) and hippocampus (c,d). Sections were incubated first with rabbit anti-cysteine sulfinate decarboxylase antiserum and mouse anti-glial fibrillary acidic protein monoclonal antibodies, and then with appropriate fluorescein- and rhodamine-labelled secondary antibodies. Observation was performed with a fluorescein filter (cysteine sulfinate decarboxylase) on the left and a rhodamine filter (glial fibrillary acidic protein) on the right. Most cysteine sulfinate decarboxylase-immunolabelled cells are also positive for glial fibrillar acidic protein. Some examples are indicated (arrow). GL: granular layer. WM: white matter; H: hilus; GFAP: glial fibrillary acidic protein; CSD: cysteine sulfinate decarboxylase.

cysteine sulfinate (0.25 mM) by around 75%, while homocysteine sulfinate (10 mM) did not elicit significant inhibition. Incubation with 25 mM of cysteine sulfinate or cysteate led to the formation of hypotaurine and taurine. The amount of taurine was about 40% that of hypotaurine. In contrast, the amount of 3-aminopropane sulfonic acid, the decarboxylated reaction product of homocysteate, was less than 0.1% of the amount of hypotaurine formed from cysteine sulfinate under the same conditions (Fig. 5). These results show that cysteine sulfinate decarboxylase has a high specificity for cysteine

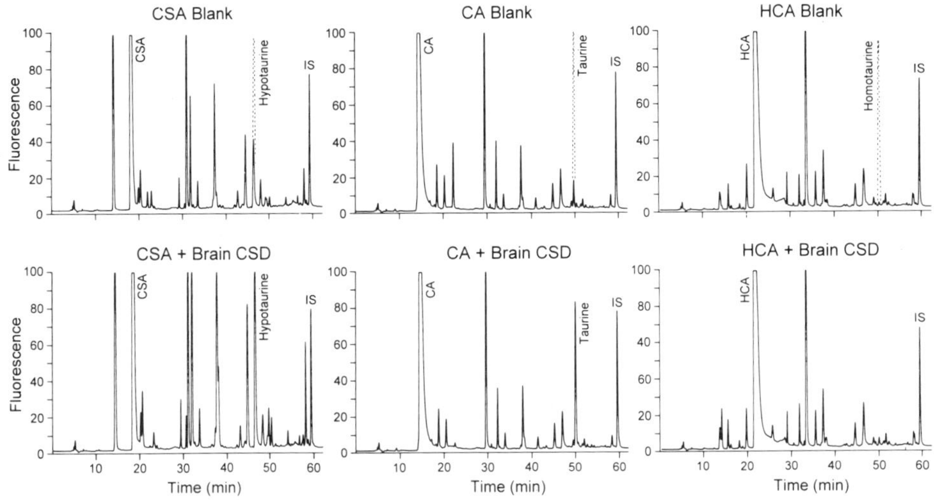

Figure. 5. HPLC analysis of the reaction products formed from cysteine sulfinate, cysteate and homocysteate by cysteine sulfinate decarboxylase immunotrapped from brain extracts. Chromatograms are shown of the o-phthaldehyde derivatives following incubation of cysteine sulfinate, cysteate or homocysteate with the Protein A-Sepharose pellets preincubated with brain extracts and antiserum or preimmune serum (blank). Spiking of the blank with addition of the corresponding decarboxylated reaction product, hypotaurine, taurine or homotaurine respectively, is indicated by a dotted line. While hypotaurine was formed from cysteine sulfinate and taurine from cysteate, no homotaurine could be detected following incubation with homocysteate. CSA: cysteine sulfinate; CA: cysteate; CSD: cysteine sulfinate decarboxylase; HCA: homocysteate.

sulfinate and cysteate which are precisely the sulfur-containing amino acids involved in the biosynthesis of taurine. Homocysteate is a weak inhibitor of cysteine sulfinate decarboxylase but does not appear to be a substrate for cysteine sulfinate decarboxylase in vitro. Accordingly, cysteine sulfinate decarboxylase is unlikely to play a role in the metabolism of homocysteate in vivo.

THE CELLULAR LOCATION OF TAURINE VERSUS CYSTEINE SULFINATE DECARBOXYLASE

While our results clearly establish that cysteine sulfinate decarboxylase is localized in astrocytes in the cerebellum and hippocampus, previous immunocytochemical investigations showed that taurine was primarily located in nerve cells at this level (6,10,20,26,27,28,34,35,48). Electron microscopic observations indicated that some taurine immunolabelling was also present in glial cells (35). The discrepancy between the cellular location of taurine and of its putative biosynthetic enzyme cysteine sulfinate decarboxylase may be explained in several ways. It may result from methodological problems concerning taurine immunocytochemistry. In spite of careful control of the lack of cross-reactivity with related compounds, one cannot entirely rule out the possibility of false positive findings. Thus, it has been reported that a taurine antiserum can label a taurine derivative (46). Another explanation could be that the cysteine sulfinate pathway is of minor importance for the synthesis of taurine in brain tissue. In peripheral tissues, several possible pathways have been proposed, the physiological significance of which remains to be established (For review see 22). The so-called cysteamine pathway appeared to be responsible for the synthesis of taurine in tissues such as heart which are devoid of cysteine sulfinate decarboxylase activity (7). Since no metabolic intermediates could be detected, this pathway is unlikely to occur in the brain. Another possibility could be that a substantial part of taurine in brain may be supplied by blood. In some species such as cat, it is well-known that taurine concentration in brain decreases following a sustained taurine-free diet (43). Finally, the discrepancy between the cellular location of taurine and cysteine sulfinate decarboxylase could be linked to the unusual metabolic half-life of taurine, which is around five days (11). As a result, the cellular distribution of taurine could primarily reflect the intercellular flux of taurine through release, uptake and passive diffusion after its synthesis in astrocytes. According to this hypothesis it is plausible that the cells that store taurine could be different from the cells that synthesize taurine.

POSSIBLE PHYSIOLOGICAL SIGNIFICANCE OF THE LOCATION OF CYSTEINE SULFINATE DECARBOXYLASE IN ASTROCYTES

The location of cysteine sulfinate decarboxylase in astrocytes indicates that taurine is synthetized in these glial cells and suggest that taurine could be released by stimulation of the receptors present on these cells (29). Indeed, stimulation of ß-adrenergic receptors has been reported to stimulate the release of taurine from astrocytes (30,41). Taurine released in such a way could play various physiological roles (For review see 19). Since taurine is known to affect the electrical activity of nerve cells (12,33), taurine could be a neuroactive signal of glial origin which modulates the activity of neurons in the immediate vicinity; i.e. it could be a possible gliotransmitter. But taurine might play other roles as a protective or regulatory agent released from astrocytes. Thus, it could be a trophic agent involved in the organisation of some neural network (42,43). Taurine could also participate to possible regulatory functions linked to its capacity to bind to membranne phospholipids (18). Finally taurine might be an osmoregulator in the central nervous system (37,38) as it has been established in fish erythrocytes (16).

CONCLUSION

Using a well-characterized specific anti-cysteine sulfinate decarboxylase antiserum and double quantitative immunofluorescence, we report that the cells immunolabelled for cysteine sulfinate decarboxylase are astrocytes in the cerebellum and the hippocampus. These results strongly disprove previous reports about the location of cysteine sulfinate decarboxylase in nerve cells obtained using an antiserum, the specificity of which was in no way established. Cysteine sulfinate decarboxylase specificity appears to be restricted to the substrates cysteine sulfinate and cysteate, which are involved in the biosynthesis of taurine. Our findings provide further evidence for a new emerging hypothesis which considers taurine to be a general metabolic regulator of glial origin in the central nervous system.

ACKNOWLEDGEMENTS

This study was supported by grants from the Institut National de la Santé et de la Recherche Médicale (INSERM U 171), the Centre National de la Recherche Scientifique (CNRS URA 1195), the Université Claude Bernard Lyon I and the Swiss National foundation (Grant 31-27822-89). The authors thank B. Barbagli and B. Benz for their valuable technical assistance.

REFERENCES

1. Almarghini, K., Remy, A. and Tappaz, M., 1991, Immunocytochemistry of the taurine biosynthesis enzyme, cysteine sulfinate decarboxylase (CSD), in the cerebellum: evidence for a glial localization, *Neuroscience*, 43: 111-119.
2. Almarghini K, Barbagli, B. and Tappaz, M., 1994, Production and characterization of a new specific antiserum against the taurine putative biosynthetic enzyme cysteine sulfinate decarboxylase, *J. Neurochem.*, In press.
3. Baba, A., Yamagami, S., Mizuo, H. and Iwata, H., 1980, Microassay of cysteine sulfinic acid by an enzymatic cycling method, *Anal. Biochem.*, 101: 288-293.
4. Bergeret, B. and Chatagner, F., 1954, Sur la présence d'acide cysteine sulfinique dans le cerveau du rat normal, *Biochim. Biophys. Acta.*, 14: 297.
5. Bignami, A., Eng, L. F., Dahl, D. and Uyeda, C. T., 1972, Localization of the glial fibrillary acidic protein in astrocytes by immunofluorescence , *Brain Res.*, 43: 429-435.
6. Campistron, G., Geffard, M. and Buijs, R. M., 1986, Immunological approach to the detection of taurine and immunocytochemical results, *J. Neurochem.*, 46: 862-68.
7. Cavallini, D., Gaull, G.E. and Zappia, V., 1980, Natural sulfur compounds: novel biochemical and structural aspects. Plenum Press, New-York.
8. Chan-Palay, V., Lin, C. T., Palay, S., Yamamoto, M. and Wu, J. Y., 1982, Taurine in the mammalian cerebellum: demonstration by autoradiography with 3H taurine and immunocytochemistry with antibodies against the taurine-synthesizing enzyme, cysteine sulfinic acid decarboxylase, *Proc. Natl.Acad. Sci. USA.*, 79: 2695-2699.
9. Chan-Palay, V., Palay, S. L. and Wu, J. Y., 1982, Sagittal cerebellar microbands of taurine neurons: immunocytochemical demonstration by using antibodies against the taurine-synthesizing enzyme cysteine sulfinic acid decarboxylase, *Proc. Natl. Acad. Sci. USA.*, 79: 4221-4225.
10. Clements, J. R., Magnusson, K. R. and Beitz, A. J., 1989, Ultrastructural description of taurine-like immunoreactive cells and processes in the rat hippocampus, *Synapse*, 4: 70-79.
11. Collins, G. G. S., 1974, The rates of synthesis, uptake and disappearance of (14C)-taurine in eight areas of the rat central nervous system, *Brain Res.*, 76: 447-459.
12. Curtis, D. R. and Watkins, J. C., 1965, The pharmacology of amino acids related to gamma-aminobutiric acid, *Pharmacol. Rev.*, 17: 347-391.
13. Davison, A. N., 1956, Amino acid decarboxylases in rat brain and liver, *Biochim Biophys. Acta.*, 19: 66-73.
14. De La Rosa, J. and Stipanuk, M. H., 1985, Evidence for a rate-limiting role of cysteine sulfinate decarboxylase activity in taurine biosynthesis in vivo, *Comp. Biochem. Physiol.*, 81B: 565-571.
15. Eng, L. F., 1982, The glial fibrillary acidic protein: the major protein constituent of glial filaments, *Scand. J. Immunol.*, 15 Suppl. 9: 41-51.
16. Fugelli, K. and Thoroed, S., 1986, Taurine transport associated with cell volume regulation in flounder erythrocytes under anisosmotic conditions, *J. Physiol. - London*, 374: 245-261.
17. Grandes, P., Do, K. D., Morino, P., Cuenod, M. and Streit, P., 1991, Homocysteate, an excitatory transmitter candidate localized in glia, *Eur. J. Neurosci.*, 3: 1370-1373.
18. Huxtable, R. J. and Sebring, L. A., 1986, Towards a unifying theory for the actions of taurine, *Trends Pharmacol. Sci.*, 7: 481-485.
19. Huxtable, R. J., 1990, Physiological actions of taurine, *Physiol. Rev.*, 72: 101-163.
20. Ida, S., Kuriyama, K., Tomida, Y. and Kimura, H., 1987, Antisera against taurine: quantitative characterization of the antibody specificity and its application to immunohistochemical study in the rat brain, *J. Neurosci. Res.*, 18: 626-631.
21. Jacobsen, J. G., Thomas, L. L. and Smith L. H. Jr., 1964, Properties and distribution of mammalian L-cysteine sulfinate carboxylyases, *Biochim. Biophys. Acta.*, 85: 103-116.
22. Jacobsen, J. G. and Smith L.H. Jr., 1968, Biochemistry and physiology of taurine and taurine derivatives, *Physiol. Rev.*, 48: 424-511.
23. Kuriyama, K., Ida, S. and Okhuma, S., 1984, Alteration of cerebral taurine biosynthesis in spontaneously hypertensive rats, *J. Neurochem.*, 42: 1600-1606.
24. Legay, F., Weise, V. K., Oertel, W. H. and Tappaz, M. L., 1987, Taurine biosynthesis in rat brain: a new and sensitive microassay of cysteine sulfinic acid decarboxylase (CSDI) activity through a selective immunotrapping and its use for distribution studies, *J. Neurochem.*, 48: 345-351.
25. Legay, F., Lecestre, D. and Tappaz, M. L., 1987, Taurine biosynthesis in rat brain in vivo: lack of relationship with cysteine sulfinic acid decarboxylase/glutamate decarboxylase associated activity (GAD/CSDII), *J. Neurochem.*, 48: 340-344.

26. Madsen, S., Ottersen, O. P. and Storm-Mathisen, J., 1985, Immunocytochemical visualization of taurine: neuronal localization in the rat cerebellum, *Neurosci. Lett.*, 60: 255-260.
27. Magnusson, K. R., Madl, J. E., Clements, J. R., Wu, J. Y., Larson, A. A. and Beitz, A. J., 1988, Colocalization of taurine- and cysteine sulfinic acid decarboxylase-like immunoreactivity in the cerebellum of the rat with monoclonal antibodies against taurine, *J. Neurosci.*, 8: 4551-4564.
28. Magnusson, K. R., Clements, J. R., Wu, J. Y. and Beitz, A. J., 1989, Colocalization of taurine- and cysteine sulfinic acid decarboxylase-like mmunoreactivity in the hippocampus of the rat, *Synapse*, 4: 55-69.
29. Martin, D. L., 1992, Synthesis and release of neuroactive substances by glial cells, *Glia*, 5: 81-94.
30. Martin, D. L. and Shain, W., 1993, beta-Adrenergic-agonist stimulated taurine release from astroglial cells is modulated by extracellular $[K^+]$ and osmolarity, *Neurochem. Res.*, 18: 437-444.
31. Oertel, W. H., Tappaz, M., Kopin, I. J., Ranson, D. H. and Schmechel, D. E., 1980, Antiserum to rat brain glutamate-cysteine sulfinate decarboxylase, *Brain Res. Bull.*, 5: 713-719.
32. Oertel, W. H., Schmechel, D. E., Weise, V. K., Ranson, D. H., Tappaz, M., Krutzsch, H. C. and Kopin, I. J., 1981, Comparison of cysteine sulphinic acid decarboxylase isoenzymes and glutamic acid decarboxylase in rat liver and brain, *Neuroscience*, 6: 2701-2714.
33. Okamoto, K., Kimura K, and Sakai, Y., 1983, Taurine-induced increase of the Cl-conductance of cerebellar Purkinje cell dendites in vitro, *Brain Res.*, 259: 319-323.
34. Ottersen, O. P., Madsen, S., Meldrum, B. S. and Storm-Mathisen, J., 1985, Taurine in the hippocampal formation of the senegalese baboon, Papio papio: an immunocytochemical study with an antiserum against conjugated taurine, *Exp. Brain Res.*, 59: 457-462.
35. Ottersen, O. P., 1988, Quantitative assessment of taurine-like immunoreactivity in different cell types and processed in rat cerebellum: an electronmicroscopic study based on a postembedding immunogold labelling procedure, *Anat. Embryol.*, 178: 407-421.
36. Pasantes-Morales, H., Chatagner, F. and Mandel, P., 1980, Synthesis of taurine in rat liver and brain in vivo, *Neurochem. Res.*, 5: 441-451.
37. Pasantes-Morales, H. and Schousboe, A., 1988, Volume regulation in astrocytes: a role for taurine as an osmoeffector, *J. Neurosci. Res.*, 20: 505-509.
38. Pasantes-Morales, H. and Schousboe, A., 1989, Release of taurine from astrocytes during potassium-evoked swelling, *Glia*, 2: 45-50.
39. Peck, E. J. and Awapara, J., 1967, Formation of taurine and isothionic acid in rat brain, *Biochim. Biophys. Acta.*, 141: 499-506.
40. Remy, A., Henry, S. and Tappaz, M., 1990, Specific antiserum and monoclonal antibodies against taurine biosynthesis enzyme cysteine sulfinate decarboxylase (CSD): identity of brain and liver enzyme, *J. Neurochem.*, 54: 870-879.
41. Shain, W. G. and Martin, D. L., 1984, Activation of beta-adrenergic receptors stimulates taurine release from glial cells, *Cell. Mol. Neurobiol.*, 4: 191-196.
42. Sturman, J. A., 1993, Taurine in development, *Physiol. Rev.*, 73: 119-147.
43. Sturman, J. A., Moretz, R. C., French, J. H. and Wisniewski, H. M., 1985, Taurine deficiency in the developing cat: persistence of the cerebellar external granule cell layer, *J. Neurosci. Res.*, 13: 405-416.
44. Taber, K. H., Lin, C. T., Liu, J. W., Thalman, R. H. and Wu, J. Y., 1986, Taurine in hippocampus: localization and postsynaptic ation, *Brain Res.*, 386: 113-121.
45. Tappaz, M., Almarghini, K., Legay, F. and Remy, A., 1992, Taurine biosynthesis enzyme cysteine sulfinate decarboxylase (CSD) from brain - the long and tricky trail to identification, *Neurochem. Res.*, 17: 849-859.
46. Tomida, Y. and Kimura, H., 1987, Immunohistochemical and biochemical studies of substances with taurine-like immunoreactivity in the brain, *Acta Histochem. Cytochem.*, 20: 31-40.
47. Wu, J. Y., 1982, Purification and characterization of cysteic acid and cysteine sulfinic acid decarboxylase and L-glutamate decarboxylase from bovine brain, *Proc. Natl. Acad. Sci. (USA).*, 79: 4270-4274.
48. Yoshida, M., Karasawa, N., Ito, M., Sakai, M. and Nagatsu, I., 1986, Demonstration of taurine-like immunoreactive structures in the rat brain, *Neurosci. Res.*, 3: 356-363.
49. Zhang, N. and Ottersen, O. P., 1992, Differential cellular distribution of two sulphur-containing amino acids in rat cerebellum - an immunocytochemical investigation using antisera to taurine and homocysteic acid, *Exp. Brain Res.*, 90: 11-20.

RELATIONS OF TAURINE RELEASE AND INFLUX TO CELL VOLUMES IN CEREBRAL CORTICAL SLICES

Simo S. Oja and Pirjo Saransaari

Tampere Brain Research Center,
Department of Biomedical Sciences,
University of Tampere, Box 607,
SF-33101 Tampere, Finland

INTRODUCTION

Taurine has been assigned two types of putative physiological functions in the central nervous system. On the one hand, taurine has been shown to cause hyperpolarization, to increase membrane chloride conductance and to inhibit neuronal activity (14). These effects have initiated hypotheses of taurine acting as an inhibitory neurotransmitter (2) or - more likely - a neuromodulator (15). On the other hand, already a number of years ago taurine was recognized as an osmoregulator in marine animals (27), and recently it has been strongly advocated to possess the same role also in the brains of terrestrial animals, including mammals and man (20, 32, 33). According to this thinking the stimulus-evoked release of taurine from nervous tissue is only a consequence of excitation-coupled cell swelling (25).

It has not been too easy to decide which one of the two postulated roles of taurine is more likely or whether they both contain some element of truth. On one hand, during the last two decades evidence has been mounting of the inhibitory actions of taurine in different preparations from the central nervous system (2, 4, 14, 19). On the other, during the last few years many studies have indicated, mainly on cultured astrocytes and neurons (5, 17, 26) but in some cases also with microdialysis in vivo (28, 29), that taurine is released from the cells by hypoosmolarity, while hyperosmolarity blocks or attenuates the stimulus-evoked taurine release. It should also be kept in mind that excitation of nervous tissue invariably tends to be associated with an increase in cell volumes (34). We have now endeavored to shed light on the two postulated roles of taurine in the brain by analyzing the properties of taurine release and influx under conditions in which the ionic composition and osmolarity of the incubation media were varied.

MATERIAL AND METHODS

Influx Studies

Slices 0.4 mm thick were manually cut from the cerebral cortices of young adult 3-month-old white NMRI mice of both sexes. The standard incubation medium used was Hepes-buffered Krebs-Ringer solution, containing (in mM) NaCl 127, KCl 5, $CaCl_2$ 0.8, $MgSO_4$ 1.3, Hepes 15, NaOH 11, and D-glucose 10 (pH 7.4), modified in most experiments as indicated elsewhere. The slices were first preincubated for 30 min in plain medium, then [^{3}H]taurine was added together with varying amounts of unlabeled taurine and the incubations continued for 20 min. During this period the influx rate was approximately constant. At the end the slices were extracted with 5% trichloracetic acid solution and the extracts subjected to scintillation counting. The amount of labeled taurine penetrating into the intracellular spaces of the slices was calculated by subtracting the amount of label in the extracellular spaces and medium adherent to slices from the total (13).

The kinetic parameters for [^{3}H]taurine influx were estimated by nonlinear regression analysis using the simplex algorithm and inversely weighted experimental data. The estimation was also checked by the method of Marquardt with equally, statistically and proportionally weighted data. The fitting was done with equations containing one or two terms conforming to saturable (Michaelis-Menten) kinetics together with one term conforming to nonsaturable (diffusional) kinetics.

Efflux Studies

The slices were preloaded for 30 min with 10 µM [^{3}H]taurine in the above Krebs-Ringer-Hepes medium and subsequently superfused with the same medium for 50 min (8). The ionic composition of medium was altered for the last 20 min of superfusion as defined later. The efflux rate constants for taurine were computed from the amounts of label released in the course of superfusion as described in Kontro and Oja (8). The superfusion period of 20 to 30 min represented the baseline prestimulation release of labeled taurine (rate constant k_1). In order to minimize random experimental variation the calculated efflux rate constants k_2 (stimulated release) for the superfusion period of 34 to 50 min were generally calculated and given as per cent of the constant k_1 (20-30 min) of each individual slice.

The release of endogenous taurine was studied in a number of experiments. The slices were then preincubated and superfused as above but omitting labeled taurine. Taurine in the slice extracts and superfusate fractions was determined by high-pressure liquid chromatography (7).

Estimation of Cell Volumes

The extracellular spaces in the slices were estimated by adding 2 g/l [^{3}H]inulin to medium for the last 20 min of experiments (11). The total volume (weight) minus the inulin space was assumed to represent the intracellular space in the slices. Total swelling or shrinking of the slices during experiments was determined by measuring the fresh and dry weights of the slices in all different experimental conditions (31). Intracellular swelling or shrinking of the slices during the experiments could then be calculated with the aid of estimated alterations in the extracellular spaces and volumes (weights).

RESULTS

The release of both endogenous and preloaded labeled taurine tended to decline slightly during superfusions (Fig. 1). The relative K^+ stimulation was somewhat greater with endogenous than with labeled taurine, but both release processes had a similarly prolonged time course. When the K^+ stimulation was carried out in hyperosmotic medium without any concomitant reduction in Na^+, the response was fairly similar, nor did it markedly differ in magnitude from the response to 50 mM K^+ in isoosmotic medium. A 25 mM reduction in NaCl likewise evoked a significant release of endogeous and labeled

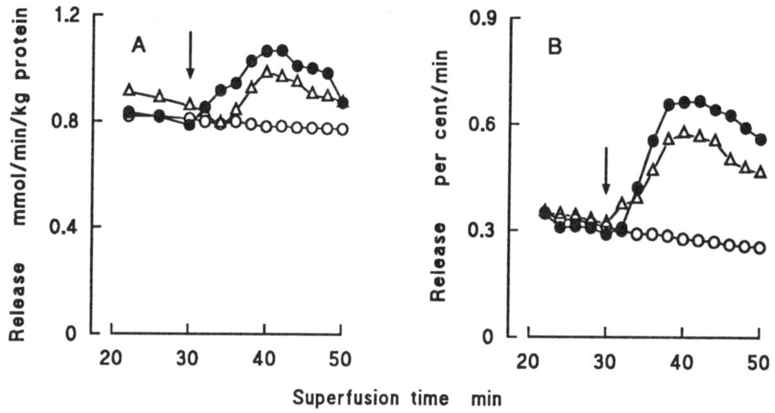

Figure 1. Release of endogeous (A) and preloaded labeled (B) taurine from superfused cerebral cortical slices. At 30 min (indicated by the arrows) the medium was left unchanged (O) or changed for 20 min to a new medium in which 50 mM Na^+ was replaced by equimolar K^+ (●) or to which 50 mM KCl was added (▵). The results, given per protein content of incubated slices, are mean values from 4-8 experiments with SEM varying from 0.5 to 9.4 per cent of the mean.

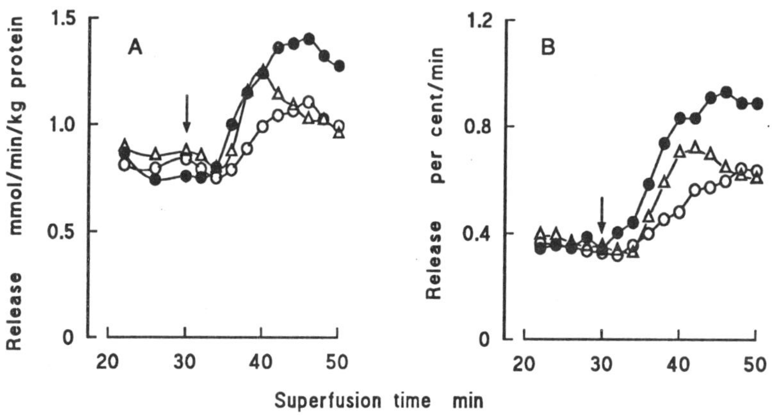

Figure 2. Release of endogenous (A) and preloaded labeled (B) taurine from superfused cerebral cortical slices. At 30 min (indicated by the arrows) the medium was changed for 20 min to a new medium from which 25 mM NaCl was omitted (▵) or all Cl^- was replaced by equimolar acetate (●) or gluconate (O). The results, given per protein content of incubated slices, are mean values from 4-6 experiments with SEM varying from 0.9 to 10.3 per cent of the mean.

taurine from the slices (Fig. 2). Total omission of Cl^- from the medium caused an even greater release of endogenous and labeled taurine. When Cl^- was replaced by the permeant anion acetate the response was greater than when an impermeant anion, gluconate, was used as replacer, but in both cases there was a pronounced enhancement of taurine release.

As an example, Fig. 3 shows how the influx of taurine was strikingly reduced when 50 mM Na^+ were replaced by equimolar K^+, particularly at the low concentrations of taurine. When 50 mM KCl was added to medium, the reduction in taurine influx was almost equal in magnitude. The omission of Cl^- inhibited the influx as well, but not so markedly as did an increase in K^+. Hypoosmotic medium (25 mM reduction in NaCl) was distinctly less inhibitory and addition of 50 mM sucrose even tended to enhance the influx (data not shown).

The results on [^{3}H]taurine influx were analyzed with computer programs which fitted nonlinear equations to the data. All results were fitted with five different equations containing 0-2 terms of the type $VS/(K_m+S)$, possibly together with the term K_dS, in which S is the concentration of taurine in medium, V the maximal velocity of transport, K_m the transport constant (equivalent to the Michaelis constant in enzyme kinetics) and K_d the

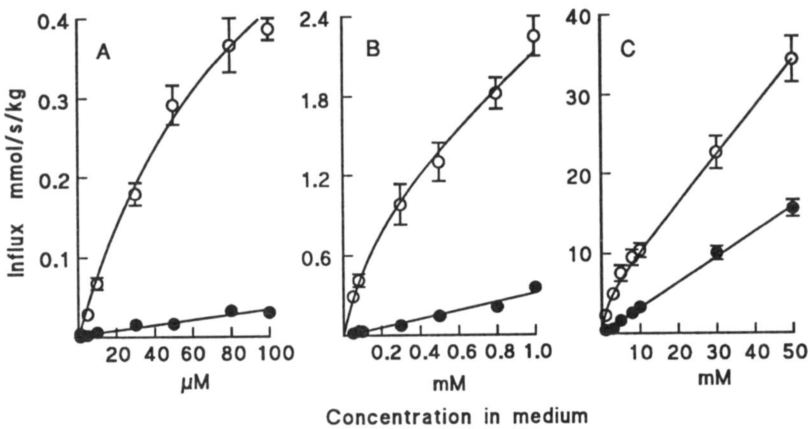

Figure 3. Influx of [^{3}H]taurine into cerebral cortical slices. The concentration of taurine varied from 1 µM to 50 mM in the medium. The 20-min experiments were carried out either in standard medium (O) or in the medium in which 50 mM Na^+ was replaced by equimolar K^+ (●). Panels A, B and C show the mean influx rates (± SEM), given per wet weight of incubated slices, within the low, middle and high concentration ranges of taurine, respectively.

Table 1. Kinetic parameters of taurine influx into mouse cerebral cortical slices under varying incubation conditions

Experiment	K_{m1} µM	V_1 µmol/s/kg	K_{m2} mM	V_2 mmol/s/kg	K_d 1/s
Control	46 ± 13	255 ± 69	2.72 ± 0.55	6.36 ± 0.43	0.620 ± 0.110
+ 50 mM K^+	-	-	-	-	0.318 ± 0.006*
+ 50 mM KCl	270 ± 67*	274 ± 40	-	-	0.360 ± 0.016*
- 25 mM NaCl	-	-	1.64 ± 0.57	3.62 ± 0.49*	0.503 ± 0.012
+ 50 mM sucrose	115 ± 33	289 ± 40	3.14 ± 0.49	9.13 ± 0.48*	0.421 ± 0.070
- Cl^-, acetate	-	-	3.69 ± 0.80	0.93 ± 0.08*	0.251 ± 0.012*
- Cl^-, gluconate	-	-	1.67 ± 0.41	1.31 ± 0.09*	0.147 ± 0.014*

Mean values ± SEM from 4-8 experiments are shown. Statistical significance of differences from the control: *P<0.01.

proportionality coefficient for nonsaturable penetration. Table 1 shows which equation fitted the data best in each case by listing the best-fit parameters from the iterative optimizations.

In the control experiments the influx could be divided into two saturable components, high- and low-affinity, combined to nonsaturable penetration (Table 1). The replacement of 50 mM Na^+ by equimolar K^+ completely abolished all saturable transport and only nonsaturable penetration into slices prevailed. The proportionality constant K_d for this penetration was also significantly diminished. When 50 mM KCl was added to medium, the low-affinity transport of taurine was abolished and there was only one saturable component left which could possibly be classified as high-affinity, but with a relatively high K_m. When the medium was rendered hyperosmotic no marked changes were observed in the transport parameters, only V for the low-affinity saturable transport was significantly increased. On the other hand, reduction of 25 mM NaCl abolished the high-affinity component and significantly diminished the V constant for the low-affinity saturable transport. The high-affinity saturable influx was entirely abolished after omission of Cl^- (Table 1). When gluconate was used as replacer only the maximal velocity of low-affinity transport was significantly diminished, whereas in the case of acetate both K_m and V were affected. In both cases the proportionality coefficient K_d for nonsaturable penetration was significantly diminished.

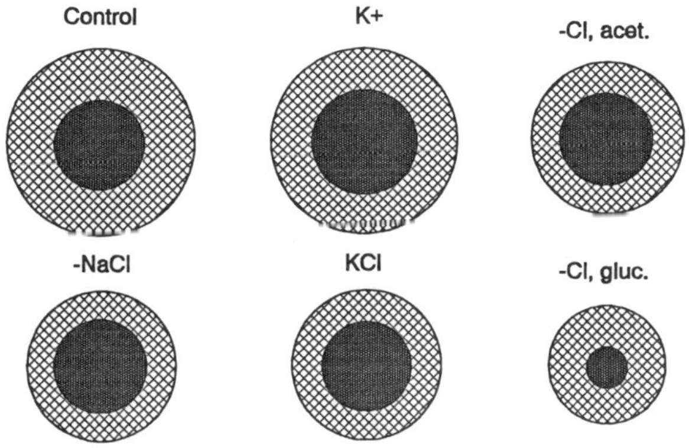

Figure 4. Schematic illustration of the changes in the mean intra- (dark circle) and extracellular (cross-hatched corona) volumes in cerebral cortical slices incubated in the varying incubation conditions shown in Table 1.

The intracellular volume changes in the slices superfused in the above varying conditions are schematically summarized in Fig. 4. Under normal standard incubation conditions there invariably occurred some swelling of slices, due to an increase in both intra- and extracellular spaces. Upon K^+ stimulation the slices swelled even more. In particular, the intracellular spaces expanded. This expansion was less pronounced when the high-K^+ medium was hyperosmotic. In hypoosmotic medium the total slice volume did not change so much, but there was a great expansion of the mean intracellular space. The situation was similar in Cl^--free, acetate-containing medium, but in contrast, in Cl^--free, gluconate-containing medium particularly the intracellular spaces shrank dramatically.

Fig. 5A is a correlation plot of the stimulated release of taurine, as expressed by the fractional efflux rate constants k_2, to the mean intracellular swelling or shrinking of the slices during the stimulation period (superfusion from 30 to 50 min). In addition to the incubation conditions described above, the experiments in Cl^--free media with K^+

stimulation (with acetate and gluconate as replacers) and the experiments in which the ionic product $K^+ \times Cl^-$ was kept constant (50 mM K^+ and 13 mM Cl^-) upon K^+ stimulation or in which Cl^- was only partially (50 mM) replaced by acetate or gluconate, were also included. The correlation tended to be positive but the coefficient was only 0.39 which did

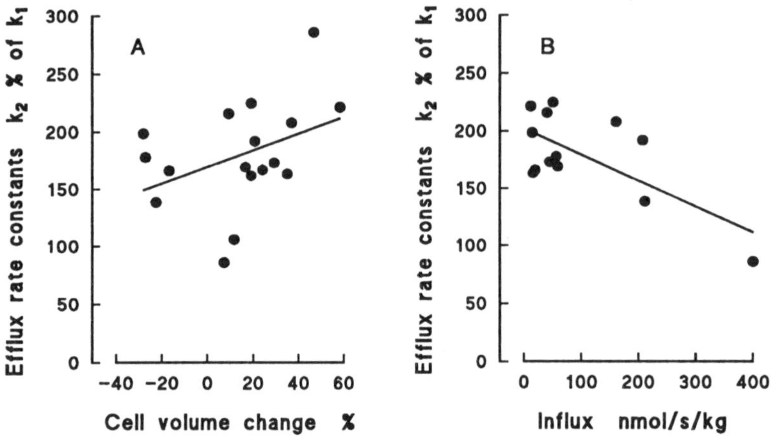

Figure 5. Correlation between the relative efflux rate constants k_2 (in per cent of basal prestimulated efflux rate k_1) and the mean intracellular volume changes in cerebral cortical slices incubated under varying experimental conditions (panel A) and correlation between the relative efflux rate constants k_2 and the influx rates of 10 μM [^{3}H]taurine (panel B).

not significantly differ from zero. We also assessed the correlation of the efflux to the influx under the same experimental conditions (Fig. 5B). In this case the correlation tended to be negative. This time the coefficient was -0.68, which significantly ($P<0.05$) differed from zero.

DISCUSSION

In cultured cells of neural origin, particularly in astrocytes (5, 20) but also in neurons (25), hypoosmotic media evoke release of taurine. Increasing concentrations of K^+ have the same effect, though accompanied by cell swelling (26, 30). The K^+-evoked release from astrocytes (22) and cerebellar granule neurons (26) is prevented when the incubation medium is rendered hyperosmotic. Similar results have been obtained in our own laboratory (17). The present data on cerebral cortical slices, however, are partially at variance with the results on cultured cells. In slices and cultured cells hypoosmotic media enhance and hyperosmotic media inhibit the release of taurine, but in slices the hyperosmotic medium could not prevent K^+ stimulation. Moreover, in cultured astrocytes the release of taurine has been reported to be unaffected by omission of Cl^- (20), and in cultured cerebellar granule neurons the K^+-stimulated release to be strictly Cl^--dependent (26), whereas in slices omission of Cl^- evokes release of taurine and does not abolish K^+ stimulation (17). In this respect the data on slices resemble more the in vivo results showing that extracellular taurine is much increased in microdialysis fluids low in Cl^- (28).

A part of taurine release in hypoosmotic medium apparently results from reduction in Na^+, not from hypoosmolarity per se. At least in slices the reduction or omission of Na^+ strongly stimulates the release of taurine (9). A critical perusal of the earlier studies on cultured cells shows, however, that this effect cannot account for the major part of the release in hypoosmotic media. The possibility that reuptake from medium of the labeled taurine released from the cells could have influenced results was not considered in the studies in question. The uptake of taurine is strictly Na^+-dependent in cerebral slices (6), neuroblastoma cells (3) and cultured astrocytes (22), the transporter not yet being saturated at the physiological extracellular Na^+ concentration. Hypoosmotic media could thus diminish taurine reuptake due to a concomitant reduction in NaCl, and in this way seemingly increase the release. Experimental conditions determine whether or not this mechanism constitutes any possible major source of misintepretation. Nevertheless, this possibility must be seriously considered with brain slices, in which the taurine molecules released from intracellular spaces must invariably traverse the extracellular space in order to have access to medium (18). Although there is a more or less stationary layer of medium also attached to cultured cells, the actual extracellular space present both in vivo and in tissue slices in vitro is missing from them.

Hypoosmolarity per se is probably a major factor affecting the release of taurine also in brain slices and probably in vivo as well (12, 29). Under certain isolated conditions there obtains a very good positive correlation between the release rates and cell swelling in cerebral cortical slices obtained from adult mice; e.g., when glutamate agonists are used as stimulators (23). A gradual reduction of osmolarity has the same effect (17). On the other hand, omission or reduction of Cl^- with gluconate as replacer for Cl^- yields totally different results. The correlation is then negative; i.e., the more intracellular shrinking occurs upon incubation, the more taurine is released (16). In addition to osmolarity, other factors are also obviously involved in the regulation of taurine release (24).

It may be noted that rather large changes in osmolarity are necessary to bring forth any measurable enhancement of taurine release from cultured cells (20, 26). It is probable that such changes are almost never encountered in the mammalian brain in vivo. They occur only in osmoconformers; i.e., in species that effectively acclimatize to changes between altered external salinity and internal salt concentration. In one such vertebrate species, the western toad *Bufo boreas*, taurine has been thought to have an osmoregulatory function in hypoosmotic conditions (1), even though the taurine levels in the brain of this toad may be too low for a major contribution to osmoregulation.

In brain slices the K^+ stimulation of taurine release and its persistence in hyperosmotic conditions can be partially attributed to inhibition of taurine influx. Such an inhibition also augments the release of taurine in low-Cl^- and Cl^--free media, as well as the release induced by hypoosmolarity. The release thus represents to some extent an 'overflow' from slices; i.e., only those taurine molecules which are not trapped by the reuptake emerge in the superfusion medium collected for analyses. However, correlation analysis indicates that only a part of the release is explicable in this way. Moreover, our results show that the capacities of taurine influx systems are too low considering the large amounts of taurine liberated from the slices already under resting conditions and even more so during stimulation. The close similarities in the properties of release of endogenous and preloaded labeled taurine witness that we have been dealing in all the present experiments with a real release of endogenous taurine, not merely with an outflow of newly accumulated radioactive molecules.

The fairly large release of taurine in neural tissue has interesting implications, irrespective of whether the release occurs as a direct response to membrane depolarization or indirectly due to cell swelling. Taurine quite generally inhibits neural excitation, not being confined to some specific brain structures and not being fast in its actions (14). An effective inhibition of reuptake under circumstances which induce release of taurine potentiates and prolongs its inhibitory effects. This phenomenon may be an important factor in the prolonged course of stimulus-evoked release of taurine observed in many brain preparations (8, 10, 21) and in the slow onset and prolonged duration of its inhibitory effects (4, 14). Taurine is thus an agent with a propensity to prevent in the neural machinery all 'overheating' phenomena, of whatever origin.

CONCLUSIONS

The present studies show that the responses of taurine release to hypo- and hyperosmolarity are similar, but the responses to K^+ stimulation and Cl^- omission dissimilar to those in cultured neurons and astrocytes. There obviously occurs a taurine release from slices which is due to cell volume adjustments. The apparent release measured is also partly attributable to an inhibition of reuptake. However, some evoked release seems to result from membrane depolarization. We do not necessarily imply that exocytosis is involved, since cell membrane carriers operating in an outward direction are also likely to contribute.

ACKNOWLEDGEMENTS

The technical assistance of Miss Paula Kosonen and Mrs Oili Pääkkönen, the secretarial aid of Mrs Irma Joukainen and the financial support of the Academy of Finland are gratefully acknowledged.

REFERENCES

1. Baxter, C.F., Baldwin, R.A., Lu, P., Imaki, H., and Sturman, J.A., 1993, Taurine in toad brain and blood under different conditions of osmolality - an immunohistochemical study, *Neurochem. Res.*, 18: 425-435.
2. Hanretta, A.T., and Lombardini, J.B., 1987, Is taurine a hypothalamic neurotransmitter?: a model of the differential uptake and compartmentalization of taurine by by neuronal and glial cell particles from the rat hypothalamus, *Brain Res. Rev.*, 12: 167-201.
3. Holopainen, I., Kontro, P., and Oja, S.S., 1984, Taurine and hypotaurine transport in neuroblastoma cells: effects of cations, *Neurochem. Int.*, 6: 217-222.
4. Huxtable, R.J., 1989, Taurine in the central nervous system and the mammalian actions of taurine, *Prog. Neurobiol.*, 32: 471-533.
5. Kimelberg, H.K., Goderie, S.K., Higman, S., Pang, S., and Waniewski, R.A., 1990, Swelling-induced release of glutamate, aspartate, and taurine from astrocyte cultures, *J. Neurosci.*, 10: 1583-1591.
6. Kontro, P., 1982, Effects of cations on taurine, hypotaurine and GABA uptake in mouse brain slices, *Neurochem. Res.*, 7: 1391-1401.
7. Kontro, P., and Oja, S.S., 1987, Taurine and GABA binding in mouse brain: effects of freezing, washing and Triton X-100 treatment on membranes, *Int. J. Neurosci.*, 32: 881-889.
8. Kontro, P., and Oja, S.S., 1987, Taurine and GABA release from mouse cerebral cortex slices: potassium stimulation releases more taurine than GABA from developing brain, *Dev. Brain Res.*, 37: 277-291.
9. Korpi, E.R., and Oja, S.S., 1983, Characteristics of taurine release from cerebral cortex slices by sodium-deficient medium, *Brain Res.*, 289: 197-204.
10. Korpi, E.R., Kontro, P., Nieminen, K., Marnela, K.-M., and Oja, S.S., 1981, Spontaneous and depolarization-induced efflux of hypotaurine from mouse cerebral cortex slices: comparison with taurine and GABA, *Life Sci.*, 29: 811-816.
11. Laakso, M.-L., and Oja, S.S., 1976, Factors influencing the inulin space in cerebral cortex slices from adult and 7-day-old rats, *Acta Physiol. Scand.*, 97: 486-494.
12. Lehmann, A., 1989, Effects of microdialysis-perfusion with anisoosmotic media on extracellular amino acids in the rat hippocampus and skeletal muscle, *J. Neurochem.*, 53: 525-535.
13. Oja, S.S., and Kontro, P., 1980, Hypotaurine uptake by brain slices from adult and 8-day-old mice, *J. Neurochem.*, 35: 1303-1308.
14. Oja, S.S., and Kontro, P., 1983, Taurine, *in:* "Handbook of Neurochemistry", Vol. 3, 2nd edn, A. Lajtha, ed., Plenum Press, New York, pp. 501-533.
15. Oja, S.S., and Kontro, P., 1990, Neuromodulatory and trophic actions of taurine, *Prog. Clin. Biol. Res.*, 351: 69-76.
16. Oja, S.S., and Saransaari, P., 1992, Cell volume changes and taurine release in cerebral cortical slices, *Adv. Exp. Med. Biol.*, 315: 369-374.

17. Oja, S.S., and Saransaari, P., 1992, Taurine release and swelling of cerebral cortex slices from adult and developing mice in media of different ionic compositions, *J. Neurosci. Res.*, 32: 551-561.
18. Oja, S.S., and Vahvelainen, M.-L., 1975, Transport of amino acids in brain slices, *in:* "Research Methods in Neurochemistry", Vol. 3, N. Marks, and R. Rodnight, eds., Plenum Press, New York, pp. 67-137.
19. Oja, S.S., Kontro, P., and Lähdesmäki, P., 1977, Amino acids as inhibitory neurotransmitters, *Prog. Pharmacol.*, 1/3: 1-119.
20. Pasantes-Morales, H., Morán, J., and Schousboe, A., 1990, Volume-sensitive release of taurine from cultured astrocytes: properties and mechanism, *Glia*, 3: 427-432.
21. Philibert, R.A., Rogers, K.L., and Dutton, G.R., 1989, Stimulus-coupled taurine efflux from cerebellar neuronal cultures: on the roles of Ca^{++} and Na^{+}, *J. Neurosci. Res.*, 22: 167-171.
22. Sánchez-Olea, R., Morán, J., and Pasantes-Morales, H., 1992, Changes in taurine transport evoked by hyperosmolarity in cultured astrocytes, *J. Neurosci. Res.*, 32: 86-92.
23. Saransaari, P., and Oja, S.S., 1991, Excitatory amino acids evoke taurine release from cerebral cortex slices from adult and developing mice, *Neuroscience*, 45: 451-459.
24. Saransaari, P., and Oja, S.S., 1992, Release of GABA and taurine from brain slices, *Prog. Neurobiol.*, 38: 455-482.
25. Schousboe, A., and Pasantes-Morales, H., 1992, Role of taurine in neural cell volume regulation, *Can. J. Physiol. Pharmacol.*, 70: S356-S361.
26. Schousboe, A., Morán, J., and Pasantes-Morales, H., 1990, Potassium-stimulated release of taurine from cultured cerebellar granule neurons is associated with cell swelling, *J. Neurosci. Res.*, 27: 71-77.
27. Simpson, J.W., Allen, K., and Awapara, J., 1959, Free amino acids in some aquatic invertebrates, *Biol. Bull.*, 117: 371-381.
28. Solis, J.M., Herranz, A.S., Herreras, O., Lerma, J., and Martin del Rio, R., 1988, Low chloride-dependent release of taurine by a furosemide-sensitive process in the *in vivo* rat hippocampus, *Neuroscience*, 24: 885-891.
29. Solis, J.M., Herranz, A.S., Herreras, O., Menéndez, N., and Martin del Rio, R., 1990, Weak organic acids induce taurine release through and osmotic-sensitive process in in vivo rat hippocampus, *J. Neurosci. Res.*, 26: 159-167.
30. Tigges, G.A., Philibert, R.A., and Dutton, G.R., 1990, K^{+}- and temperature-evoked taurine efflux from hypothalamic astrocytes, *Neurosci. Lett.*, 119: 23-26.
31. Vahvelainen, M.-L., and Oja, S.S., 1972, Kinetics of influx of phenylalanine, tyrosine, tryptophan, histidine and leucine into slices of brain cortex from adult and 7-day-old rats, *Brain Res.*, 40: 477-488.
32. van Gelder, N.M., 1989, Brain taurine content as a function of cerebral metabolic rate: osmotic regulation of glucose derived water production, *Neurochem. Res.*, 14: 495-497.
33. Wade, J.V., Olson, J.P., Samson, F.E., Nelson, S.R., and Pazdernik, T.L., 1988, A possible role of taurine in osmoregulation within the brain, *J. Neurochem.*, 51: 740-745.
34. Walz, W., and Hinks, E.C., 1985, Carrier-mediated KCl accumulation accompanied by water movements is involved in the control of physiological K^{+} levels by astrocytes, *Brain Res.*, 343: 44-51.

TAURINE RELEASE FROM MOUSE HIPPOCAMPAL SLICES: EFFECTS OF GLUTAMATERGIC SUBSTANCES AND HYPOXIA

Pirjo Saransaari and Simo S. Oja

Tampere Brain Research Center
Department of Biomedical Sciences
University of Tampere, Box 607
SF-33101 Tampere, Finland

INTRODUCTION

The hippocampus is involved in many important functions, including generation of long-term potentiation, memory formation, learning, arousal, emotions and regulation of autonomic functions. Since the major part of excitatory innervation uses glutamate as transmitter in the hippocampus (8), this region contains a high density of glutamate receptors (26). The activity of this excitatory network is modulated by GABA, the main inhibitory transmitter released by local interneurons (11, 12). The structural analogue of GABA, taurine, found at high concentrations in the hippocampus (15, 21, 31), inhibits CA1 pyramidal neurons, increasing chloride conductance and causing hyperpolarization (42). Taurine-like immunoreactivity has been localized in interneurons, pyramidal neurons and dentate granule cells in the rat hippocampus (5, 22). The taurine-synthesizing enzyme cysteine sulfinate decarboxylase has also been identified in pyramidal basket interneurons (42). Despite the above findings no specific function has been assigned for taurine in the hippocampus.

Microdialysis studies in vivo on adult animals have demonstrated that taurine is released from the hippocampus by the glutamate agonists N-methyl-D-aspartate (NMDA) and kainate (17, 18, 19, 24), indicating interaction of taurine with hippocampal glutamatergic systems. This interaction could be of importance in view of the protective effects of taurine in epilepsy (27, 28), hypoxia, ischemia (39) and excitotoxicity (10), all of which may involve excitatory amino acids. For example, the extracellular levels of both taurine and glutamate are increased during ischemia in the rat hippocampus (2, 16, 20). Hippocampal pyramidal cells are known to be among the neurons most vulnerable to transient ischemia and subsequent cell death (34). On the other hand, it has been shown that the hippocampus in the newborn is particularly sensitive to excitotoxic agents (7), but more resistant to hypoxic insults than in the adult (4, 9). We have now studied the hippocampal taurine release in vitro in developing and adult mice using a superfusion system focusing particularly on the actions of glutamate receptor agonists on the release. The effects of hypoxic superfusion conditions were also investigated.

Taurine in Health and Disease, Edited by R. Huxtable
and D.V. Michalk, Plenum Press, New York, 1994

EXPERIMENTAL PROCEDURES

Material: NMRI mice of both sexes, aged 7 days and 3 months (adults), were used throughout. [1,2-^{3}H]Taurine (specific activity 1.07 PBq/mol) was obtained from Amersham International, U.K. MK-801 was a gift from Merck, Sharp & Dohme (Rahway, NJ). Other reagents were from common commercial sources.

Efflux experiments: Slices 0.4 mm thick weighing 15-20 mg were prepared with a Stadie-Riggs tissue slicer from mouse hippocampi and used immediately in efflux experiments. In studies on the release of exogenous taurine the slices were preloaded for 30 min with 10 µM (50 MBq/l) [^{3}H]taurine in preoxygenated Krebs-Ringer-Hepes-glucose medium under O_2 and superfused with the same medium as described in detail in Oja and Kontro (29). The medium was pooled during the first 20 min of superfusion, whereafter 2-min fractions (0.5 ml) were collected. At 30 min the medium was changed to other modified media. In a number of experiments the preloading medium was gassed with N_2 and the superfusion experiments carried out under N_2 (hypoxic conditions). After superfusion the slices were weighed, homogenized in ice-cold 5 % (w/v) trichloracetic acid solution and centrifuged, and the clear supernatants used for scintillation counting or taurine assays by high-performance liquid chromatography as described in Kontro and Oja (13). The effluent samples were subjected to the same analyses.

Estimation of efflux rate constants for exogenous taurine: Desaturation curves of labeled taurine from the slices were plotted as a function of time on the basis of the radioactivities remaining in the slices after superfusion and recovered in the collected superfusate fractions (14). The efflux rate constants of taurine for the time intervals of 20 to 30 min (k_1) and 34 to 50 min (k_2) were computed as negative slopes for the regression lines of the logarithm of radioactivity remaining in the slices vs. superfusion time.

Statistical calculations: The presence of statistically significant differences between sample means was detected by variance analysis. Comparisons of individual means were by Hartley's sequential method of testing.

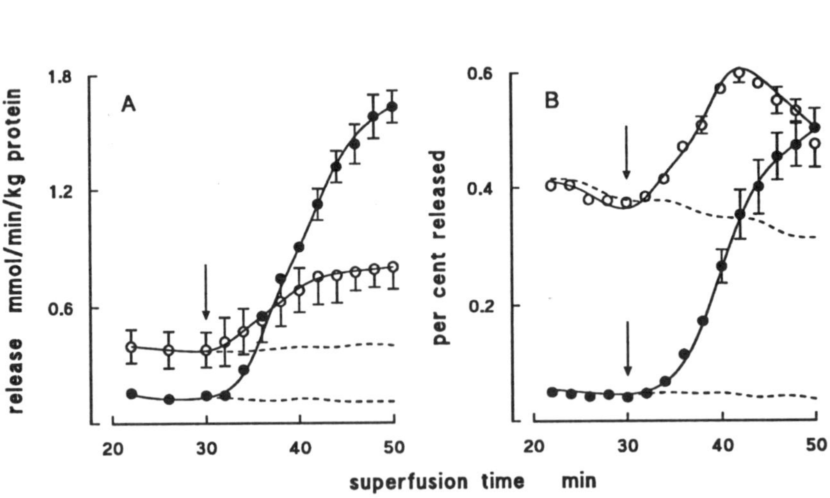

Figure 1. Release of endogenous (A) and preloaded labeled (B) taurine from hippocampal slices from 7-day-old (-●-) and 3-month-old (-O-) mice. The slices were superfused for 50 min with Krebs-Ringer-Hepes medium which contained 50 mM K$^+$ from 30 min onwards, as indicated by the arrow. The results are mean values of 3-7 separate experiments with SEMs. The dashed lines indicate spontaneous, unstimulated efflux.

RESULTS

The basal unstimulated release of endogenous taurine from the hippocampus of 7-day-old mice was low. Stimulation with 50 mM K^+ enhanced the release about 14-fold, the release steadily increasing during the whole stimulation period of 30-50 min (Fig. 1A). The basal release was higher in the adults but the response to K^+ stimulation only about 2-fold (Fig. 1A). The results on the release of [^{3}H]taurine were strikingly similar to these (Fig. 1B). The basal release of [^{3}H]taurine was greatly enhanced when Ca^{2+} was omitted

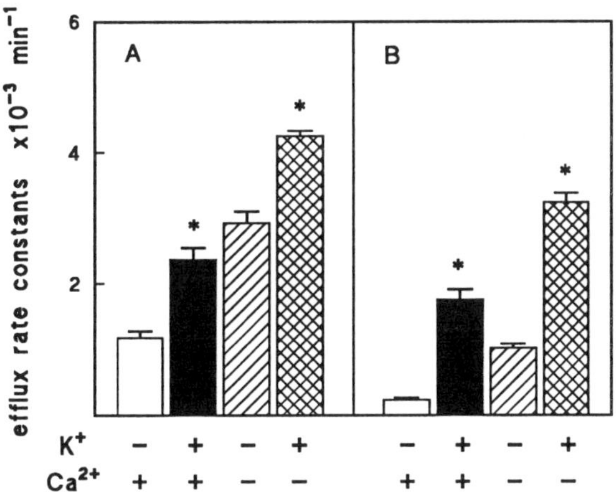

Figure 2. Release of [^{3}H]taurine from hippocampal slices from 3-month-old (A) and 7-day-old (B) mice. The graph shows efflux rate constants k_2 ± SEM (34-50 min) in the absence and presence of 50 mM K^+ and 0.8 mM Ca^{2+} as indicated. The asterisk indicates statistically significant ($P<0.01$) K^+ stimulation.

Table 1. Effects of glutamatergic substances on [^{3}H]taurine release from hippocampal slices from 3-month- and 7-day-old mice

Substance in medium from 30-50 min	Efflux rate constants ± SEM k_2 (34-50 min), % of k_1 3-month-old	7-day-old
None (control)	82.9 ± 6.8	88.2 ± 1.2
Kainate	124.6 ± 6.6**	259.8 ± 25.1**
+ CNQX	113.9 ± 4.9**	126.9 ± 5.6**
Quisqualate	90.9 ± 2.1	307.0 ± 37.4**
+ CNQX	90.1 ± 6.7	150.8 ± 10.8**
AMPA	102.7 ± 5.2	497.7 ± 46.5**
+ CNQX	106.6 ± 3.6*	226.7 ± 17.0**
NMDA	186.8 ± 8.4**	744.1 ± 77.2**
+ MK-801	94.8 ± 7.4	107.5 ± 6.1*
Tetrazolylglycine	138.5 ± 12.2**	937.6 ± 94.7**
+ MK-801	97.6 ± 3.7	188.7 ± 6.3**
CNQX	83.8 ± 5.5	82.5 ± 5.4
MK-801	91.8 ± 1.7	80.6 ± 2.8

The slices were preloaded for 30 min with 10 µM [^{3}H]taurine in Krebs-Ringer-Hepes-glucose medium, pH 7.4, and then superfused for 50 min; from 30 min onwards together with the above effectors (concentration: S0.1 mM). The results are given as percentages of the rate constant k_1 (20-30 min) for the basal efflux of each slice. The k_1 constants were $1.906 \pm 0.053 \times 10^{-3}$ min^{-1} (mean ± SEM, n=52) and $0.315 \pm 0.018 \times 10^{-3}$ min^{-1} (n=52) for 3-month-olds and 7-day-olds, respectively. The number of independent experiments was four. Significance of differences from the corresponding controls: *$P<0.05$, **$P<0.01$.

and 2.0 mM EDTA added to superfusion medium in both 7-day- and 3-month-old mice (Fig. 2). The K^+ stimulation was, however, clearly evident in the Ca^{2+}-free medium at both ages, but the relative stimulation was diminished, particularly in the immature hippocampus (Fig. 2).

The glutamate receptor agonists kainate and NMDA (both 0.1 mM) greatly stimulated [^{3}H]taurine release from hippocampal slices from both developing and adult mice, the affects being considerably greater in the former (Table 1, Fig. 3). Moreover, (RS)-α-amino-3-hydroxy-5-methylisoxazole-4-propionate (AMPA) and quisqualate (both 0.1 mM) were ineffective in adults but stimulated the release in 7-day-old mice (Table 1). (RS)-(Tetrazol-5-yl)glycine (0.1 mM) also potentiated taurine release in both age group (Table 1, Fig. 4). The antagonist 6-cyano-7-nitroquinoxaline-2,3-dione (CNQX) (0.1 mM) significantly ($P<0.01$) inhibited the potentiations by kainate (Fig. 3, Table 1), quisqualate

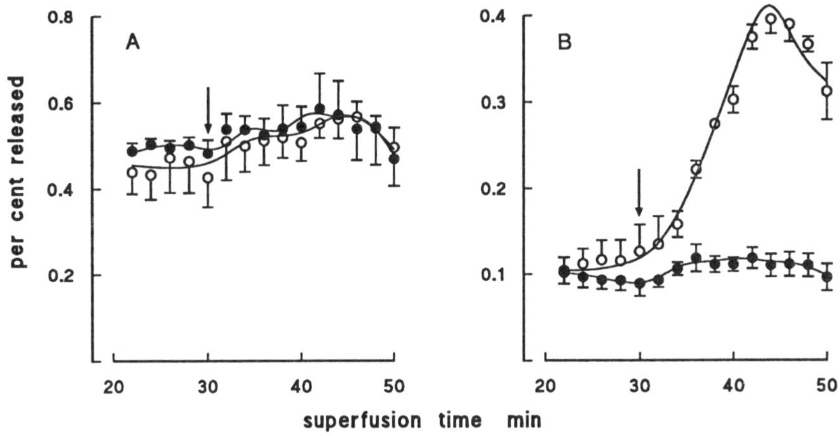

Figure 3. Release of [^{3}H]taurine from hippocampal slices from 3-month-old (A) and 7-day-old (B) mice in the presence of 0.1 mM kainate (-O-) and 0.1 mM kainate together with 0.1 mM CNQX (-●-). The results are mean values ± SEM of 4 separate experiments. Kainate and CNQX were added to the superfusion medium at 30 min as indicated by the arrow.

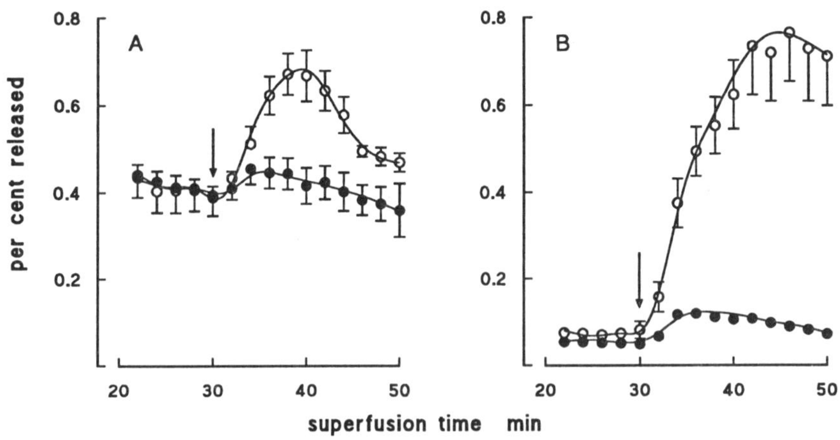

Figure 4. Release of [^{3}H]taurine from hippocampal slices from 3-month-old (A) and 7-day-old (B) mice in the presence of 0.1 mM tetrazolylglycine (-O-) and 0.1 mM tetrazolylglycine together with 0.1 mM MK-801 (-●-). The results are mean values ± SEM of 4 separate experiments. Tetrazolylglycine and MK-801 were added to the superfusion medium at 30 min, as indicated by the arrow.

and AMPA in 7-day-old mice but had no effect on the kainate stimulation in the adults. The enhancements by NMDA and tetrazolylglycine were also significantly (P<0.01) reduced by the antagonist (+)-5-methyl-10,11-dihydro-5H-dibenzo(a,d)-cyclohept-5,10-imine hydrogen maleate (MK-801) in both developing and adult animals (Table 1, Fig. 4). CNQX and MK-801 (both 0.1 mM) alone had no effect on taurine release. The specific agonist of metabotropic glutamate receptors, 1-aminocyclopentane-1,3-dicarboxylate (ACPD) alone or together with the antagonist 2-amino-3-phosphonopropionate (AP3) (both 0.1 mM) had no effect on taurine release at either age. Neither did the agonist 1-aminocyclopentane-trans-1,3-dicarboxylate (trans-ACPD) (data not shown).

When the hippocampal slices were superfused under N_2 from the beginning of superfusions with medium previously gassed with N_2, the basal release in the adults increased about 2-fold. The release steadily increased with time in contrast to the release

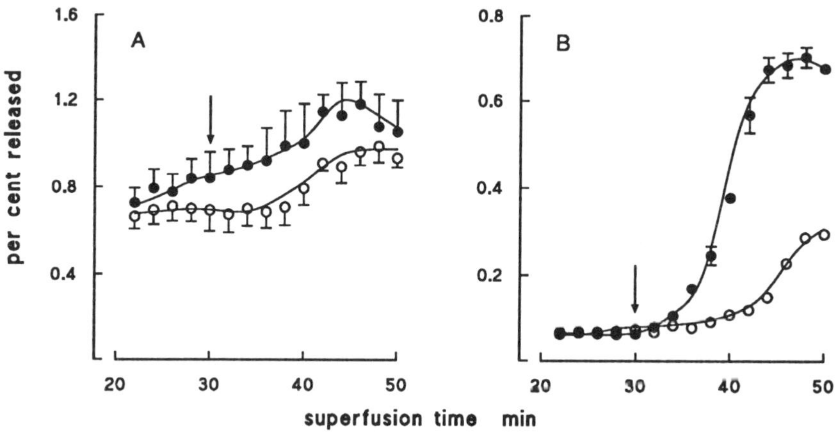

Figure 5. Spontaneous (-O-) and K^+-stimulated (-●-) release of [^{3}H]taurine under N_2 from hippocampal slices from 3-month-old (A) and 7-day-old (B) mice. The whole superfusion was carried out under N_2. K^+ (50 mM) was added to medium at 30 min, as indicated by the arrow. The results are mean values ± SEMs of 4 separate experiments.

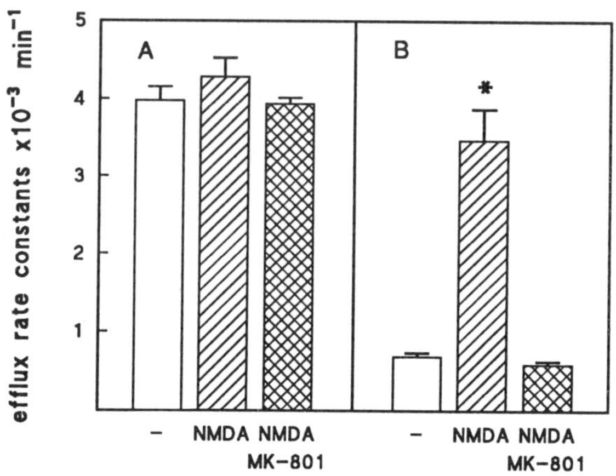

Figure 6. Effects of NMDA and MK-801 (both 0.1 mM) on [^{3}H]taurine release from hippocampal slices from 3-month-old (A) and 7-day-old (B) mice superfused under N_2. The graph shows efflux rate constants k_2 ± SEM (34-50 min). The asterisk indicates statistically significant (P<0.01) stimulation.

in normal aerobic conditions (cf. Figs 1 and 5). There was no significant stimulation by K^+. In 7-day-olds the basal efflux likewise increased in hypoxic conditions, but the relative K^+ stimulation of the release was of the same order of magnitude as in aerobic conditions (Figs 1 and 5). The potentiation of release by 0.1 mM NMDA in hypoxic conditions was 508 % in 7-day-olds but insignificant (only 108 %) in adults (Fig. 6). MK-801 was still able to abolish the NMDA-stimulated release.

DISCUSSION

Two consistent features of K^+-stimulated taurine release are the slow time course of stimulation and the markedly large release from immature brain tissue (36). These were also now clearly evident in the hippocampus. This pattern is in striking contrast, for example, to the release of glutamate from hippocampal synaptosomes; K^+ depolarization releases relatively small amounts of glutamate in young animals (6). The Ca^{2+} dependency of stimulated taurine release was now similar to that in cerebral cortical slices (14), being only partially Ca^{2+}-dependent in both mature and immature hippocampus. However, the latter seems to be more sensitive to Ca^{2+} omission.

The potentiation of taurine release by glutamate agonists in the hippocampus has earlier been demonstrated in adults in vivo (18, 19, 24, 25) and in vitro (23). We show here for the first time that the stimulation of release by NMDA, kainate, quisqualate and AMPA is more pronounced in the immature than the mature hippocampus, as is also the case in the cerebral cortex (35, 36). In agreement with the studies of Magnusson et al. (23), we observed that NMDA was the most powerful agonist in potentiating the release of taurine. Immature nervous tissue appears to be more sensitive to depolarization by the agonists and K^+ than mature tissue. The effects of kainate, quisqualate and AMPA were considerably reduced by the non-NMDA antagonist CNQX, indicating that both kainate and AMPA/quisqualate receptors are involved in the evoked release of taurine in the developing hippocampus. This regulation appears to be lost during maturation, since CNQX was unable to affect the kainate-evoked taurine release significantly in the adults. Moreover, the stimulation by NMDA was inhibited by the specific antagonist MK-801 in both adult and developing mice. The regulation thus appears to be operating also in adulthood. This inference is further strengthened by our results with the new potent NMDA receptor agonist tetrazolylglycine (38), which enhanced taurine release similarly to NMDA at both ages studied. These effects were both antagonized by MK-801 in the same way. On the other hand, we could not confirm the in vivo results demonstrating that non-NMDA receptors are also involved in the glutamate agonist-evoked release of taurine in adults (24), though the kainate potentiation was evident.

The ability of NMDA to evoke substantial release of taurine in the immature hippocampus tallies with the overexpression of the NMDA subtype of glutamate receptors in hippocampal formation during postnatal development (days 4-13) (43). An increased number of receptors correlates with a development of the afferent input and elaboration of dendrites in the hippocampus and with the developmental onset of long-term potentiation (32, 33). At the same time the number of kainate-binding sites remains relatively low in the rat hippocampus (1). The participation of metabotropic glutamate receptors in taurine release is unlikely in either adult or developing hippocampus, since the agonists of this receptor type, ACPD and trans-ACPD, and the antagonist AP3 (37) were without any effects. The stimulation caused by quisqualate in the immature hippocampus could be due to its potency on AMPA receptors, even though it also activates metabotropic glutamate receptors (37).

It has been suggested that glutamate agonist-evoked release of taurine in the hippocampus does not occur during normal neurotransmission but as a result of excitotoxicity and cellular swelling (23, 24). Moreover, a replacement of Cl^- by permeant anions in microdialysis fluids greatly increases extracellular taurine in the rat hippocampus in vivo (40), probably by an osmotically sensitive mechanism (41). We have previously demonstrated with the same superfusion technique as used now that the evoked release of taurine from the cerebral cortex of adult mice may be at least partially related to cell volume regulation in the context of excitation-coupled ion and water movements across plasma membranes, whereas this inference is less warranted with respect to 3-day-old mice (35). On the other hand, the hippocampal release of taurine evoked by NMDA in vivo has been shown to occur independently of cell swelling (25). The pronounced glutamate receptor-activated release of taurine could be of great importance in counteracting excitotoxic effects, particularly in immature nervous tissue.

In agreement with other studies (3, 4, 7) we noted that the immature hippocampus is more resistant to a hypoxic insult than the mature hippocampus, as judged from the remaining stimulation by K^+ and NMDA of the release of taurine under N_2. Ischemia causes a large increase in the release of both excitatory and inhibitory amino acids from the hippocampus in adults (3, 20, P. Saransaari, unpublished results). GABA and taurine could counteract the depolarizing effects of excitatory amino acids by increasing membrane chloride conductance (30). Indeed, it has been shown that taurine improves the recovery of neuronal function in hippocampal slices after hypoxia in vitro (39). Taurine-containing neurons within the hippocampus are likewise less susceptible than other hippocampal neurons to damage induced by ischemia in vivo (44). In our present studies the basal taurine release in the adults under N_2 increased so much that both K^+ and NMDA could evoke much less additional release than in normal conditions. Hypoxia per se seems thus to provoke so much release of taurine that other mechanisms, e.g., K^+ and NMDA stimulation, are almost overshadowed. On the other hand, the NMDA-evoked release, antagonized by MK-801, was still discernible, particularly in the immature hippocampus, indicating that the NMDA receptor-stimulated release of taurine is still operating in hypoxic conditions.

CONCLUSIONS

K^+ and glutamate agonists enhanced taurine release more in the immature mouse hippocampus than in the adult, NMDA being the most potent. An activation of the kainate, AMPA and NMDA subtypes of glutamate receptor appears to be involved in taurine release in 7-day-old animals, whereas only NMDA receptors affect the release in adults. The glutamate agonist-evoked taurine release could counteract excitotoxic effects, particularly in immature nervous tissue. Hypoxic conditions per se induce taurine release, which could contribute to recovery from ischemic damage.

ACKNOWLEDGEMENTS

The skilful technical assistance of Miss Paula Kosonen and Mrs Irma Rantamaa and the financial support of the Academy of Finland are gratefully acknowledged.

REFERENCES

1. Ben-Ari, Y., Tremblay, E., Berger, M., and Nitecka, L., 1984, Kainic acid seizure syndrome and binding sites in developing rats, *Dev. Brain Res.*, 14: 284-288.

2. Benveniste, H., Drejer, J., Schousboe, A., and Diemer, N.H, 1984, Elevation of the extracellular concentrations of glutamate and aspartate in rat hippocampus during transient cerebral ischemia monitored by intracerebral microdialysis, *J. Neurochem.*, 43: 1369-1374.
3. Cherici, G., Alesiani, M., Pellegrini-Giampietro, D.E., Moroni, F., 1991, Ischemia does not induce the release of excitotoxic amino acids from the hippocampus of newborn rats, *Dev. Brain Res.*, 60: 235-240.
4. Cherubini, E., Ben-Ari, Y., and Krnjevic, K., 1989, Anoxia produces smaller changes in synaptic transmission, membrane potential and input resistance in immature rat hippocampus, *J. Neurophysiol.*, 62: 882-895.
5. Clements, J.R., Magnusson, K.R., and Beitz, A.J., 1989, Ultrastructural description of taurine-like immunoreactive cells and processes in the rat hippocampus, *Synapse*, 4: 70-79.
6. Collard, K.J., Edwards, R., and Liu, Y., 1993, Changes in synaptosomal glutamate release during postnatal development in the rat hippocampus and cortex, *Dev. Brain Res.*, 71: 37-43.
7. Cook, T.M., and Crutcher, K.A., 1986, Intrahippocampal injection of kainic acid produces significant pyramidal cell loss in neonatal rats, *Neuroscience*, 18: 79-92.
8. Davies, S.N., and Collingridge, G.L., 1989, Role of excitatory amino acid receptors in synaptic transmission in area CA1 of rat hippocampus, *Proc. R. Soc. Lond. B*, 236: 373-384.
9. Ferriero, D.M., Arcavi, L.J., and Simon, R.P., 1990, Ontogeny of excitotoxic injury to nicotinamide adenine dinucleotide phosphate diaphorase reactive neurons in the neonatal rat striatum, *Neuroscience*, 36: 417-424.
10. French, E.D., Vezzani, A., Whetsell, W.O., Jr., and Schwarcz, R., 1986, Antiexcitotoxic actions of taurine in the rat hippocampus studied in vivo and in vitro, *Adv. Exp. Med. Biol.*, 203: 349-362.
11. Freund, T.F., and Buzsáki, Gy., 1988, Alterations in excitatory and GABAergic inhibitory connections in hippocampal transplants, *Neuroscience*, 27: 373-385.
12. Frotscher, M., Léránth, Cs., Lübbers, K., and Oertel, W.H., 1984, Commissural afferents innervate glutamate decarboxylase immunoreactive non-pyramidal neurons in the guinea-pig hippocampus, *Neurosci. Lett.*, 46: 137-143.
13. Kontro, P., and Oja, S.S., 1987, Taurine and GABA binding in mouse brain: effects of freezing, washing and Triton X-100 treatment on membranes, *Int. J. Neurosci.*, 32: 881-889.
14. Kontro, P., and Oja, S.S., 1987, Taurine and GABA release from mouse cerebral cortex slices: potassium stimulation releases more taurine than GABA from developing brain, *Dev. Brain Res.*, 37: 277-291.
15. Kontro, P., Marnela, K.-M., and Oja, S.S., 1980, Free amino acids in the synaptosome and synaptic vesicle fractions of different bovine brain areas, *Brain Res.*, 184: 129-141.
16. Korf, J., Klein, H.C., Venema, K., and Postema, F., 1988, Increases in striatal and hippocampal impedance and extracellular levels of amino acids by cardiac arrest in freely moving rats, *J. Neurochem.*, 50: 1087-1096.
17. Lehmann, A., Isacsson, H., and Hamberger, A., 1983, Effects of *in vivo* administration of kainic acid on the extracellular amino acid pool in the rabbit hippocampus, *J. Neurochem.*, 40: 1314-1320.
18. Lehmann, A., Lazarewicz, J.W., and Zeise, M., 1985, N-Methylaspartate-evoked liberation of taurine and phosphoethanolamine in vivo: site of release, *J. Neurochem.*, 45: 1172-1177.
19. Lehmann, A., Hagberg, H., Lazarewicz, J.W., Jacobson, I., and Hamberger, A., 1986, Alterations in extracellular amino acids and Ca^{2+} following excitotoxin administration and during status epilepticus, *Adv. Exp. Med. Biol.*, 203: 363-373.
20. Lekieffre, D., Callebert, J., Plotkine, M., and Boulu, R.G., 1992, Concomitant increases in the extracellular concentrations of excitatory and inhibitory amino acids in the rat hippocampus during forebrain ischemia, *Neurosci. Lett.*, 137: 78-82.
21. Lombardini, J.B., 1976, Regional and subcellular studies on taurine in the rat central nervous system, *in:* "Taurine", R. Huxtable and A. Barbeau, eds., Raven Press, New York, pp. 311-326.
22. Magnusson, K.R., Clements, J.R., Wu, J.-Y., and Beitz, A.J., 1989, Colocalization of taurine- and cysteine sulfinic acid decarboxylase-like immunoreactivity in the hippocampus of the rat, *Synapse*, 4: 55-69.
23. Magnusson, K.R., Koerner, J.F., Larson, A.A., Smullin, D.H., Skilling, S.R., and Beitz, A.J., 1991, NMDA-, kainate- and quisqualate-stimulated release of taurine from electrophysiologically monitored rat hippocampal slices, *Brain Res.*, 549: 1-8.
24. Menéndez, N., Herreras, O., Solis, J.M., Herranz, A.S., Martin del Rio, R., 1989, Extracellular taurine increase in rat hippocampus evoked by specific glutamate receptor activation is related to the excitatory potency of glutamate agonists, *Neurosci. Lett.*, 102: 64-69.

25. Menéndez, N., Solis, J.M., Herreras, O., Herranz, A.S., Martin del Rio, R., 1990, Role of endogenous taurine on the glutamate analogue-induced neurotoxicity in the rat hippocampus in vivo, *J. Neurochem.*, 55: 714-717.
26. Monaghan, D.T., Holets, V.R., Toy, D.W., and Cotman, C.W., 1983, Anatomical distributions of four pharmacologically distinct ^{3}H-L-glutamate binding sites, *Nature*, 306: 176-179.
27. Oja, S.S., and Kontro, P., 1983, Taurine, *in:* "Handbook of Neurochemistry", Vol. 3, 2nd edn, A. Lajtha, ed., Plenum Press, New York, pp. 501-533.
28. Oja, S.S., and Kontro, P., 1983, Free amino acids in epilepsy: possible role of taurine, *Acta Neurol. Scand.*, 67, Suppl. 93: 5-20.
29. Oja, S.S., and Kontro, P., 1989, Release of endogenous taurine and γ-aminobutyric acid from brain slices from the adult and developing mouse, *J. Neurochem.*, 52: 1018-1024.
30. Oja, S.S., Korpi, E.R., and Saransaari, P., 1990, Modification of chloride flux across brain membranes by inhibitory amino acids in developing and adult mice, *Neurochem. Res.*, 15: 797-804.
31. Palkovits, M., Elekes, I., Lang, T., and Patthy, A., 1986, Taurine levels in discrete brain nuclei of rats, *J. Neurochem.*, 47: 1333-1335.
32. Pokorny, J., and Yamamoto, T., 1981, Postnatal ontogenesis of hippocampal CA1 area in rats. I. Development of dendritic arborization in pyramidal neurons, *Brain Res. Bull.*, 7: 113-120.
33. Pokorny, J., and Yamamoto, T., 1981, Postnatal ontogenesis of hippocampal CA1 area in rats. II. Development of ultrastructure in striatum lacunosum and moleculare, *Brain Res. Bull.*, 7: 121-130.
34. Pulsinelli, W.A., Brierley, J.B., and Plum, F., 1982, Temporal profile of neuronal damage in a model of transient forebrain ischemia, *Ann. Neurol.*, 11: 491-498.
35. Saransaari, P., and Oja, S.S., 1991, Excitatory amino acids evoke taurine release from cerebral cortex slices from adult and developing mice, *Neuroscience*, 45: 451-459.
36. Saransaari, P., and Oja, S.S., 1992, Release of GABA and taurine from brain slices, *Prog. Neurobiol.*, 38: 455-482.
37. Schoepp, D., Bockaert, J., and Sladeczek, F., 1990, Pharmacological and functional characteristics of metabotropic excitatory amino acid receptors, *Trends Pharmacol. Sci.*, 11: 508-515.
38. Schoepp, D.D., Smith, C.L., Lodge, D., Millar, J.D., Leander, J.D., Sacaan, A.I., and Lunn, W.H.W., 1991, D,L-(Tetrazol-5-yl) glycine: a novel and highly potent NMDA receptor agonist, *Eur. J. Pharmacol.*, 203: 237-243.
39. Schurr, A., Tseng, M.T., West, C.A., and Rigor, B.M., 1987, Taurine improves the recovery of neuronal function following cerebral hypoxia: an in vitro study, *Life Sci.*, 40: 2059-2066.
40. Solis, J.M., Herranz, A.S., Herreras, O., Lerma, J., and Martin del Rio, R., 1988, Low chloride-dependent release of taurine by a furosemide-sensitive process in the *in vivo* rat hippocampus, *Neuroscience*, 24: 885-891.
41. Solis, J.M., Herranz, A.S., Herreras, O., Menéndez, N., and Martin del Rio, R., 1990, Weak organic acids induce taurine release through an osmotic-sensitive process in *in vivo* rat hippocampus, *J. Neurosci. Res.*, 26: 159-167.
42. Taber, K.H., Lin, C.-T., Liu, J.-W., Thalmann, R., and Wu, J.-Y., 1986, Taurine in hippocampus: localization and postsynaptic action, *Brain Res.*, 386: 113-121.
43. Tremblay, E., Roisin, M.P., Represa, A., Charriant-Marlangue, C., and Ben-Ari, Y., 1988, Transient increased density of NMDA binding sites in the developing rat hippocampus, *Brain Res.*, 461: 393-396.
44. Wu, J.-Y., Johansen, F.F., Lin, C.-T., and Liu, J.-W., 1987, Taurine system in the normal and ischemic rat hippocampus, *Adv. Exp. Med. Biol.*, 217: 265-274.

TAURINE NEURONS IN RAT HIPPOCAMPAL FORMATION ARE RELATIVELY INERT TO CEREBRAL ISCHEMIA

Jang-Yen Wu[1], Chin-Tarng Lin[3], Flemming Fryd Johansen[2] and Jie- Wen Liu[1]

[1]Department of Physiology and Cell Biology
University of Kansas
Lawrence, Kansas 66045-2106
[2]Institute of Neuropathology
University of Copenhagen
Copenhagen, Denmark
[3]Institute of Biomedical Sciences
Academia Sinica
Taipei, Taiwan

INTRODUCTION

It is well-known that transient cerebral ischemia causes selective loss of neurons in the hippocampus, striatum and cortical layers (14,24). It has been suggested that the difference in nerve cell vulnerability to ischemic damage is probably due to differences in the nature of their transmitters (14,22,27). For instances, GABAergic neurons in the striatum are more susceptible to transient ischemia than the cholinergic neurons (7) and dopaminergic neurons are more sensitive to ischemic damage than the GABAergic neurons in the gerbil brain (26). Since taurine has been demonstrated to have protective effects on the photoreceptor in the cat (12) and on the light-induced disruption of isolated frog rod outer segments (23), we decided to investigate the influence of transient cerebral ischemia on the taurine neurons in rat hippocampus. Taurine neurons were identified immunocytochemically with antibodies against cysteine sulfinic acid decarboxylase, the rate-limiting enzyme of taurine biosynthesis. Cysteine sulfinic acid decarboxylase has previously been used as a specific marker for identification of taurine neurons in the retina (10,17-19), cerebellum (3,4) and hippocampus (10,26,28).

Taurine in Health and Disease, Edited by R. Huxtable
and D.V. Michalk, Plenum Press, New York, 1994

MATERIALS AND METHODS

Transient Ischemia

Twenty min of transient cerebral ischemia was induced in eight male adult Wistar rats (200-300g) by means of 4-vessel occlusion (13,24). Another four rats were sham-operated but had no ischemia. Briefly, rats were deprived of food and anesthethized with 2% halothane in 98% oxygen prior to electrocauterization of the vertebral arteries. A Martin Electrotom 30 (setting 4) and a 0.5 mm thick cauterization needle was used. The animals were kept food-deprived with free access to water until the next day. They were reanesthetized with halothane in oxygen, and the common carotid arteries were gently exposed before anesthesia was switched off. Two minutes later the common carotids were clamped with metal clasps for 20 min. The rat body temperature was maintained at 37°C with an incandescent lamp. After ischemia the animals survived for 7 days.

Production and Characterization of Antiserum Against Cysteine Sulfinic Acid Decarboxylase

The methods for purification of cysteine sulfinic acid decarboxylase, and production and characterization of its antiserum have been described previously (29,31). Briefly, the decarboxylase was purified from mammalian brain by a combination of ammonium sulfate fractionation, column chromatographies on sephadex G-200, hydroxylapatite, DEAE cellulose and preparative polyacrylamide gel electrophoresis. The purity of the cysteine sulfinic acid decarboxylase preparation was judged by polyacrylamide gel electrophoresis, gradient gel electrophoresis and isoelectric focusing in which a single protein band co-migrated with the enzyme activity. The purified enzyme preparation was injected into rabbits for the production of polyclonal antibodies. The specificity of anti-cysteine sulfinic acid decarboxylase serum was established by immunodiffusion, immunoelectrophoresis and immunoprecipitation tests (29,31). The well-characterized anti-decarboxylase has been used extensively for the identification of taurine neurons in various parts of the central nervous system including cerebellum (3,4), retina (10,17-19) and hippocampus (10,26,28).

Immunohistochemistry

The immunohistochemical procedure used in the present study was similar to the method described previously (17,26,28,30). Rats from sham-operated and ischemic experiments were anesthetized with ether and perfused through the left cardiac ventricle with 200 ml of fixative (4% paraformaldehyde and 0.01% glutaraldehyde in 0.1 M phosphate buffer, pH 7.4). After perfusion, the brain was removed from the skull and the hippocampus was dissected. Tissue blocks from the central portion of the hippocampus were immersed overnight in 2% paraformaldehyde in phosphate buffer. The blocks were rinsed in phosphate buffer and immersed in the same buffer containing 0.025 M hydroxylamine for 30 min to block excess aldehyde groups. Tissue sections (50 μm) were then cut with a vibratome. The sections were incubated with normal goat IgG (0.1 mg/ml) for 10 min, followed by incubation with rabbit antiserum against cysteine sulfinic acid decarboxylase for 16 hr at 4°C (1:500 dilution). After two buffer washings (10 min each), sections were further incubated with biotinylated goat anti-rabbit IgG (1:200 dilution), followed by additional washing and incubation in avidin-biotin-peroxidase complex solution for 30 min (Vector Laboratories, Burlingame, CA). Sections were washed again in buffer and refixed in 2.5% glutaraldehyde for 5 min. After extensive washing, sections were incubated with 0.05% 3,3'- diaminobenzidine-4 HCl in 0.2 M Tris HCl buffer (pH 7.2) plus 0.01% H_2O_2 for 2 min. Sections were then washed, counterstained in 0.01% OsO_4 for 1 min, rinsed in distilled water and mounted on glass slides with 50% glycerine in buffer. Control sections were incubated with either preimmune rabbit serum or antiserum which had been preabsorbed with excess antigen.

RESULTS

CA-1 Region

In the sham-operated animals, cysteine sulfinic acid decarboxylase-immunoreactivity was found in all strata of CA1 including stratum oriens, stratum pyramidale and stratum radiatum (Figs. 1A and 1B). Certain larger stained perikarya in the stratum pyramidale could be the basket neurons (Figs. 1A and 1B). Many smaller decarboxylasepositive interneurons were seen in all strata. In addition, reaction product was localized in the neuronal processes in each layer. These findings were also seen in CA2 and CA3. In the ischemic animals, a massive degeneration of pyramidal neurons was found in the CA-1 stratum pyramidale and no clear neuronal profile could be seen. However, most stained interneurons in the stratum pyramidale, stratum oriens, and stratum radiatum were similar

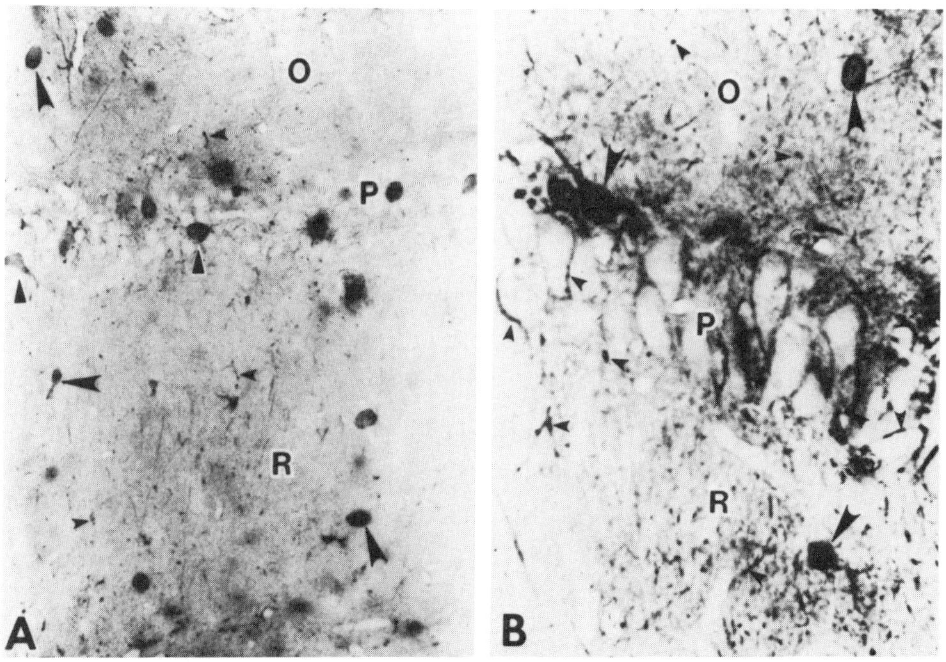

Figure 1 A. Immunohistochemical localization of cysteine sulfinic acid decarboxylase in sham-operated hippocampal formation area CA1. Immunoreactivity to cysteine sulfinic acid decarboxylase is seen in some interneurons (large black arrowheads) in the strata oriens (O) and radiatum (R) and in the somata of some pyramidal shaped interneurons in stratum pyramidale (P) (black-and-white arrowheads). Some stained neuronal processes are also found in those areas (small arrowheads) (X 300). **B.** Immunohistochemical localization of cysteine sulfinic acid decarboxylase in sham- operated hippocampal formation area CA1. A higher magnification showing stained interneurons (large arrowheads) in strata oriens (O), pyramidale (P) and radiatum (R). Some stained neuronal processes are also seen in the three layers (small arrowheads) (X 600).

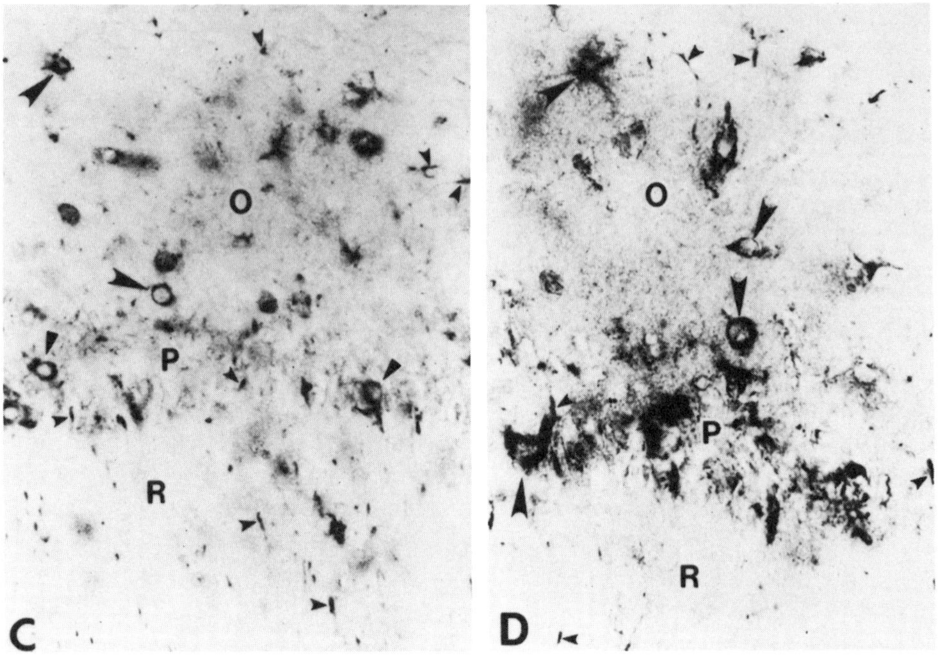

Figure 1 C. Immunohistochemical localization of cysteine sulfinic acid decarboxylase in ischemic hippocampal formation area CA1. Immunoreactivity to cysteine sulfinic acid decarboxylase is still seen in some interneurons (large arrowheads) in strata oriens (O) and radiatum (R) and in the pyramidal shaped interneurons in stratum pyramidale (P) (black-and-white arrowheads). The morphology of the stained neurons appears to be normal. Some stained neuronal processes are still visible in the three layers (small black arrowheads). The unstained pyramidal neurons show severe degeneration (X 300). **D.** Immunohistochemical localization of cysteine sulfinic acid decarboxylase in ischemic hippocampal formation area CA1. A higher magnification showing normal stained interneurons in the strata oriens (O) and pyramidale (P) (large arrowheads). The unstained pyramidal neurons show marked degeneration. Some stained neuronal processes are still present (small arrowheads). R:stratum radiatum (X 600).

to the sham-operated brain and appeared normal morphologically (Figs. 1C and 1D). Numerous decarboxylase-positive neuronal processes could still be seen in each layer although the staining was somewhat weaker than in the sham- operated group. The control section from ischemic animal treated with preimmune serum showed no specific staining.

CA-4 Region

Similar results were obtained in the CA4 region (Figs. 2A-B). Although cysteine sulfinic acid decarboxylase-positive interneurons in the CA4 region of the sham-operated animals were not as numerous as those in the CA1 region, several cysteine sulfinic acid decarboxylase-positive neurons and neuronal processes surrounding the pyramidal cells could easily be identified (Fig. 2A). Again, ischemia caused destruction of neurons in the CA4 region (Fig. 2B). Although it was difficult to identify the precise neuronal profiles in the CA4 region after ischemia, some cysteine sulfinic acid decarboxylase-positive neuronal processes could be identified (Fig. 2B). Sections treated with preimmune serum as control showed no specific staining (data not shown).

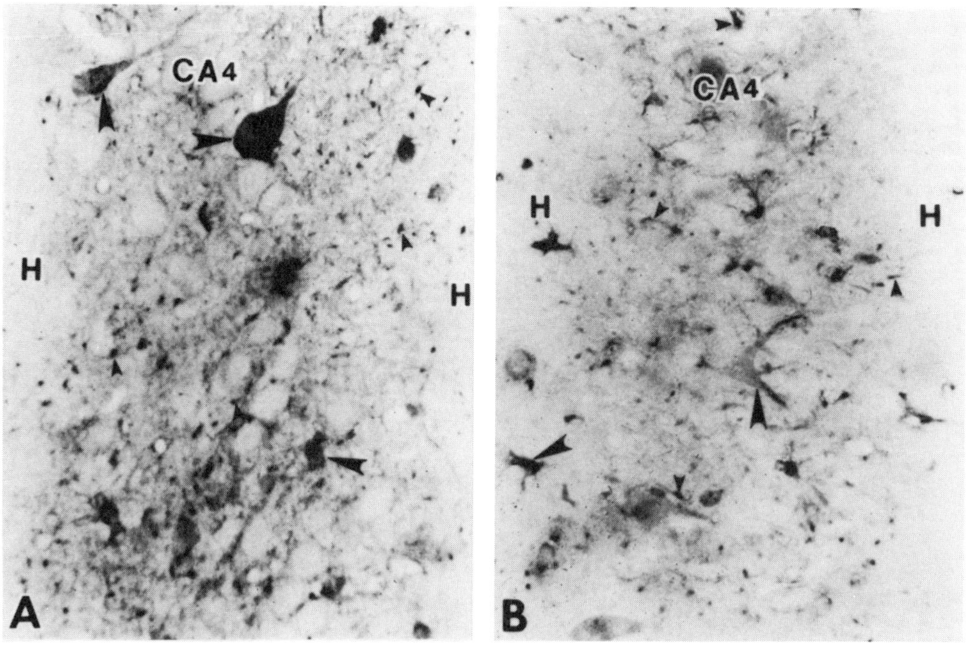

Figure 2 A. Immunohistochemical localization of cysteine sulfinic acid decarboxylase in sham-operated hippocampal formation area CA4. Immunoreactivity to cysteine sulfinic acid decarboxylase is found in some somata of pyramidal basket interneurons (large arrowheads) and in some neuronal processes (small arrowheads). H:dentate hilus (X 600). **B.** Immunohistochemical localization of cysteine sulfinic acid decarboxylase in ischemic hippocampal formation area CA4. Immunoreactivity to cysteine sulfinic acid decarboxylase is shown in some interneurons (large arrowhead) and neuronal processes (small arrowheads). Some pyramidal neurons are degenerated. H:dentate hilus (X 600).

Dentate Gyrus

In sham-operated animals, immunoreactivity to cysteine sulfinic acid decarboxylase was seen in the somata of some interneurons in the molecular layer, within the hilus in some presumably pyramidal basket interneurons and in some granule cell-like neurons (Figs. 3A, 3B, 4A). In the stained cells, immunoreactivity was found distributed throughout the cytoplasm. Some stained neuronal processes derived from the pyramidal basket interneurons in the hilus were seen passing through the granule cell layer and extended into the molecular layer (Fig. 3B Inset). No clear punctate reaction product surrounding the granule cell was shown. Similarly, in the molecular layer and hilus, numerous stained neuronal processes in the tangential sections were observed. The stained and unstained interneurons in the molecular layer and hilus received no punctate immunoreaction product. When the ischemic dentate gyrus was stained with anti-cysteine sulfinic acid decarboxylase, decarboxylase-positive interneurons in the molecular layer, granule cell layer and hilus were visible (Figs. 3A, 3B, 4B) and their morphology appeared to be normal. The stained neuronal processes in the hilus, granule cell layer and molecular layer appeared not to be affected, although the intensity of staining was somewhat weaker than the sham-operated animals. The control sections showed no specific staining.

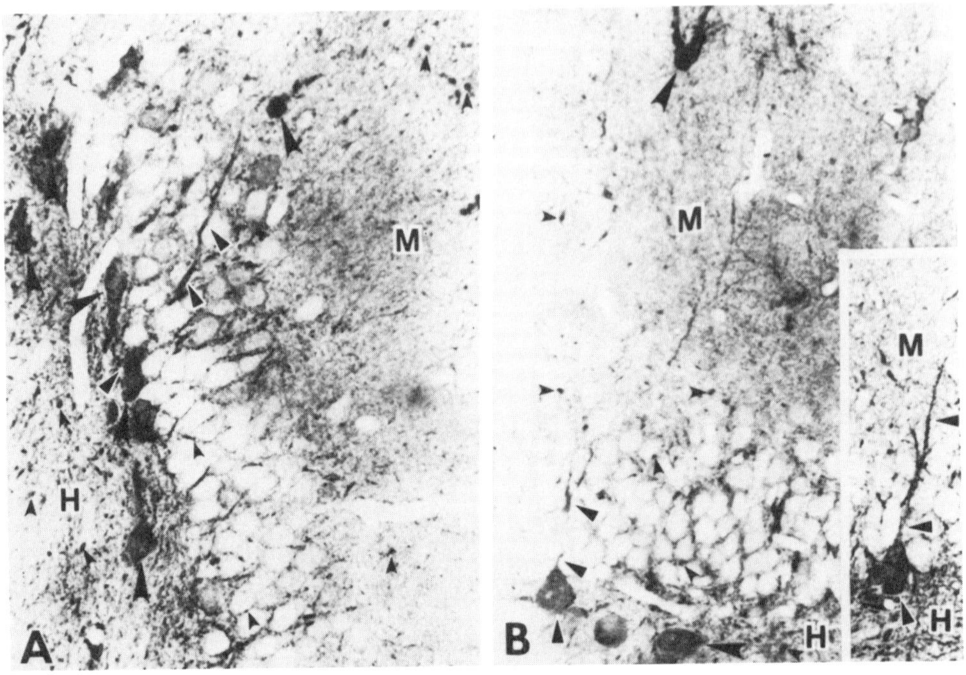

Figure 3 A. Immunohistochemical localization of cysteine sulfinic acid decarboxylase in sham-operated dentate gyrus. Immunoreactivity to cysteine sulfinic acid decarboxylase is seen in some somata of interneurons (large black arrowheads) in the hilus (H), the molecular layer (M), and in the granule cell layer. In the granule cell layer, some stained interneuronal processes are seen to pass through the entire granule cell layer and extended into the molecular layer (black-and-white arrowheads) (X 600). **B.** Immunohistochemical localization of cysteine sulfinic acid decarboxylase in sham- operated dentate gyrus. Some somata of interneurons (larger arrowheads) containing immunoreactivity to cysteine sulfinic acid decarboxylase are seen in the molecular layer (M) and in the hilus (H). A stained presumably granule basket interneuron (black-and-white arrowhead) is noted in the hilus close to the granule cell layer. The stained neuronal processes are also visible in the entire area (small arrowheads) (X 600). Inset: A stained presumably granule basket interneuron with its long neuronal process extended into the molecular layer (M) is shown (black-and-white arrowheads). H:hilus (X 600).

DISCUSSION

The distribution of immunoreactivity cysteine sulfinic acid decarboxylase in hippocampal formation in the sham-operated animal is similar to that previously reported (26). The staining included the somata and their neuronal processes of some pyramidal shaped interneurons in the pyramidal cell layer and of some interneurons in strata oriens and radiatum. There was no clear punctates surrounding the pyramidal cell bodies but only the oblique or cross sections of neuronal processes. Our previous immunoelectron microscopic observations have shown that most of the staining which surround pyramidal cell bodies are dendritic processes (26). In the dentate gyrus, many interneurons and their neuronal processes in the hilus contained Immunoreactivity. In the granule cell layer, not only some presumably pyramidal basket interneurons and their neuronal processes, but also a few granule cell-like neurons contained immunoreactivity to cysteine sulfinic acid decarboxylase, suggesting that perhaps there are subtypes of granule cells which are decarboxylase-positive similar to the Purkinje cells observed in the cerebellum (4,5). Similarly, immunoreactivity was also noted in some interneurons and their neuronal processes in the molecular layer. However, there was no punctate immunoreactivity surrounding the stained or unstained

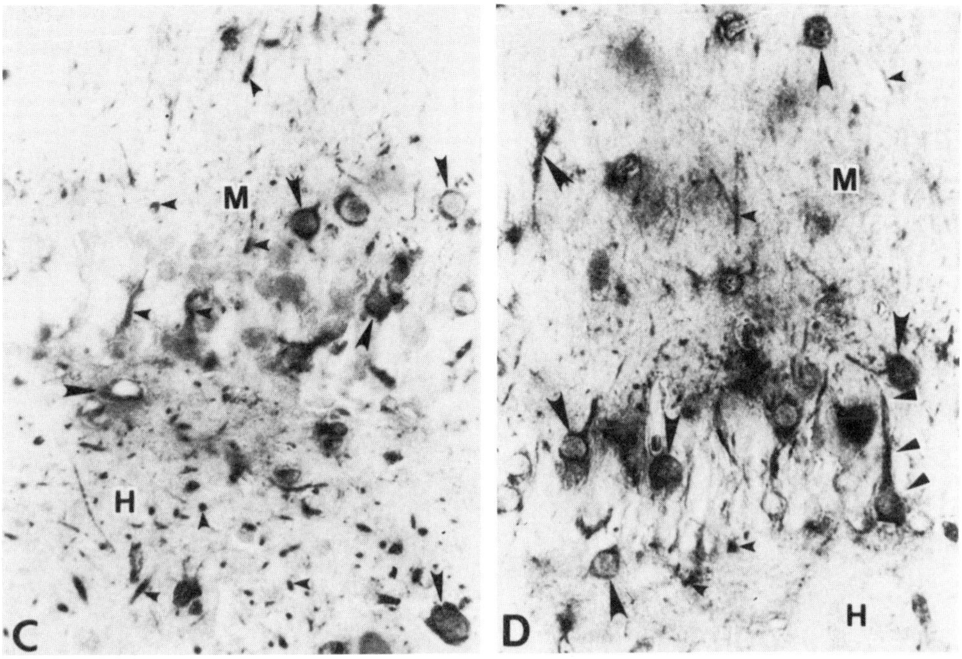

Figure 3 C. Immunohistochemical localization of cysteine sulfinic acid decarboxylase in the ischemic dentate gyrus. Immunoreactivity to cysteine sulfinic acid decarboxylase is shown in the somata of interneurons in the hilus (H) and in the granule cell layer (large arrowheads) and also in the neuronal processes (small arrowheads)(X 600). **D.** Immunohistochemical localization of cysteine sulfinic acid decarboxylase in the ischemic dentate gyrus. Most stained interneurons in the granule cell layer and in molecular layer (M) show normal morphology (large black arrowheads). A stained granule basket interneuron (black-and-white arrowhead) is shown (X 600).

interneurons in the hilus and in the molecular layer. It is not clear whether the stained neuronal processes surrounding the granule cell somata is the axonal terminals or dendritic processes. Further studies using immunoelectron microscopy are needed.

Although transient ischemia caused severe degeneration of CA-1 pyramidal cells, the decarboxylase-stained interneurons (including some pyramidal shaped neurons) revealed normal morphology. Similarly, the cysteine sulfinic acid decarboxylase-positive granule cell-like neurons as well as the decarboxylase-positive presumably pyramidal basket neurons were relatively well preserved. Also the decarboxylase-positive interneurons and their neuronal processes in the hilus and in the molecular layers were resistant to the ischemic insult. Thus, neurons which contain the taurine synthesizing enzyme (presumably taurine neurons) show normal morphology after transient ischemia. These results suggest that taurine in the cysteine sulfinic acid decarboxylase-positive neurons may play an important role in protecting the neurons from ischemic insult.

The mechanism underlining this protective function is not clear. Taurine possibly parcipitates in the regulation of calcium metabolism (6,15,21). In addition, taurine has been shown to be involved in maintaining the structural integrity of the cell membrane. The membrane stabilizing property of taurine is suggested by experiments demonstrating retinal degeneration in cats with taurine deficiency (2,12). Schurr et al. (25) reported that taurine can protect neurons from hypoxic damage. When brain slices were kept in a bathing medium containing taurine, neurons showed normal electrophysiological responses and normal morphology. But if the brain slices were kept in a medium without taurine the neurons became severely damaged and lost electrophysiological responses. This protective function of taurine may be due to its modulation of Ca^{2+} fluxes across neuronal membranes (6,20,21). This notion is supported by the observation that hypoxia-induced Ca^{2+} influx is believed to play a major role in producing irreversible damage in neuronal tissue (11). It has also been shown that during ischemia, there is an increased release of taurine (9) in concordance with excessive release of excitotoxins (1). Since taurine inhibits cellular calcium uptake it may have protective effect against cytotoxicity elicted by excitatory amino acids (16). The relative concentration of the protective inhibitory transmitters (e.g. taurine, GABA) and the excitotoxic transmitters (e.g. glutamate, aspartate) in the microenvironment of a particular neuron may determine its vulnerability towards ischemia. In addition to its

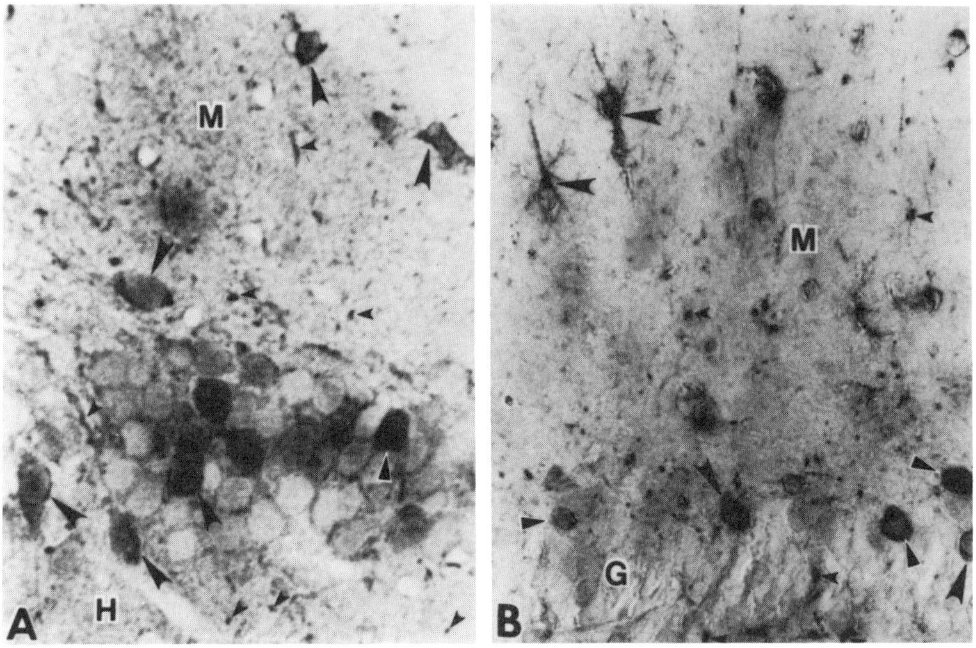

Figure 4 A. Immunohistochemical localization of cysteine sulfinic acid decarboxylase in the sham- operated dentate gyrus. Stained interneurons (large black arrowheads) are seen in the hilus (H), in the granule cell layer and in the molecular layer (M). One granule cell-like neuron in the granule cell layer also contains immunoreactivity to cysteine sulfinic acid decarboxylase (black-and-white arrowhead). The stained neuronal processes are also found in the entire area (small black arrowheads) (X 600). **B.** Immunohistochemical localization of cysteine sulfinic acid decarboxylase in ischemic dentate gyrus. Stained interneurons (large arrowheads) which appear to be normal are seen in the granule cell layer (G) and molecular layer (M). Two granule cell-like neurons containing immunoreactivity to cysteine sulfinic acid decarboxylase appear to be morphologically well preserved neurons (black-and-white arrowheads). Stained neuronal processes are visible in the molecular layer and granule cell layer although the staining intensity has decreased (small arrowheads) (X 600).

protective function in the brain, taurine has also been shown to protect myocardial tissue from calcium overload, hypoxic and reoxygenation stress (8).

Another possible mechanism involved in the protective function of taurine is its role as an inhibitory neurotransmitter. This notion is supported from our previous observations that GABA neurons which contain GABA, another major inhibitory transmitter in the brain, are also less vulnerable to ischemic damage than other neurons in the hippocampus (13).

Our recent findings of a specific receptor for taurine in the brain (32) may provide an alternative mechanism related to the function of taurine as a protective agent.

ACKNOWLEDGEMENTS

This work was supported in part by NIH grant, NS20978, NSF grant, BNS-8820581, and American Heart Association-Kansas Affiliate (JYW) and Weiman Foundation Copenhagen (to FFJ). The authors thank Jan Elder, Sharon Hopkins and Judy Wiglesworth for typing the manuscript.

REFERENCES

1. Beneveniste, H., Drejer, J., Schousboe, A., and Diemer, N.H. 1984, Elevation of the extracellular concentrations of glutamate and aspartate in rat hippocampus during transient cerebral ischemia monitored by intracerebral microdialysis, *J.Neurochem.*, 43:1369-1374.
2. Berson, E.L., Hayes, K.C., Rabin, A.R., Schmidt, S.Y., and Watson, G. 1976, Retinal degeneration in cat fed casein. II. supplementation with methionine, cysteine or taurine, *Invest.Ophthalmol.*, 15:52-58.
3. Chan-Palay, V., Lin, C.-T., Palay, S., Yamamoto, M., and Wu, J.-Y. 1982, Taurine in the mammalian cerebellum: demonstration by autoradiography with (]3]H)taurine and immunocytochemistry with antibodies against the taurine-synthesizing enzyme, cysteine-sulfinic acid decarboxylase, *Proc.Natl.Acad.Sci.USA*, 79:2695-2699.
4. Chan-Palay, V., Palay, S.L., and Wu, J.Y. 1982, Sagittal cerebellar microbands of taurine neurons -immunocytochemical demonstration by using antibodies against the taurine synthesizing enzyme cysteine sulfinic decarboxylase acid, *Proc.Natl.Acad.Sci.USA*, 79:4221-4225.
5. Chan-Palay, V., Wu, J.-Y., and Palay, S.L. 1979, Immunocytochemical localization of gamma aminobutyric acid transaminase at cellular and ultrastructure levels, *Proc.Natl.Acad.Sci.USA*, 76:2067-2071.
6. Dolara, P., Agresti, A., Giotti, A., and Pasquini, G. 1973, Effect of taurine on calcium kinetics of guinea pig heart, *Eur.J.Pharmacol.*, 21:352-358.
7. Francis, A. and Pulsinetti, W. 1982, The response of GABAergic and cholinergic neurons to transient cerebral ischemia, *Brain Res.*, 243:271-278.
8. Franconi, F., Stendardi, I., Matucci, R., Bennardini, F., Baccaro, C., and Giotti, A. 1985, A protective effect of taurine on "hypoxic" and "reoxygenated" guinea pig heart, *in:* "Taurine: biological actions and clinical perspectives", Oja, S.S., Ahtee, L., Kontro, P. and Paasonen, M.K., eds., Alan R. Liss, New York, pp. 177-182.
9. Hagberg, H., Lehmann, A., Sandberg, M., Nystrom, B., Jacobsen, I., and Hamberger, A. 1985, Ischemia-induced shift of inhibitory and excitatory amino acids frm intra- and extracellular compartments, *J.Cereb.Blood Flow Metab.*, 5:413-419.
10. Hanretta, A.T. and Lombardini, J.B. 1987, Is taurine a hypothalamic neurotransmitter?: a model of the differential uptake and compartmentalization of taurine by neuronal and glial cell particles from the rat hypothalamus, *Brain Res.Rev.*, 12:167-201.
11. Hansen, A.J. 1985, Effect of anoxia on ion distribution in the brain, *Physiol.Rev.*, 65:101-148.
12. Hayes, K.C., Carey, R.E., and Schmidt, S.Y. 1975, Retinal degeneration associated with taurine deficiency in the cat, *Science*, 188:949-951.
13. Johansen, F., Lin, C.T., Schousboe, A., and Wu, J.-Y. 1989, An immunocytochemical investigation of L-glutamate decarboxylase in the rat hippocampal formation: The influence of cerebral ischemia, *J.Comp.Neurol.*, 281:40-53.
14. Jorgensen, M.B. and Diemer, N.H. 1982, Selective neuron loss after cerebral ischemia in the rat: possible role of transmitter glutamate, *Acta Neurol.Scand.*, 66:536-546.
15. Kuriyama, K., Muramatsu, M., Nakagawa, K., and Kakita, K. 1978, Modulating role of taurine on release of neurotransmitters and calcium transport in excitable tissues, *in:* "Taurine and Neurological Disorders", Barbeau, A. and Huxtable, R.J., eds., Raven Press, New York, pp. 201-216.
16. Lehmann, A., Hagberg, H., and Hamberger, A. 1984, A role for taurine in the maintenance of homeostasis in the central nervous system during hyperexcitation? *Neuroscience Letters*, 52:341-346.
17. Lin, C.-T., Li, H.-Z., and Wu, J.-Y. 1983, Immunocytochemical localization of L-glutamate

decarboxylase, gamma-aminobutyric acid transaminase, cysteine sulfinic acid decarboxylase, aspartate aminotransferase and somatostatin in rat retina, *Brain Res.*, 270:273-283.

18. Lin, C.-T., Song, G.-X., and Wu, J.-Y. 1985, Ultrastructural demonstration of L-glutamate decarboxylase and cysteine sulfinic acid decarboxylase in rat retina by immunocytochemistry, *Brain Res.*, 331:71-80.
19. Lin, C.-T., Song, G.-X., and Wu, Y.-Y. 1985, Is taurine a neurotransmitter in rabbit retina? *Brain Res.*, 337:293-298.
20. Lombardini, J.B. 1978, High-affinity transport of taurine in the mammalian central nervous system, *in:* "Taurine and Neurological Disorders", Barbeau, A. and Huxtable, R.J., eds., Raven Press, New York, pp. 119-136.
21. Lop'z-Colom', A.M. 1982, Taurine receptors in CNS membranes: binding studies, *in:* "Taurine in Nutrition and Neurology", Huxtable, R.J. and Pasantes-Morales, H., eds., Plenum Press, New York, pp. 293-310.
22. Nemoto, E.M. 1979, Studies on the pathogenesis of ischemic brain damage and its amelioration by barbiturate therapy, *in:* "Brain and Heart Infarct", Zulch, K.J. and Kaufman, W., eds., Springer Verlag, pp. 306-317.
23. Pasantes-Morales, H., Ademe, R.M., and Quesada, O. 1981, Protective effect of taurine on the light-induced disruption of isolated frog rod outer segments, *J.Neurochem.Res.*, 6:337-348.
24. Pulsinelli, W.A. and Brierly, J.B. 1979, A new model of bilateral hemispheric ischemia in the unanesthetized rat, *Stroke*, 10:267-272.
25. Schurr, A., Tsen, M.T., and Rigor, B.M. 1986, Protection against hypoxia by taurine - study using the in vitro hippocampal slice, *Neurosci.Abstracts*, 12:1526.(Abstract)
26. Taber, K.H., Lin, C.T., Liu, J.W., Thalmann, R.H., and Wu, J.Y. 1986, Taurine in hippocampus: localization and postsynaptic action, *Brain Res.*, 386:113-121.
27. Weinberger, J. and Cohen, C. 1982, Differential effect of ischemia on active uptake of dopamine, GABA, and glutamate, *J.Cereb.Blood Flow Metab.*, 2:265a.
28. Wu, J., Johansen, F.F., Lin, C., and Lui, J. 1987, Taurine system in the normal and ischemic rat hippocampus, *in:* "The biology of taurine: methods and mechanisms", Huxtable, R.J., Franconi, F. and Giotti, A., eds., Plenum Press, New York, pp. 265-274.
29. Wu, J.-Y. 1982, Purification and characterization of cysteic acid and cysteine sulfinic acid decarboxylase and L-glutamate decarboxylase from bovine brain, *Proc.Natl.Acad.Sci.USA*, 79:4270-4274.
30. Wu, J.-Y. and Lin, C.-T. 1985, Immunocytochemical techniques, *in:* "Neuromethods, Vol. 3: Amino Acids", Boulton, A.A., Baker, G.B. and Wood, J.D., eds., Humana Press, New Jersey, 115-178.
31. Wu, J.-Y., Lin, C.-T., Hwang, B., Wei, S., and Lin, H.S. 1986, Antibodies against enzymes synthesizing amino acid neuro-transmitters, *in:* "Neurohistochemistry, Modern Methods and Applications", Panula, P., Paivarinta, H. and Soinila, S., eds., Alan R. Liss, Inc., New York, pp. 21-47.
32. Wu, J.-Y., Tang, X.W., and Tsai, W.H. 1992, Taurine receptor: Kinetic analysis and pharmacological studies, *Adv.Exp.Med.Biol.*, 315:263-268.

CALCIUM AND TAURINE INTERACTION IN MAMMALIAN BRAIN METABOLISM

Giampietro Sgaragli,[1] Maria Frosini,[1] Mitri Palmi,[1]
Loria Bianchi,[2] and Laura Della Corte[2]

[1]Istituto di Scienze Farmacologiche
Università di Siena, Italy
[2]Dipartimento di Farmacologia
Università di Firenze, Italy

INTRODUCTION

Taurine is present in relatively high (mM) concentrations in mammalian tissues, including the brain. Despite extensive studies of its synthesis, release in response to nerve stimulation and possible neuronal or extraneuronal uptake there is still no consensus about its functions (9,8,12). It has variously been designated as a neurotransmitter, neuromodulator, a modulator of calcium fluxes, and also as an osmoregulator or membrane stabilizer and an antioxidant. These effects may, at least in some cases, be interlinked. Thus, for example, the "neuroprotective" effects of taurine may be related to the osmoregulatory, antioxidant and calcium ion modulating functions of taurine.

With regard to the integrated neurovegetative functions that are controlled by the central nervous system (CNS), much attention in the last few years has been devoted to clarifying the possible role of taurine in thermoregulation. Regulation of body temperature involves a delicate balance between the production and loss of heat. The hypothalamus regulates the set point at which body temperature is then maintained. One of the most striking and consistent effects of central (ICV) administration of taurine in mammals is hypothermia. The primary action of taurine in thermoregulatory effectors is expressed peripherally mainly by the promotion of heat dissipation which takes place through a reduction in peripheral vasomotor tone and by the inhibition of shivering (17). Furthermore, a decrease in metabolic heat production and an increase in evaporative heat loss - both under serotonergic control - may also play a role in taurine-induced hypothermia.

In fever, where the thermoeffectors work in just the opposite direction, the hypothalamic set point is raised , and can be brought back to normal by aspirin-like drugs. Fever is caused by various conditions that have, as a common feature, the enhanced formation of cytokines such as interleukin 1 or tumor necrosis factor, which will in turn induce the synthesis of prostaglandin E_2 in vascular organs of the preoptic hypothalamic area

Taurine in Health and Disease, Edited by R. Huxtable
and D.V. Michalk, Plenum Press, New York, 1994

(5). Prostaglandins act within the hypothalamus to raise body temperature by processes that are still unclear. Aspirin-like drugs reduce fevers caused by interleukin-1 and various cytokine enhancers but do not reduce fever caused by prostaglandins (13), thus suggesting that these compounds act by inhibiting prostaglandin E_2 synthesis. Although there is agreement over the nature of fever mediators, the precise neurochemical mechanisms have not yet been understood.

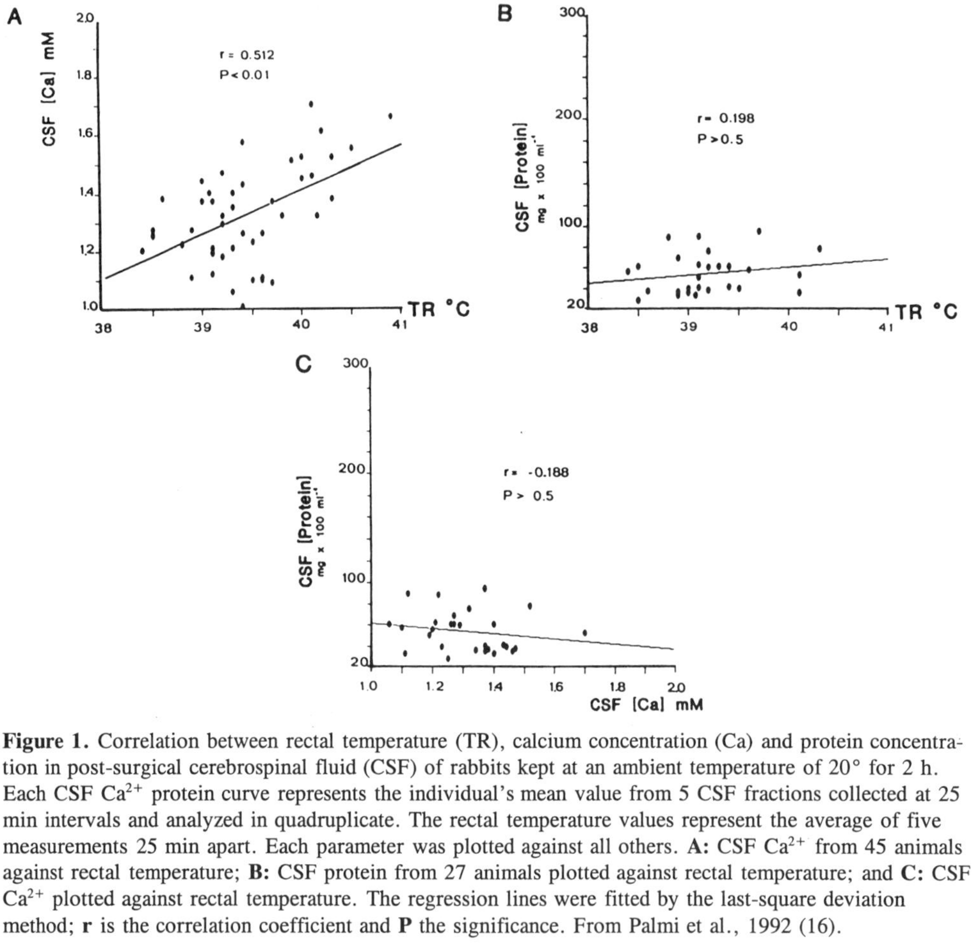

Figure 1. Correlation between rectal temperature (TR), calcium concentration (Ca) and protein concentration in post-surgical cerebrospinal fluid (CSF) of rabbits kept at an ambient temperature of 20° for 2 h. Each CSF Ca^{2+} protein curve represents the individual's mean value from 5 CSF fractions collected at 25 min intervals and analyzed in quadruplicate. The rectal temperature values represent the average of five measurements 25 min apart. Each parameter was plotted against all others. **A:** CSF Ca^{2+} from 45 animals against rectal temperature; **B:** CSF protein from 27 animals plotted against rectal temperature; and **C:** CSF Ca^{2+} plotted against rectal temperature. The regression lines were fitted by the last-square deviation method; **r** is the correlation coefficient and **P** the significance. From Palmi et al., 1992 (16).

Early experimental evidence of Myers et al. (14,15) states that the set point for body temperature in mammals is regulated by the antagonistic action of Ca^{2+} and Na^+ ions in the extracellular fluid within the posterior hypothalamus. We have recently shown that when the normothermic conditions of animals were altered, as in post-surgical fever or *E. coli* endotoxin and interleukin-1 fever, there is a consistent and highly significant upward shift of cerebrospinal fluid Ca^{2+} content concomitant with an increase of rectal temperature

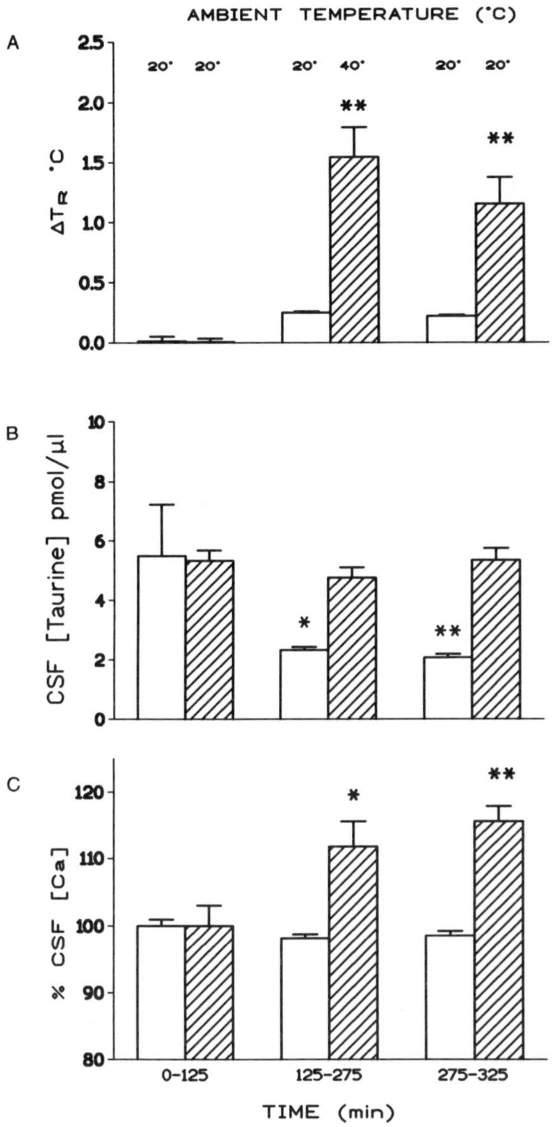

Figure 2. Effect of heat stress on core temperature, cerebrospinal fluid calcium and taurine concentrations. Open bars: control rabbits maintained at an ambient temperature of 20°C. Hatched bars: rabbits exposed to 40° C for a 40-min period starting at 125 min (Middle bar: at 125 min, the temperature in the thermostatted chamber began to rise, reaching 40°C at approximately 175 min. This temperature was maintained for 40 min and then reduced back to 20°C over the next 50 min). * $p < 0.05$, ** $p<0.01$ vs their own pre-treatment (0-125) values. n= 5 animals

(16). As shown in Fig. 1, increased calcium levels are linearly correlated to the degree of temperature gain. These findings suggest that brain calcium metabolism plays a role in the development of fever.

One may conceive that whenever there is an imbalance of temperature regulation, even under hypothermic conditions, changes in brain calcium metabolism may follow giving rise to modifications of cerebrospinal fluid calcium concentrations. To test the hypothesis that taurine might be involved in thermoregulation by affecting calcium metabolism in the brain, two experimental approaches were undertaken, a physiological and a pharmacological one. The first approach was to study whether hyperthermia induced by exposing animals to high ambient temperature could elicit changes in the cerebrospinal fluid taurine and calcium content. The aim of the second approach was to study how taurine and a series of taurine analogs injected ICV would modify cerebrospinal fluid calcium content as well as the thermal and neurological behaviour.

METHODS

Adult, male New Zealand albino rabbits weighing 2.0-2.5 kg were supplied with food and water ad libitum and individually caged at an ambient temperature of 15-20°C with natural night and day cycles. Under sodium pentobarbital anaesthesia, a cannula guide was chronically implanted by means of a stereotaxic apparatus into the cisterna magna for the withdrawal of cerebrospinal fluid, as previously described (16). For the ICV administration of drugs, an inflow cannula guide was inserted stereotaxically into a lateral cerebral ventricle. Samples of cerebrospinal fluid drawn from the cisterna magna through a polyethylene tubing connected to a peristaltic pump at a constant flow rate of 5 μl/min were collected into plastic tubes at 25 min fractions (125 μl). Fractions were refrigerated at 4 °C and analysed on the same day as in the experimental session. The rate of cerebrospinal fluid withdrawal was well below 9.6 $\pm$ 0.9 (S.D.) μl/min, which is the value reported for cerebrospinal fluid formation in conscious, restrained rabbits (2). Compounds dissolved in pyrogen-free water were administered ICV in 10 μl volume by a micrometer syringe.

Conscious rabbits were individually housed in a thermostated chamber and then restrained in stainless steel cages. The ambient temperature was set at 20 °C or - for some experiments - at 40 °C. The core temperature was measured every 5 min by a thermocouple thermometer (Columbus Instruments, Columbus, OH) which was inserted 10 cm into the rectum and connected to a personal computer with an isothermex program. Core body temperature was monitored for at least 1 h before the experimental session.

Cerebrospinal fluid samples were analysed for amino acid content using reversed-phase HPLC with the *O*-pthalaldehyde pre-column derivatization procedure as previously described (4). The complete separation of aspartate, glutamate, taurine and GABA was obtained with a total running time of 20 min using a 3 step gradient (mobile phase : methanol/potassium acetate 0.1 M , pH 5.5; flow rate: 1.5 ml/min). This method proved to be extremely sensitive, with detection limits between 7 and 16 fmol/μl, and reliable as evident from the basal mean values $\pm$ S.E.M. (n=4) of aspartate, glutamate, taurine and GABA in rabbit cerebrospinal fluid samples which were: 4.18 $\pm$ 0.37, 23.95 $\pm$ 3.45, 5.33 $\pm$ 0.36 and 1.16 $\pm$ 0.14 pmol/μl.

Cerebrospinal fluid Na^+, K^+, Mg^{2+} and Ca^{2+} contents were determined using an HPLC method with conductimetric detection as already described (6).

Values are expressed as means $\pm$ S.E.M. The significance of differences between pre- and post-treatment data was checked by analysis of variance (ANOVA). A p value less than 0.05 was considered significant.

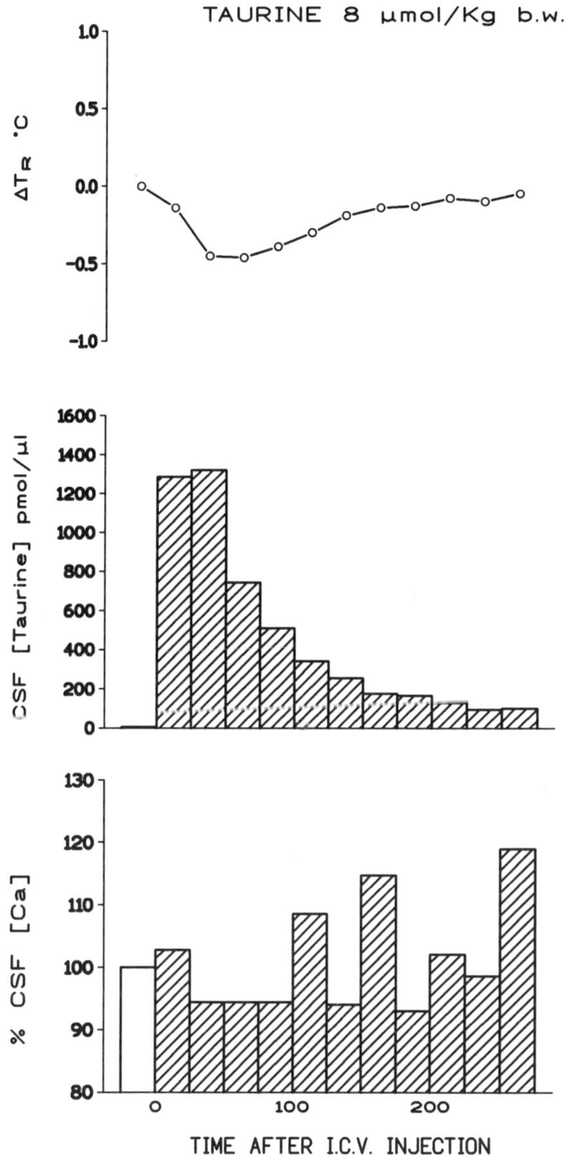

Figure 3. Effects of ICV injection of taurine on core temperature and cerebrospinal fluid calcium concentration in one rabbit. The kinetic of cerebrospinal fluid taurine concentration in cisterna magna is also represented

RESULTS

Effects of Heat Stress on Cerebrospinal Fluid Taurine and Calcium Content

Rabbits exposed to ambient temperature of 40°C for a period of 40 min developed hyperthermia. This started after a 20 min lag period, plateaued in about 1 h and persisted for about 4 h after the animals had been exposed once again to an ambient temperature of 20°C. (Fig. 2, panel A). In control animals, withdrawal of cerebrospinal fluid from cisterna magna was accompanied by a significant reduction in taurine concentration, averaging 60% (panel B). Among the other amino acids analysed only GABA concentration exhibited a behaviour comparable to that of taurine, decreasing to about 20% (data not shown). Noticeably, in animals made hyperthermic by exposure to high temperature, the cerebrospinal fluid taurine concentration did not show significant changes. The same was observed with GABA contents (data not shown). Calcium concentrations, which were unchanged in control animals (panel C), significantly increased in hyperthermic animals by 15 - 20 % and remained elevated for the entire observation period. Na^+, K^+ and Mg^{2+} levels, however, did not show any significant changes (data not shown).

Effects of of Taurine and Some of Its Analogues on Cerebrospinal Fluid Calcium Concentration and on Thermal and Neurological Behaviour

As shown in Fig. 3, ICV administration of taurine induced hypothermia in a dose-related manner. Concomitantly, cerebrospinal fluid calcium concentrations increased significantly, even with the lowest dose of taurine. This effect was already maximal with the dose of 8 μmol/kg and amounted to an increase of 25-30% over the basal conditions.

Calcium concentrations were still elevated at the time when rectal temperatures had returned to basal values. The kinetics of cerebrospinal fluid taurine and Ca^{2+} concentrations following ICV administration of taurine (8 μmol/kg) to one rabbit are reported in Fig. 4. cerebrospinal fluid taurine concentration in cisterna magna had already peaked in the first fraction; i.e. 25 min following injection into the lateral ventricle. The kinetic data of the taurine decay from cisternal cerebrospinal fluid fitted a two-exponential model with half-lives of 20 and 90 min respectively.

The degree of selectivity of the effects induced by ICV taurine administration was investigated by comparing the effects elicited by some compounds structurally related to taurine. The results are summarized in Table 1. Guanidinoethane sulfonic acid, well known as a taurine uptake inhibitor, induced the same effects as taurine, whereas 6-aminomethyl-3-methyl-4H-1,2,4-benzothiadiazine-1,1-dioxide (TAG), reported in the literature as a taurine antagonist, behaved in the opposite way. The two cyclic taurine derivatives, piperidine-3-sulfonic acid and pyridine-3-sulfonic acid, behaved as a taurine agonist and antagonist, respectively. Finally, the phosphonic analogue of taurine only induced a weak hyperthermia and slight excitation.

DISCUSSION

This study demonstrates that when thermoregulation is set towards the promotion of heat dissipation - as it happens when animals undergo hypothermia following ICV injection of taurine or develop hyperthermia under heat stress - calcium metabolism in the brain is altered so that cerebrospinal fluid Ca^{2+} concentration increases. Although this finding gives support to the possible involvement of calcium in the central mechanisms of thermoregulation, it does not address the issue of the source of cerebrospinal fluid calcium.

Table 1. Effect of taurine and related compounds on gross motor behavior and on rectal temperature and calcium concentration in cerebrospinal fluid

Compound	Action	Rectal Temperature		CSF [Ca^{2+}]		Gross Motor Behavior
		0 - 100 min	**100 - 300 min**	**0 - 100 min**	**100 - 300 min**	
Taurine		hypothermia	return to basal value	no effect	↑	-
Piperidine-3-sulfonate	taurine agonist?	hypothermia	hyperthermia	no effect	no effect	-
Pyridine-3-sulfonate	taurine antagonist?	hyperthermia	hyperthermia	↑	↑	+ + +
Guanidinoethane sulfonate	taurine uptake inhibitor	hypothermia	hyperthermia	no effect	↑	-
2-Aminoethane phosphonate		no effect	hyperthermia	no effect	no effect	+
TAG	taurine agonist	hyperthermia	hyperthermia	↑	return to basal value	+ +

- = sedation and sleepiness
\+ = alertness
++ = alertness, laboured breathing
+++ = alertness, laboured breathing, episodical convulsion
↑ = increase

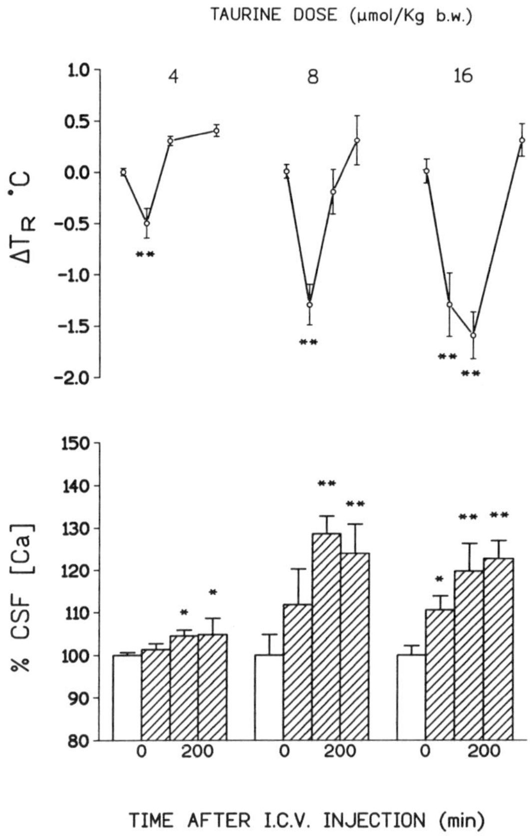

Figure 4. Effects of ICV injection of taurine on core temperature and cerebrospinal fluid calcium concentration. Dose-response relationship. Three-five rabbits were injected with each dose. * $p<0.05$, ** $p<0.01$ vs their own pre-treatment control values

The high plasma concentration of calcium - twice that of cerebrospinal fluid (10, 6) - however, does not allow one to exclude the blood as a possible source of calcium. calcium content in cerebrospinal fluid of mammals is known to be maintained within a relatively limited range even in the presence of marked changes in Ca^{2+} levels in plasma and body fluids (11,7). It has been shown that at the site of passive permeation of Ca^{2+} in the choroid plexus a calcium pump, which can be blocked by ruthenium red, recycles the filtered calcium back into the capillary circulation. The activity of this pump increases in proportion to the concentration of cerebrospinal fluid calcium (1). It is possible that under the experimental conditions of the present study the mechanism of Ca^{2+} outflow from cerebrospinal fluid was inhibited to such an extent as to allow cerebrospinal fluid Ca^{2+} concentration - and subsequently extracellular fluid - to increase to levels which may critically influence neuronal activity.

An increase in cerebrospinal fluid Ca^{2+} concentration might originate also from the depletion of calcium stores in brain cells. Morphine has been reported to induce calcium depletion in discrete brain regions of the rat along with hypothermia (3). Cold stress has been associated in rats and mice with hypothermia and reduced calcium contents in several brain areas (18). In principle, we cannot exclude that the two above-mentioned mechanisms may work jointly to promote the increase in cerebrospinal fluid calcium concentration, although the latter seemed independent of the direction in which peripheral thermoeffectors operated.

Withdrawal of cerebrospinal fluid from cisterna magna is an unphysiological procedure which, as shown in the present study in control animals, gives rise to a progressive decay in cerebrospinal fluid of taurine (and GABA) concentration, whilst leaving unaffected aspartate and glutamate concentrations. Noticeably, heat stress did induce a change in brain taurine (and GABA) metabolism so that taurine (and GABA) cerebrospinal fluid concentration remained constant with time as if its output into the extracellular fluid were stable or even increased; just the opposite to what would have happened to animals kept at a neutral ambient temperature. If this hypothesis is correct, then heat stress, by causing an increased availability of taurine in the extracellular space, would have mimicked the pharmacological approach in which the supply of taurine to cerebrospinal fluid or extracellular space was brought about by an ICV injection.

Finally, the finding that some taurine analogs (pyridine-3-sulfonic acid and TAG) do promote the increase of cerebrospinal fluid Ca^{2+} concentration in a taurine-like manner, in spite of their opposite effects on thermal and gross neurological behaviour, suggests the existence in the brain of at least two sites that are sensitive to taurine; one dealing with the regulation of cerebrospinal fluid Ca^{2+} content - interleukin-1 sensitive site (16) - and the other connected to the synaptic mechanisms of thermoregulation.

ACKNOWLEDGEMENTS

This work was supported by MURST and CNR, Roma, Italy and BIOMED 1 BMH1 CT-94-1402 (European Community). The authors thank B. Gorelli, A. Benocci and R. Borghi for technical assistance, and M.L. Valacchi for typing the manuscript. Nathalie Adams is acknowledged for improving the manuscript.

REFERENCES

1. Barkai, A.I., and Meltzer, H.L., 1982, Regulation of calcium entry into the extracellular environment of the rat brain, *J. Neurosci.* 2:1322-1328.
2. Barkai, A.I., Budek, M., Brown, D.L., and Fieve, R.R., 1978, Application of an improved cerebroventricular perfusion technique in the rabbit: Effect of pentobarbital or haloperidol on monoamine metabolites and proteins in the perfusate, *Neuropharmacology* 17:409-414.

3. Cardenas, H.L., and Ross, D.H., 1975, Morphine induced calcium depletion in discrete regions of rat brain, *J. Neurochem.* 24:487-493.
4. Della Corte L., Bianchi, L., Federico, S., and Michelassi, S., 1991, In vivo HPLC estimation of extracellular aspartate, glutamate, taurine and GABA in rat striatum: Improved methodology, *Eur. J. Neurosci.* suppl. 4, p.151.
5. Dinarello, C.A., and Bernheim, H.A., 1981, Ability of human leucocyte pyrogen to stimulate brain prostaglandin synthesis in vitro, *J.Neurochem.* 37: 702-708.
6. Frosini, M., Gorelli, B., Matteini, M., Palmi, M., Valoti, M., and Sgaragli, G.P., 1993, HPLC determination of cation levels in cerebrospinal fluid and plasma of conscious rabbits, *J. Pharmacol. Toxicol. Meth.* 29: 99-104.
7. Graziani, L.J., Kaplan, R.K., Escriva, A., and Katzman, R., 1967, Calcium flux into cerebrospinal fluid during ventricular and ventriculocisternal perfusion, *Am. J. Physiol.* 213:629-636.
8. Huxtable,R.J., 1989, Taurine in the central nervous system and the mammalian actions of taurine, *Prog. Neurobiol.* 32:471-533.
9. Huxtable,R.J., 1992, Physiological actions of taurine, *Physiol. Rev.* 72:101-159.
10. Katzman, R., and Pappius, H.M., 1973, "Brain Electrolytes and Fluid Metabolism", Williams & Wilkins, Baltimore.
11. Katzman, R., Graziani, L., and Ginsberg, S., 1968, Cation exchange in blood, brain and cerebrospinal fluid, *in:* "Brain Barrier Systems", Lajhta, A., and Ford, D.H., eds., Elsevier, New York, pp. 283-296.
12. Lombardini,J.B., 1991, Taurine and retinal function, *Brain Res.* 616:471-533.
13. Milton,A.S., 1982, Prostaglandins in fever and the role of action of antipyretic drugs, *in:* "Pyretics and Antipyretics. Handbook of Experimental Pharmacology", Milton, A.S., ed., Springer Verlag, Berlin, pp. 257-303.
14. Myers, R.D., 1980, Hypothalamic control of thermoregulation: Neurochemical mechanism, *in:* "Handbook of the Hypothalamus", Morgane, P., and Panksepp, J., eds., Marcel Dekker, New York, pp. 83-210.
15. Myers, R.D., and Veale, W.L., 1970, Body temperature: Possible ionic mechanism in the hypothalamus controlling the set-point, *Science,* 170: 95-97.
16. Palmi, M., Frosini, M., and Sgaragli, G.P,, 1992, Calcium changes in rabbit cerebrospinal fluid during endotoxin, IL-1 , and prostaglandin fever, *Pharmacol. Biochem. Behav.* 43:1253-1262.
17. Sgaragli,G.P., and Palmi, M., 1985, The role and mechanisms of action of taurine in mammalian thermoregulation, *in:* "Taurine: Biological Actions and Clinical Perspectives", Oja, S.S., Ahtee, L., Kontro, P., and Paasonen, M.K., eds., Alan R. Liss, Inc., New York, pp. 343-357.
18. Watanabe, H.K., Ho,I.K., and Hoskins, B., 1987, Effects of cold stress on brain regional calcium content in rats and mice, *Brain. Res. Bull.* 19:407-409.

REGULATION OF PROTEIN KINASE C ACTIVITY BY TAURINE AND ß-ALANINE DURING EXCITOTOXICITY IN CAT AND MOUSE CEREBELLAR CULTURES

Ekkhart Trenkner, Dongjic Liu, Carolyn Harris and John Sturman

Department of Developmental Biochemistry
Institute for Basic Research in Developmental Disabilities
1050 Forest Hill Road, Staten Island, NY 10314, USA

INTRODUCTION

Taurine can prevent excitotoxic neuronal cell death in early postnatal mouse cerebellar cultures but appears to enhance excitotoxicity in cat cerebellar cultures. This study provides data that could explain these species-specific differences.

Protein kinase C is involved in the regulation of basic neuronal functions through signal transduction pathways (1). It phosphorylates proteins in ion channels (2) and transmitter receptor proteins, and activates enzymes, as well as phosphorylating cytoskeletal proteins (3). As a result, it controls in part neurite outgrowth (4). Depending on the state of development, high levels of protein kinase C are distributed in distinct areas of the brain and the enzyme is believed to play an important role in the regulation of synaptic function (13).

In neurons, protein kinase C is activated in part by the activation of Ca^{2+} channels and increased amounts of intracellular Ca^{2+} that activate the inositol-phospholipase C cascade. Elevation of intracellular Ca^{2+} has been shown to overactivate protein kinase C as well as the signal transduction cascade in rodents. It has been suggested that this leads to abnormal phosphorylation and neuronal cell death. Therefore, inhibition or down-regulation of protein kinase C activity should prevent abnormal phosphorylation and cell death, as has been shown for neuronal cells of rodents (6,10).

NMDA-type glutamate receptors, when stimulated, induce Ca^{2+} influx in a concentration-dependent fashion. It has been suggested that, under neurodegenerative conditions in vivo and in vitro, excitotoxic concentrations of glutamate are released into the extracellular space. The glutamate depolarize glutamate receptor-bearing neuronal cells and possibly overstimulates protein kinase C activity in neurons. This may lead to an increased ion flux, particularly of Ca^{2+}, and neuron cell death (5,11,12). Thus, the control of intracellular Ca^{2+} can determine the activity and overactivity of Ca^{2+}-dependent enzyme

Taurine in Health and Disease, Edited by R. Huxtable
and D.V. Michalk, Plenum Press, New York, 1994

reactions, such as protein kinase C or proteases, that are crucial steps in the development and maintenance of neuronal tissues and their regeneration after injury and degeneration.

Van Gelder (22) as well as we proposed a model that describes the balanced relationship between glutamate and taurine (17). Taurine is essential for the development and survival of neuronal cells in the mammalian central nervous system (7,16). Taurine reaches millimolar concentrations, comparable to those of glutamate. The highest concentrations of taurine occur in newborn and early postnatal stages. Taurine is readily synthesized in mouse whereas cat acquires most of its taurine through the diet. We have shown that taurine prevents excitotoxicity in early postnatal mouse cerebellar cultures (17) and have demonstrated that the taurine synthesizing enzyme, cysteinesulfinic acid decarboxylase, is induced in vitro by injury as well as by excitotoxic concentrations of glutamate or NMDA in both mouse and cat, suggesting not only that taurine is released, but also that new taurine synthesis occurs in case of need (21). The molecular mechanism of taurine's protective role, however, is only poorly understood.

The study presented here is based a) on the suggestion that abnormal phosphorylation through overstimulation of protein kinase C activity causes neuronal cell death in mouse but not in cat (6,12) and b) that taurine inhibits protein kinase C activity in synaptosomal fractions and mitochondrial preparations of rat cortex and retina (8,9), indicating a regulating role of taurine. We found significant differences between cat and mouse cerebellar cells in the down-regulation of protein kinase C activity and will discuss results which suggest that the elevation of protein kinase C activity in cat is not linked to apoptosis of these cells. In addition, we demonstrate in mouse cerebellar cells that glutamate-induced toxicity reduces energy levels in mitochondria that are restored in the presence of 10^{-2} M taurine.

MATERIALS AND METHODS

Materials

Glutamate, taurine, ß-alanine and FCCP (Carbony cyanide p-trifluoro methoxy-phenyl-hydrazone) were obtained from Sigma. AP-5 and MK-801 were purchased from RBI Res. Biochemicals Inc.

Cerebellar Tissue Culture

Early postnatal mouse and cat cerebellar cells (P7) were cultured as described in detail elsewhere (18,19,20). Briefly, 3 X 10^5 cells were plated per well of an eight chamber Lab-tech plate, precoated with 100 μg/ml Poly-D-Lysine in BME (Gibco) supplemented with 0.1% glucose, 0.2 mg glutamine, 10% horse serum and 25 μg/ml penicillin and streptomycin. Cells were maintained for 3 days at 37°C in 5% CO_2 before treatments. The glutamate antagonist AP-5 (NMDA-receptor specific) was added at a final concentration of 10^{-4} M.

Cell Viability

Cell viability was assessed by morphological criteria: healthy neurons exhibited a smooth, even cell membrane around the soma with extended processes. Degenerating granule cells showed a rough cell surface, irregular shape and loss of neurites and lysis of cells leaving only nuclei behind. Trypan blue-exclusion correlated well with the morphological assessment. At least 400 cells in different randomly selected fields of not less than two wells per experimental condition were counted.

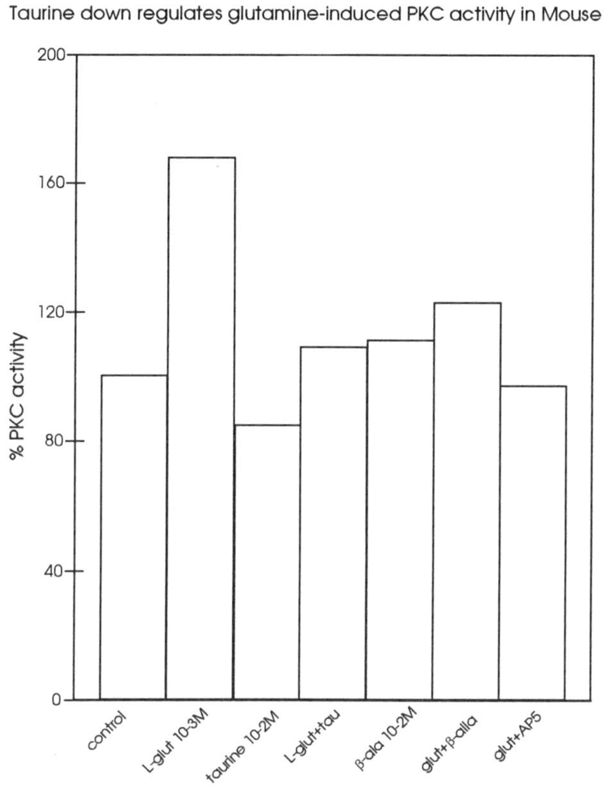

Figure 1. 3 x 10^7 seven-day-old mouse cerebellar cells were cultured for 48 h on poly D-lysine coated dishes in BME + 10% horse serum. At this time, glutamate (10^{-3} M), taurine (10^{-2} M), ß-alanine (10^{-2} M), AP-5 (10^{-4} M) or combinations of the above were added for 1 h. Cells were then scraped off the dish, homogenized in lysis buffer and protein kinase C activity determined using a synthetic substrate (GIBCO, BRL).

Mitochondrial Staining

In order to visualize energy levels of mitochondria cultures were stained with the fluorescent dye JC-1 (Molecular Probes). 10 μg/ml were dissolved in culture medium, cells incubated for 15 min. at 37°C and subsequently washed 3 times in dye free medium (14).

Protein Kinase C Activity

Protein kinase C activity was determined according to the procedure developed by GIBCO BRL, using an artificial substrate.

RESULTS AND DISCUSSION

Further evidence is reported that confirms significant differences of mouse and cat cerebellar cultures in response to ß-alanine and taurine. Protein kinase C activity was determined before and after three day-old early postnatal cerebellar cultures were treated for 1 h with an excitotoxic dose of glutamate (10^{-3} M) in combination with taurine, ß-alanine, and the NMDA antagonist AP-5. In both mouse (Fig. 1) and cat (Fig. 2) cells, glutamate induces protein kinase C activity. This glutamate-induced increase of protein

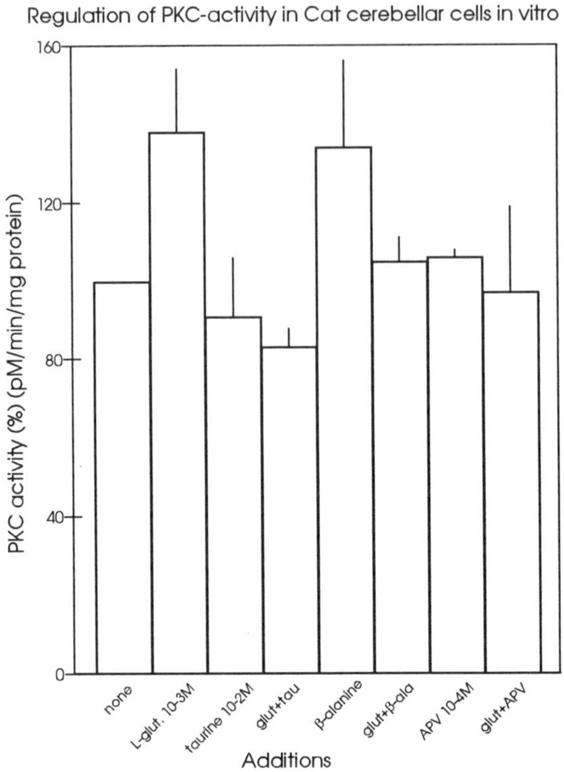

Figure 2. Regulation of protein kinase C activity in 7-day-old cat cerebellar cells. Conditions as for Fig. 1.

kinase C activity was down-regulated to normal levels by taurine in both mouse and cat cerebellar cultures. However, in mouse, excitotoxic neuronal cell death was prevented, whereas in cultures of cat cerebellum it was not. Taurine (10^{-2} M) alone reduced protein kinase C activity by approximately 20%. On the other hand, ß-alanine alone insignificantly induced protein kinase C activity in mouse, but significantly induced protein kinase C activity in cat cerebellar cells. In the presence of excitotoxic concentrations of glutamate, ß-alanine had little or no effect in mouse cultures while in cat cultures glutamate-induced protein kinase C activity was down-regulated by, ß-alanine to control levels. Considering that abnormal phosphorylation through overactivity of protein kinase C has been found to cause cell death in rodents, our results obtained in cat cerebellar cultures suggest a different mechanism. For example, the down-regulation of glutamate-induced protein kinase C activity by taurine accelerated cell death while down-regulation by ß-alanine prevented cell death. Particularly interesting is the fact that ß-alanine alone stimulated protein kinase C activity in mouse and, significantly, in cat cerebellar cells without obvious neuronal cell death, although protein kinase C activity reached levels as high as those observed in the presence of glutamate. These results clearly demonstrate that elevated protein kinase C activity is not necessarily toxic. They also indicate that glutamate might induce protein kinase C isozymes different from those induced by ß-alanine. This is supported by the fact that the glutamate-induced protein kinase C activity was down-regulated by both ß-alanine and taurine.

As expected, the glutamate antagonist AP-5 as well as MK-801 (not shown) blocked the glutamate-dependent induction of protein kinase C activity, as shown in Figures 1 and 2.

Since protein kinase C activity is directly related to the activity of the mitochondrial electrochemical gradient, we analyzed the activity of this in gradient the presence of

Table 1. Taurine restores the mitochondrial electrochemical gradient and prevents apoptosis

Additions			% Cell Survival
FCCP (μM)	Glutamate (μM)	Taurine (mM)	
None			87 ± 2
Medium			86 ± 3
4			78 ± 2
40			36 ± 5
	1000		37 ± 2
	100		72 ± 3
	10		82 ± 2.5
4	1000		18.5 ± 3.5
4	100		39.5 ± 1.5
4	10		66.5 ± 5.5
40	1000		7.5 ± 7.5
40	100		13.5 ± 3.5
40	10		23.5 ± 11.5
4	1000	10	76 ± 5.5
40	1000	10	$68 + 4$

In cerebellar cultures of early postnatal mouse cerebellum reduction of energy levels of mitochondria in neurons through uncoupling of the mitochondrial electrochemical gradient by FCCP, raises neuronal vulnerability towards glutamate. Taurine prevented this process, maintained the integrity of the electrochemical gradient, and thus prevented neuronal cell death.

glutamate by labelling mitochondria with a dye, JC-1 (Molecular Probes), that allows the energy levels of mitochondria to be visualized. Mouse cerebellar cultures were labelled with JC-1 after treatment for 1 h with glutamate (10^{-3} M) alone, or glutamate with 10^{-2} M taurine. As seen in Figure 3, mitochondria of glutamate-treated migrating presumptive granule cells appeared grey and black (original color green) indicating low energy levels. Only a few mitochondria in glial processes appear light, indicating a high energy level. Cell shape and viability of migrating neurons as well as neurons in reaggregates showed signs of excitotoxicity. This reduction of mitochondrial energy levels coincided with an increase of protein kinase C activity. This increase can be seen as a result of the breakdown of the electrochemical gradient that sets Ca^{2+} free, leads to an increase of Ca^{2+}-dependent and protein kinase C activities. However, when taurine was added together with glutamate, conditions that result in the down-regulation of protein kinase C activity and prevention of excitotoxicity (Fig. 1), energy levels of mitochondria in neuronal cells (migrating or assembled in reaggregates) were maintained, expressed here as light colored (original color yellow-orange), and were indistinguishable from control cultures (not shown) (Fig. 3). Thus, taurine might modulate Ca^{2+} influx and transport into mitochondria (9) and thereby not only restore neuronal migration and cell survival in an excitotoxic environment (17) but also maintain the integrity of the mitochondrial electrochemical gradient, and thus the

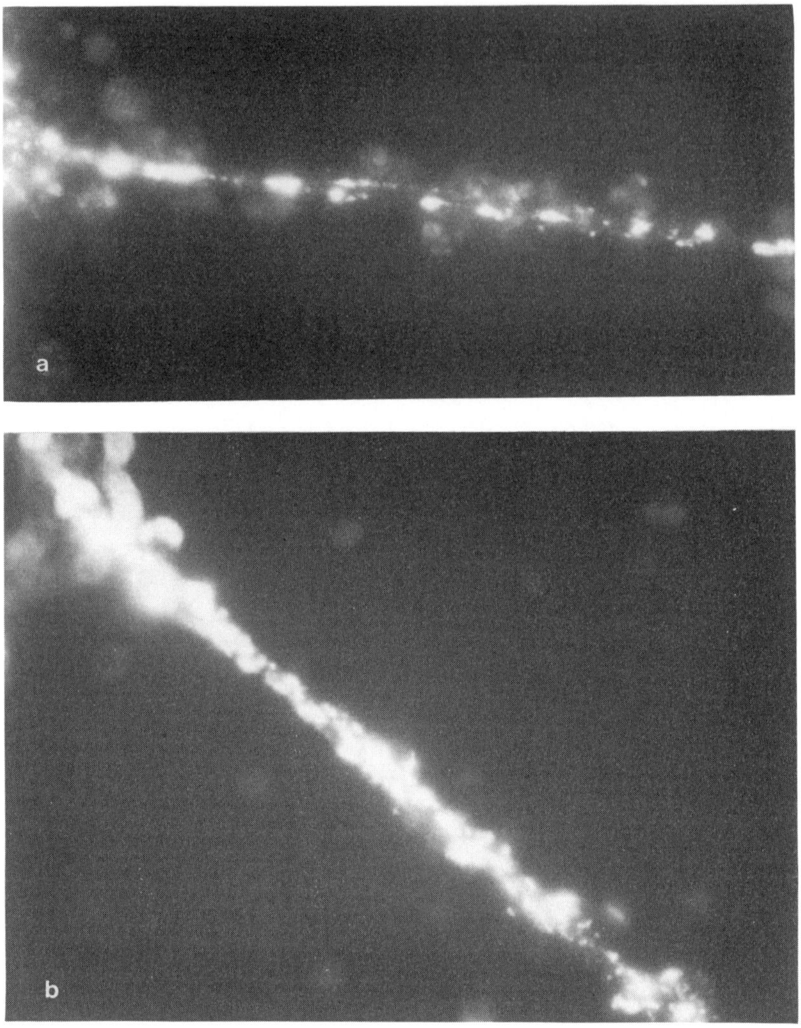

Figure 3. Mouse P7 cerebellar cells (5 x 10^4) were cultured for three days in Labtech dishes without poly D-lysine coating before treatment with **A:** glutamate (10^{-3} M) or **B:** glutamate and taurine (10^{-3} M) for 1 h. Subsequently cultures were stained with the fluorescent dye JC-1 (10 μg/ml) in order to visualize the energy levels of mitochondria (light: high energy, grey-black low energy). The figures show migrating cells (x 1600). Low energy levels are observed in cultures treated with 10^{-3} M glutamate. Mitochondria in presumptive glial processes remained unaffected by glutamate depolarization. However, in the presence of 10^{3} M taurine the energy levels of glutamate-treated cultures are indistinguishable from control cultures.

energy levels of neuronal cells. This might be a plausible mechanism for the protective action of taurine.

In order to investigate the role of mitochondria further, mouse cerebellar cells were cultured in the presence of an uncoupler of the proton pump, FCCP, which reduces the energy levels of cells. As shown in Table 1, when cultured with low concentrations of FCCP, neurons seem more vulnerable to glutamate excitotoxicity; suboptimal concentrations of FCCP together with physiological concentrations of glutamate lead to excitotoxic cell death whereas these compounds alone showed little or no effect on the survival of neuronal cells in vitro. In the presence of taurine, however, cell survival reached control levels, suggesting that taurine can overcome the effects of FCCP and restore energy levels of mitochondria. Experiments are in progress to determine how energy levels are regulated in cerebellar cultures of kitten.

In conclusion, the presented data confirm our previous results and suggest a possible mechanism for the role of taurine during excitotoxic reactions. A detailed analysis of phosphorylated proteins under the conditions described above will enable us to determine whether taurine in mouse and ß-alanine in cat inhibit phosphorylation of subcellular proteins as described by Lombardini (19), whether glutamate-induced increase of protein kinase C activity leads to "over phosphorylation" and cell death, and whether the presence of taurine or ß-alanine and the observed down-regulation of protein kinase C activity will actually produce a "normal" pattern of phosphorylated proteins and normal mitochondrial energy levels in vulnerable neurons. Only then can we call taurine "the rescue molecule".

ACKNOWLEDGEMENTS

This research was supported by NSF grant # BNS-8910218. We are grateful to Ann Parese for secretarial assistance.

REFERENCES

1. Browning, M.D., Huganir, R., and Greengard, P., 1985, Protein phosphorylation and neuronal function, *J. Neurochem.* 45: 11-23.
2. Barban, J.M., Snyder, S.H., and Alger, B.E., 1985, Protein kinase C regulates ionic conductance in hippocampal pyramidal neurons: Electrophysiological effects of phorbol esters, *Proc. Natl. Acad. Sci.*. USA 82: 2538-2542.
3. Baudier, J., and Cole, R.D., 1987, Phosphorylation of tau proteins to a state like that in Alzheimer's brain is catalyzed by a calcium/calmodulin-dependent kinase and modulated by phospholipids, *J. Biol. Chem.*. 36: 17,577-17,583.
4. Cambray-Deakin, M.A., Adu, J., and Burgoyne, R.D., 1990, Neuritogenesis in cerebellar granule cells *in vitro:* a role of protein kinase C, *Dev. Brain Res.*. 53: 40-46.
5. Choi, D.W., 1988, Glutamate neurotoxicity and disease of the nervous system, *Neuron.* 1: 623-634.
6. Felipo, V., Minana, M.D., and Grisolia, S., 1993, Inhibitors of protein kinase C prevent the toxicity of glutamate in primary neuronal cultures, *Brain Res.* 604: 192-196.
7. Huxtable, R.J., 1992, The physiological action of taurine, *Physiol. Rev.*. 72: 101163.
8. Li, Y.P., and Lombardini, J.B., 1991, Inhibition by taurine of phosphorylation of specific synaptosomal proteins in the rat cortex: effects of taurine on the stimulation of calcium uptake in mitochondria and inhibition of phosphorinositide turnover, *Brain Res.* 553: 89-96.
9. Lombardini, J.B., 1988, Effects of taurine and mitochondria metabolic inhibitors on ATP-dependent Ca^{2+} uptake in synaptosomal and mitochondrial subcellular fractions of rat retina, *J. Neurochem.*. 51: 200-205.
10. Mattson, M.P., 1991, Evidence for the involvement of protein kinase C in neurodegenerative changes in cultured human cortlcal neurons, *Exp. Neurologv.* 112: 93-103.
11. Mattson, M.P., 1990, Second messengers in neuronal growth and degeneration, in: "Current Aspects of the Neurosciences", N. Osborne, ed., MacMillan Co., New York, Vol. 2, pp. 1-48.
12. Mattson, M.P., Rychlik, B., You, J.S., and Sisken, J.E., 1991, Sensitivity of cultured human embryonic cerebral cortical neurons to excitatory amino acid induced calcium influx and neurotoxicity, *Brain Res.*. 542: 97-106.
13. Schwartz, K.J., Merritt, A., Beau, B.P., and Lovinger, D.M., 1993, Protein kinase C modulates glutamate receptor inhibition of Ca^{2+} channels and synaptic transmission, *Nature* 361: 165-168.
14. Smiley, S.T., Reers, M., Mottola-Hartshorn, C., Liu, M., Chen, A., Smith, T.W., Steele, G.D., and Chen, L.B., 1991, Intracellular heterogeneity in mitochondrial membrane potentials revealed by J-aggregate-forming lipophilic cation JC-1, *PIVAS*, USA 88: 3671-3675.
16. Sturman, J.A., 1993, Taurine in development, *Physiol. Rev.,* 73: 119-147.
17. Trenkner, E., 1990, The role of taurine and glutamate during early postnatal cerebellar development of normal and weaver mutant mice, in: "Excitatory Amino Acids and Neuronal Plasticity", Y. Ben-Ari, ed., Plenum Press, Adv. Exp. Med. and Biol. 268: pp. 239-244.

18. Trenkner, E., 1991, Cerebellar cells in culture, in: Culturing Nerve Cells, G. Banker and K. Goslin, eds., pp. 283-307.
19. Trenkner, E., and Sidman, R.L., 1977, Histogenesis of mouse cerebellum in microwell cultures: cell reaggregation and migration, fiber- and synapse formation, *J. Cell Biol.*. 75: 915-940.
20. Trenkner, E., and Sturman, J.A., 1991, The role of taurine in the survival and function of cerebellar cells in cultures of early postnatal cat, *Int. J. Devl. Neurosci.*. 9: 77-88.
21. Trenkner, E., Gargano, A., Scala, P., and Sturman, J.A., 1992, Taurine synthesis in cat and mouse *in vivo* and *in vitro,* in: "Adv. Exp. Med. and Biol.", Plenum Press, Vol. 315, pp. 7-14.
22. Van Gelder, N.M., 1990, Neuronal discharge hypersynchrony and the intercranial water balance in relation to glutamic acid and taurine redistribution: migraine and epilepsy, *Progr. Clin. Biol. Res.*, 351:1-20.

VOLUME REGULATION IN CULTURED NEURONS: PIVOTAL ROLE OF TAURINE

Herminia Pasantes-Morales, Elizabeth Chacón,
Roberto Sánchez-Olea and Julio Morán

Institute of Cell Physiology
National University of Mexico
Mexico City, Mexico

INTRODUCTION

The ability of cells to adapt by active processes to changes in external osmolarity has been known for a long time (3). Two new concepts have emerged in recent years: first, that the adaptive response is present in cells which grow and live in media of strictly controlled osmolarity, and secondly, that a number of organic molecules are importantly involved in volume regulation in vertebrates (5,8).

The widespread occurrence of mechanisms of cell volume control suggests that volume regulation is a fundamental property of animal cells, from which nervous tissue cells are not excluded. The ability of brain to adjust to changes in osmolarity has been known since the work of Cserr et al. (1) who showed that brain behaves as an imperfect osmometer in physiological or pathological conditions leading to changes in plasma osmolarity. These studies also showed that although marked changes in the concentration of inorganic ions (Na^+, K^+ and Cl^-) occur during osmotic adaption, they are not sufficient to account for the extent of the regulatory response and the difference was attributed to the involvement of "idiogenic" osmolytes. Subsequent work has suggested that amino acids, and particularly taurine, represent an important part of these idiogenic osmolytes together with other organic molecules such as polyols and amines.

The involvement of taurine in volume regulation in nervous tissue has been demonstrated in various preparations ranging from the whole brain in vivo (17) or in situ experiments (6,15,18) to in vitro systems including brain slices (7), cells in culture (10,12) and synaptosomes (11). In all these studies it was clear that a decrease in osmolarity in the extracellular milieu caused by experimental hyponatremia or directly by reducing the osmolarity in the medium resulted in an increased efflux of taurine and consequently in a decrease of tissue taurine concentration.

Examination of morphological changes occurring in brain during hyponatremia as well as in other pathological situations causing brain edema suggested that astrocytes may be the

Taurine in Health and Disease, Edited by R. Huxtable
and D.V. Michalk, Plenum Press, New York, 1994

most affected cells, as they show marked swelling in these conditions whereas neurons appear less altered (4). These observations point to a different response of neurons and astrocytes, which may have various interpretations. Either neurons are somehow protected from swelling or they may have a more effective regulatory response. An interesting observation in this respect is reported by Lehmann et al. in this volume, concerning changes in the taurine content of cerebellar cells during hyponatremia. These workers show that neurons, but not glial cells, swell in hypernatremia while concomitantly taurine is mobilized from neurons to glia. These observations appear to support the interpretation that neurons have an effective mechanism for volume regulation which involves the efflux of taurine, and that astrocytes, by accumulating the released taurine and perhaps also other osmolytes, have a key role in the homeostasis of nervous tissue during brain edema.

To clarify these points further, it is important to characterize the mechanisms of volume regulation in neurons. Cell cultures represent a convenient model for this purpose as these preparations contain a high proportion (95%) of granule neurons.

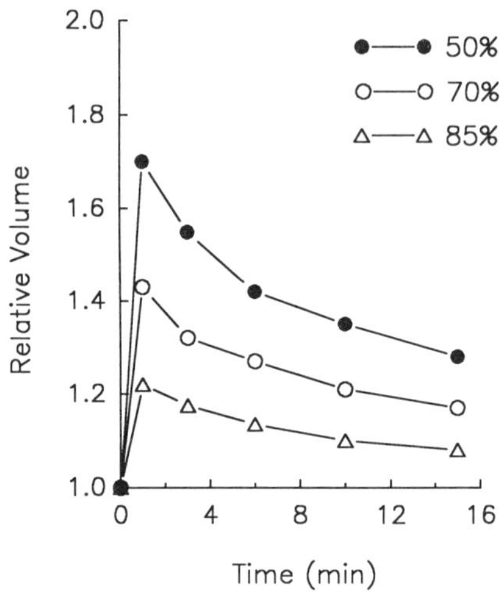

Figure 1. Changes in cell volume of cerebellar granule neurons in hyposmotic conditions. Cultured cerebellar granule neurons were detached by treatment with Ca^{2+}-free saline solution containing 0.01% trypsin. Cells were exposed to media made hyposmotic (50%, 70% or 85% of the isosmotic solutions) by correspondingly decreasing NaCl. Cell volume was measured by electronic sizing as described by Pasantes-Morales et al. (9). Results are expressed as relative volume considering cell volume in isosmotic conditions as 1. Results are means of 4-14 experiments. SEM values do not extend beyond the symbols.

VOLUME REGULATION IN CEREBELLAR GRANULE NEURONS IN CULTURE

Cerebellar granule neurons were obtained from 8-day old rats and cultured as previously described (9). Cell volume changes were measured by electric sizing in neurons detached in calcium-free media by mild trypsination (9).

Upon exposure to hyposmotic solutions (50%, 70%, 85%), neurons first swell, with increases in cell volume from 1.0 (taken as relative volume in isosmotic conditions) to 1.68,

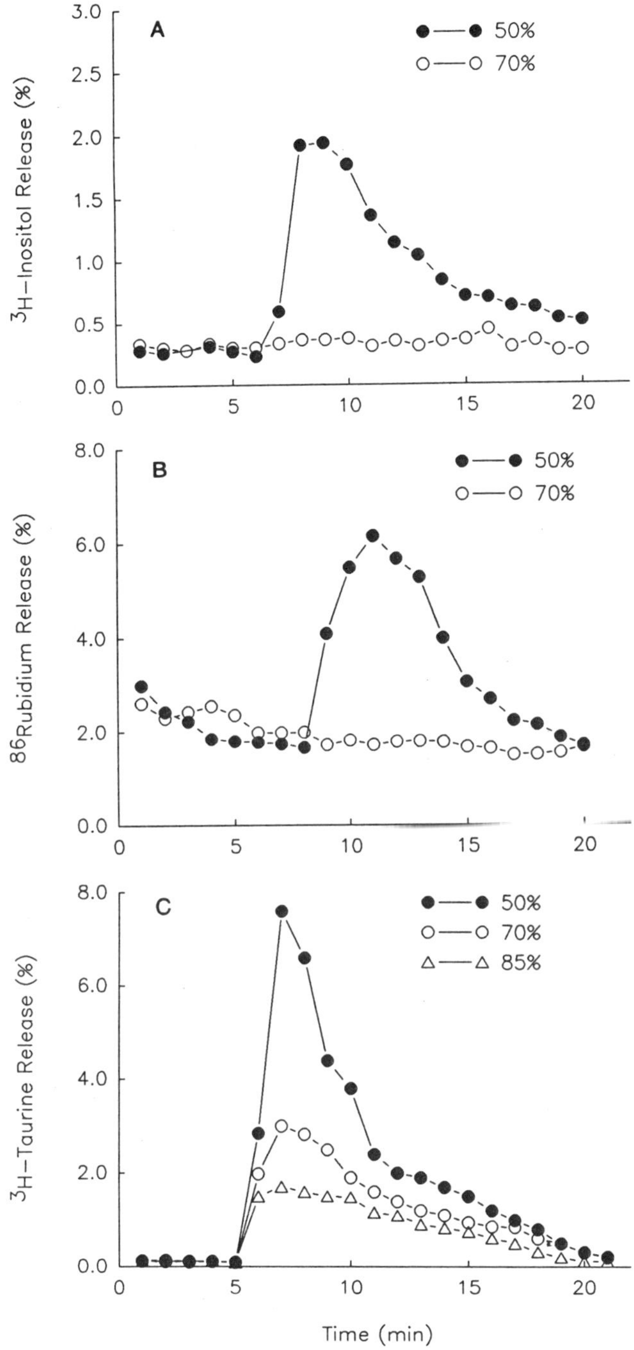

Figure 2. Hyposmolarity-induced release of taurine, inositol and Rb^+ from cerebellar granule cells in hyposmotic conditions. Cells were loaded with (A) [^{3}H]inositol, (B) $^{86}Rb^+$ or (C) [^{3}H]taurine during 30-60 min, then washed and superfused with isosmotic medium for 8 min. After attaining a baseline efflux, cells were superfused with media of decreased osmolarity as indicated. Results are expressed as fractional release; i.e. the radioactivity released in each fraction as the percentage of the total radioactivity accumulated by cells during loading. Results are means of 4-8 experiments.

1.41 and 1.20, according to the decrease in external osmolarity (Fig. 1). The peak of swelling is attained within 1 min after the hyposmotic stimulus. At this time an active regulatory phase starts which occurs despite the persistence of the hyposmotic conditions, and cell volume returns towards normal. Volume regulation at the longest time registered in this work (15 min) is only partial, however, as the volume remains higher than that occurring under isosmotic conditions: 1.32, 1.18 and 1.06 in solutions of 50%, 70% and 85%, respectively. The efficiency of volume regulation, defined as the percentage of recovery of cell volume between min 1 (maximal volume) and min 15, is also found to be dependent on the decrease of osmolarity, being 53%, 59% and 70%, respectively, for hyposmotic solutions of 50%, 70% and 85% (Fig. 1).

TAURINE AND VOLUME REGULATION IN CEREBELLAR GRANULE NEURONS

The general mechanism underlying volume regulatory processes in animal cells involves the activation of transmembrane fluxes of osmotically active solutes in the necessary direction to correct the change in cell water. Consequently, during a regulatory decrease in cell volume, intracellular osmolytes leave the cell.

In previous reports we described an efflux of taurine and other amino acids from cerebellar granule neurons in response to hyposmotic conditions (12). Taurine is present in high concentrations in granule neurons (92 nmol/mg protein), being exceeded only by glutamate and glutamine (13). The time course of this taurine efflux was found to parallel the time course of cell volume decrease, as observed comparing Fig. 1 and 2.

In the present study, the volume-associated efflux of taurine from cerebellar granule neurons was compared to that of other osmolytes presumably involved in the volume regulatory process; namely K^+ (using $^{86}Rb^+$ as a tracer) and inositol. A marked difference was observed in the sensitivity of the efflux among these three osmolytes. As shown in Fig. 2, the three osmolytes are released upon reductions in osmolarity of 50% or more. However, only taurine efflux is activated when osmolarity is reduced by less than 50%. Furthermore, the volume-associated taurine release was more sensitive than the efflux of other amino acids as more than 80% of the taurine pool is released upon hyposmotic (50%) conditions whereas for glycine, aspartate and glutamate this proportion is 40-60% and it is much lower for alanine, serine, histidine, arginine and lysine (7).

MECHANISM OF TAURINE EFFLUX ASSOCIATED TO CELL SWELLING

Previous studies from our laboratory have provided evidence supporting the notion that the diffusional component of taurine transport which is consistently observed in most cells, including granule neurons, could be the mechanism activated during the efflux (14). This diffusional component increases markedly in hyposmotic conditions, in clear contrast with the saturable, energy-dependent component, which is unaffected. A more direct demonstration of the diffusional nature of taurine fluxes activated by hyposmolarity comes from the observation that increasing extracellular taurine concentration to an extent that exceeds the intracellular level leads to influx rather than to efflux of the amino acid during cell swelling until the intracellular and the extracellular concentration are in equilibrium. Obviously, under these conditions the influx of taurine counteracts the efflux of water carried by the fluxes of other osmolytes and cell volume regulation does not occur (Fig.3).

This observation was used as an experimental tool to characterize the permeability properties of the taurine pathway to other amino acids. The rationale for these experiments is that any compound able to permeate through the pathway opened by hyposmolarity will lead to an impairment of cell volume regulation. A variety of amino acids were examined

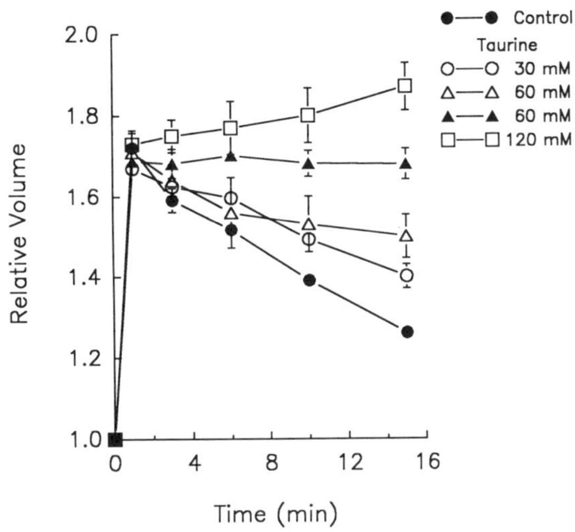

Figure 3. Effect of extracellular taurine on volume regulation of cerebellar granule neurons in hyposmotic conditions. Cell volume was measured as described in Fig. 1. Cells were resuspended in 50% hyposmotic medium containing the indicated concentrations of taurine replacing equiosmolar concentrations of NaCl. Results are expressed as relative volume considering cell volume in isosmotic conditions as 1. Results are means ± SEM of 4-8 experiments.

including neutral amino acids (GABA, alanine, glycine, glutamine,) basic amino acids (lysine and arginine) and the acidic amino acids, glutamate and aspartate. In these experiments the concentration of amino acids in the medium was 90 mM (90 mOsmol) which replaced equiosmolar amounts of NaCl.

Fig. 4 shows that glycine and taurine were the most permeable of the amino acids tested, as they completely abolished volume regulation and even caused cell swelling (expressed as negative change). Alanine was less effective. In the presence of GABA the first phase of the volume regulatory process was unaffected whereas a blockade was observed at the last phase. Glutamine did not affect cell volume. Glutamate and aspartate were permeable through the pathway as K^+ salts, but not as Na^+ salts, as predicted by previous results showing that K^+ (and Rb^+) but not Na^+ can permeate through this pathway. Aspartate was somewhat more permeable than glutamate (Fig. 4). The basic amino acids, arginine and lysine, showed a completely different effect. In the presence of these amino acids cells did not swell but rather a decrease in cell volume was observed. These observations indicate a permeability order for amino acids of glycine > taurine > aspartate > glutamate > alanine >> glutamine. The basic amino acids, as mentioned, have opposite effects, accelerating the regulatory process or/and reducing the swelling phase. There is no obvious explanation at present for the effects of the basic amino acids.

It is interesting to note that the ability of amino acids to permeate through the pathway opened by hyposmolar conditions in cerebellar granule neurons closely parallels the sensitivity of the efflux process in these cells. Upon reduction in osmolarity, taurine and glycine efflux was larger (80% and 60%, respectively of the endogenous pool) than for glutamate and aspartate (30%-40%) and notably less for other amino acids (10%-15%) (13). Also, in good agreement with the present results, essentially no efflux of glutamine is observed in response to cell swelling (7).

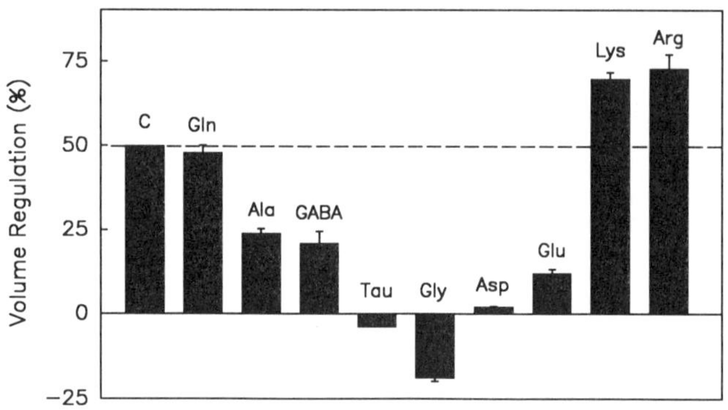

Figure 4. Effect of extracellular amino acids on volume regulation in cerebellar granule neurons. Cell volume was measured in 50% hyposmotic medium containing 90 mM of the indicated amino acids replacing equiosmolar concentrations of NaCl. Cell volume was measured as in Figs. 1 and 3 and volume regulation in the presence of the amino acids is expressed as the percentage of recovery of cell volume between min 1 (maximal volume) and min 15. Results are means of 6-8 experiments with SEM indicated by vertical bars.

It is of interest that some correlation is observed between the amino acid release and the permeability properties of the hyposmolarity-activated pathway and the decrease in tissue levels of amino acids in brain in vivo after acute hyponatremia. Taurine and glutamate levels markedly decreased under these conditions whereas glutamine levels were unchanged (2). After long term hyponatremia essentially all the taurine pool is exhausted as well as the glutamate pool. Glutamine concentration is somewhat decreased (16).

All these results emphasize the importance of organic osmolytes, and notably taurine, in the maintenance of osmotic equilibrium of neurons during hyponatremia. It seems clear that this adaptive response represents a mechanism of survival in face of pathological conditions. The question arises of whether this is a mechanism operating only in conditions of stress or if it has some physiological significance. During the normal cell function numerous situations can be envisaged which entail changes in cell volume requiring adjustment. Gross changes in cell volume occur during growth and proliferation of the cell. Also microosmotic gradients originate associated with secretory processes, intracellular movement, synthesis and degradation of macromolecules and changes in membrane potential. In all these conditions an inert molecule, such as taurine, having osmoprotectant properties may be the preferred osmolyte.

ACKNOWLEDGEMENTS

The excellent technical assistance of Ms. Claudia Peña and Claudia Rodriguez is gratefully acknowledged. This work was supported in part by DGAPA-UNAM (grant No.IN-202093).

REFERENCES

1. Cserr, H.F., DePasquale, M., and Patlak, C.S. 1987, Regulation of brain water and electrolytes during acute hyperosmolality in rats, *Am.J.Physiol.*, 253:F522-F529.
2. Gullans, S.R. and Verbalis, J.G. 1993, Control of brain volume during hyperosmolar and hypoosmolar conditions, *Annu.Rev.Med.*, 44:289-301.
3. Hoffmann, E.K. and Simonsen, L.O. 1989, Membrane mechanisms in volume and pH regulation in vertebrate cells, *Physiol.Rev.*, 69:315-382.
4. Kimelberg, H.K. and Ransom, B.R. 1986, Physiological and pathological aspects of astrocytic swelling, *in:* "Astrocytes, Vol. II", Fedoroff, F. and Vernadakis, A., eds., Academic Press, New York, pp. 129-136.
5. Law, R.O. and Burg, M.B. 1991, The role of organic osmolytes in the regulation of mammalian cell volume, *in:* "Advances in Comparative and Environmental Physiology", Gilles, R., ed., Springer--Verlag, Berlin, pp. 189-225.
6. Lehmann, A. 1989, Effects of microdialysis-perfusion with anisoomotic media on extracellular amino acids in the rat hippocampus and skeletal muscle, *J.Neurochem.*, 53:525-535.
7. Oja, S.S. and Saransaari, P. 1992, Taurine release and swelling of cerebral cortex slices from adult and developing mice in media of different ionic compositions, *J.Neurosci.Res.*, 32:551-561.
8. Pasantes-Morales, H. and Del Río, R.M. 1990, Taurine and mechanisms of cell volume regulation, *in:* "Taurine: Functional Neurochemistry, Physiology, and Cardiology", Pasantes-Morales, H., Martin, D.L., Shain, W. and del Río, R.M., eds., Wiley-Liss, New York, pp. 317-328.
9. Pasantes-Morales, H., Maar, T.E., and Morán, J. 1993, Cell volume regulation in cultured cerebellar granule neurons, *J.Neurosci.Res.*, 34:219-224.
10. Pasantes-Morales, H. and Schousboe, A. 1988, Volume regulation in astrocytes: A role for taurine as an osmoeffector, *J.Neurosci.Res.*, 20:505-509.
11. Sánchez Olea, R. and Pasantes-Morales, H. 1990, Chloride dependence of the K^+-stimulated release of taurine from synaptosomes, *Neurochem.Res.*, 15:535-540.
12. Schousboe, A., Moran, J., and Pasantes-Morales, H. 1990, Potassium-stimulated release of taurine from cultured cerebellar granule neurons is associated with cell swelling, *J.Neurosci.Res.*, 27:71-77.
13. Schousboe, A. and Pasantes-Morales, H. 1992, Role of taurine in neural cell volume regulation, *Can.J.Physiol.Pharmacol.*, 70 Suppl.:S356-S361.
14. Schousboe, A., Sánchez Olea, R., Morán, J., and Pasantes-Morales, H. 1991, Hyposmolarity-induced taurine release in cerebellar granule cells is associated with diffusion and not with high-affinity transport, *J.Neurosci.Res.*, 30:661-665.
15. Solís, J.M., Herranz, A.S., Herreras, O., Lerma, J., and Del Río, R.M. 1988, Does taurine act as an osmoregulatory substance in the rat brain, *Neuroscience Letters*, 91:53-58.
16. Sterns, R.S., Baer, J., Ebersol, S., Thomas, D., Lohr, J.W., and Kamm, D.E. 1993, Organic osmolytes in acute hypernatremia, *Am.J.Physiol.*, 264:F833-F836.
17. Thurston, J.H., Hauhart, R.E., and Dirgo, J.A. 1980, Taurine: a role in osmotic regulation of mammalian brain and possible clinical significance, *Life Sci.*, 26:1561-1568.
18. Wade, J.V., Olson, J.P., Samson, F.E., Nelson, S.R., and Pazdernik, T.L. 1988, A possible role for taurine in osmoregulation within the brain, *J.Neurochem.*, 51:740-745.

TAURINE AS AN ORGANIC OSMOLYTE IN THE INTACT BRAIN:

IMMUNOCYTOCHEMICAL AND BIOCHEMICAL STUDIES

Erlend A. Nagelhus[1], Mahmood Amiry-Moghaddam[1],
Anders Lehmann[2], and Ole P. Ottersen[1]

[1]Department of Anatomy
Institute of Basic Medical Sciences
University of Oslo
P.O. Box 1105
Blindern, N-0317 Oslo, Norway
[2]Department of Anatomy and Cell Biology
University of Göteborg
Sweden (Present address: Astra Hässle AB
S-431 83 Mölndal, Sweden)

INTRODUCTION

The brain must possess mechanisms that enables it to adjust to changes in plasma osmolality. Such changes can be quite pronounced in a number of conditions, including diabetes and certain renal and gastrointestinal disorders (31), and would have had deleterious consequences more often were it not for the brain's capacity to counteract the accompanying volume changes. It is clear that the brain responds to osmotic stress by regulating its electrolyte contents (1). However, as large fluctuations in intracellular electrolyte concentrations may interfere with cell function in various ways, the ability to use nonperturbing, organic osmolytes would represent a definite advantage. Several organic compounds show a decreased concentration in brain following experimentally induced hypo-osmolality, suggesting that they are involved in osmoregulation (13). The largest concentration changes are exhibited by amino acids such as taurine, glutamate, and glutamine. But other compounds including creatine also show a conspicuous decrease. There is also evidence that the major organic osmolytes of the kidney are present in the brain and act in a similar capacity there (35).

Of the above-mentioned compounds, taurine has received particular attention as a possible regulator of cell volume in the central nervous system (7). Studies of cultured neurons or glial cells have shown that this amino acid is released subsequent to manipulations that tend to increase cell volume (10,20,23,24,25,27,28,32). Taurine has properties that make it well suited as an osmoregulator: it occurs in relatively high

Taurine in Health and Disease, Edited by R. Huxtable
and D.V. Michalk, Plenum Press, New York, 1994

concentrations and does not participate in essential metabolic pathways. Possible drawbacks are the low ability of a number of species to biosynthesize taurine, and the reported ionotropic effects of taurine which could come into play upon an efflux to the brain's extracellular space.

Assumptions concerning the importance of taurine in brain volume regulation have mainly relied on data from cell cultures. Comparatively few investigations have dealt with the intact brain. Studies employing in vivo microdialysis (14,30,36) have demonstrated an increased concentration of taurine in the extracellular space following an hypo-osmotic challenge, and the work of Trachtman et al. (33) suggests that depletion of taurine directly affects the animal's resistance to osmotic stress.

Many questions remain unresolved, however. Most notably, there has so far been no direct demonstration of an acute, osmotically dependent loss of taurine from mammalian brain cells in situ. Such data are required, as the microdialysis studies referred to above made no attempts to identify the sources of the taurine overflow. Hence an influx from blood could not be ruled out. Furthermore, if hypo-osmotic stress does induce an acute cellular efflux of taurine, the question arises whether it occurs indiscriminately from all cell types or whether the use of taurine as an osmoregulator is restricted to select populations of cells. It must be emphasized here that immunocytochemical analyses of the distribution of taurine (3,8,17,18,19,22,37) have revealed a substantial cellular and regional heterogeneity that has not been duly considered in the interpretation of the data obtained from cell cultures.

In a recent study of the toad brain (2), light microscopic immunocytochemistry displayed a redistribution of taurine between intra- and extracellular compartments following experimentally induced hypo-osmolarity.

The present paper reviews our attempts to shed light on the cellular handling of taurine in the intact rat brain during acute hypo-osmotic stress. To this end we have used in vivo microdialysis and semiquantitative immunocytochemical procedures at the light and electron microscopic levels.

IS ACUTE HYPO-OSMOTIC STRESS ASSOCIATED WITH A FLUX OF TAURINE FROM BLOOD TO BRAIN?

Many cell types have been shown to release taurine in response to hypo-osmotic swelling (7), and it is possible that the increase in extracellular taurine in the brain occurring after systemic water injection is derived not only from neural cells but also from blood. To address this question a 15% decrease in plasma osmolality was induced in urethane-anaesthetized rats by intraperitoneal injections of distilled water (150 ml/kg). This led to a 25-fold elevation of taurine in plasma and a 5-fold increase in cerebrospinal fluid (sampled from the cisterna magna). The changes had a rapid onset and the maximum cerebrospinal fluid concentration was reached about 2 h after water injection (Fig. 1). The changes in taurine concentration in dialysate, sampled from a microdialysis probe in the hippocampus, paralleled the changes in the cerebrospinal fluid but the absolute values were smaller. To examine the patency of the blood-cerebrospinal fluid-barrier we measured the albumin concentration in the cerebrospinal fluid. This was transiently increased and reached a peak level of about 70% above basal at 90-120 min after the water injection (Fig. 1), indicating a temporary leakiness of the barrier (15).

These data raise the possibility that an influx of taurine from blood contributes to the extracellular accumulation of taurine during an acute hypo-osmotic challenge. This influx may be driven by the pronounced increase in the plasma level of taurine and facilitated by the transient opening of the blood-cerebrospinal fluid barrier.

Since leukocytes and erythrocytes are known to release taurine during hypo-osmotic swelling (5,6), and since hemolysis was observed in the rats administered water, experiments were carried out to assess the contribution of blood cells to the elevation of taurine in plasma. Blood was diluted with distilled water in vitro to simulate the in vivo situation, and plasma amino acids were determined. There was a 2-fold increase in taurine concentration which indicates that blood cells contribute to the elevation in circulating taurine in vivo (15). However, the effect observed in vitro accounted for less than 10% of the in vivo response, and consequently, peripheral cells other than blood cells must be the main contributors of taurine during generalized hypo-osmolality.

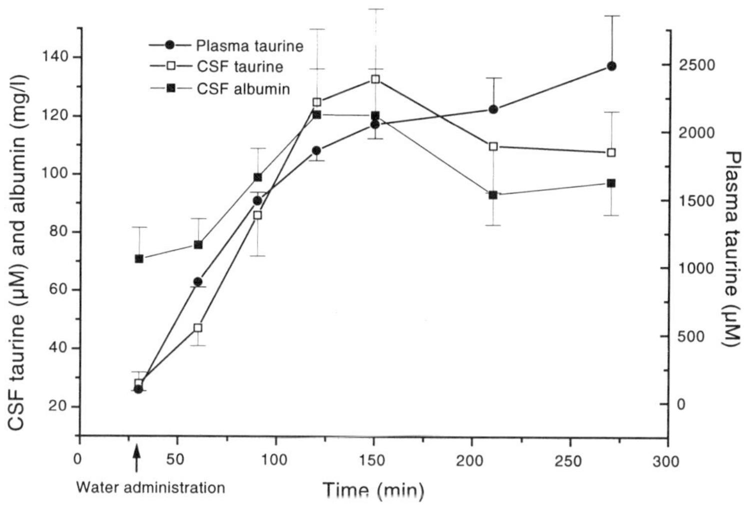

Figure 1. Changes in taurine and albumin in plasma and cerebrospinal fluid after intraperitoneal injection of distilled water (150 ml/kg) in rats. Cerebrospinal fluid was sampled from a catheter inserted into the cisterna magna, and blood was collected from the femoral artery. Standard error of the mean is indicated; n = 6. CSF: Cerebrospinal fluid. Adapted from Lehmann et al., 1991.

CELLULAR REDISTRIBUTION OF TAURINE DURING ACUTE HYPO-OSMOTIC STRESS

Although an influx from the blood may be responsible for part of the increase in extracellular taurine concentration, ample evidence from culture studies predicts that acute hypo-osmotic stress also induces an efflux of taurine from brain cells. In addition, intracerebral microdialysis-perfusion with hypo-osmotic medium greatly stimulates taurine release (14,30,36). The cellular handling of taurine was investigated by use of antibodies raised against taurine-glutaraldehyde-carrier protein. These antibodies (no. 20; 18; no. 286; 38) selectively recognize the fixation products of taurine that are formed in tissue exposed to glutaraldehyde and were employed for postembedding immunocytochemistry at the light and electron microscopic levels (21,22). Semiquantitative data were obtained from both types of preparation, by optical densitometry and computer-assisted calculation of gold particle densities, respectively. Plasma hypo-osmolality was induced by water loading (Table 1), as outlined in the preceding paragraph.

The immunocytochemical analyses were focused on the cerebellum. This part of the central nervous system exhibits a differentiated pattern of taurine distribution with

Table 1. Effects of water-loading on cerebellar taurine concentration, cerebellar specific gravity and plasma osmolality

Groups of animals	Taurine concentration (μmol/g wet weight)	Specific gravity	Plasma osmolality (mmol/kg)
C_4	5.65 ± 0.12 (6)	1.0482 ± 0.0002 (6)	294 ± 1 (6)
W_4	4.45 ± 0.10 (6) *	1.0441 ± 0.0003 (6) *	250 ± 2 (6) *
C_8	8.66 ± 0.25 (5)	1.0488 ± 0.0003 (5)	288 ± 2 (5)
W_8	5.37 ± 0.21 (5) ††	1.0460 ± 0.0003 (5) ††	247 ± 3 (5) ††
R_8	6.31 ± 0.17 (5) †	1.0492 ± 0.003 (5) ††	292 ± 3 (5) ††

Animals were treated with 150 ml/kg isotonic saline (controls, C; killed 4 or 8 h after injection), distilled water (W_4; 4 h survival, and W_8, 8 h survival), or by water followed 4 h later by hypertonic saline ("reversed" animals; R_8). Values are given as mean ± standard error of the mean (number of animals in parenthesis). Asterisks indicate values significantly different from corresponding values in group C_4 (*, $P < 0.001$; Student's t-test, two tails). Daggers indicate values significantly different from corresponding value in preceding group (†, $P < 0.05$; ††, $P < 0.01$ or better; ANOVA, Dunnett's t-test). The data were from two series of animals, represented by the first two groups and the last three groups, respectively. The data for C_4 and C_8 are not directly comparable since they were obtained in different experiments. The amino acid concentrations (determined by HPLC) were not corrected for swelling. Modified from Nagelhus et al., 1993.

particularly high concentrations in the Purkinje cells and substantially lower concentrations in glial cells (22).

The light- and electron microscopic data revealed a pronounced loss of taurine from Purkinje cells in animals that had been killed 4 h after the water injection (Fig. 2). The loss amounted to about 50%, as judged from the calibrated gold particle counts (Table 2). The cross section area of the Purkinje cell somata was no different from that of controls (Table 2), suggesting that these cells had maintained their original volume in spite of the osmotic perturbation (21).

The loss of immunoreactivity from the Purkinje cells was associated with an *increased* immunoperoxidase labelling of the surrounding glial cells (Fig. 2). This enhanced immunolabelling reflected an increase in the total contents of taurine in these cells (see legend to Table 2). However, due to their marked swelling the intracellular *concentration* of taurine in glial cells was actually lower than in controls, as assessed from the quantitative immunogold data (Table 2).

Some of the water-loaded animals received an intraperitoneal injection of hypertonic saline after 4 h which re-established iso-osmolality. These animals were allowed to survive for additional 4 h and then subjected to immunocytochemical analysis. The changes in taurine distribution were now reversed: Purkinje cells appeared strongly labelled and glial cells much less so, as in controls (Fig. 2).

Adjacent sections to those analyzed above were immunolabelled for GABA. The cellular distribution of the latter amino acid was not visibly affected by the water or saline injections.

These results (21) suggest that acute hypo-osmolality causes a cellular *redistribution* of taurine, and not only a diffuse loss of this amino acid from all cell compartments. Specifically in the cerebellar cortex there seems to be a reversible exchange of taurine between Purkinje cells - a neuronal compartment - and adjacent glial cells. The mechanism for these changes remains to be established, but extrapolating from cell culture studies (10,20,23,24,25,27,28,32) it appears likely that diffusion through volume-sensitive channels could be involved. It should be recalled that even if the glial cells display an increased total contents of taurine following water loading their intracellular taurine concentration remains lower than that in the Purkinje cells (as judged by the quantitative electron microscopic data). The changes observed here in the water-loaded animals could thus be explained simply by assuming that taurine channels open in Purkinje cells as well as in glia, and that taurine passively diffuses through these channels along its concentration gradient. It is implicit in this explanation that volume regulation of the Purkinje cells will be achieved at the expense of a swelling in the glial cells (provided that compensatory mechanisms do not

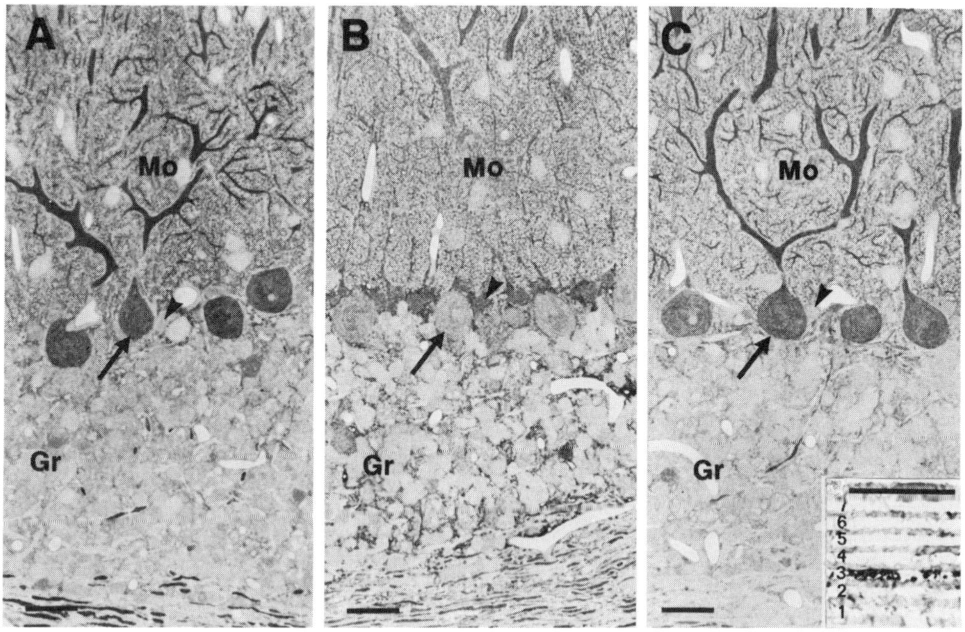

Figure 2. Redistribution of taurine-like immunoreactivity following water loading. Photomicrographs are of semi-thin sections incubated with taurine antiserum 286 diluted 1:3000. The animals were treated either with **(A)** 150 ml/kg isotonic saline or **(B)** distilled water 4 h prior to sacrifice, or with **(C)** water followed by hypertonic saline, 8h and 4 h prior to sacrifice, respectively. Note severe loss of taurine immunoreactivity from Purkinje cells (arrows) and increased glial labelling (arrowheads) after water loading (B), and re-establishment of the normal pattern of labelling after reversal with hypertonic saline (C). A test sandwich incubated together with the tissue sections (inset C) displays highly selective labelling of taurine. The conjugates were prepared from GABA (1), glutamate (2), taurine (3), glycine (4), none (5), aspartate (6) and glutamine (7). Abbreviations: Gr, granule cell layer; Mo, molecular layer. Sagittal sections from vermis anterior (lobulus V). Bars: 25 μm. Adapted from Nagelhus et al., 1993

intervene), as the glial cells will have to act as acceptors for the extruded taurine. In other words, wherever the above mechanism applies, the compartment with the highest taurine level will profit from an opening of its taurine channels whereas any neighbouring compartment with low taurine concentrations risks an increased swelling by doing this.

The re-establishment of the normal pattern of labelling following reversal with hypertonic saline must, in contrast, depend on active processes as taurine will have to be moved against its concentration gradient. One must postulate that active uptake mechanisms exist on Purkinje cells and that these mechanisms determine the direction of flux in the

Table 2. Redistribution of taurine in water-loaded animals assessed by quantitative immunogold cytochemistry

Group of animal	Gold particle density over Purkinje cells (particles/μm^2)	Cross sectional area of Purkinje cell somata (μm^2)	Gold particle density over glial cells (particles/μm^2)	Areal density of mitochondria in glial cells (number/μm^2)
C_4	65.0 ± 3.0 (17)	431 ± 8 (75)	27.3 ± 1.7 (21)	1.19 ± 0.26 (19)
	51.4 ± 1.9 (17)	435 ± 8 (66)	20.7 ± 1.1 (17)	0.83 ± 0.18 (17)
	54.9 ± 2.1 (25)	414 ± 6 (73)	25.0 ± 1.7 (16)	0.97 ± 0.21 (18)
	45.7 ± 1.9 (20)	415 ± 7 (59)	18.9 ± 0.8 (25)	0.88 ± 0.13 (26)
	74.4 ± 3.0 (15)	389 ± 8 (56)	29.7 ± 2.0 (19)	0.65 ± 0.10 (17)
	54.7 ± 1.6 (20)	394 ± 7 (52)	22.4 ± 1.2 (21)	0.94 ± 0.16 (22)
Mean ± SEM	57.7 ± 4.2	413 ± 8	24.0 ± 1.6	0.91 ± 0.07
W_4	21.9 ± 1.7 (15)	409 ± 8 (64)	12.1 ± 0.9 (32)	0.26 ± 0.06 (31)
	25.2 ± 1.8 (19)	397 ± 7 (68)	12.2 ± 0.9 (27)	0.30 ± 0.12 (27)
	19.8 ± 1.4 (16)	427 ± 6 (84)	11.2 ± 1.0 (25)	0.26 ± 0.07 (23)
	29.0 ± 1.2 (19)	364 ± 9 (51)	17.0 ± 1.0 (30)	0.24 ± 0.07 (26)
	24.3 ± 1.6 (17)	392 ± 7 (73)	13.2 ± 0.7 (33)	0.34 ± 0.06 (36)
Mean ± SEM	24.0 ± 1.6 *	398 ± 9	13.1 ± 1.0 *	0.28 ± 0.02 *

Groups of animals are denoted as in Table 1 (C, control animals; W, water-loaded animals; R, "reversed" animals; subscripts represent time of survival after first injection). The gold particle density was calculated over Purkinje cell bodies and adjacent cell bodies and processes of Golgi epithelial cells. The same glial profiles were also analysed with respect to the areal density of mitochondria. The decrease in mitochondrial density after water loading indicated a glial swelling of about 225%. Using this correction factor the glial cell contents of taurine was found to *increase* by 77% following water loading. The electron micrographs had a final magnification of 34500. A Zeus Image Analysis System (Pixel Works, Bergen, Norway) was used for assessment of cross-sectional area of Purkinje cell bodies in adjacent semithin sections. Mean ± standard error of the mean is indicated for each type of observation in each animal (number of observations in parenthesis) and for each group of animals. Asterisks, significantly different from group C_4 ($p < 0.001$; Student's t-test, two tails). Taurine antiserum no. 286 diluted 1:500. Modified from Nagelhus et al., 1993.

reversal phase. This issue should now be amenable to more definitive studies as several taurine transporters have been isolated and cloned in the last couple of years (9,16,29).

A neuronal-glial exchange of taurine in acute hypo-osmotic stress adds to the list of known service functions that glial cells maintain vis-à-vis adjacent neurons (4). However, one must be careful not to generalize from the cerebellum to all other brain regions. Thus immunocytochemical preparations from the hippocampus, obtained from the same series of animals as above, did not show a strong enrichment of taurine in glial cells following water loading, and the pyramidal cell bodies remained enriched in taurine as in the control animals (Nagelhus, Lehmann and Ottersen; unpublished observations). Although quantitative analyses have not yet been done and modest changes cannot be excluded, the data from the hippocampus suggest that opening of taurine channels in response to osmotic stress might not be a property common to all cells and regions in the brain. It should also be remembered that the distribution of taurine varies among central nervous system regions; while a neuronal labelling predominates in the cerebellum and hippocampus, most of the brain stem and thalamus display a stronger labelling in glial cells than in neurons (17). This must have consequences for the cellular redistribution of taurine during osmotic stress.

The immunocytochemical data described above indicate that the role of taurine as an osmoregulator is far more complex than hitherto appreciated from studies of cultured cells. This complexity follows from the cellular and regional heterogenity in the distribution of taurine, and the indirect evidence of a nonuniform concentration and/or behaviour of volume-sensitive taurine channels.

So far we have been dealing exclusively with the acute effects of an alteration in plasma osmolality. An immunocytochemical analysis was also performed of rats in late pregnancy, which have a chronic, physiologic hypo-osmolality (12). These rats show a perfectly normal pattern of taurine immunolabelling in the cerebellar cortex in so far as the Purkinje cells are more intensely stained than the glial cells (Nagelhus, Lehmann, and Ottersen; unpublished observations). Biochemical studies have reported, however, that rats in late pregnancy display a 50 % reduction in the brain level of taurine, which must reflect a nonselective loss of taurine from all cell compartments (12). A reduced tissue level of taurine was also noted in our acute experiments (Table 1) (21). Taken together, our data indicate that a reduction in plasma osmolality induces an acute redistribution of taurine that is reversible and superimposed on a nonselective loss of taurine from all compartments. The latter type of change is seen also for other amino acids and may represent the major adaptive change in *chronic* aniso-osmotic conditions.

The cells of certain aquatic invertebrates contain more taurine than K^+, and taurine may be the main osmoregulator of these cells (11). It has been argued that although taurine behaves as an osmolyte in the mammalian brain, the tissue level of the amino acid is so low that even if the cells were depleted of their taurine content during hypo-osmotic swelling, the loss would contribute very little to the total loss of osmolytes. However, immunocytochemical data suggest that taurine is heterogeneously distributed in the brain and that it may reach very high concentrations in some cells including the Purkinje cells. In such cells, taurine may contribute substantially to volume regulation.

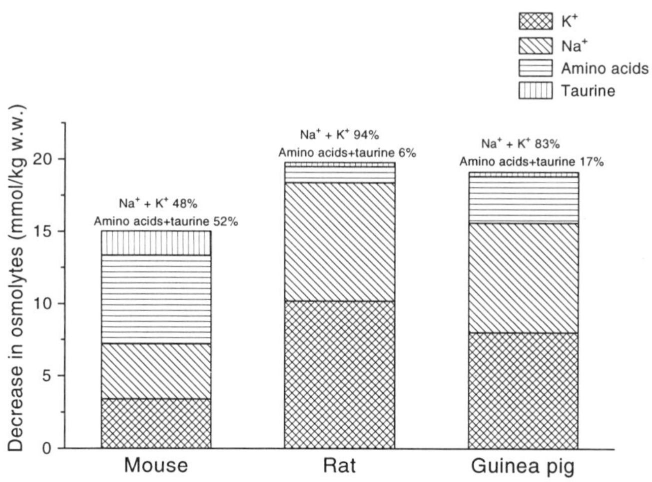

Figure 3. Relative contribution of inorganic cations and taurine and other amino acids to the downregulation of cerebrocortical osmolytes during water intoxication. The animals were administered 150 ml/kg distilled water and killed 4 h later. The decrease in osmolytes is expressed in relation to saline controls. n = 5 (guinea pig), n = 6 (rat and mouse). Adapted from Puka et al., 1991

SPECIES DIFFERENCES

The significance of taurine as an osmoregulator may not only be restricted to certain cell populations, but may also vary among species. A study on the utilization of inorganic cations versus amino acids in brain volume regulation during acute plasma hypo-osmolality revealed that taurine and other amino acids are quantitatively more important than Na^+ and K^+ in the mouse (26). In contrast, the rat and the guinea pig primarily down-regulate inorganic ions in response to acute osmotic stress (according to measurements in cerebrocortical tissue; Fig. 3). These observations were based on biochemical analyses and thus do not exclude cellular redistributions of taurine of the kind presently reported for the cerebellum.

The well-known species differences with respect to cerebral taurine concentrations have been proposed to relate to disparities in brain glucose consumption (34). Alternatively, the differences in taurine levels may reflect adaptation to species-specific osmotic pressure (26). Thus, the mouse has high cerebrocortical taurine concentrations, low cortical water content and high blood osmolality as compared to the guinea pig. The rat occupies an intermediate position with respect to these parameters (Fig. 4). Thus the positive correlation between plasma osmolality and cerebral taurine level applies not only to the individual animal subjected to an osmotic challenge, but also across different species.

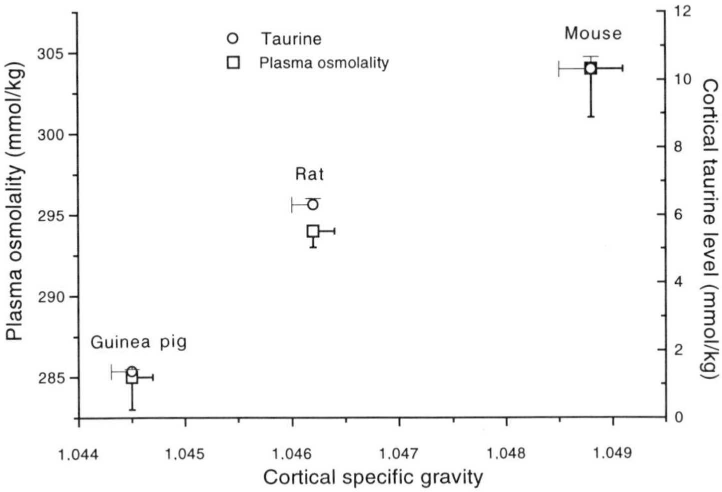

Figure 4. Relationship between plasma osmolality, cortical specific gravity and cortical taurine levels in the guinea pig (n = 5), rat (n = 6) and mouse (n = 6). Osmolality was measured with a vapor pressure osmometer, taurine was quantitated with HPLC and specific gravity was measured with a density gradient column. Standard error of the mean is indicated. Modified from Puka et al., 1991

ACKNOWLEDGEMENTS

The expert technical assistance of A.T. Bore, K.M. Gujord, J. Knutsen, T. Nordby, B. Riber, and T.A. Slyngstad is gratefully acknowledged. We also wish to thank S. Madsen for preparing the taurine no. 20 antiserum and J. Storm-Mathisen for continuous help and support.

This work was supported by the Norwegian Council for Science and the Humanities, the Swedish Medical Research Council (grant B92-12x-09053-03A), European Science Foundation (Twinning grant), Bergliot and Sigurd Skaugen's Research Fund, Lise and Arnfinn Heje's Fund, H.P. Pettersen's Fund, Alette and Kristian Schreiner's Fund, and Skipsreder Fredriksen's Fund.

REFERENCES

1. Ballanyi, K., and Grafe, P., 1988, Cell volume regulation in the nervous system, *Renal Physiol. Biochem.* 3-5:142-157.
2. Baxter, C.F., Baldwin, R.A., Lu, P., Imaki, H., and Sturman, J.A., 1993, Taurine in toad brain and blood under different conditions of osmolality: an immunocytochemical study, *Neurochem. Res.* 18:425-435.
3. Campistron, G., Geffard, M., and Buijs, R.M., 1986, Immunological approach to the detection of taurine and immunocytochemical results, *J. Neurochem.* 46:862-868.
4. Erecinska, M., and Silver, I.A., 1990, Metabolism and role of glutamate in mammalian brain, *Prog. Neurobiol.* 35:245-296.
5. Fugelli, K., and Thoroed, S.M., 1986, Taurine transport associated with cell volume regulation in flounder erythrocytes under anisosmotic conditions, *J. Physiol.* 374:245-261.
6. García, J.J., Sánches Olea, R., and Pasantes-Morales, H., 1991, Taurine release associated to volume regulation in rabbit lymphocytes, *J. Cell. Biochem.* 45:207-212.
7. Huxtable, R.J., 1992, The physiological actions of taurine, *Physiol. Rev.* 72:101-163.
8. Ida, S., Kuriyama, K., Tomida, Y., and Kimura, H., 1987, Antisera against taurine: Quantitative characterization of the antibody specificity and its application to immunohistochemical study in the rat brain, *J. Neurosci. Res.* 18:626-631.
9. Jhiang, S.M., Fithian, L., Smanik, P., McGill, J., Tong, Q., and Mazzaferri, E.L., 1993, Cloning of the human taurine transporter and characterization of taurine uptake in thyroid cells, *Fed. Eur. Bio. Soc.* 318: 139-144.
10. Kimelberg, H.K., Goderie, S.K., Higman, S., Pang, S., and Waniewski, R.A., 1990, Swelling-induced release of glutamate, aspartate and taurine from astrocyte cultures, *J. Neurosci.* 10:1583-1591.
11. Lange, R., 1964, The osmotic adjustment in the echinoderm, Strongylocentrotus droebachiensis. *Comp. Biochem, Physiol.* 13:205-216.
12. Law, R.O., 1989, Effects of pregnancy on the contents of water, taurine, and total amino nitrogen in rat cerebral cortex, *J. Neurochem.* 53:300-302.
13. Law, R.O., 1991, Amino acids as volume-regulatory osmolytes in mammalian cells, *Comp. Biochem. Physiol.* 99:263-277.
14. Lehmann, A., 1989, Effects of microdialysis-perfusion with anisoosmotic media on extracellular amino acids in the rat hippocampus and skeletal muscle, *J. Neurochem.* 53:525-535.
15. Lehmann, A., Carlström, C., Nagelhus, E.A., and Ottersen, O.P., 1991, Elevation of taurine in hippocampal extracellular fluid and cerebrospinal fluid of acutely hypoosmotic rats: contribution by influx from blood? *J. Neurochem.* 56:690-697.
16. Liu, Q.-R., Lopez-Corcuera, B., Nelson, H., Mandiyan, S., and Nelson, N., 1992, Cloning and expression of a cDNA encoding the transporter of taurine and ß-alanine in mouse brain, *Proc. Natl. Acad. Sci. USA* 89:12145-12149.
17. Madsen, S., 1990, Immunocytochemical visualization of taurine-containing and taurine-synthesizing cells, *in:* "Taurine: Functional Neurochemistry, Physiology, and Cardiology", Martin, D.L., Shain, W., and Martin del Rio, R., eds., Wiley-Liss, New York, pp. 21-28.
18. Madsen, S., Ottersen, O.P., and Storm-Mathisen, J., 1985, Immunocytochemical visualization of taurine: neuronal localization in the rat cerebellum, *Neurosci. Lett.* 60:255-260.
19. Magnusson, K.R., Madl, J.E., Clements, J.R., Wu, J.-Y., Larson, A.A., and Beitz, A.J., 1988, Colocalization of taurine- and sulfinic acid decarboxylase-like immunoreactivity in the cerebellum of the rat with monoclonal antibodies against taurine, *J. Neurosci.* 8:4551-4564.
20. Martin, D.L., Madelian, V., Seligmann, B. and Shain, W., 1990, The role of osmotic pressure and membrane potential in K^+-stimulated taurine release from cultured astrocytes and LRM55 cells, *J. Neurosci.* 10:571-577.
21. Nagelhus, E.A., Lehmann, A., and Ottersen, O.P., 1993, Neuronal-glial exchange of taurine during hypo-osmotic stress: a combined immunocytochemical and biochemical analysis in rat cerebellar cortex, *Neuroscience* 54:615-631.

22. Ottersen, O.P., 1988, Quantitative assessment of taurine-like immunoreactivity in different cell types and processes in rat cerebellum: an electronmicroscopic study based on a postembedding immunogold labelling procedure, *Anat. Embryol.* 178:407-421.
23. Pasantes-Morales, H., and Schousboe, A., 1988, Volume regulation in astrocytes: A role for taurine as an osmoeffector, *J. Neurosci. Res.* 20:505-509.
24. Pasantes-Morales, H., and Schousboe, A., 1989, Release of taurine from astrocytes during potassium-evoked swelling, *Glia* 2:45-50.
25. Pasantes-Morales, H., Morán, J., and Schousboe, A., 1990, Volume-sensitive release of taurine from cultured astrocytes: properties and mechanism, *Glia* 3:427-432.
26. Puka, M., Sundell, K., Lazarewicz, J.W., and Lehmann, A., 1991, Species differences in cerebral taurine concentrations correlate with brain water content, *Brain Res.* 548:267-272.
27. Sánches-Olea, R., Morán, J., Schousboe, A., and Pasantes-Morales, H., 1991, Hyposmolarity-activated fluxes of taurine in astrocytes are mediated by diffusion, *Neurosci. Lett.* 130:233-236.
28. Schousboe, A., Sánchez Olea, R., Morán, and Pasantes-Morales, H., 1991, Hyposmolarity-induced taurine release in cerebellar granule cells is associated with diffusion and not with high-affinity transport, *J. Neurosci. Res.* 30:661-665.
29. Smith, K.E., Borden, L.A., Wang, C.-H. D., Hartig, P.R., Branchek, T.A., and Weinshank, R.L., 1992, Cloning and expression of a high affinity taurine transporter from rat brain, *Molec. Pharm.* 42:563-569.
30. Solis, J.M., Herranz, A.S, Herreras, O., Lerma, J. and Martín del Rio, R., 1988, Does taurine act as an osmoregulatory substance in the rat brain? *Neurosci. Lett.* 91:53-58.
31. Strange, K., 1992, Regulation of solute and water balance and cell volume in the central nervous system, *J. Am. Soc. Nephrol.* 3:12-27.
32. Tigges, G.A., Philibert, R.A., and Dutton, G.R., 1990, K^+- and temperature-evoked taurine efflux from hypothalamic astrocytes, *Neurosci. Lett.* 119:23-26.
33. Trachtman, H., del Pizzo, R., and Sturman, J.A., 1990, Taurine and Osmoregulation. III. Taurine deficiency protects against cerebral edema during acute hyponatremia, *Pediatr. Res.* 27: 85-88.
34. van Gelder, 1990, Neuronal discharge hypersynchrony and the intracranial water balance in relation to glutamatic acid and taurine redistribution: Migraine and epilepsy, *in:* "Taurine: Functional Neurochemistry, Physiology, and Cardiology", Pasantes-Morales, H., Martin, D.L., Shain, W., and Martin del Rio, R., eds., Wiley-Liss, New York, pp. 1-20.
35. Verbalis, J.G., and Gullans, S.R., 1991, Hyponatremia causes large sustained reductions in brain content of multiple organic osmolytes in rats, *Brain Res.* 567:274-282.
36. Wade, J.V., Olson, J.P., Samson, F.E., Nelson, S.R., and Pazdernik, T.L., 1988, A possible role for taurine in osmoregulation within the brain, *J. Neurochem.* 51:740-745.
37. Yoshida, M., Karasawa, N., Ito, M., Sakai, M., and Nagatsu, I., 1986, Demonstration of taurine-like immunoreactive structures in the rat brain, *Neurosci. Res.* 3:356-363.
38. Zhang, N., and Ottersen, O.P., 1992, Differential cellular distribution of two sulphur-containing amino acids in rat cerebellum, *Exp. Brain Res.* 90:11-20.

ALTERATIONS OF VENTRICULAR CONTRACTILITY AND MYOFIBRIL LOSS IN TAURINE-DEFICIENT HEARTS

Norma Lake

Departments of Physiology and Ophthalmology
McGill University
Montreal, Quebec, Canada

INTRODUCTION

Although the sulfur-containing amino acid, taurine, is the major constituent of the free amino acid pool of mammalian myocardium (10), its function remains elusive. Taurine is not a substrate for Krebs cycle metabolism, and it does not participate in protein synthesis. There is little biosynthesis of taurine within the heart; instead its millimolar intracellular levels are determined by active transmembrane transport by a specific high affinity carrier. Normal plasma levels of taurine are in the micromolar range. Although the cDNA sequences of several taurine transporters have been elucidated for other tissues (11,30), the isoform(s) in the heart has not been isolated. The liver is one of the major organs for taurine synthesis, but in many species including humans and cats, the liver capacity is limited, and the diet is a major source of taurine. Experimental taurine depletion can be induced in cats, foxes and young monkeys by feeding taurine-free diets. In animals such as rats and mice, whose liver taurine synthesis is much greater than in cats or humans (10), rendering them less susceptible to low taurine diets, specific taurine depletion of many tissues can be achieved by in vivo treatment with compounds, such as β-alanine or guanidinoethane sulfonate, that are antagonists of taurine transport systems (9).

TAURINE DEFICIENCY PRODUCES DEFICITS IN VENTRICULAR CONTRACTILE FORCE

Depletion of tissue taurine levels can have many deleterious effects. Attention here is focussed on alterations of ventricular contractility. A number of studies have used echocardiography to measure heart chamber dimensions at different stages of the cardiac cycle. Observers of cats and foxes that had other clinical signs of "spontaneous" heart

Taurine in Health and Disease, Edited by R. Huxtable and D.V. Michalk, Plenum Press, New York, 1994

failure noted that decrements of the ventricular shortening fraction were associated with low plasma taurine levels (23,25). Novotny and colleagues (24) studied normal cats maintained on taurine-free diets that were taurine-supplemented or taurine-deficient (unsupplemented). They examined ventricular contractile dynamics by observing pressure-volume relationships in the perfused and paced left ventricle of isolated hearts. A balloon was inflated within the ventricle to set the end diastolic pressure and the peak systolic pressure developed was monitored. They reported two major findings for hearts from taurine-deficient cats: (i) active systolic function was depressed, and (ii) passive diastolic chamber compliance or distensibility was increased.

My own studies of force generation began with small papillary muscles excised from hearts of control rats or those depleted of 75% of myocardial taurine using treatment with a transport antagonist in vivo (18). There were no changes in DNA; i.e. no cell death occurred. However, preparations from taurine-deficient rats showed reduced contractility; in particular, a large deficit in generation of isometric force (18). More recently I have been using small ventricular trabeculae (5,14). Their thinness (60 to 200 μ) permits the use of a laser diffraction technique to monitor sarcomere length simultaneously with force, to ensure that force generation is evaluated at comparable sarcomere lengths. This is crucial when one is interested in mechanisms at the level of the sarcomere. An eyepiece micrometer was used to measure total muscle dimensions, and the force generated in response to electrical pacing was expressed per unit cross sectional area to allow comparisons between muscles. In taurine-depleted rats the yield of trabeculae and their dimensions were not different from those of untreated litter mates who served as the controls.

Over the range of sarcomere lengths at which cardiac muscle operates, the force (in milliNewtons /mm^2) developed by the taurine-depleted trabeculae was less than 60% of that developed by the control trabeculae at any given sarcomere length. In addition the passive force (to length changes) was significantly lower, indicating greater compliance of taurine-deficient muscles. Note the similarities between these data from guanidinoethane sulfonate-treated rats and those obtained by Novotny et al. (24) from cats on taurine-deficient diets: reduced active force, and increased passive distensibility. In addition, when fibres from normal rat hearts were chemically skinned to remove the permeability barrier of the sarcolemma, applications of guanidinoethane sulfonate up to 50 mM had no effects on the force generated or on sensitivity to calcium (6), thus making it unlikely that direct effects of guanidinoethane sulfonate were causing the deficits seen in the muscles from rats treated with guanidinoethane sulfonate to produce taurine depletion.

WHAT IS THE ORIGIN OF THE FORCE DEFICIT?

Force–Length Relation

To obtain information concerning the origin of the force deficits observed in taurine-depleted muscles, we analyzed the relationships between force and its major determinants. Relative force was computed: force was expressed as percent of maximum developed (separately for control or depleted muscles). There was no difference in the curves of relative force and sarcomere length, indicating no change in the basic sarcomere length-active tension relationship in ventricular cells. Furthermore, for both groups the curve was convex towards the ordinate in buffers containing 1.6 mM calcium and convex towards the abscissa for those containing 0.4 mM calcium. This suggests that the reduction of force in depleted muscles did not result from reduced activation of myofibrils, since previous work has shown that this would indeed alter the force-sarcomere length relation (12).

Force–Calcium Relation

Curves for the two groups were also very similar when examining the relationships between relative force and calcium concentration in the perfusate at a fixed sarcomere length. The slope of the relation (Hill coefficient) and the minimum concentration required for maximal force (1.6 mM calcium) were the same for both groups. There was a slight reduction in sensitivity to calcium since the depleted muscles required 0.6 mM calcium (versus 0.5 mM for the control) to generate half maximal force. It is doubtful that this small change has physiological significance, especially at the usual extracellular calcium levels, and it certainly would not be sufficient to explain the deficit of 40% of developed force at 1.6 mM calcium.

Excitation-Contraction Coupling

Calcium is required in the vicinity of the myofibrillar proteins to activate contraction. Our current understanding of excitation-contraction coupling, eg. in rat ventricle, is that "trigger" calcium, which enters through calcium channels during the action potential, releases a much larger amount of calcium from the sarcoplasmic reticulum which then activates the myofibrils. Trans-sarcolemmal extrusion, and re-sequestration of calcium by the sarcoplasmic reticulum allows relaxation to proceed and re-primes the sarcoplasmic reticulum for the next release. The electrophysiology I have done indicates that reduced force was probably <u>not</u> a result of reduced calcium entry, since the duration of the action potential is actually prolonged in taurine-depleted muscles compared to control (14,16). Specific stimulation protocols were employed to examine if calcium entry, extrusion or recycling through the sarcoplasmic reticulum were altered by taurine depletion. The force-interval relation probes the effect on subsequent force development of interposing a variable rest interval (from 0.3 to 300 s) into the basic pacing rate. Experiments have shown that drugs that modify calcium channels alter the effects of short rest intervals, while modifying calcium extrusion changes the responses to longer rest intervals (27). The curves of control and taurine-depleted muscles were superimposable, which indicates that the transport of calcium across the sarcolemma membrane is not altered in taurine-depleted rats (5). There was only a slight (nonsignificant) decrease in the recycling of calcium through the sarcoplasmic reticulum, examined by use of the decay of post extra systolic potentiation. It thus appears that the force deficit in taurine-depleted muscle is <u>not</u> dependent on alterations in calcium handling.

Calsequestrin

To see if force was being limited in taurine-deficient muscles by a reduced store of releasable calcium in the sarcoplasmic reticulum, we measured calsequestrin (the primary calcium-binding protein of sarcoplasmic reticulum). The total amount of extractable sarcoplasmic reticulum protein was not different, and western blots (immunoblots) of polyacrylamide gel separations of heart extracts indicated that the quantity of calsequestrin also was not different from control, suggesting that the sarcoplasmic reticulum calcium store size was not changed (5). Other experiments are required to investigate if the calcium binding capacity or affinity of the calsequestrin is altered by taurine deficiency.

Myosin Polymorphism

Three myosin isozymes have been seen in ventricular muscle (7). In normal adult rat ventricle V_1 is the predominant isoform. Conditions such as hypothyroidism, diabetes, excess sympathetic activity or work overload shift the isozyme pattern to V3. Our physiological studies did <u>not</u> detect reduced maximal velocity of shortening (a correlate of isoform switching) after 6 weeks on guanidinoethane sulfonate (18), and this was confirmed by <u>no</u> change in the relative distribution of the V1 and V3 isoforms (5). Both control and taurine-depleted muscles expressed predominantly the V1 isoform, so isoform switching did not contribute to the force deficit.

Density of Contractile Elements

Since we could not find changes in myofibril activation level, calcium sensitivity, calcium stores, calcium handling, or myosin isoforms, sufficiently large enough to account for the large force deficit in taurine-deficient muscles, we then tested the possibility that there was a reduced population of force generators. Strontium is a calcium ion analogue that enters the cell through calcium channels and directly activates the contractile elements. That its action is directly on the contractile elements, and not via release of calcium from the sarcoplasmic reticulum, is shown by the marked activation of contraction by strontium in the presence of ryanodine, a compound that blocks sarcoplasmic reticulum calcium release and calcium-activated contractions (13,22). A key observation in our studies of trabeculae from taurine-depleted hearts was that the large deficit in force developed/ cross-sectional area persisted during strontium activation (5,14). This finding suggested a reduction in the number of contractile elements, and prompted my studies of ventricular ultrastructure and contractile protein content.

VENTRICULAR ULTRASTRUCTURE IN TAURINE-DEPLETED RATS

In my studies, fixation was by retrograde perfusion (Langendorff preparation) or by immersion using 4–5% glutaraldehyde in cacodylate buffer containing 0.5% alcian blue as recommended by Sommer (3) and including 50 mM lysine to preserve actin filaments from the destructive disorganization produced by glutaraldehyde and osmium (4). This particular cocktail was found to be superior to several other less complex conventional pathology protocols with respect to the preservation of the myofibrillar lattice. I have found striking differences between control and taurine-depleted tissue in the integrity of the myofibrillar bundles. Many of the myocytes in depleted tissue contained myofibrillar bundles in various stages of degeneration. Lipid droplets were frequent and the bundles ranged from normal to disorganized, with dropout of thick filaments from the usual crystalline array (in transverse sections), to small areas of loss of myofibrillar bundles and replacement by amorphous debris. In some regions there were small remnants of myofibrillar bundles surrounded by amorphous material, and a disproportionate number of mitochondria that themselves appeared fairly normal, by cursory assessment. The most severe degeneration was usually localized close to the sarcolemma; I never saw any myocytes with their entire population of myofibrils degenerated. There did not appear to be an increase in collagen or fibrosis. Examples of some of these findings are illustrated in the electron micrographs of Figure 1, and in Lake (15).

All of the rats used for the initial study (15) were male. The most severe myofibril damage accompanying taurine deficiency that I have found in preliminary studies was in hearts from two lactating dams consuming taurine-free diets. This situation may correspond to peripartum cardiomyopathy (1,8). During pregnancy and postnatal development when the rat pups are suckled, the mother provides much of the very high taurine levels found in developing tissues, and she is highly vulnerable to taurine deficiency. Law (20) has reported the loss of 48% of brain taurine in pregnant rats which he interprets as a response to preserve brain volume during the plasma hypoosmolality of pregnancy. Stipanuk et al.

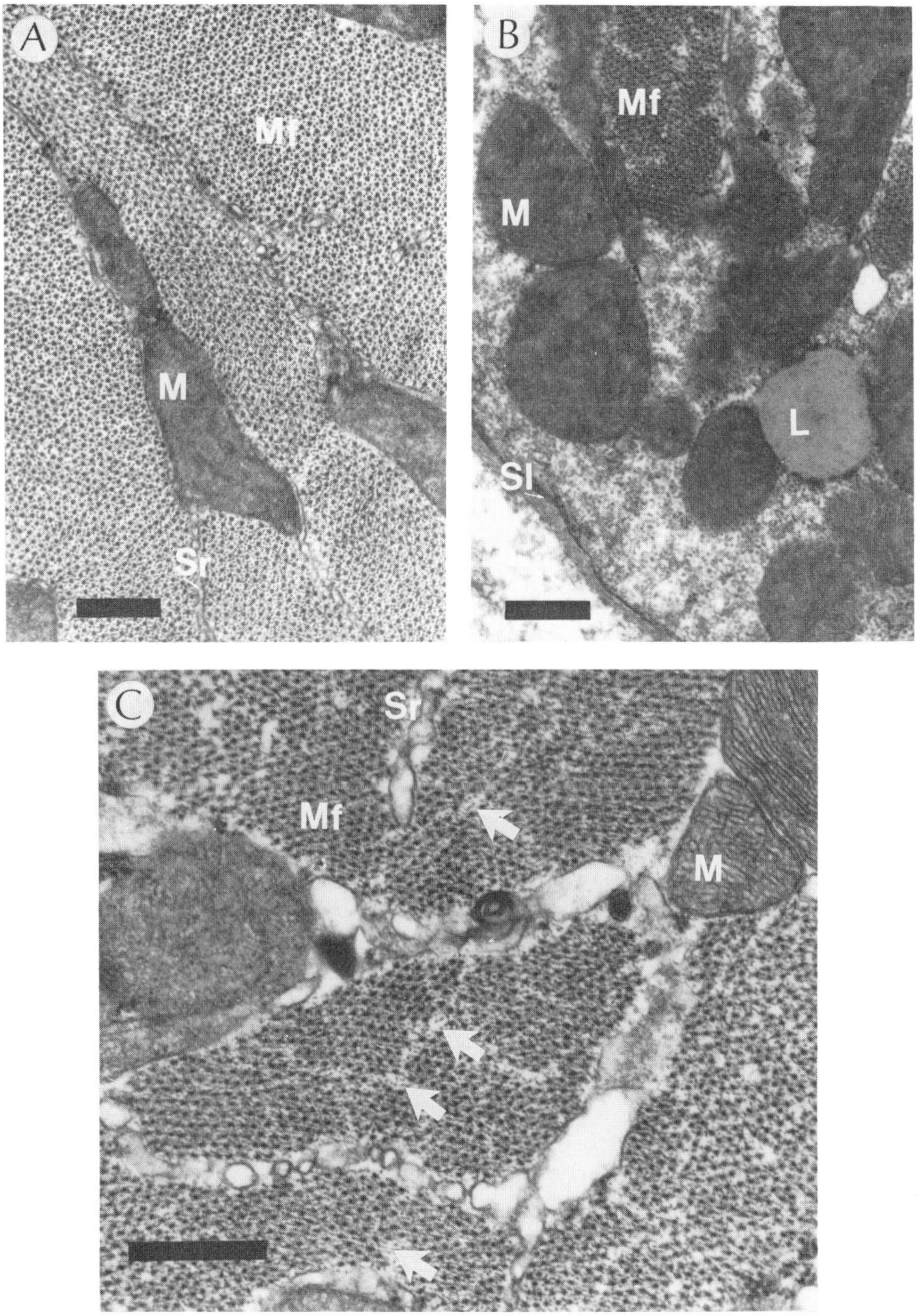

Figure 1. Electron micrographs of transverse sections of ventricular tissue from control (**A**) and taurine-depleted (**B,C**) rats. **A**: Note the crystal-like array of thick and thin myofilaments (Mf); other components visible are the irregular-shaped mitochondria (M) and tubular sacs of the sarcoplasmic reticulum (Sr). A highly disrupted region of a myocyte is shown in **B**; note the large lipid granule (L) and the amorphous debris that has replaced the myofibrillar bundles lying close to the sarcolemma (Sl). **C** is a higher power view of a less disrupted region. Note the disorganization of the myofibrillar lattice and in some places (->) the apparent loss of thick filaments, while thin filaments remain. Scale bars represent 0.5 μ.

[Fig. 2 of (28)] have described a significant fall in heart taurine content in untreated mother rats at the time of parturition and early lactation. Consistent with increased vulnerability, we have observed that the degenerative effects of guanidinoethane sulfonate-induced taurine deficiency on the retina and optic nerve are most marked in lactating rat dams (17,19), and anecdotal evidence for a similar conclusion exists for cats.

BIOCHEMICAL ESTIMATES OF MAJOR SARCOMERIC PROTEINS

In order to quantitate the losses of contractile fibres apparent in the micrographs from taurine-depleted animals, I used polyacrylamide gel electrophoresis (SDS-PAGE) to separate the various sarcomeric proteins. Initially I concentrated on myosin and actin since they are the most abundant sarcomeric proteins and easily identified by molecular weight. Left ventricular samples were homogenized and extracted on ice in a buffer containing 5 mM EGTA, the antiprotease PMSF, 4 mM dithiothreitol and 50 mM sodium pyrophosphate to solubilize myosin [modified from (7)]. After centrifugation the supernatant was mixed with ice cold glycerol before plating or storing at -2°C, to facilitate myosin solubility and protect the extracted proteins from denaturation. Figure 1D of Lake (15) is a gel typical of extracts from control and guanidinoethane sulfonate-treated rats, run on a Biorad Mini Protean apparatus. The gels were fixed, stained with Coomassie blue and destained before quantitation using a BioImage system (Millipore), that scans the gels, digitizes the images and gives the integrated area times optical density (OD) of the separated bands in OD units calibrated with external OD wedge standards. The results indicated that although equal amounts of protein were extracted from the control and taurine-deficient ventricles, the myosin heavy chain content was reduced by 24% from control (range 14-33%, $P < .01$). The actin content was unchanged or slightly reduced (not statistically significant), and the myosin heavy chain: actin OD ratio (per lane) was significantly reduced.

TAURINE DEPLETION REDUCES THE NUMBER OF VENTRICULAR FORCE GENERATORS

In summary, I have four pieces of evidence for the hypothesis that the substantial deficits in force developed by taurine-depleted muscles reflect a reduced population of force generators. (i) When strontium was added to the perfusate to directly activate the contractile proteins, the substantial force deficit persisted, indicating that the <u>number</u> of contractile elements was substantially less than in controls; (ii) The ultrastructural studies show a partial loss of thick filaments and myofibrils in myocytes from taurine-depleted rats; (iii) I have found a significant decrease of contractile protein content, with losses in myosin heavy chain that confirm the loss of myofibrils seen in the electron micrographs; (iv) Consistent with a loss of elastic elements, it was noted that compliance or distensibility of the taurine-depleted muscles was significantly higher [as has been also reported for the cat (24)]. Since the thin filaments, that are in large part responsible for the parallel elasticity of muscle, are anchored to myosin (29,31), if the number of myosin filaments is decreased, then the density of the thin filaments would also be expected to be decreased, and compliance increased.

SPECIES DIFFERENCES?

I have pointed out the similarities, in terms of the functional effects on ventricular contractile activity, of taurine depletion induced by guanidinoethane sulfonate treatment in rats, and by taurine-deficient diets in cats. Whilst it would be satisfying to demonstrate an underlying mechanism common to both species, this is not immediately apparent since various claims have been made that no specific tissue pathology has been seen in the ventricles of such cats. For example, Pion et al. (26) report that although myocardial tissue lesions were found in taurine-depleted cats, they did not provide an explanation of the heterogeneity of ventricular function alteration by taurine depletion. In fact, the authors were unable to provide an explanation of the heterogeneity. In contrast, in my studies using guanidinoethane sulfonate the extent of taurine depletion and the effects on ventricular function and pathomorphology were rather uniform amongst animals treated for 6-7 weeks, while the two lactating dams (no guanidinoethane sulfonate) had lower taurine content and more numerous lesions. I plan to examine systematically the correlation between these parameters by varying the duration and type of depletion protocol. Against the hypothesis that the effects I observe are produced solely by guanidinoethane sulfonate are the findings that guanidinoethane sulfonate had no direct effects on force developed or sensitivity to calcium in skinned myocytes, and that pathomorphology was similar in the pilot study of lactating female rats not exposed to guanidinoethane sulfonate. However further data are needed to establish this point.

HOW DOES TAURINE DEFICIENCY LEAD TO MYOSIN LOSSES?

The intervening steps between taurine deficiency and myosin heavy chain losses also need to be elucidated. Two possible mechanisms whose alteration could affect myosin heavy chain turnover are myosin heavy chain gene regulation and degradation. Some aspects of the former can be readily evaluated since the relevant species and isoform specific mRNA probes exist. Specific lysis of myosin is more difficult to explore in view of the relative lack of knowledge about protein degradation in heart muscle. Recently Bertini et al. (2) have described cardiomyopathy in one patient with selective lysis of thick filaments (mechanism unknown), while Margossian et al. (21) have provided evidence that in human idiopathic dilated cardiomyopathy there is activation (cause unknown) of a protease that specifically breaks down a myosin light chain (LC2), which contributes to contractile failure.

ACKNOWLEDGMENTS

This work was supported by the Canadian Medical Research Council and the Faculty of Medicine, McGill University.

REFERENCES

1. $Adler, A.K., and Davis, M.R., 1986, Peripartum cardiomyopathy: two case reports and a review, *Obstetrical and Gynecological Survey* 41:675-682.
2. Bertini, E., Bosman, C., Salviati, G., Boldrini,S. Servidei, E. Ricci, F., Del Nonno, F., Gagliardi, M.G., and Bevilacqua, M., 1993, Myopathy and hypertrophic cardiomyopathy with selective lysis of thick filaments, *Virchows Arch. A Pathol. Anat Histopathol.* 422:327-331.
3. Bossen, E.H., Sommer, J.R., and Waugh, R.A., 1978, Comparative stereology of the mouse and finch left ventricle, *Tissue Cell* 10:773-784.
4. Boyles, J., Anderson, L., and Hutcherson, P., 1985, A new fixative for the preservation of actin filaments: fixation of pure actin filament pellets, *J. Histochem. Cytochem.* 33:1116-1128.
5. Eley, D.W., Lake, N., and ter Keurs, H.E.D.J., 1993, Taurine depletion and excitation-contraction coupling in rat myocardium, *Circ. Res.* submitted.
6. Godt, R.E., 1992, personal communication.

7. Hoh, J.F.Y., McGrath, P., and Hale, P.T., 1977, Electrophoretic analysis of multiple forms of rat cardiac myosin: effects of hypophysectomy and thyroxine replacement, *J. Molec. Cell Cardiol.* 10:1053-1076.
8. Homans, D.C., 1985, Peripartum cardiomyopathy, *New England J. Med.* 312:1432-1437.
9. Huxtable, R.J., Laird, H.E., and Lippincott, S., 1979, The transport of taurine in the heart and the rapid depletion of tissue taurine content by guanidinoethyl sulfonate, *J. Pharmacol. Exp. Ther.* 211:465-472.
10. Jacobsen, J.G., and Smith, L.L.H., 1968, Biochemistry and physiology of taurine and taurine derivatives, *Physiol. Rev.* 48:424-511.
11. Jhiang, S.M., Fithian, L., Smanik, P., McGill, J., Tong, Q., and Mazzaferri, E.L., 1993, Cloning of the human taurine transporter and characterization of taurine uptake in thyroid cells, *FEBS Lett.* 318:139-144.
12. Kentish, J.C., ter Keurs, H.E.D.J., Ricciardi, L., Bucx, J., and Noble, M., 1986, Comparison between the sarcomere length force relations of intact and skinned trabeculae from rat right ventricle. Influence of calcium concentration on these relations, *Circ. Res.* 58:755-768.
13. Kondo, N., 1987, Can strontium replace calcium as an activator of internal calcium release in cardiac muscles? A study on a dual action of A23187, *J. Molec. Cell Cardiol.* 19:391-397.
14. Lake, N., 1992, Effects of taurine deficiency on arrhythmogenesis and excitation-contraction coupling in cardiac tissue, *Adv. Exp. Med. Biol.* 315:173-179.
15. Lake, N., 1993, Loss of cardiac myofibrils: Mechanism of contractile deficits induced by taurine deficiency, *Am. J. Physiol. Heart Circ. Physiol.* 264: H1323-H1326.
16. Lake, N., De Roode, M., and Nattel, S., 1987, Effects of taurine depletion on rat cardiac electrophysiology: in vivo and in vitro studies. *Life Sci.* 40:997-1005.
17. Lake, N., Malik, N., and De Marte, L., 1988, Taurine depletion leads to loss of optic nerve axons, *Vis. Res.* 28:1071-1076.
18. Lake, N., Splawinski, J., Juneau, C., and Rouleau, J.L., 1990, Effects of taurine depletion on intrinsic contractility of rat ventricular muscles, *Can. J. Physiol. Pharmacol.* 68:800-806.
19. Lake, N., and Verdone-Smith, C., 1989, Immunocytochemical localization of taurine in the mammalian retina, *Curr. Eye Res.* 8:163-173.
20. Law, R.O., 1989, Effects of pregnancy on the contents of water, taurine, and total amino nitrogen in rat cerebral cortex, *J. Neurochem.* 53:300-302.
21. Margossian, S.S., White, H.D., Caulfield, J.B., Norton, P., Taylor, S., and Slayter, H.S., 1992, Light chain 2 profile and activity of human ventricular myosin during dilated cardiomyopathy: Identification of a causal agent for impaired myocardial function, *Circulation* 85:1720-1733.
22. Mitchell, M.R., Powell, T., Terrar, D.A., and Twist, V.W., 1987, Electrical activity and contraction in cells isolated from rat and guinea-pig ventricular muscle: a comparative study, *J. Physiol.* 391:527-544.
23. Moise, N.S., Pacioretty, L.M., Kallfeiz, F.A., and Gilmour, R.F., 1989, Association between low plasma taurine and dilatative cardiomyopathy in silver foxes, *FASEB J.* 3:6040A.(abstract)
24. Novotny, M.J., Hogan, P.M., Paley, D.M., and Adams, H.R., 1991, Systolic and diastolic dysfunction of the left ventricle induced by dietary taurine deficiency in cats, *Am. J. Physiol. Heart. Circ. Physiol.* 261:H121-H127.
25. Pion, P.D., Kittleson, M.D., Rogers, Q.R., and Morris, J.G., 1987, Myocardial failure in cats associated with low plasma taurine: a reversible cardiomyopathy, *Science* 237:764-768.
26. Pion, P.D., Kittleson, M.D., Skiles, M.L., Rogers, Q.R., and Morris, J.G., 1992, Dilated cardiomyopathy associated with taurine deficiency in the domestic cat: Relationship to diet and myocardial taurine content, *Adv. Exp. Med. Biol.* 315:63-73.
27. Schouten, V.J.A., Van Deen, J.K., De Tombe, P.P., and Verveen, A.A., 1987, Force-interval relationship in heart muscle of mammals: a calcium compartment model, *Biophys. J.* 51:13-26.
28. Stipanuk, M.H., Kuo, S.-M., and Hirschberger, L.L., 1984, Changes in maternal taurine levels in response to pregnancy and lactation, *Life Sci.* 35:1149-1155.
29. Trinick, J., 1991, Elastic filaments and giant proteins in muscle, *Curr. Opin. Cell Biol.* 3:112-118.
30. Uchida, S., Kwon, H.M., Yamauchi, A., Preston, A.S., Marumo, F., and Handler, J.S., 1992, Molecular cloning of the cDNA for an MDCK cell Na^+- and Cl^--dependent taurine transporter that is regulated by hypertonicity, *Proc. Natl. Acad. Sci. U.S.A.* 89:8230-8234.
31. Wang, K., McCarter, R., Wright, J., Beverly, J., and Ramirez-Mitchell, R., 1991, Regulation of skeletal muscle stiffness and elasticity by titin isoforms: A test of the segmental extension model of resting tension, *Proc. Natl. Acad. Sci. U.S.A.* 88:7101-7105.

DEVELOPMENTAL EFFECTS OF TAURINE DEPLETION ON SYNAPTOSOMAL PHOSPHOLIPIDS IN THE RAT

Ryan J. Huxtable, John Murphy and Pierre-Louis Lleu

Department of Pharmacology
College of Medicine, University of Arizona
Tucson, Arizona 85724

INTRODUCTION

No one working with taurine doubts its biological significance. Evidence continues to accumulate as to its fundamental involvement in phenomena as diverse as calcium homeostasis, osmoregulation, and membrane stabilization. The actions of taurine on these phenomena are, perhaps, interrelated. It continues to be a puzzle, however, as to why, if taurine is important for cell function, its concentration in cognate structures varies so much from species to species, or even temporally within a given species. Taurine concentrations in the heart, for example, can vary, depending on the species, over a tenfold range (10). Synaptosomal taurine concentrations in the rat can vary sixfold, depending on the developmental stage of the brain (14,23).

We have considered that the key to these variations might lie in the phospholipid composition of the membrane (11,12,19,20). This also varies enormously in cognate structures from species to species, or within the same structure, depending on developmental stage. Furthermore, many of the functions of membrane phospholipids seem to be related to the actions of taurine - as with calcium regulation and osmoregulation, to mention two important membrane functions. For both of these functions, the actions of taurine probably involves a modulation of membrane properties. For example, we have published on the ability of taurine to modify the binding of Ca^{2+} to membrane phospholipids (16,17,30). The effect of taurine on the K_m and B_{max} of Ca^{2+} binding to artificial phospholipid membranes, and the modulatory actions of other cations on Ca^{2+} binding, is quite strongly dependent on the ratio of neutral phospholipids in the membrane (11,12,17-19).

We considered, therefore, that membrane phospholipid composition and function might be modified cellular taurine concentration. We have begun a test of this hypothesis in brain synaptosomes (P_2B fraction). Synaptosomal membranes can be prepared in a relatively homogenous state. We report herein the consequences of taurine depletion induced by guanidinoethane sulfonate on the phospholipid composition and biochemistry on rat brain synaptosomal membranes.

Taurine in Health and Disease, Edited by R. Huxtable and D.V. Michalk, Plenum Press, New York, 1994

MATERIALS AND METHODS

[^{3}H-Methyl]methionine was purchased from New England Nuclear, Boston, Mass. Guanidinoethane sulfonate was synthesized as previously described (15). Sprague-Dawley rats with timed pregnancies were maintained on Wayne rat chow and were kept in pans on a 12 h photo-period for the duration of the study. Animals were taurine-depleted during development by maintaining the mothers during pregnancy and lactation on drinking water containint 1% guanidinoethane sulfonate. Litter sizes were adjusted at birth to eight male pups.

Preparation of Synaptosomal P_2B Fraction

Synaptosomal P_2B fractions were isolated from 7, 14, 21, 28, 56 and 119 day-old rats on discontinous sucrose gradients, as previously described (23). Protein concentrations were determined by the Lowry procedure (24). P_2B preparations were also prepared using a continous sucrose gradient from control and guanidinoethane sulfonate-treated rats for comparision with the preparations described by Whittaker and co-workers (6,35). Briefly, gradients were prepared overnight at room temperature by successive stacking of 0.8, 0.9, 1.0, 1.1, and 1.2 M sucrose layers. Synaptosomal P_2 fractions were resuspended in 0.32 M sucrose and layered on the top of the gradient and centrifuged at 53,000 g for 2 h. Pelleted subcellular fractions were fixed in 2% glutaraldehyde in O.1 M PO_4^{3-} followed by 2% OsO_4 in the same buffer and examined by transmission electron microscopy (Fig. 1). Mitochondrial contamination of the synaptosomal fraction was evaluated by determining succinate dehydrogenase activity according to the method of Green et al. (7).

Incorporation of [^{3}H-Methyl]Methionine in Proteins and Phospholipids

At 7, 14, 21, 28 and 56 days of age, male rat pups were randomly selected, one from each litter, weighed and injected i.p. with 300 μCi/kg of [^{3}H-methyl]methionine (specific activity 200 mCi/mmol). Nine hours later, rats were decapitated, and their brains removed and weighed. Brains were pooled such that approximately 2 g were used for subcellular fractionation.

Synaptosomal protein (0.2 ml) was precipitated with 0.2 ml of 7% sulfosalicylic acid. After centrifugation, the precipitated protein was digested with a tissue solubilizer (NEN, Solvable), scintillation cocktail added (ICN, Ecolyte) and the sample counted. Soluble radioactivity was determined on a separate aliquot.

Lipids were extracted using a modification of the Christiansen method (2). P_2B synaptosomal fraction (1 ml) was homogenized in a Wheaton glass homogenizer tube with 20 ml of 2:1 $CHCl_3$:CH_3OH containing 0.05% butylated hydroxytoluene. The homogenate was filtered, and the homogenizer tube was rinsed once with an additional 15 ml of the same solvent and once with 15 ml of 1:2 $CHCl_3$:CH_3OH containing 0.05% butylated hydroxytoluene. The rinsings were filtered and added to the original filtrate. The solvent was removed on a rotary evaporator and the residue taken up in 1 ml of 19:1 $CHCl_3$:CH_3OH and frozen for determination of radioactivity and phosphorus. For the former, aliquots (0.75 ml) were dried in a scintillation vial, scintillation cocktail added (ICN, cytoscint) and the samples counted.

Phospholipid Separation and Analysis

Phospholipid analysis was performed on extracts of P_2B fractions from 7, 14, 21, 28, 56 and 119 day-old rats. TLC separation of the major classes of phospholipids was carried out using two sequential solvent systems on silica gel plates (31). TLC plates were first

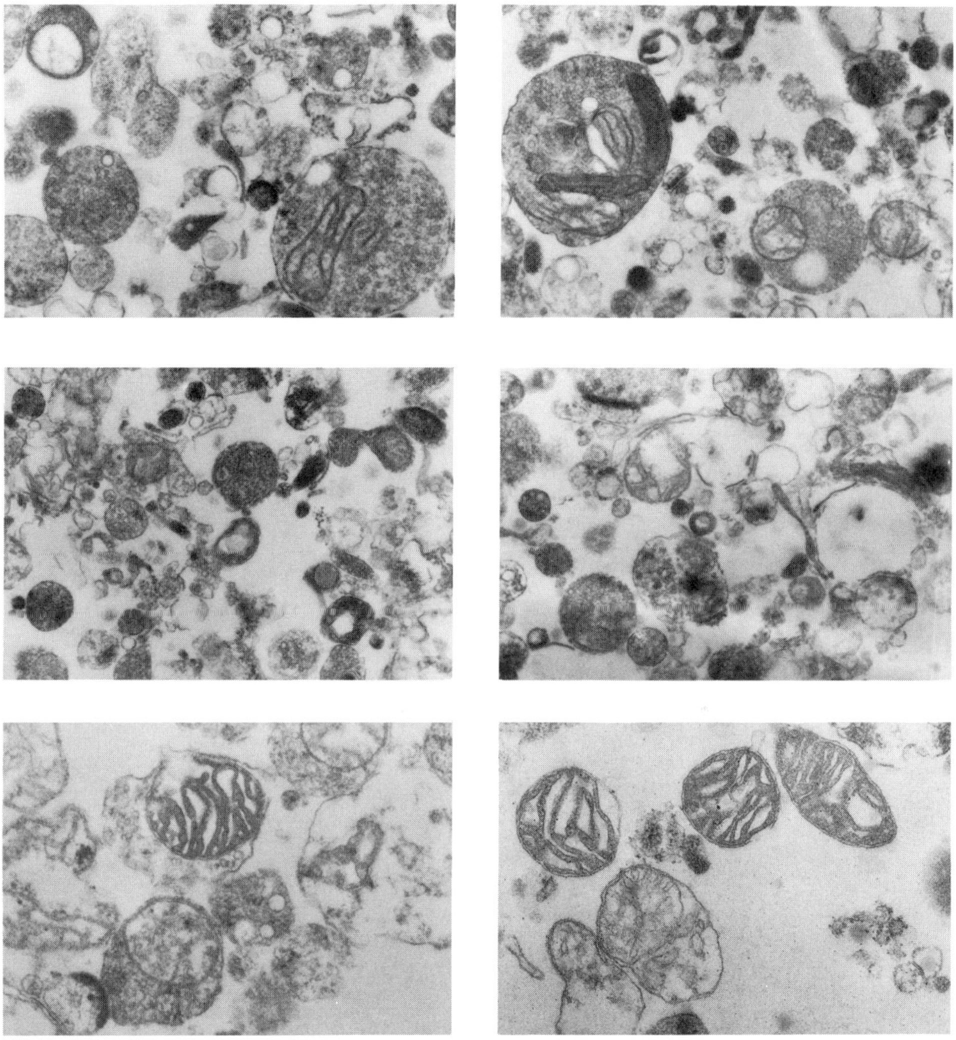

Figure 1. Electron micrographs of fractions from brains of control and taurine-depleted rats: Light P_2B fraction, **A:** control; **B:** taurine-depleted; Heavy P_2B fraction, **C:** control; **D:** taurine-depleted; Mitochondrial fraction, **E:** control; **F:** taurine-depleted. All micrographs have an original magnification of 33,800.

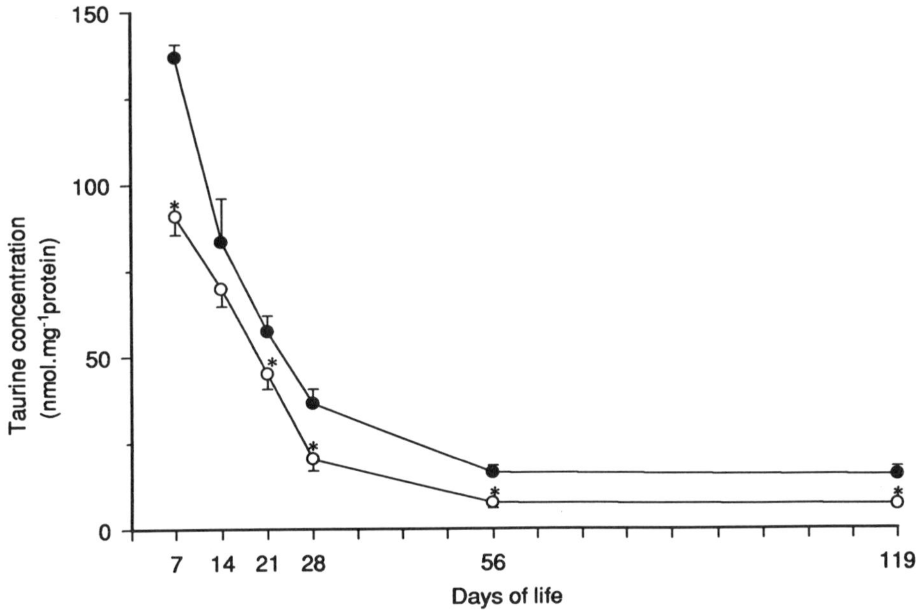

Figure 2. Taurine concentration in synaptosomes of developing rat brain. Filled circles: Control; Open circles: Synaptosomes from guanidinoethane sulfonate-exposed rats. Data are means ± SD for 4 preparations per point (3 for day 119). *p < .05 relative to control value.

developed in a solvent system of chloroform, methanol, acetic acid, formic acid and water (70:40:12:4:2 v/v) at 4°C for 1 h. After being dried, the TLC plates were further developed in a solvent system of hexane, isopropyl ether, and acetic acid (65:35:2 v/v) at room temperature for 40 min (14). TLC plates were visualised using iodine and the phospholipid bands assigned using standards (Sigma).

The phosphorus analysis was based on the method of Parker and Peterson (29). Phospholipid-containing areas were scratched from the plate and digested in concentrated sulfuric acid (0.1 ml) for 90 min at 250°C in tubes placed in heating blocks. Then the tubes were cooled and 50 μl of 30% H_2O_2 added. Samples were oxidized for 1 h at 155°C. Deionized phosphate-free water (2 ml) was added to the cooled tubes. A 1:1 mixture (0.8 ml) of a 10.1 mM ammonium molybdate solution and 280 mM fresh ascorbic acid solution was added to the samples (9), which were then heated in boiling water for 7 min. Tubes were cooled and centrifuged. The optical density of samples was read at 797 nm. Standards tubes contained known solution of potassium dihydrogen phosphate.

HPLC Analysis of Taurine and Methionine

Aliquots of P_2B fraction (0.4 ml) were deproteinized with 3.5% sulfosalicylic acid. Taurine and methionine were determined on separate analyses because of their large differences in concentration. Deproteinized samples were filtered and 20 μl aliquots were mixed with 80 μl of *o*-phthaldehyde solution (prepared by mixing 9 ml of solution of 343 mg of sodium tetraborate decahydrate with a solution of 54 mg OPA in 1 ml HPLC-grade methanol and 50 μl of mercaptoethanol). One minute after the addition of OPA, 20 μl of sample was injected onto an Altex ultrasphere ODS C18 reverse phase column (4.6 mm x 25 cm) equipped with a Beckman model 157 fluorescence detector. The column was developed with a mixture of two solvents. Solvent A consisted of 80% 40 mM sodium acetate (adjusted to pH 6.8 with acetic acid) and 20% methanol plus 1% tetrahydrofuran. Solvent B consisted of 20% 10 mM sodium acetate adjusted to pH 6.8 with acetic acid and 80% methanol. A mixture of 80% solvent A and 20% solvent B was run for 10 min, then a linear gradient from 20 to 50% B was run over the next 12 min, followed by a gradient from 50 to 100% B over the next 2 min. Taurine and methionine eluted at 19.4 and 29.2 min, respectively.

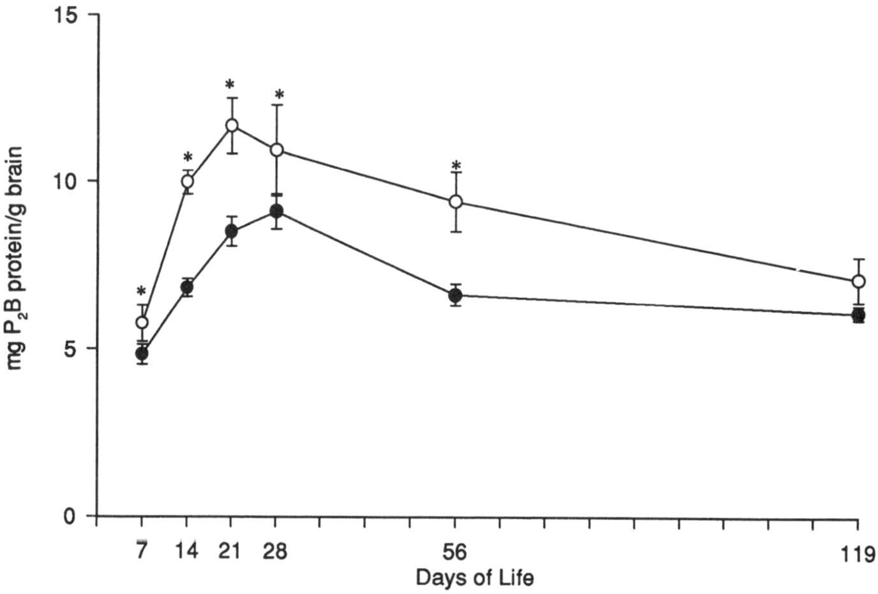

Figure 3. Synaptosomal density (mg protein/g wet weight of tissue) in developing brains of control and taurine-depleted rats. Symbols as for Fig. 2.

RESULTS AND DISCUSSION

Guanidinoethane Sulfonate Treatment of Rats

Pregnant rats were maintained on drinking water containing 1% guanidinoethane sulfonate for the final two weeks of pregnancy, and throughout lactation. The pups were weaned onto the same drinking water. This treatment had no effect on the number of stillbirths or on litter size. No significant effect was observed on body weight gain or on brain weight in the pups, although both parameters were slightly lower in the guanidinoethane sulfonate-exposed animals (data not shown).

Morphology and Composition of Synaptosomes from Taurine-Depleted Rats

Synaptosomes from taurine-depleted animals showed a continual fall in taurine concentration with development, as has been previously reported for rats (14). However, taurine levels were significantly lower than those in control animals at all time points examined, extending through the 119th day of life (Fig. 2). At this time, synaptosomes from animals exposed to guanidinoethane sulfonate had taurine levels only 46% of those found in control animals. No effect was seen on other amino acids.

Synaptosomal density (i.e. mg P_2B protein/g brain) was significantly higher in taurine-depleted animals at all timepoints examined except for 119 days (Fig. 3). This raised concerns as to how comparable the P_2B fractions were from the two groups of animals. We compared the fractions for three parameters: separation pattern on sucrose density gradients; succinate dehydrogenase activity; and electron microscopic appearance.

Table 2. Succinate dehydrogenase activity in brain subcellular fractions of 119-day-old control and taurine-depleted rats

Band	Major component	Succinate dehydrogenase activity (μmol/min/mg protein)			
		Control		Depleted	
I	Myelin	0.01 ± 0.01	(0.2%)	0 ± 0	(0%)
II	Synaptosomes (light)	0.49 ± 0.12	(8.3%)	0.47 ± 0.06	(8.4%)
III	Synaptosomes (heavy)	0.73 ± 0.01	(10.4%)	0.58 ± 0.10	(10.6%)
IV	Mitochondria	2.06 ± 0.18	(81.0%)	1.92 ± 0.15	(81.0%)

Enzyme activities are reported as means ± SD for three preparations per point. The percent of total activity contained in a given fraction is shown in parentheses.

No differences in band pattern were seen on either discontinuous or continuous sucrose density gradients. On the latter, P_2 (crude synaptosomal) preparations from both control and guanidinoethane sulfonate-exposed brains showed four bands, corresponding to myelin, light synaptosomal, heavy synaptosomal, and mitochondrial fractions. The median depths of these bands, measured from the top of the gradient, were 10 mm, 63 mm, 75 mm and 90 mm, respectively.

The specific activity of succinate dehydrogenase, a marker enzyme for mitochondria, did not differ significantly between control and taurine-depleted preparations in any of the bands (Table 1). Furthermore, the percent distribution of enzyme activity along the gradient was the same in the two preparations (Table 1). Most of the activity (81%) was found in the mitochondrial fraction. The remaining activity is distributed between the two synaptosomal layers, representing mitochondria present in nerve endings.

Finally, electron microscopy showed no difference between the preparations (Fig. 1). Regardless of the preparation, the light synaptosomal fraction contained larger synaptosomal fragments than the heavy fraction, with only a few free mitochondria being visible. Most of these are found in pinched-off nerve endings.

We conclude from this that the preparations are essentially comparable, and that the increased density of synaptosomal protein (Fig. 3) found in taurine-deficient brains is not due to a fundamental difference in the nature of the fraction being isolated.

Phospholipid Composition of Synaptosomes from Taurine-depleted and Control Brains

Phospholipid compositions are reported on Table 2 for days 7, 14, 21, 28, 56 and 119 of life. In view of the correlation we have found between taurine concentration and neutral phospholipid ratio (14,22,23), the most interesting effect seen in taurine-depleted

Table 2. Phospholipid concentrations in brain P_2B fractions from control and taurine-depleted rats during development

Days of life	Phosphatidylethanolamine	Phosphatidylserine	Phosphatidylinositol	Phosphatidylcholine	Sphingomyelin	Total phospholipids
Control						
7	164 ± 14	39.8 ± 6.3	19.4 ± 5.2	260 ± 31	13.4 ± 6.5	497 ± 45
14	206 ± 18	39.0 ± 9.1	22.3 ± 6.4	297 ± 32	8.3 ± 3.0	575 ± 55
21	181 ± 7	60.2 ± 3.7	22.6 ± 4.9	225 ± 11	14.3 ± 1.9	504 ± 23
28	216 ± 40	61.8 ± 14.9	26.2 ± 11.1	248 ± 46	24.4 ± 6.0	598 ± 90
56	122 ± 7	47.1 ± 3.5	22.5 ± 7.8	132 ± 6	21.2 ± 3.1	349 ± 18
119	100 ± 10	40.1 ± 4.0	12.1 ± 4.1	102 ± 10	12.8 ± 0.2	267 ± 23
Depleted						
7	156 ± 9	37.6 ± 13.6	22.0 ± 1.6	247 ± 15	17.8 ± 11.6	481 ± 22
14	186 ± 14	49.4 ± 14.3	30.4 ± 9.1	285 ± 17	17.4 ± 13.9	565 ± 17
21	176 ± 3	60.7 ± 3.9	23.4 ± 4.6	225 ± 10	17.5 ± 3.0	495 ± 21
28	197 ± 18	69.3 ± 7.5	23.0 ± 8.4	239 ± 26	23.7 ± 4.7	553 ± 43
56	94 ± 3	34.3 ± 1.6	9.1 ± 2.5	121 ± 6	14.4 ± 6.6	259 ± 30
119	88 ± 1	32.7 ± 2.6	21.9 ± 2.6	106 ± 2	24.4 ± 3.2	273 ± 3

Phospholipid concentrations are expressed in nmol phosphorus/mg protein. Values are means ± SD for 4 determinations per set.

synaptosomes is the phosphatidylethanolamine/phosphatidylcholine ratio during development (Fig. 4). Although no difference from control synaptosomes is seen during the developmental period, adult rats (days 56 of life onwards) exhibit a marked decrease in the ratio of these two neutral phospholipids in the synaptosomal fraction. The lowered synaptosomal taurine concentrations occurring in the developmental period, therefore, are not correlated with any change in neutral phospholipid ratio in this period.

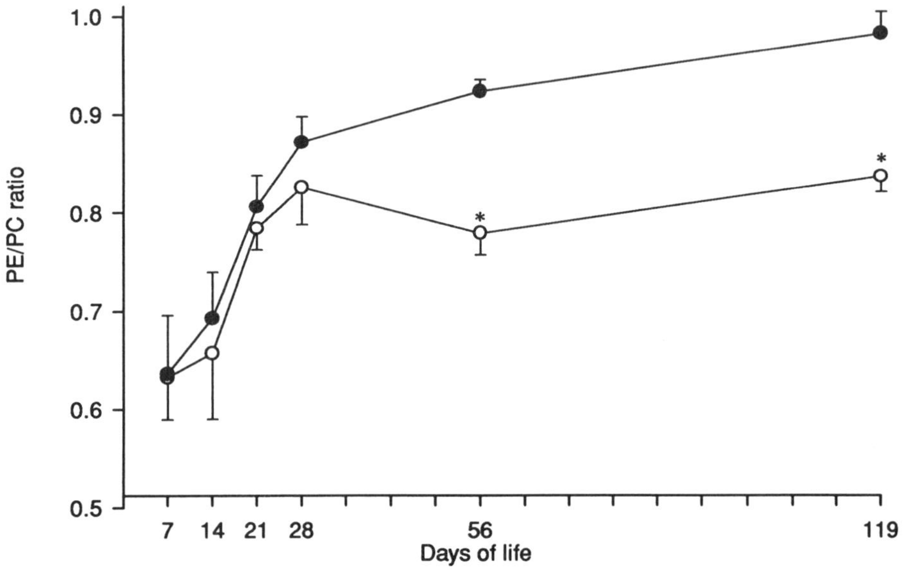

Figure 4. Neutral phospholipid ratio (phosphatidylethanolamine to phosphatidylcholine) in synaptosomal P_2B fraction of developing brain from control and taurine-depleted rats. Symbols as for Fig. 2.

Phospholipid Methylation of Synaptosomes from Taurine-depleted and Control Brains

In a quantitative sense, the major pathway of biosynthesis of phosphatidylcholine is the CDP choline route. However, phosphatidylcholine can also be biosynthesized directly from phosphatidylethanolamine by the phospholipid methylation route. These neutral phospholipids have opposing effects on membranes: phosphatidylcholine stabilizing lipid bilayers, and phosphatidylethanolamine destabilizing (33). The methylation route is being increasingly recognized as an important regulator of numerous membrane functions (3). These functions include Ca^{++} channel opening (25), altered Ca^{++} binding and Ca^{++} ATPase activities (27), inhibition of Na^{+} and Ca^{++} exchange in the heart (5,28), histamine release and various receptor activities (26). Thus, N-methylation unmasks cryptic ß-adrenergic receptors (32,34).

The phospholipid methylation route starts with the adenosylation of methionine to S-adenosylmethionine. The enzyme catalyzing this conversion is methionine adenosyltransferase (EC 2.5.1.6). The second enzyme is phospholipid methyltransferase I. This enzyme transfers a methyl group from S-adenosylmethionine to phosphatidylethanolamine, yielding monomethylphosphatidylethanolamine (4). Methyltransferase I activity is high in brain synaptosomes, where it is localized on the inner (cytoplasmic) aspect of the membrane. Further transfer of methyl groups from two additional molecules of S-adenosylmethionine yield phosphatidylcholine. The enzyme carrying out these methylations is phospholipid methyltransferase II. This enzyme is localized in synaptosomes on the outer aspect of the membrane.

Taurine concentration in the brain may be another regulator of phospholipid methylation. Schaffer et al. (8) have recently reported that in the isolated, perfused rat heart, taurine reverses the increase in phospholipid methylation induced by the perfusion of methionine (and also reverses the negative inotropy of methionine). These workers also found that 10 mM taurine inhibited the incorporation of [^{3}H]methyl groups into phospholipids on incubation of cardiac sarcolemmal membranes with 300 μM [^{3}H-methyl]methionine. Earlier, Cantafora et al. (1) had reported that the feeding of a 0.4% taurine-containing diet to guinea pigs led to decreased levels of phosphatidylcholine in liver membranes without change in phosphatidylethanolamine levels. In the brain, however, a positive correlation is observed between taurine concentration and phosphatidylcholine:phosphatidylethanolamine ratio (14).

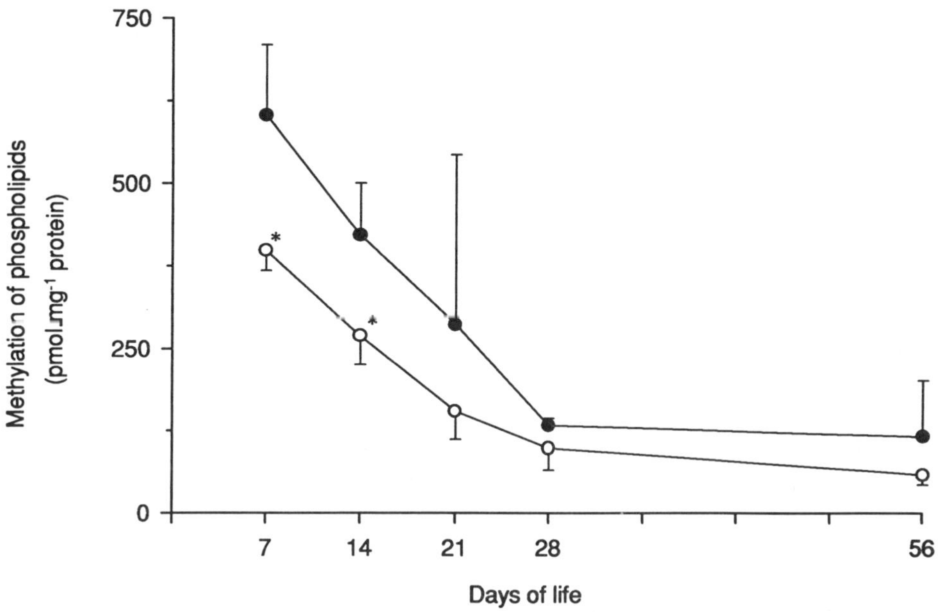

Figure 5. Incorporation of [^{3}H-methyl]methionine into synaptosomal phospholipids during brain development in control and taurine-depleted rats. Symbols as for Fig. 2.

Phospholipid methylation rates were determined by injection [^{3}H-methyl]methionine in vivo followed by isolation of synaptosomal P_2B fraction and determination of label incorporation into the phospholipid fraction. Although the levels of free methionine in synaptosomes fell with development, no difference was seen in methionine content between control and taurine-depleted preparations (data not shown). Phospholipid methylation rate fell with development (Fig. 5). At all time points examined, however, synaptosomes from taurine-depleted brains showed lower methylation rates than those from control brains. These methylation rates correlated well with taurine concentration (Fig. 6); i.e. the lower the taurine concentration, the lower the methylation rate. Methylation data from the taurine-depleted preparations fall on the same curve as those from the control preparations, strongly suggesting that taurine has a direct stimulatory effect on methylation. This is an opposite effect to that reported for taurine in the heart (8).

It appears, therefore, that, although taurine depletion has only an indirect effect on the neutral phospholipid ratio, it may have a direct effect on phospholipid methylation. In in vitro experiments, we have been unable to demonstrate an effect of taurine on methyl transferase activity. We postulate, therefore, that if taurine affects phospholipid methylation, it does so via an effect on the membrane environment rather than by a direct stimulatory action on the enzymes responsible for methylating phosphatidylethanolamine.

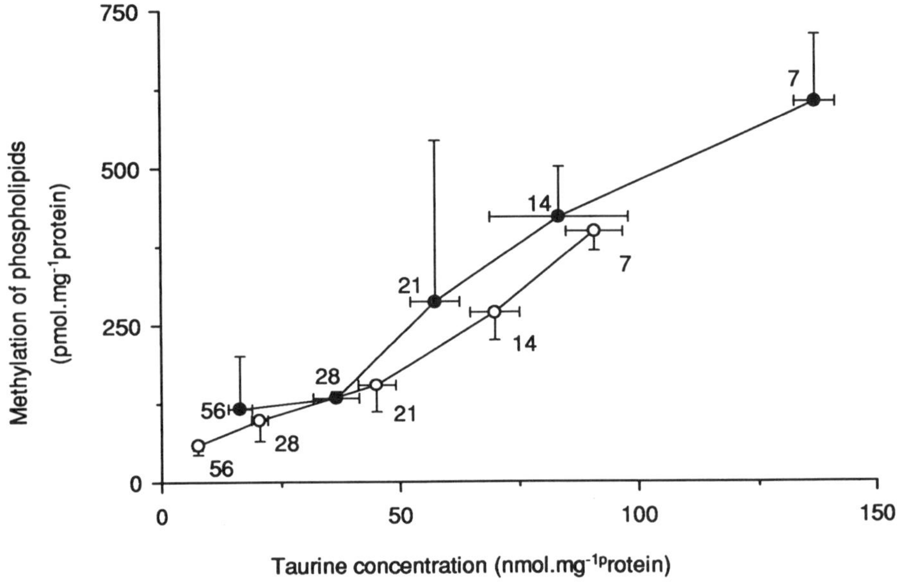

Figure 6. Relationship between phospholipid methylation and taurine concentration in synaptosomes from developing brain of control and taurine-depleted rats. Numbers by each point indicate the day of life. Symbols as for Fig. 2.

CONCLUSION

The taurine-depleted rat and cat (21) are quite different models. The former involves depletion achieved by means of the taurine transport antagonist, guanidinoethane sulfonate. In the cat, depletion is achieved by dietary deprivation. In the rat, there is a deficiency of taurine throughout development. Taurine depletion in the cat, on the other hand, was initiated after the developmental period. These models concur, however, in showing that artificially induced taurine depletion can lead to marked disturbances in membrane phospholipid composition. This composition is a major and fundamental characteristic determining the biological properties of the membrane.

An enantiostatic function has been proposed for taurine (13). Enantiostasis implies that taurine helps maintain constancy of function in a changing biochemical system. In brain synaptosomes, an induced permanent reduction in taurine content has resulted in a reciprocal permanent change in membrane composition and, in the case of the rat, phospholipid methylation rates. We can conceive that the net result of such reciprocal changes under physiological conditions helps minimize disturbances in synaptosomal function. This, in turn, reflects an enantiostatic function of taurine.

REFERENCES

1. Cantafora, A., Mantovani, A., Masella, R., Mechelli, L., and Alvaro, D. 1986, Effect of taurine administration on liver lipids in guinea pigs, *Experientia*, 42:407-408.
2. Christiansen, K. 1975, Lipid extraction procedure for in vitro studies of glyceride synthesis with labeled fatty acid, *Anal.Biochem.*, 66:93-99.
3. Crews, F. 1985, Phospholipid methylation and membrane function, *in:* "Phospholipids and Cellular Regulation, Vol. I", Kuo, K.F., ed., CRC Press, Boca Raton, Florida, pp. 131-158.
4. Crews, F.T., Hirata, F., and Axelrod, J. 1980, Identification and properties of methyltransferases that synthesize phosphatidylcholine in rat brain synaptosomes, *J.Neurochem.*, 34:1491-1498.
5. Dyer, J.R. and Greenwood, C.E. 1988, Evidence for altered methionine methyl-group utilization in the diabetic rat's brain, *Neurochem.Res.*, 13:517-523.
6. Gray, E.G. and Whittaker, V.P. 1962, The isolation of nerve endings from brain: an electron microscopic study of cell fragments derived by homogenization and centrifugation, *J.Anat.*, 96:79-88.
7. Green, D.E., Mii, S., and Kohout, P.M. 1955, Studies on the terminal electron transport system: I. Succinate dehydrogenase, *JBC*, 217:551-567.
8. Hamaguchi, T., Azuma, J., and Schaffer, S. 1991, Interaction of taurine with methionine: Inhibition of myocardial phospholipid methyltransferase, *J.Cardiovasc.Pharmacol.*, 18:224-230.
9. Huxtable, R. and Bressler, R. 1973, Determination of orthophosphate, *Anal.Biochem.*, 57:604-608.
10. Huxtable, R.J. 1976, Metabolism and function of taurine in the heart, *in:* "Taurine", Huxtable, R.J. and Barbeau, A., eds., Raven Press, New York, pp. 99-119.
11. Huxtable, R.J. 1987, From heart to hypothesis: a mechanism for the calcium modulatory actions of taurine, *in:* "The Biology of Taurine: Methods and Mechanisms", Huxtable, R.J., Franconi, F. and Giotti, A., eds., Plenum Press, New York, pp. 371-388.
12. Huxtable, R.J. 1990, The interaction between taurine, calcium and phospholipids: Further investigations of a trinitarian hypothesis, *in:* "Taurine: Functional Neurochemistry, Physiology, and Cardiology", Pasantes-Morales, H., Martin, D.L., Shain, W. and del Río, R.M., eds., Wiley-Liss, New York, pp. 185-196.
13. Huxtable, R.J. 1992, Physiological actions of taurine, *Physiol.Rev.*, 72:101-163.
14. Huxtable, R.J., Crosswell, S., and Parker, D. 1989, Phospholipid composition and taurine content of synaptosomes in developing rat brain, *Neurochem.International*, 15:233-238.
15. Huxtable, R.J., Laird, H.E., and Lippincott, S.E. 1979, The transport of taurine in the heart and the rapid depletion of tissue taurine content by guanidinoethyl sulfonate, *Journal of Pharmacology and Experimental Therapeutics*, 211:465-471.
16. Huxtable, R.J. and Peterson, A. 1988, The effect of taurine on calcium binding to brain synaptosomes, *Pharmacologist*, 30:A86.(Abstract)
17. Huxtable, R.J. and Peterson, A. 1989, Altered calcium binding induced by taurine in brain synaptosomal membranes, *J.Neurochem.*, 52 (Supp.):S161 B.(Abstract)
18. Huxtable, R.J. and Sebring, L. 1987, Modulation by taurine of calcium binding to phospholipid vesicles and cardiac sarcolemma, *Proc.Western Pharmacol.Soc.*, 30:153-155.
19. Huxtable, R.J. and Sebring, L.A. 1986, Towards a unifying theory for the action of taurine, *Trends Pharmacol.Sci.*, 7:481-485.
20. Huxtable, R.J. and Sebring, L.A. 1989, Taurine and the heart: the phospholipid connection, *in:* "Taurine and the Heart", Iwata, H., Lombardini, J.B. and Segawa, T., eds., Kluwer Academic Press, Boston, pp. 31-42.
21. Lleu, P.-L., Bowers, R., Gargano, A., Messing, J.M., Sturman, J., and Huxtable, R.J. 1994, Taurine depletion and synaptosomal phospholipid content in cat brain, *in:* "Taurine in Health and Disease", Huxtable, R.J. and Michalk, D., eds., Plenum Press, New York, pp. in press.
22. Lleu, P.-L., Croswell, S., and Huxtable, R.J. 1992, Phospholipids, phospholipid methylation and taurine content in synaptosomes of developing rat brain, *in:* "Taurine: Nutritional Value and Mechanisms of Action", Lombardini, J.B., Schaffer, S.W. and Azuma, J., eds., Plenum Publishing, New York, pp. 221-228.
23. Lleu, P.-L. and Huxtable, R.J. 1992, Phospholipid methylation and taurine content of synaptosomes from cerebral cortex of developing rat, *Neurochem.International*, 21:109-118.

24. Lowry, O.J., Rosebrough, N.J., Farr, A.L., and Randall, R.J. 1951, Protein measurement with folin phenol reagent, *J.Biol.Chem.*, 193:265-275.
25. McGivney, A., Crews, F.T., Hirata, F., Axelrod, J., and Siraganian, R.R. 1981, Rat basophilic leukemia cells defective in phospholipid methyltransferase enzyme, Ca^{2+} influx and histamine release: Reconstruction by hybridization, *Proc.Natl.Acad.Sci.USA*, 78:6176-6180.
26. Okumura, K., Panagia, V., Beamish, R.E., and Dhalla, N.S. 1987, Biphasic changes in the sarcolemmal phosphatidylethanolamine N-methylation activity in catecholamine-induced cardiomyopathy, *J.Mol.Cell.Cardiol.*, 19:356-366.
27. Panagia, V., Elimban, V., Ganguly, P.K., and Dhalla, N.S. 1987, Decreased Ca2+ binding and Ca2+ ATPase activities in heart sarcolemma upon phospholipid methylation, *Mol.Cell.Biochem.*, 78:65-71.
28. Panagia, V., Makino, N., Ganguly, P.K., and Dhalla, N.S. 1987, Inhibition of Na^{+}-Ca^{2+} exchange in heart sarcolemmal vesicles by phosphatidylethanolamineN-methylation, *Eur.J.Biochem.*, 166:597-603.
29. Parker, F. and Peterson, N.F. 1965, Quantitative analysis of phospholipids and phospholipid fatty acids from silica gel thin layer chromatograms, *J.Lipid Res.*, 6:455-460.
30. Sebring, L.A. and Huxtable, R.J. 1986, Low affinity binding of taurine to phospholiposomes and cardiac sarcolemma, *Biochim.Biophys.Acta*, 884:559-566.
31. Skipski, V.P., Peterson, R.F., and Barclay, M. 1964, Quantitative analysis of phospholipids by thin-layer chromatography, *Biochem.J.*, 90:374-378.
32. Strittmatter, W.J., Hirata, F., and Axelrod, J. 1979, Phospholipid methylation unmasks cryptic β-adrenergic receptors in rat reticulocytes, *Science*, 204:1207-1209.
33. Sun, G.Y. and Sun, A.Y. 1985, Ethanol and membrane lipids, *Alcohol Clin.Exp.Res.*, 9:164-180.
34. Tallman, J.F.J., Henneberry, R.C., Hirata, F., and Axelrod, J. 1979, Control of β-adrenergic receptors in Hela cells, *in:* "Catecholamines: Basic and Clinical Frontiers, Vol. I", Usdin, E., Kopin, I.J. and Barchas, J., eds., Pergamon Press, pp. 489-491.
35. Whittaker, V.P. and Barker, L.A. 1972, The subcellular fractionation of brain tissue with special reference to the preparation of synaptosomes and their component organelles, *in:* "Methods in Neurochemistry, Volume II", Fried, R., ed., Marcel Dekker, NY, pp. 1-52.

TAURINE DEPLETION AND SYNAPTOSOMAL PHOSPHOLIPID CONTENT IN CAT BRAIN

P.-L. Lleu[1], R.J. Bowers[1], A. Gargano[2], J. Sturman[2] and R.J. Huxtable[1]

[1]Department of Pharmacology
College of Medicine, University of Arizona
Tucson, Arizona 85724
[2]Department of Developmental Biochemistry
Institute for Basic Research in Developmental Disabilities
1050 Forest Hill Road
Staten Island, NY 10314

INTRODUCTION

Unlike the rat, the cat is dependent upon a dietary source of taurine. Cats maintained on taurine-deficient diets show profound and rapid falls in plasma taurine concentrations, followed by slower falls in tissue concentrations. Taurine deficiency in cats is associated with a large number of problems, including retinal degeneration and blindness (1), dilated cardiomyopathy (9), and reproductive (11) and developmental disorders (10,12).

In view of the parallels found between taurine content and phospholipid composition and biochemistry (2,3,6,7) we have examined whether the profound effects of taurine deprivation in the cat are associated with changes in the phospholipid composition of brain synaptosomes.

MATERIALS AND METHODS

The taurine-deficient adult cats used in this study had been maintained on a taurine-deficient diet for at least 2 years. The taurine-supplemented and taurine-deficient diets have been described previously, along with the conditions under which the cats were maintained (11).

The synaptosomal preparation and the isolation, separation and quantitation of phospholipids are reported elsewhere in this volume (5).

Taurine in Health and Disease, Edited by R. Huxtable and D.V. Michalk, Plenum Press, New York, 1994

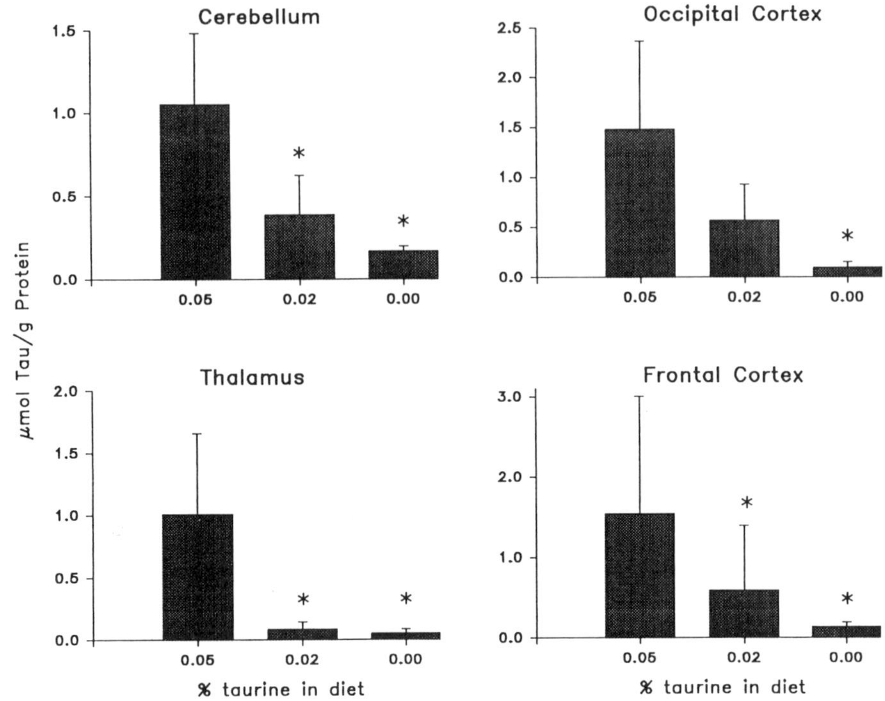

Figure 1. Synaptosomal (P_2B) taurine concentrations in cats maintained for at least two years on a normal taurine diet (0.05% taurine), a taurine-deficient diet (0.02% taurine), or a taurine-free diet. * $p < .05$ relative to 0.05% diet.

RESULTS

Synaptosomal (P_2B) fractions were prepared from four regions of the cat brain, the cerebellum, occipital cortex, thalamus and frontal cortex. Taurine concentrations were much lower than in P_2B fractions of the rat brain (4,8). Compared to brains from cats maintained on a diet containing 0.05% taurine, P_2B fractions from animals that had received diets containing 0.02% and 0.0% taurine showed lower taurine concentrations (Fig. 1).

A higher yield of P_2B protein (mg protein/g brain tissue) was obtained from the occipital cortex and thalamus of cats maintained on a diet containing 0.02% taurine, as compared to those on a 0.05% diet (Fig. 2). However, these increases did not persist in P_2B fractions from animals maintained on a completely taurine-free diet.

The phospholipid content of P_2B from the four regions studied are reported on Table 1. Compared to the control (0.05% taurine) group, the 0.02% taurine group showed a number of changes that did not persist in the 0% taurine group. The total phospholipid content (per mg synaptosomal protein) is significantly reduced in P_2B fractions prepared from all areas except the frontal cortex in animals maintained on the 0.02% diet as compared to the 0.05% diet (Fig. 3). No difference is observed, however, in total synaptosomal phospholipids between the 0.05% and taurine-free diets. Among the individual classes of phospholipids, in synaptosomes from the 0.02% taurine group phosphatidylethanolamine and phosphatidylcholine are significantly lower in all areas except the frontal cortex. These differences do not persist, however, in animals maintained on a 0% taurine diet.

The neutral phospholipid ratios (phosphatidylethanolamine to phosphatidylcholine) are shown on Fig. 4. In thalamic and frontal cortical synaptosomes from cats on the 0.02% taurine diet, the ratio is significantly higher. This increase persists in the frontal cortex of

Table 1. Phospholipid composition (nMol phosphorus/mg protein) of synaptosomal P_2B fractions from cats maintained on diets of various taurine content.

Dietary Taurine (%)	Phosphatidylethanolamine	Phosphatidylserine	Phosphatidylinositol	Phosphatidylcholine	sphingomyelin	Total Phospholipid
Cerebellum						
0.05	240 ± 47	65 ± 9	30 ± 9	262 ± 46	32 ± 15	629 ± 125
0.02	161 ± 16[a]	57 ± 12	35 ± 25	216 ± 8[a]	22 ± 8	492 ± 43[a]
0.00	215 ± 51	70 ± 20	25 ± 11	258 ± 69	39 ± 5[b]	607 ± 150
Occipital lobe						
0.05	227 ± 10	47 ± 5	19 ± 3	228 ± 16	36 ± 13	556 ± 25
0.02	169 ± 34[a]	55 ± 13	15 ± 2	159 ± 28[a]	26 ± 11	424 ± 82[a]
0.00	225 ± 54	52 ± 11	17 ± 2	214 ± 48	19 ± 14	542 ± 83
Thalamus						
0.05	252 ± 28	66 ± 9	29 ± 4	249 ± 29	40 ± 8	637 ± 57
0.02	185 ± 41[a]	57 ± 14	24 ± 8	172 ± 39[a]	37 ± 12	475 ± 101[a]
0.00	274 ± 19[b]	64 ± 21	31 ± 3	244 ± 38	28 ± 4	640 ± 81
Frontal cortex						
0.05	229 ± 22	50 ± 16	20 ± 3	225 ± 15	40 ± 16	564 ± 44
0.02	205 ± 41	62 ± 13	16 ± 4	181 ± 34	37 ± 9	501 ± 89
0.00	227 ± 37	60 ± 12	7 ± 4[a,b]	197 ± 25	26 ± 5	540 ± 118

Data are means ± SD for four samples per point. [a] $p < .05$ compared to 0.05% taurine group; [b] $p < .05$ compared to 0.02% taurine group.

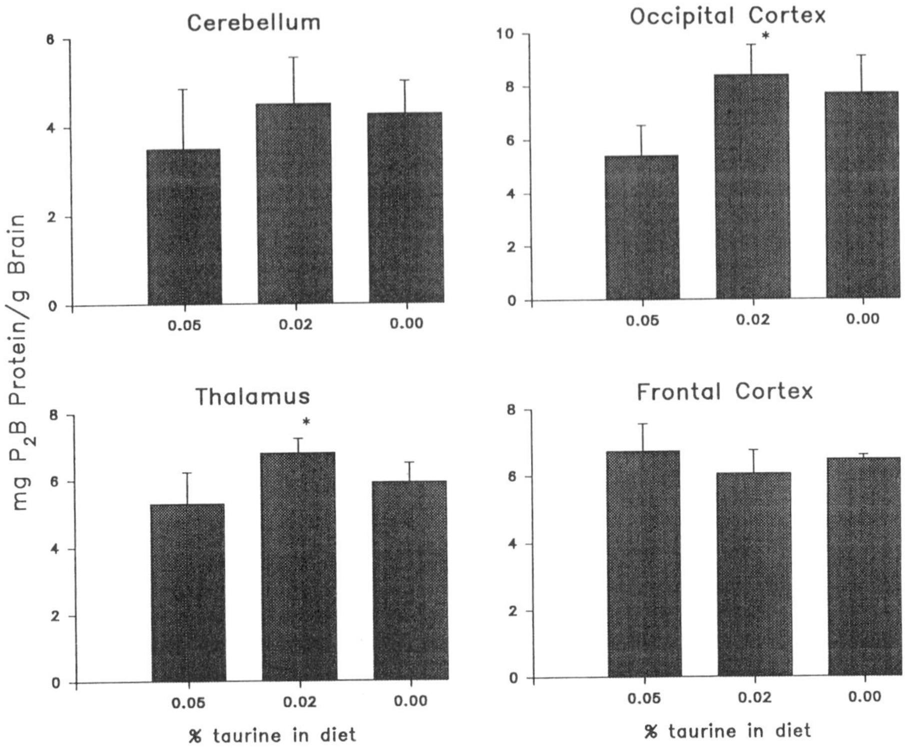

Figure 2. Synaptosomal (P_2B) protein density (mg protein/g wet weight of brain tissue) in cats maintained on diets containing 0.05%, 0.02% or 0% taurine. * $p < .05$ relative to 0.05% diet.

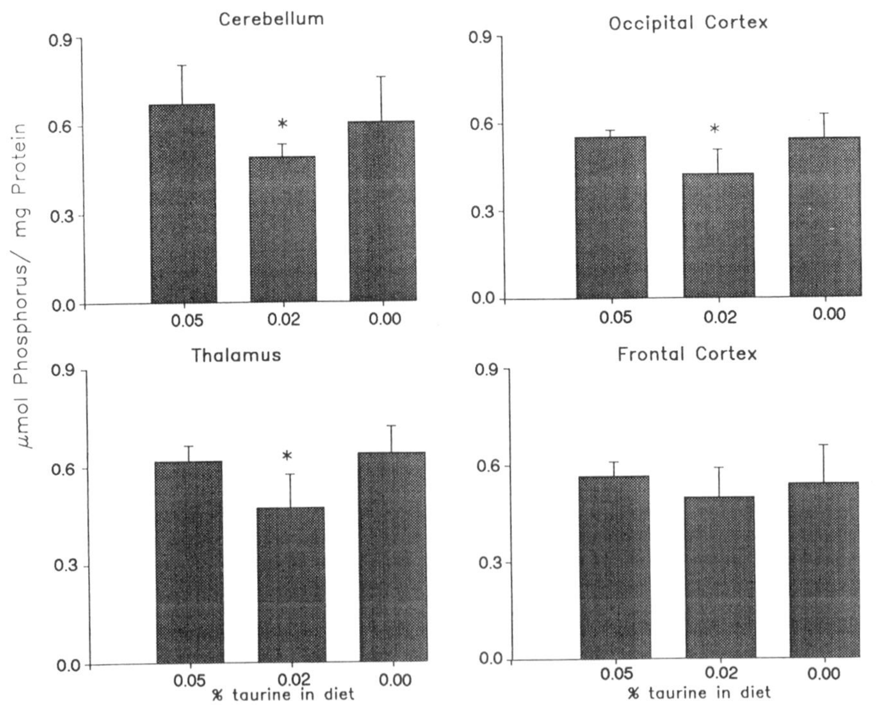

Figure 3. Total synaptosomal (P_2B) phospholipids in cats maintained on diets containing 0.05%, 0.02% or 0% taurine. * $p < .05$ relative to 0.05% diet.

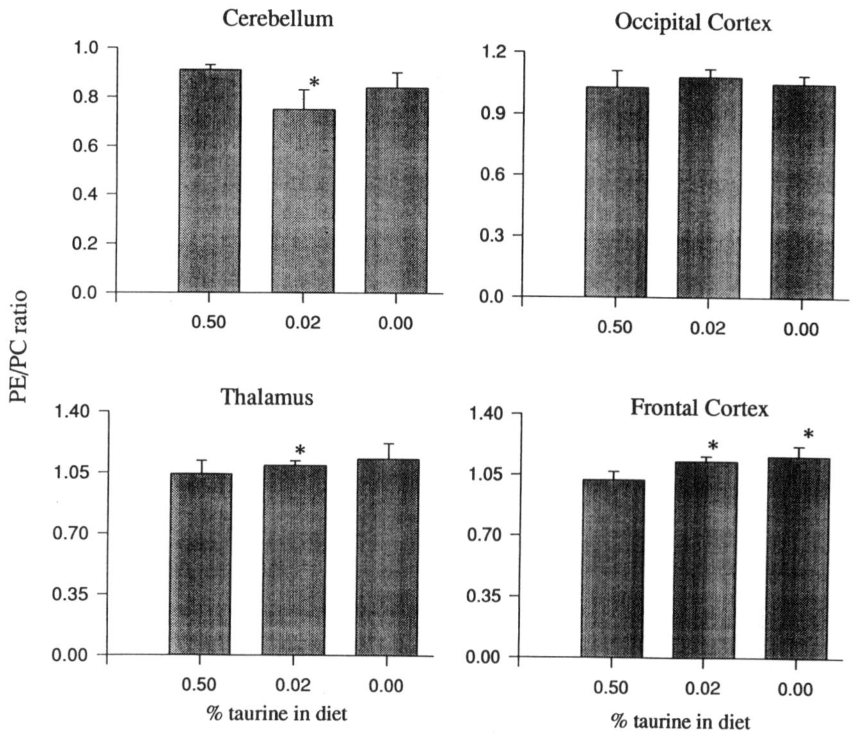

Figure 4. The neutral phospholipid ratio (phosphatidylethanolamine to phosphatidylcholine) in synaptosomal (P_2B) fractions of cats maintained on diets containing 0.05%, 0.02% or 0% taurine. * $p < .05$ relative to 0.05% diet

animals on the taurine-free diet. In the thalamus of these animals, although the ratio remains elevated, the increased standard deviation leads to the difference escaping significance. In the cerebellum, however, the neutral phospholipid ratio is significantly decreased.

DISCUSSION

The relationship between taurine and synaptosomal phospholipid content has been examined and compared in cats maintained on three dietary taurine regimens: a normal taurine diet (0.05% taurine by weight), a taurine-deficient diet (0.02% taurine), and a taurine-deprived diet.

The most marked differences in phospholipid composition were observed in the deficient group. The changes found in this group are comparable to those seen in brain synaptosomes of rats made taurine-deficient by treatment with guanidinoethane sulfonate (5). The neutral phospholipids are decreased in all four brain areas examined. The relative decreases lead to a significant decrease in the ratio of phosphatidylethanolamine to phosphatidylcholine in the cerebellum [comparable to the changes seen in brain synaptosomes of rats maintained throughout development on the taurine-depleting agent, guanidinoethane sulfonate (5)] and a significant increase in the ratio in another two areas.

These findings indicate that taurine deficiency induced in the adult cat is associated with changes in the biochemical or biophysical properties of the synaptosomal fraction. The synaptosomal P_2B fraction is enriched in pinched-off nerve endings. The differences we observe could be due either to an alteration in the type of cell fragments and organelles being isolated in the P_2B fraction or to an alteration in the biochemical characteristics of the nerve endings. Electron microscopic examination of brain P_2B fractions from control and taurine deficient rats failed to show recognizable differences in the contents of the fractions (5). However, whatever the reason for the changes, they point to taurine deficiency having profound effects on brain biochemistry. The probability is that these effects correlate with functional changes.

REFERENCES

1. Hayes, K.C., Carey, R.E., and Schmidt, S.Y. 1975, Retinal degeneration associated with taurine deficiency in the cat, *Science*, 188:949-951.
2. Huxtable, R.J. 1987, From heart to hypothesis: a mechanism for the calcium modulatory actions of taurine, *in:* "The Biology of Taurine: Methods and Mechanisms", Huxtable, R.J., Franconi, F. and Giotti, A., eds., Plenum Press, New York, pp. 371-388.
3. Huxtable, R.J. 1990, The interaction between taurine, calcium and phospholipids: Further investigations of a trinitarian hypothesis, *in:* "Taurine: Functional Neurochemistry, Physiology, and Cardiology", Pasantes-Morales, H., Martin, D.L., Shain, W. and del Río, R.M., eds., Wiley-Liss, New York, pp. 185-196.
4. Huxtable, R.J., Crosswell, S., and Parker, D. 1989, Phospholipid composition and taurine content of synaptosomes in developing rat brain, *Neurochem.International*, 15:233-238.
5. Huxtable, R.J., Murphy, J., and Lleu, P.-L. 1994, Developmental effects of taurine depletion on synaptosomal phospholipids in the rat, *in:* "Taurine in Health and Disease", Huxtable, R.J. and Michalk, D., eds., Plenum Press, New York, this volume.
6. Huxtable, R.J. and Sebring, L.A. 1986, Towards a unifying theory for the action of taurine, *Trends Pharmacol.Sci.*, 7:481-485.
7. Huxtable, R.J. and Sebring, L.A. 1989, Taurine and the heart: the phospholipid connection, *in:* "Taurine and the Heart", Iwata, H., Lombardini, J.B. and Segawa, T., eds., Kluwer Academic Press, Boston, pp. 31-42.
8. Lleu, P.-L. and Huxtable, R.J. 1992, Phospholipid methylation and taurine content of synaptosomes from cerebral cortex of developing rat, *Neurochem.International*, 21:109-118.
9. Pion, P.D., Kittleson, M.D., Rogers, Q.R., and Morris, J.G. 1987, Myocardial failure in cats associated with low plasma taurine: a reversible cardiomyopathy, *Science*, 237:764-768.
10. Sturman, J.A., Gargano, A.D., Messing, J.M., and Imaki, H. 1986, Feline maternal taurine deficiency: effect on mother and offspring, *J.Nutr.*, 116:655-667.
11. Sturman, J.A. and Messing, J.M. 1991, Dietary taurine content and feline reproduction and outcome, *J.Nutr.*, 121:1195-1203.
12. Sturman, J.A., Moretz, R.C., French, J.H., and Wisniewski, H.M. 1985, Postnatal taurine deficiency in the kitten results in a persistence of the cerebellar external granule cell layer: correction by taurine feeding, *J.Neurosci.Res.*, 13:521-528.

CELL VOLUME REGULATION IN TAURINE DEFICIENT CULTURED ASTROCYTES

Julio Morán, Thomas Maar and
Herminia Pasantes-Morales

Institute of Cell Physiology
National University of Mexico
Mexico City, Mexico

INTRODUCTION

A large variety of cells are able to regulate their cell volume under anisosmotic conditions. Cell volume regulation involves the activation of fluxes of osmotically active solutes, which have been classified as inorganic ions, Na^+, K^+ and Cl^-, and organic molecules, such as polyols and amino acids. In the nervous system, taurine seems to be the most important organic osmolyte. Cultured astrocytes and neurons regulate cell volume under hyposmotic conditions (6,14). The induced release of taurine parallels this process, suggesting that taurine is involved in the regulatory mechanism (15,18). Moreover, substantial evidence supports a similar role for taurine in brain in vivo. Brain taurine levels decrease subsequent to plasma hyposmolarity during experimental chronic hyponatremia (22). Correspondingly, increases are observed in extracellular taurine concentrations, assessed by dialysis in brain, in response to a number of conditions that induce neural cell swelling such as seizure activity, excitotoxic agents and intracellular acidosis (1,8,9). This association of taurine and cell swelling has been directly demonstrated by showing a significant release of taurine after a direct reduction in the extracellular osmolarity (7). Numerous observations in systems in vitro, including brain slices, synaptosomes, isolated retina and cultured neurons and astrocytes, also document the increased efflux of taurine associated with cell swelling, suggesting its involvement in the processes of cell volume adjustment in nervous tissue (2,12,15,17,18).

Taurine deficiency has deleterious consequences in excitable tissues of adult and developing animals. Tissue taurine depletion leads to impaired myocardial function, changes in the immune system, and problems with pregnancy and outcome (16,19,20). In nervous tissue, taurine deficiency causes a marked delay in the maturation and migration of cortical neurons and cerebellar granule cells (21). In adult animals, a reduction in taurine content leads to a severe dysfunction of retinal photoreceptors (3). The mechanisms subserving all these effects of taurine deficiency are still unknown. Swelling in cells and intracellular

Taurine in Health and Disease, Edited by R. Huxtable
and D.V. Michalk, Plenum Press, New York, 1994

structures of retina (3) and brain (13) of taurine-deficient animals has been observed, suggesting that taurine deficiency may be associated with a disability in the mechanisms of cell volume control and this in turn, may alter a number of cellular functions.

In the present work we examined a possible link between taurine deficiency and cell swelling in taurine-deficient cultured cerebellar astrocytes. Reduction in taurine levels was obtained either by chronic treatment with a blocker of taurine transport, guanidinoethane sulfonate, or by growing cells in a chemically defined medium essentially free of taurine. We have previously shown that this treatment effectively reduces taurine levels in mouse cultured astrocytes and neurons (11).

TAURINE-DEFICIENT ASTROCYTES

Astrocytes treated for 2 days with 2 mM guanidinoethane sulfonate showed a marked decrease in taurine concentration (11). The taurine content of astrocytes decreased about 65%, from 210 to 74 nmol/mg protein (Fig. 1). The concentrations of the main components

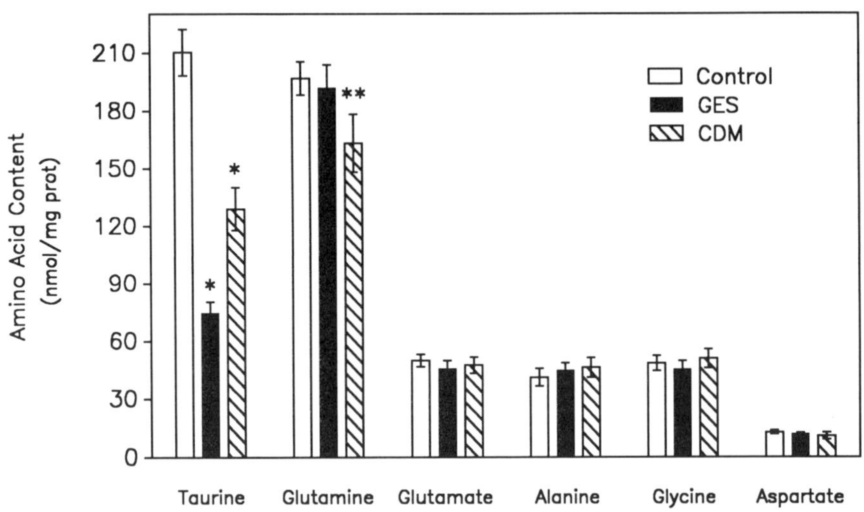

Figure 1. Effect of guanidinoethane sulfonate and a chemically-defined medium on free amino acid content of cultured astrocytes. Rat cerebellar astrocytes were cultured in a basal medium containing 10% fetal calf serum. After two weeks, 2 mM guanidinoethane sulfonate was added to the culture or medium was totally replaced by a chemically-defined medium without taurine. After 48 h of treatment, amino acid content was measured by HPLC. Results are means $\pm$ SEM of 12-18 culture dishes. Control corresponds to untreated cells. Asterisk indicates statistically significant difference from the control value: $*P < 0.001$, $**P < 0.01$.

of the free amino acid pool (aspartate, glutamate, glutamine, glycine, ß-alanine and alanine), as previously reported, did not change significantly (Fig. 1) (11). Astrocytes grown for 48 h in a chemically defined medium with no fetal serum, and, therefore, essentially free of taurine, also showed a substantial reduction in taurine concentration from 210 to

Table 1. Effect of guanidinoethane sulfonate and a chemically-defined medium on the cell volume of cerebellar astrocytes under isosmotic conditions

	Cell Volume (pl)	Swelling (%)	Significance (n)
Control	1.12 ± 0.02	0	
guanidinoethane sulfonate	1.30 ± 0.04	16	$p < 0.001$ (14)
chemically defined medium	1.22 ± 0.02	10	$p < 0.01$ (4)

Taurine-deficient cells were obtained by exposure to guanidinoethane sulfonate (2 mM) or a chemically defined medium. Cell volume was measured in an isosmotic medium as described in Figure 1. Cell volume values represent the absolute cell volume in picoliters (pl) and are mean ± SEM of the number of experiments in parenthesis. Swelling is expresed as the percent of increase in cell volume over control.

129 nmol/mg of protein. A decrease in glutamine levels of 19% was also observed (Fig. 1). Addition of 0.2 mM taurine to the chemically defined medium restored the cell levels of taurine, but not glutamine (not shown). Both, the morphology and viability of astrocytes cultured with guanidinoethane sulfonate or chemically defined medium remained unaffected. Neither of the two experimental procedures used in the present study totally depleted cells of intracellular taurine.

The reduction in taurine content in rat astrocytes was lower than that found previously for mice astrocytes. The taurine pool in astrocytes in culture is the result of a steady state between an active accumulation process, from taurine present in the fetal serum that supplements the culture medium, and a continuous leakage. When taurine uptake is effectively blocked by guanidinoethane sulfonate, the efflux of taurine continues, leading to a progressive decrease of the intracellular pool so that after 48 h only about 35% of the original taurine content remains in the cell. This remaining pool could not be depleted, even after 14 days of treatment with guanidinoethane sulfonate. It is possible that this pool originates from biosynthesis or represents a sequestered pool not participating in the influx-efflux process above described. It is not unlikely that at least a fraction of this pool corresponds to taurine bound to membranes or to intracellular organelles. The taurine content of cells grown in a chemically defined medium was higher than that of cells treated with guanidinoethane sulfonate. This difference may be due to reaccumulation by the carrier of taurine released by astrocytes through the leakage efflux.

TAURINE DEFICIENCY AFFECTS CELL VOLUME

A reduction in taurine concentration in cultured astrocytes had clear consequences in the handling of cell volume. The mean volume of cells treated with guanidinoethane sulfonate was 1.30 ± 0.04 pl, which is significantly higher (16%) than the volume of controls (1.12 ± 0.02 pl) (Table 1). Cells cultured in chemically defined medium also showed a 10% increase in volume (1.22 ± 0.02 pl) as compared to controls (Table 1). Although there is no obvious explanation for the difference found in cell volume in taurine-deficient cells, it might be attributed to a lower stability of the cell membrane in such cells. It is well known that taurine exerts membrane stabilizing effects, observed in various preparations and conditions (5). This stabilizing effect of taurine has been attributed to an interaction with membrane phospholipids (5). However, if that is the mechanism by which taurine-deficient cells are swollen, one must assume that the stabilizing effect of taurine is taking place only intracellularly because addition of 0.2 mM taurine to this preparation did not prevent the increase in basal cell volume (not shown).

VOLUME REGULATION IS IMPAIRED IN TAURINE-DEFICIENT CELLS

The effect of taurine reduction caused either by guanidinoethane sulfonate or by chemically defined medium on the ability of astrocytes to recover volume after exposure to hyposmotic conditions was examined. Astrocytes containing normal taurine concentrations exhibited the typical time course of change in cell volume after exposure to hyposmotic medium (150 mOsm). Astrocytes first swell, increasing volume by more than 50% (1.53 relative volume) within 1 min. Then the volume regulatory phase starts, leading to a cell volume decrease to a relative volume of about 1.21, within 16 min. Therefore, during this time, astrocytes exhibited a regulatory response of about 60% (Fig. 2).

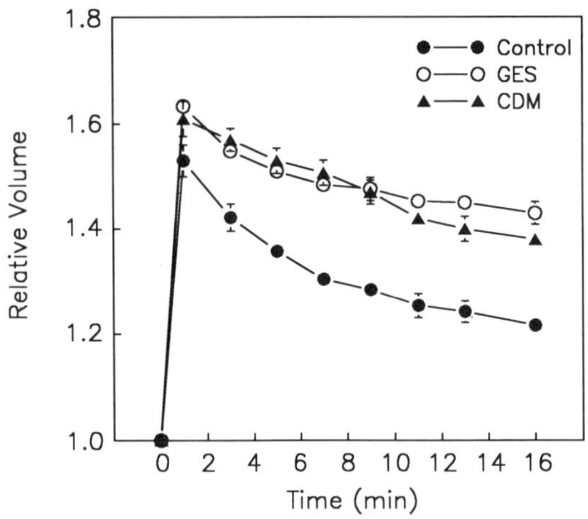

Figure 2. Changes in cell volume in taurine-deficient cerebellar astrocytes induced by a 50% hyposmotic medium. Astrocytes were cultured 2 weeks and treated the last 48 h with guanidinoethane sulfonate (GES) (2 mM) or chemically defined medium (CDM). Cells were detached by enzymatic treatment, exposed to isosmotic (300 mOsm) or 50% hyposmotic media (150 mOsm), and cell volume was measured after 16 min. Cell volume was measured using a Coulter Counter coupled to a particle analyzer device. Osmolarity was decreased by reducing the concentration of NaCl from 135 to 60 mM. Results are expressed as percentage changes relative to volume under isosmotic conditions (300 mOsm) and are mean ± SEM of 6-8 independent experiments.

The process of cell volume regulation in hyposmolar conditions was less efficient in cells with low taurine concentration. Guanidinoethane sulfonate-treated astrocytes showed two main differences with respect to taurine-containing cells. First, the maximal cell volume observed after exposure to hyposmotic medium was about 1.63, significantly higher than in controls. Secondly, the volume regulatory phase was inhibited, as cell volume at the end of the regulatory phase was 1.45; i.e. much higher than in controls. The volume regulatory

Table 2. Hyposmolarity evoked release of free amino acids from astrocytes grown with guanidinoethane sulfonate or a chemically-defined medium

	Stimulated Release (%)		
Amino Acid	Control	guanidinoethane sulfonate	defined medium
Aspartate	43.4	41.2	40.7
Glutamate	40.3	41.1	44.8
Glutamine	46.9	44.9	42.3
Glycine	51.2	49.8	53.4
Taurine	72.6	65.8	71.1
ß-Alanine	56.0	56.9	43.9
Alanine	44.8	43.9	40.5

After two weeks in vitro, astrocytes were cultured 48 h in a medium containing 2 mM guanidinoethane sulfonate or in a chemically defined medium. Culture medium was removed and cells were washed and incubated during 15 min with 50% hyposmotic media. At the end of the incubation period, free amino acids in the medium and in cells were measured by HPLC. Results correspond to free amino acid stimulated-release expressed as % of total concentration initially present in the cell. Results are means of 6-8 experiments.

response was, therefore, only 30% (Fig. 2). Astrocytes grown in chemically defined medium showed an intermediate behavior, showing a cell volume of 1.61 during the swelling phase and regulating to a final volume of 1.40 above the value in isosmotic conditions; i.e. a regulatory response of 35% (Fig. 2). Cells grown in chemically defined medium supplemented with 100 μM taurine exhibited a regulatory volume decrease similar to control cells (not shown).

Guanidinoethane sulfonate-treated astrocytes exposed to a 30% hyposmolar medium also showed a less efficient volume regulatory process. Under these conditions, taurine-deficient cells exhibited an initial swelling of 1.47 and a volume regulation of 40%, significantly different from controls, which first swell to 1.36 and showed a volume regulation of 75% (not shown). These results therefore, suggest that high concentrations of taurine as those present in cultured astrocytes, are required for the volume regulatory process to occur with normal efficiency.

POTASSIUM AND FREE AMINO ACID RELEASE IS NOT AFFECTED IN TAURINE DEFICIENT CELLS

During the volume regulatory response, astrocytes release a number of osmolytes, which include potassium and free amino acids, particularly taurine (4,10). To investigate the possibility that the volume-sensitive efflux of these other osmolytes was increased to compensate for taurine deficiency, the hyposmolarity-evoked release of free amino acids and of $^{86}Rb^+$, a tracer for potassium, was examined. This hypothesis however, was not confirmed by experimental evidence, since the volume-sensitive fluxes of $^{86}Rb^+$ and free amino acids were not increased in taurine-deficient cells. Results in Table 2 show that the efflux of all the major components of the free amino acid pool elicited by hyposmolarity was essentially similar in taurine-containing and taurine-deficient cells. An increase in the efflux in isosmotic conditions of aspartate and glutamate were observed in cells exposed to a chemically defined medium (not shown). Similarly, the hyposmolarity-induced efflux of

Table 3. Hyposmolarity evoked release of ^{86}Rb from astrocytes grown with guanidinoethane sulfonate or a chemically-defined medium

	Basal		Stimulated	
	Control	Guanidinoethane sulfonate	Control	Guanidinoethane sulfonate
^{86}Rb Release (%)	1.45	1.67	4.00	3.90

Cerebellar astrocytes were treated with guanidinoethane sulfonate (2 mM) as described in Fig.1. Cells preloaded with $^{86}Rb^+$ were superfused with isosmotic medium (300 mOsm) for 5 min and then with a 50% hyposmotic media (150 mOsm) for 15 min. The efflux of $^{86}Rb^+$ is expressed as the percent of radioactivity released per min under isosmotic conditions (min 5, basal) and at the maximal release under hyposmolar conditions (min 10, stimulated).

$^{86}Rb^+$ showed the same time course and was quantitatively similar in controls and in taurine-deficient cells (Table 3).

CONCLUDING REMARKS

The results of this work contribute to the evidence suggesting that taurine is involved in the process of cell volume control in astrocytes. The participation of taurine in the mechanisms of volume regulation is supported by several types of evidence. Taurine content in brain decreases during hyponatremia, either induced experimentally (22) or occurring as a result of physiological processes such as pregnancy. Also, the release of taurine in response to swelling has been well characterized in a variety of nervous tissue preparations including in vivo (1,8,9) and in vitro experiments in retina (2), brain slices (12), cultured cells (15,18) and isolated nerve endings (17). Taurine deficiency causes functional alterations that are generally associated with swelling of cells and organelles in preparations such as photoreceptors, cortical neurons and lymphocytes (3,13,19). Our results showing impairment in cell volume control in taurine-depleted astrocytes in culture confirm the role of taurine as an osmolyte in brain cells and offer a simple model to investigate a possible connection between this role of taurine and the problems observed in taurine-deficient animals.

ACKNOWLEDGEMENTS

This work was supported in part by a grant from DGAPA-UNAM (IN 204589). We acknowledge the technical assistance of Ms. Claudia Rodríguez. T. Maar is from the Royal Danish School of Pharmacy.

REFERENCES

1. del Río, R.M, Herranz, A.S., Herreras, O., Menéndez, N., and Solís, J.M. 1990, Possible osmoregulatory role of taurine in the cellular swelling evoked by weak organic acids in the rat hippocampus, *in*: "Taurine: Functional Neurochemistry, Physiology, and Cardiology", Pasantes-Morales, H., Martin, D.L., Shain, W. and del Río, R.M., eds., Wiley-Liss, New York, pp. 357-368.
2. Domínguez, L., Montenegro, J., and Pasantes-Morales, H. 1989, A volume-dependent, chloride-sensitive component of taurine release stimulated by potassium from retina, *J.Neurosci.Res.*, 22:356-361.

3. Hayes, K.C., Carey, R.E., and Schmidt, S.Y. 1975, Retinal degeneration associated with taurine deficiency in the cat, *Science*, 188:949-951.
4. Hoffmann, E.K. and Simonsen, L.O. 1989, Membrane mechanisms in volume and pH regulation in vertebrate cells, *Physiol.Rev.*, 69:315-382.
5. Huxtable, R.J. 1992, Physiological actions of taurine, *Physiol.Rev.*, 72:101-163.
6. Kimelberg, H.K. and Frangakis, M. 1986, Volume regulation in primary astrocyte cultures, *Adv.Biosci.*, 61:177-186.
7. Lehmann, A. 1989, Effects of microdialysis-perfusion with anisoomotic media on extracellular amino acids in the rat hippocampus and skeletal muscle, *J.Neurochem.*, 53:525-535.
8. Lehmann, A., Hagberg, H., Jacobson, I., and Hamberger, A. 1985, Effect of status epilepticus on extracellular amino acids in the hippocampus, *Brain Res.*, 359:147-151.
9. Lehmann, A., Isacsson, H., and Hamberger, A. 1983, Effects of in vivo administration of kainic acid on the extracellular amino acid pool in the rabbit hippocampus, *J.Neurochem.*, 40:1314-1320.
10. Macnight, A.D.C. 1988, Principles of cell volume regulation, *Renal Physiol.Biochem.*, 5:114-141.
11. Morán, J. and Pasantes-Morales, H. 1991, Taurine-deficient cultured cerebellar astrocytes and granule neurons obtained by treatment with guanidinoethane sulfonate, *J.Neurosci.Res.*, 29:533-537.
12. Oja, S.S. and Saransaari, P. 1992, Taurine release and swelling of cerebral cortex slices from adult and developing mice in media of different ionic compositions, *J.Neurosci.Res.*, 32:551-561.
13. Palackal, T., Moretz, R.C., Wisniewski, H.M., and Sturman, J.A. 1988, Ultrastructural abnormalities in the visual cortex of kittens from taurine-deficient mothers, *Brain Dysfunction*, 1:71-89.
14. Pasantes-Morales, H., Maar, T.E., and Morán, J. 1993, Cell volume regulation in cultured cerebellar granule neurons, *J.Neurosci.Res.*, 34:219-224.
15. Pasantes-Morales, H. and Schousboe, A. 1988, Volume regulation in astrocytes: A role for taurine as an osmoeffector, *J.Neurosci.Res.*, 20:505-509.
16. Pion, P.D., Kittleson, M.D., Rogers, Q.R., and Morris, J.G. 1987, Myocardial failure in cats associated with low plasma taurine: a reversible cardiomyopathy, *Science*, 237:764-768.
17. Sánchez Olea, R. and Pasantes-Morales, H. 1990, Chloride dependence of the K^+-stimulated release of taurine from synaptosomes, *Neurochem.Res.*, 15:535-540.
18. Schousboe, A., Moran, J., and Pasantes-Morales, H. 1990, Potassium-stimulated release of taurine from cultured cerebellar granule neurons is associated with cell swelling, *J.Neurosci.Res.*, 27:71-77.
19. Schuller-Levis, G.B. and Sturman, J.A. 1988, Immunologic consequences of taurine deficiency in cats, *FASEB Journal*, 2:A1617
20. Sturman, J.A., Gargano, A.D., Messing, J.M., and Imaki, H. 1986, Feline maternal taurine deficiency: effect on mother and offspring, *J.Nutr.*, 116:655-667.
21. Sturman, J.A., Moretz, R.C., French, J.H., and Wisniewski, H.M. 1985, Taurine deficiency in the developing cat: persistence of the cerebellar external granule cell layer, *J.Neurosci.Res.*, 13:405-416.
22. Thurston, J.H., Hauhart, R.E., and Dirgo, J.A. 1980, Taurine: a role in osmotic regulation of mammalian brain and possible clinical significance, *Life Sci.*, 26:1561-1568.

FELINE MATERNAL TAURINE DEFICIENCY: EFFECTS ON VISUAL CORTEX OF THE OFFSPRING. A MORPHOMETRIC AND IMMUNOHISTOCHEMICAL STUDY

John A. Sturman, Peimin Lu, Yue-Xiang Xu,
and Humi Imaki

Department of Developmental Biochemistry
Institute for Basic Research in Developmental
Disabilities
1050 Forest Hill Road
Staten Island, NY 10314

INTRODUCTION

Cats are dependent on an adequate dietary source of taurine to maintain their body pools because their endogenous taurine biosynthesis is limited by a low activity of cysteinesulfinic acid decarboxylase (7). Without such a dietary source, cats become taurine depleted and develop a number of abnormalities, including retinal degeneration (6) and dilated cardiomyopathy (21). Taurine deficiency in female cats has an adverse effect on feline reproduction resulting in excessive fetal wastage due to abortion or resorption, stillborn and low-birth-weight live kittens (27,28). Only 40% of kittens born alive survive to weaning at 8 weeks after birth. Kittens born alive from taurine-deficient mothers have significantly smaller brain weights than kittens from mothers fed a taurine-sufficient diet, associated with greatly reduced taurine concentrations. Those kittens from taurine-deficient mothers that survive to weaning at 8 weeks after birth also have smaller brain weights and taurine concentrations than kittens from mothers fed a taurine-sufficient diet, and have an abnormal morphology in the cerebellum (29) and visual cortex (18,19), at least.

We have extended these observations in this study by directly counting neurons and glial cells in the various layers of visual cortex of newborn and 8-week-old kittens, and by examining the taurine-like, GABA-like, and glutamate-like immunoreactivity of the cells using antibodies prepared against the individual amino acids conjugated with glutaraldehyde to bovine serum albumin.

Taurine in Health and Disease, Edited by R. Huxtable
and D.V. Michalk, Plenum Press, New York, 1994

MATERIALS AND METHODS

Animals

Female domestic cats raised in the colony at the Institute for Basic Research and vaccinated against rhinotracheitis, panleukopenia, calici virus [(FVR-C-CP (MLV) Pitman-Moore, Washington Crossing, NJ] were fed a completely defined, purified diet (taurine-free) (BioServ, Frenchtown, NJ) alone or containing 0.05% taurine for at least 6 months prior to mating (27,28). All were bred as follows: at estrus they were caged with a male for 1 week; conception was defined as the middle of this period. Male cats were fed cat chow, except for the interval of cohabitation with females during which they received the same diet as the females. Pregnancies were confirmed by palpation and in some cases also by X-ray 4-6 week after conception.

Live newborn kittens were used within 24 h of birth or were nursed by their own mothers kept in the same cage until 8 weeks after birth. Kittens were anesthetized with sodium pentobarbital (Nembutal, Abbott Labs., North Chicago, IL) prior to killing by intracardiac perfusion (1% paraformaldehyde and 2.5% glutaraldehyde in 0.1 M Sorensen's phosphate buffer, pH 7.4, for 15 min). The brains were dissected within a few hours, hemisected, and stored in the perfusion solution for 3 to 6 h in the refrigerator for immunocytochemistry, and longer for morphometry. Three kittens from each diet group at each age were used for the immunohistochemical studies and five kittens from each diet group at each age were used for quantitative morphometric studies. Other kittens were killed by exsanguination following an overdose of sodium pentobarbital, and the occipital cortex region of the brain dissected and processed for measurement of amino acid concentrations by reverse-phase HPLC, described in detail elsewhere (27). This study was approved by the Institute's Animal Welfare Committee.

Morphometric Analysis

Thin blocks of tissue, less than 1 mm in width and slightly over 1 mm in length encompassing the entire cortical thickness and some underlying white matter, were dissected from the medial wall of hemisphere between the crown of lateral (marginal) gyrus and the superior bank of splenial sulcus near the occipital pole, the area demonstrated to represent the central 10° in the contralateral visual hemi-field in the cat primary visual cortex (32), well within the cytoarchitectonically identified area 17, or striata (16,22). All the tissue blocks were rinsed overnight in 0.1 M Sorensen's buffer, postfixed in 2% OsO_4, dehydrated in an ascending series of acetone concentrations and embedded in Epon 812 or EM-bed 812 (Electron Microscopy Sciences, Ft. Washington, PA). One-half micron sections were cut perpendicular to the cortical surface, mounted on glass slides and stained with 1% toluidine blue, and light micrographs were taken on a Zeiss Photomicroscope III using Kodak Panatomic-X film.

In toluidine blue-stained sections, the nuclei of cortical cells in the newborn kittens appeared sharply defined, but precise identification of cell types proved virtually impossible due to the transient state of their differentiation. Therefore, nuclei of neurons and glial cells were combined to obtain overall cell density for a given layer. In toluidine blue-stained sections of 8-week-old kittens identification of cell types into neurons, astrocytes, oligodendrocytes and microglia was achieved according to the criteria described for the rat brain by Ling et al. (11). Determination of the position and limits of cortical layers was made by comparing the distributions of neurons in toluidine blue-stained sections with those in Nissl-stained sections of the kitten visual cortex (Figures 1 and 2), which in turn were compared with Nissl-stained area 17 of the (old World) monkeys (3,13), in addition to directly comparing toluidine blue-stained sections of kitten visual cortex with those of the

3-month-old monkeys (17). For laminar analysis of 8-week-old kitten visual cortex, layer II-III was further divided into 3 equal parts, Layer IVB, into 2 equal parts, and layers IVC, V and VI were separated into 2 sublayers each as in the monkey visual cortex (15,17).

For morphometric analysis, a stereological method was used to determine the numerical density of cells for each layer of visual cortex following the procedure used by O'Kusky and Colonnier (17), for the adult monkey visual cortex. For this purpose, stained sections were visualized under a microscope with a 100x objective lens and a 10x ocular lens fitted with a micrometer grid divided into 100 squares, each corresponding to 100 μm^2 on the section. Using camera lucida the perimeters of nuclei identified within the grid and those lying on the margins, except for forbidden lines, each marked for the cell types, were drawn on a roll of thermosensitive printing paper (Morrell Instrument Co., Melville, NY), after aligning the pial surface of cortex and the end of the roll flush with a margin of grid. This procedure was repeated for the rest of the cortical thickness by sliding the section and paper grid by grid until the white matter was reached. This resulted in a sample probe which usually was composed of 17 to 20 transfers each representing a 100 μm wide section. From each section 2 such probes, non-overlapping with one another, were prepared from each of 5 blocks to obtain 10 probes from each animal, yielding a total of 100 probes from 5 control and 5 taurine-deficient kittens at each age.

After all the cortical layers and sublayers were marked, nuclear profiles on the probe were traced by a stylus on an electromagnetic tablet, attached to a semi-automatic image analyzer, Videoplan (Morrell Instrument Co., Melville, NY), and NEC Pinwriter (NEC Technologies, Inc., Boxborough, MA). Measurements of the nuclei were fed separately for each cell type layer into the Videoplan which calculated numerical density of nuclei/mm^3 of tissue adopting the formula by O'Kusky and Colonnier (15) using nuclei as test objects, following the proposal by Weibel (34) and Weibel and Bolender (35). Since the tissue processing employed in this study was basically the same as that in O'Kusky and Colonnier (15), the same shrinkage factors, 16% for linear, 29% for areal and 41% for volumetric measurements, were used to calculate original (intact) values, and the group means of numerical density, thickness of individual layers and the number of cells under 1 mm^2 of surface were computed using an IBM Personal Computer with a program developed for a similar study with young rhesus monkeys (17).

Immunohistochemistry

Immunogens were synthesized by the method of Campistron et al. (4), starting with 100 mg of taurine (or GABA, glutamate, β-alanine, etc.), and coupling with glutaraldehyde to bovine serum albumin or poly-L-lysine (MW 30,000-70,000, Sigma Chemical Co., St. Louis, MO).

Rabbits were immunized by first injecting 500 μg of immunogen emulsified in Freund's complete adjuvant, followed every 4 weeks by injection of 500 μg immunogen emulsified in Freund's incomplete adjuvant. Injections alternated between amino acid-glutaraldehyde-bovine serum albumin and amino acid-glutaraldehyde-polylysine. Two rabbits were injected subcutaneously and two intramuscularly. Blood was collected 7 and 21 days after each injection and serum frozen in 1 ml batches.

The serum was characterized for titer and cross-reactivity using the ELISA method described by Lake and Verdone-Smith (8). High, stable titers of reactivity against amino acid-glutaraldehyde-bovine serum albumin were reached after 3 injections (> 1:64,000 dilution). The anti-taurine serum had a small cross-reactivity to GABA-glutaraldehyde-bovine serum albumin and to β-alanine-glutaraldehyde-bovine serum albumin. This reactivity was removed by preadsorption with both conjugates and resulted in little loss of taurine-glutaraldehyde-bovine serum albumin activity. Antibodies to the other immunogens were similarly treated prior to use.

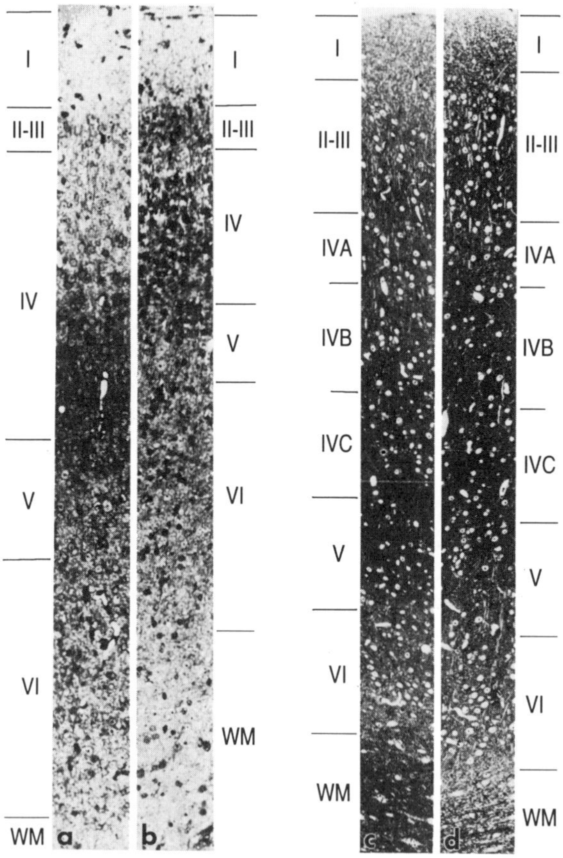

Figure 1. Light micrographs of the visual cortex of newborn (a,b) and 8-week-old (c,d) kittens showing the distribution of cells in the cortical layers: a,c, from taurine-supplemented mothers; b,d from taurine-deficient mothers. 0.5 μm sections stained with toluidine blue. a,b, X 128; c,d X 61.

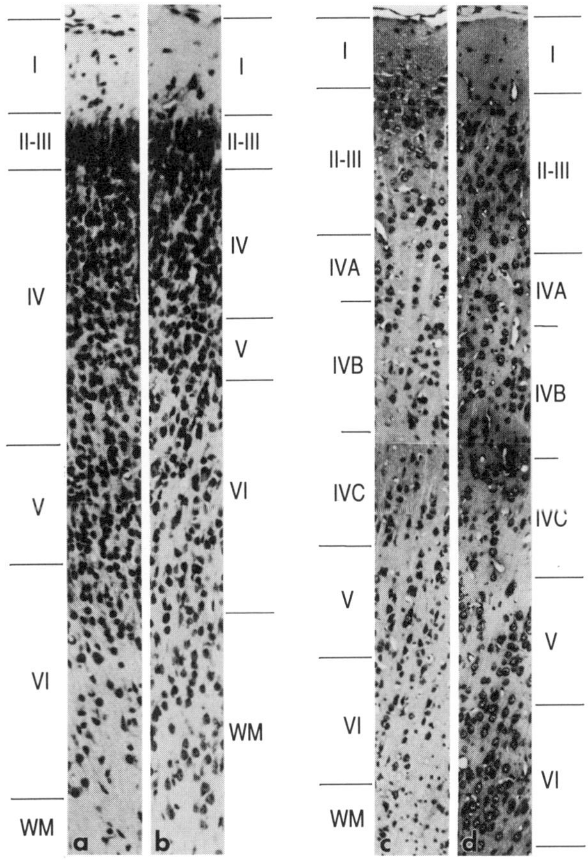

Figure 2. Light micrographs of the visual cortex of newborn (a,b) and 8-week-old (c,d) kittens showing the distribution of cells in the cortical layers: a,c, from taurine-supplemented mothers; b,d from taurine-deficient mothers. 6 μm sections stained with cresyl violet. a,b X 128; c,d X 61.

Tissue was embedded in paraffin and serial 6 μm coronal sections cut and mounted on glass slides. Sections were deparaffined and hydrated by warming at 60°C for 65 min in a histoclear bath, followed by graded ethanol washes. Staining with the antisera (1 to 2000 dilution) was carried out at 4°C overnight and visualized using the peroxidase-conjugated avidin/DAB method (Dako Corp., Carpinteria, CA) (33). Control sections were included in every batch, those in which the antisera was preadsorbed with the immunogen prior to processing, and others replacing the primary antiserum with preimmune serum at similar dilutions. Control slides showed no visible staining after processing, whereas slides prepared with the antisera showed brown reaction products. Rabbit anti-cow glial fibrillary acidic protein antibody was used for localizing kitten glial fibrillary acidic proteine (obtained from Dako Corp., code number 2334). Counterstaining, when employed, was with hematoxylin. Photomicrographs were taken on a Zeiss Axiophot using Kodak 40 color reversal film.

Table 1. Thickness of laminae in kittens from taurine-supplemented and taurine-deficient mothers

Layer	Supplemented	Deficient
Newborn		
I	112 ± 5	105 ± 11
II-III	107 ± 7	81 ± 6[1]
IV	279 ± 17	202 ± 33[2]
V	141 ± 8	107 ± 6[3]
VI	341 ± 27	278 ± 39[4]
Total	980 ± 29	773 ± 61[3]
8-Week-old		
I	170 ± 30	170 ± 17
II-III	354 ± 44	363 ± 25
IV-A	163 ± 22	162 ± 19
IV-B	303 ± 33	286 ± 29
IV-C	263 ± 20	267 ± 32
V	291 ± 41	289 ± 25
VI	304 ± 44	330 ± 26
Total	1850 ± 161	1868 ± 138

Mean laminar thickness ± SD (in μm) from five kittens in each group. [1] $P < 0.0003$; [2] $P < 0.002$; [3] $P < 0.0001$; [4] $P < 0.02$

RESULTS

The visual cortex of newborn and 8-week-old kittens from taurine-supplemented and taurine-deficient mothers are depicted in Figs. 1 and 2 using standard histological stains. Clear differences are evident in the distribution of cells both between groups and between ages. The total thickness of the visual cortex and of all individual laminae, with the exception of layer I, is significantly smaller in newborn kittens from taurine-deficient mothers, but in 8-week-old kittens there are no such differences (Table 1). The brain weight at birth of kittens from taurine-deficient mothers is 70% of kittens from taurine-supplemented mothers, whereas at 8 weeks after birth it is more than 90% (27). The numerical density of cells is significantly greater in all laminae of newborn kittens from taurine-deficient mothers, except layer I, whereas there are few differences in 8-week-old kittens (Table 2). At birth, it is difficult to distinguish between neurons and glial cells, so only the total was measured, whereas at 8 weeks after birth, measures were made separately of neurons, oligodendrocytes, astrocytes and microglia. The only differences found are a lower numerical density of oligodendrocytes in layers IV-B.1 and V-2 of taurine-deficient kittens and a greater numerical density of astrocytes throughout the visual cortex, with significance ($P < 0.05$) reached in layers II-III.3 and IV-B.1 (Table 3). The number of cells under 1 mm^2 of visual cortex surface is not different in newborn kittens (Table 4) nor in 8-week-old kittens for neurons, oligodendrocytes or microglia, although there are significantly more astrocytes in layers II-III and V of taurine-deficient kittens (Table 5). These numerical data are supported by an increased number of cells positive for glial fibrillary acidic protein, presumbly reactive astrocytes, in the visual cortex of 8-week-old taurine-deficient kittens (Fig. 3).

Table 2. Numerical density of cells (neurons plus glia) in newborn kittens from taurine-supplemented and taurine-deficient mothers

Layer	Supplemented	Deficient
I	198 ± 24	217 ± 29
II-III	548 ± 63	719 ± 90^{1}
IV	392 ± 64	535 ± 107^{2}
V	275 ± 47	333 ± 52
VI	269 ± 31	342 ± 56^{3}

^{1} $P < 0.008$; ^{2} $P < 0.03$; ^{3} $P < 0.04$.
Each value represents the mean $\pm$ SD (thousands/mm^2) from five kittens in each group.

The visual cortex from both groups of kittens at birth and at 8 weeks after birth was examined for immunoreactivity to taurine, GABA and glutamate using antibodies prepared against the amino acids conjugated to bovine serum albumin with glutaraldehyde. The only obvious difference in newborn kittens is a decreased taurine-like immunoreactivity in taurine-deficient kittens, including the band of staining near the pial surface, which had been noted previously in newborn kitten cerebellum (12). A more detailed examination of this region by electron microscopy shows it to be packed with dendrites, axons and glial processes (Fig. 4). The use of immunogold labelling with the taurine antibody shows that virtually all of the taurine-like immunoreactivity is located within these processes (Fig. 5). At 8 weeks after birth, there is little difference in taurine-like and glutamate-like immunoreactivity between taurine-supplemented and taurine-deficient kittens, but there is considerably more GABA-like immunoreactivity in the visual cortex of taurine-deficient kittens, especially in layer VI and in the white matter (Fig. 6). Direct measurement of tissue concentrations of amino acids shows differences only in taurine between kittens from taurine-supplemented and taurine-deficient mothers (Fig. 7).

DISCUSSION

Interest and information concerning the biological roles of taurine have burgeoned during the past two decades and much research has pointed to an important role for taurine in a variety of developmental processes (26). We have previously reported that kittens from taurine-deficient mothers have an impaired migration of cerebellar granule cells which can be prevented by feeding postnatal taurine supplements (29,30). The visual cortex from these same kittens also displayed morphological abnormalities (18,19). The present work describes quantitative morphometric measurements of the visual cortex from newborn and 8-week-old kittens from taurine-supplemented and taurine-deficient mothers, along with immunohistochemical and biochemical measurements.

Table 3. Numerical density of neurons and glia in 8-week-old kittens from taurine-supplemented and taurine-deficient mothers

Layer	Neurons		Oligodendrocytes	
	Supplemented	Deficient	Supplemented	Deficient
I	11.9 ± 7.9	10.2 ± 5.2	13.0 ± 7.7	12.5 ± 4.1
II-III.1	50.5 ± 8.1	54.2 ± 16.4	0.2 ± 0.4	0.2 ± 0.4
II-III.2	43.8 ± 5.6	46.2 ± 10.5	0.5 ± 0.6	0.2 ± 0.5
II-III.3	37.6 ± 8.1	45.8 ± 7.4	0	0
IV-A	39.0 ± 9.4	40.0 ± 11.6	1.2 ± 1.5	0.1 ± 0.2
IV-B.1	39.3 ± 5.1	44.8 ± 9.2	3.0 ± 1.6	0.9 ± 0.6[1]
IV-B.2	34.9 ± 5.3	39.9 ± 10.2	1.1 ± 0.9	2.0 ± 1.7
IV-α	41.1 ± 6.5	48.3 ± 10.5	1.4 ± 1.6	1.8 ± 1.5
IV-Cβ	42.2 ± 7.8	48.4 ± 11.8	2.0 ± 0.9	1.9 ± 2.1
V-1	40.7 ± 7.2	42.8 ± 16.0	3.5 ± 1.4	4.0 ± 2.2
V-2	35.6 ± 10.2	34.6 ± 6.2	3.3 ± 1.5	1.5 ± 0.2[1]
VI-A	47.5 ± 6.1	54.2 ± 15.8	3.9 ± 1.1	3.5 ± 1.5
VI-B	45.4 ± 5.3	48.7 ± 15.9	6.7 ± 2.1	7.0 ± 3.5
Mean	38.4 ± 6.0	41.8 ± 8.5	3.4 ± 0.7	2.9 ± 0.8

Layer	Astrocytes		Microglia	
	Supplemented	Deficient	Supplemented	Deficient
I	6.3 ± 4.5	10.8 ± 5.5	4.4 ± 1.9	7.3 ± 2.5
II-III.1	10.3 ± 4.1	17.0 ± 7.9	6.4 ± 1.5	7.5 ± 3.5
II-III.2	9.8 ± 3.3	12.8 ± 6.1	6.0 ± 2.6	5.9 ± 3.0
II-III.3	11.0 ± 2.9	16.8 ± 4.5[2]	5.8 ± 2.7	6.4 ± 1.3
IV-A	12.0 ± 3.4	14.5 ± 5.5	3.4 ± 1.3	4.3 ± 1.3
IV-B.1	12.7 ± 3.8	18.0 ± 2.8[2]	3.8 ± 1.6	4.7 ± 1.9
IV-B.2	12.0 ± 2.7	15.7 ± 5.4	3.8 ± 2.6	5.6 ± 2.8
IV-Cα	14.7 ± 3.8	17.0 ± 4.4	3.2 ± 1.9	4.9 ± 2.3
IV-Cβ	11.6 ± 2.0	14.4 ± 4.7	4.9 ± 2.9	5.2 ± 3.8
V-1	12.6 ± 4.2	17.6 ± 4.1	5.1 ± 2.8	4.8 ± 2.9
V-2	11.8 ± 3.1	16.5 ± 5.6	3.6 ± 1.6	7.1 ± 3.8
VI-A	14.6 ± 4.5	17.0 ± 2.9	6.0 ± 3.1	7.1 ± 4.0
VI-B	14.5 ± 6.6	17.8 ± 6.1	3.0 ± 1.8	6.0 ± 2.7
Mean	12.2 ± 3.8	15.6 ± 4.8	4.5 ± 1.3	5.8 ± 2.4

[1] $P < 0.03$; [2] $P < 0.04$

Each value represents the mean % SD (in thousands/mm^3) from five kittens in each group.

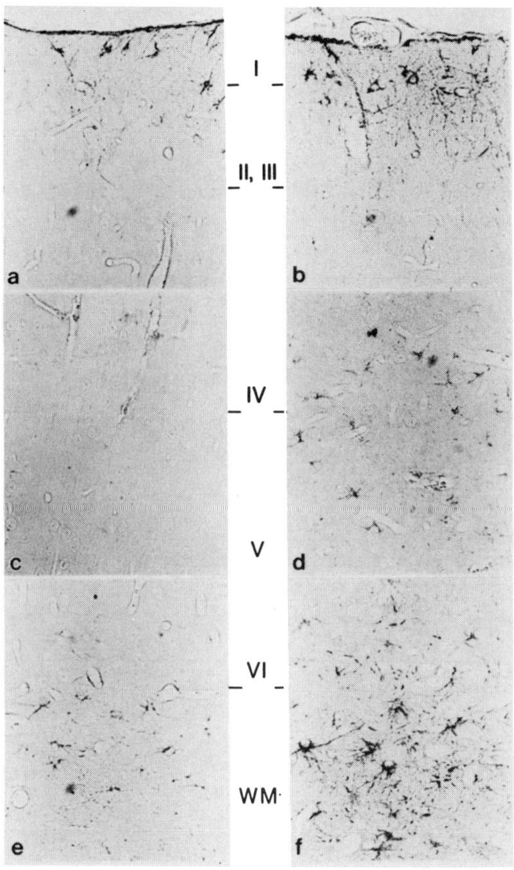

Figure 3. Portions of the visual cortex of 8-week-old kittens born to and nursed by mothers fed a taurine-supplemented diet (a,c,e), or a taurine-free diet (b,d,f) showing the distribution of cells reacting positively to an antibody to glial fibrillary acidic protein. X 254. From reference 37.

Table 4. Number of cells (neurons plus glia) beneath 1 mm^2 of surface in newborn kittens from taurine-supplemented and taurine-deficient mothers

Layer	Supplemented	Deficient
I	22.3 ± 3.4	22.7 ± 2.0
II-III	58.8 ± 9.0	57.7 ± 6.2
IV	109.3 ± 17.2	105.2 ± 6.5
V	38.5 ± 6.2	35.4 ± 3.7
VI	91.9 ± 14.0	95.0 ± 21.2
Total	320.8 ± 37.7	316.0 ± 31.5

Table 5. Number of neurons and glia under 1 mm^2 of surface in 8-week-old kittens from taurine-supplemented and taurine-deficient mothers

Layer	Neurons		Oligodendrocytes	
	Supplemented	Deficient	Supplemented	Deficient
I	1.9 ± 1.0	1.7 ± 1.0	2.3 ± 1.3	2.1 ± 0.7
II-III	15.6 ± 2.4	17.5 ± 3.2	0.1 ± 0.1	0.04 ± 0.1
IV-A	6.3 ± 1.3	6.4 ± 1.3	0.2 ± 0.2	0.01 ± 0.0[3]
IV-B	11.3 ± 1.8	11.9 ± 1.3	0.6 ± 0.4	0.4 ± 0.3
IV-C	10.9 ± 1.4	12.7 ± 1.3	0.5 ± 0.2	0.5 ± 0.3
V	11.1 ± 2.3	11.1 ± 2.1	1.0 ± 0.3	0.8 ± 0.5
V-1	14.1 ± 1.9	16.8 ± 3.9	1.6 ± 0.5	1.7 ± 0.8
Total	71.1 ± 7.4	78.0 ± 12.5	6.3 ± 1.0	5.4 ± 2.5

Layer	Astrocytes		Microglia	
	Supplemented	Deficient	Supplemented	Deficient
I	1.1 ± 0.9	1.8 ± 0.7	0.7 ± 0.2	1.2 ± 0.4
II-III	3.7 ± 1.1	5.6 ± 1.5[1]	2.2 ± 0.7	2.4 ± 0.6
IV-A	1.9 ± 0.4	2.3 ± 0.8	0.5 ± 0.2	0.7 ± 0.2
IV-B	3.7 ± 0.8	4.7 ± 0.9	1.4 ± 0.5	1.4 ± 0.4
IV-C	4.3 ± 2.3	4.1 ± 0.6	1.3 ± 0.6	1.3 ± 0.4
V	3.5 ± 0.7	4.9 ± 1.0[1]	1.2 ± 0.4	1.7 ± 0.8
V-1	4.3 ± 1.5	5.7 ± 1.1	1.3 ± 0.6	2.1 ± 0.9
Total	22.6 ± 4.2	29.1 ± 5.3	8.4 ± 1.2	10.8 ± 2.8

[1] $P < 0.05$. Each value represents the mean ± SD (thousands) from five kittens in each group.

Much greater differences in quantitative measures are found in newborn kittens than in 8-week-old kittens. The thickness of all laminae, except for Layer I, and total thickness are significantly reduced in newborn taurine-deficient kittens. The 21% reduction of total cortical thickness is similar to that of 10-day-old rats undernourished from birth (1). As found in this work, the reduced thickness does not result from a reduced number of neurons, but rather from a reduced dendritic growth and a reduced cortical volume, which is reflected in a greater numerical cell density (5,9,10,25). By 8 weeks after birth, there are no longer such great differences between the groups, with changes limited to small

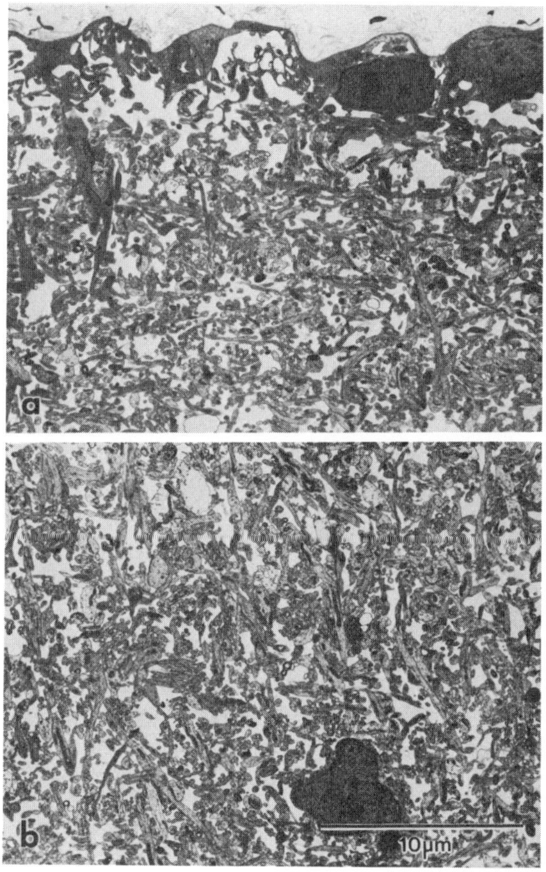

Figure 4. Low magnification electron micrographs of the visual cortex of a newborn kitten born to a mother fed a taurine-supplemented diet, showing (a) the superficial, and (b) the intermediate levels of layer I, consisting of dendritic shafts, axon terminals and occasional cell bodies and processes of astrocytes near the pial surface. From reference 36.

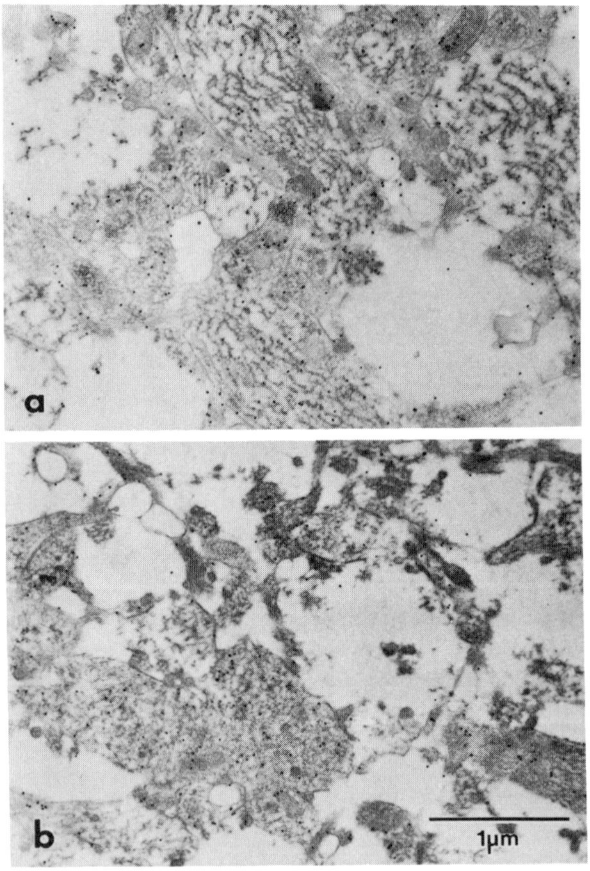

Figure 5. Electron micrographs of visual cortex of a newborn kitten from a taurine-supplemented mother, showing that 15 nm gold particles bound to the taurine antibody are localized over a loosely organized plexus of dendrites, axons and astrocytic processes in (a) the superficial, and (b) the intermediate levels of layer I. From reference 36.

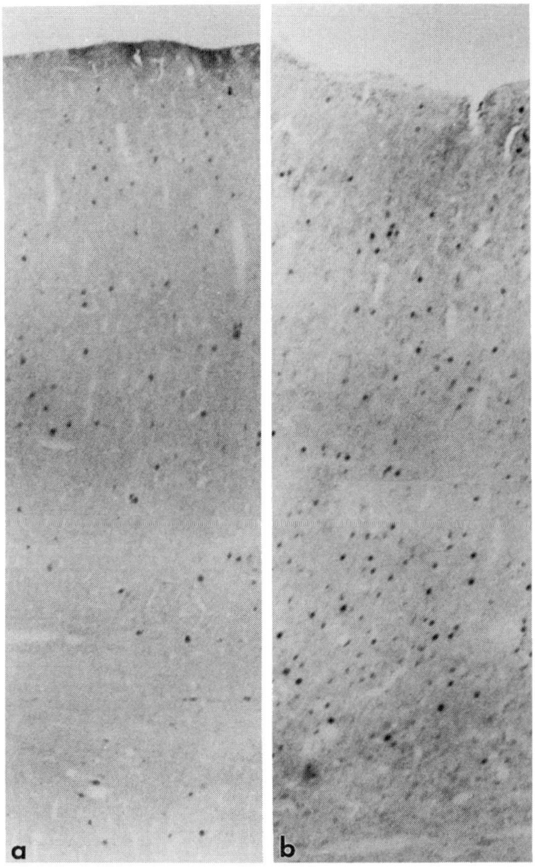

Figure 6. Visual cortex of 8-week-old kittens immunostained with GABA antiserum. (a) Kitten from a taurine-supplemented mother; (b) kitten from a taurine-deficient mother. X 104.

decreases in the numerical density of oligodendrocytes in two laminae and small increases in astrocytes in two laminae. These results suggest that the taurine-deficient milieu throughout pregnancy has a greater impact on the development of the visual cortex than the reduced taurine conconcentrations in the post-natal brain. It is also likely that the most severely affected live newborn kittens do not survive to weaning, since only 40% of those kittens born live are still alive at 8 weeks, and we are studying the more moderately taurine-deficient kittens. It is of note that there are relatively few quantitative changes in the visual cortex of rhesus monkeys deprived of taurine from birth for 3,6 or 12 months (17,20).

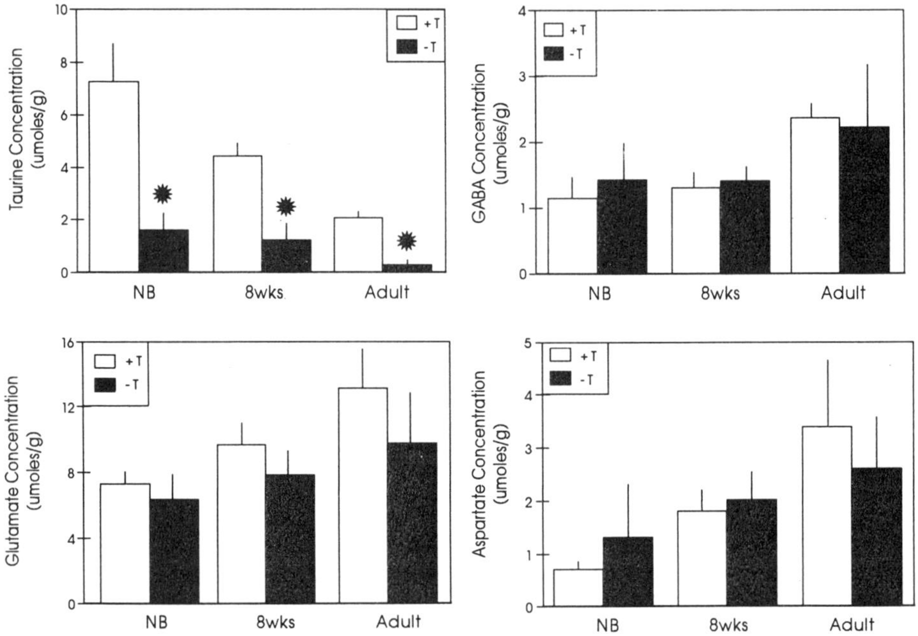

Figure 7. Concentrations of inhibitory and excitatory amino acids in occipital lobe of taurine-supplemented (+T) and taurine-deficient (-T) cats and kittens (NB: newborn; 8 wks: 8-week-old). Each value represents the mean from six animals ± SD. ☼ Significantly smaller, $P < 0.0001$.

The major observations from immunohistochemical studies are the presence of a band of staining near the pial surface of newborn kittens, unique to taurine and decreased in taurine-deficient kittens, which is localized in the mass of neuronal and glial processes in this region, and an increased number of neurons showing GABA-like immunoreactivity in 8-week-old kittens from taurine-deficient mothers. A similar result was noted in the visual cortex of cats that had been dark-reared for 4-5 months from birth (14). Such cats also had increased numbers of neurons compared to normal light-reared cats, unlike the present study. A number of reports suggest that the GABA neurotransmitter and receptor system have a special importance to the ontogeny of the kitten visual cortex (2,23,24,31) raising the intriguing possibility that some of the abnormalities in taurine-deficient kittens result from the taurine imbalance interacting with the GABA system.

ACKNOWLEDGEMENTS

This research was supported by NIH grant HD-16634 and by the New York State Office of Mental Retardation and Developmental Disabilities. We are grateful to Alice Gargano and Jeffrey Messing for technical assistance and to Ann Parese for secretarial assistance.

REFERENCES

1. Bass, N.H., Netsky, M.G., and Young, G., 1970, Effect of neonatal malnutrition on the developing cerebrum. I. Microchemical and histologic study of cellular differentiation in the rat, *Arch. Neurol.* 23:289-302.
2. Beaulieu, C., and Cynader, M., 1990, Effect of the richness of the environment on neurons in cat visual cortex. I. Receptive field properties, *Dev. Brain Res.* 53:71-81.
3. Brodmann, K., 1905, Beitrage zur histologischen Lokalisation der Grosshirnrinde. III. Die Rindenfelder der niederen Affen, *J. Psychol. Neurol.* 4:177-226.
4. Campistron, G.M., Geffard, and Buijs, R.M., 1986, Immunological approach to the detection of taurine and immunocytochemical results, *J. Neurochem.* 46:862-868.
5. Cragg, B.G., 1972, The development of cortical synapses during starvation in the rat, *Brain* 95:143-150.
6. Hayes, K.C., Carey, R.E., and Schmidt, S.Y., 1975, Retinal degeneration associated with taurine deficiency in the cat, *Science* 188:949-951.
7. Knopf, K., Sturman, J.A., Armstrong, M., and Hayes, K.C., 1978, Taurine: an essential nutrient for the cat, *J. Nutr.* 108:773-778.
8. Lake, N., and Verdone-Smith, C., 1989, Immunocytochemical localization of taurine in the mammalian retina, *Curr. Eye Research* 8:163-173.
9. Leuba, G., and Rabinowicz, T., 1979, Long-term effects of postnatal undernutrition and maternal malnutrition on mouse cerebral cortex. I. Cellular densities, cortical volume and total numbers of cells, *Expl. Brain Res.* 37:283-298.
10. Leuba, G., and Rabinowicz, T., 1979, Long-term effects of postnatal undernutrition and maternal malnutrition on mouse cerebral cortex. II. Evolution of dendritic branchings and spines in the visual region, *Expl. Brain Res.* 37:299-308.
11. Ling, E.A., Paterson, J.A., Privat, A., Mori, S., and Leblond, C.P., 1973, Investigation of glial cells in semithin sections. I. Identification of glial cells in the brain of young rats, *J. Comp. Neurol.* 149:43-72.
12. Lu, P., Schuller-Levis, G., and Sturman, J.A., 1991, Distribution of taurine-like immunoreactivity in cerebellum of kittens from taurine-supplemented and taurine-deficient mothers, *Int. J. Devl. Neurosci.* 6:621-629.
13. Lund, J.S., 1973, Organization of neurons in the visual cortex, area 17, of the monkey (Macaca mulatta), *J. Comp. Neurol.* 147:455-495.
14. Mower, G.D., Rustad, R., and White, W.F., 1988, Quantitative comparisons of gamma-aminobutyric acid neurons and receptors in the visual cortex of normal and dark-reared cats, *J. Comp. Neurol.* 272:293-302.
15. O'Kusky, J., and Colonnier, M., 1982, A laminar analysis of the number of neurons, glia and synapses in the visual cortex (area 17) of adult macaque monkeys, *J. comp. Neurol.* 210:278-290.
16. O'Leary, J.L., 1941, Structure of the area striata of the cat, *J. Comp. Neurol.* 75:131-64.
17. Palackal, T., Kujawa, M., Moretz, R., Neuringer, M., and Sturman, J., 1991, A laminar analysis of the number of neurons, astrocytes, oligodendrocytes and microglia in the visual cortex (area 17) of 3-month-old rhesus monkeys fed a human infant soy-protein formula with or without taurine supplementation from birth, *Dev. Neurosci.* 13:20-33.
18. Palackal, T., Moretz, R., Wisniewski, H., and Sturman, J., 1986, Abnormal visual cortex development in the kitten associated with maternal dietary taurine deprivation, *J. Neurosci. Res.* 15:223-239.
19. Palackal, T., Moretz, R.C., Wisniewski, H.M., and Sturman, J.A., 1988, Ultrastructural abnormalities in the visual cortex of kittens from taurine-deficient mothers, *Brain Dysfunct.* 1:71-89.
20. Palackal, T., Neuringer, M., and Sturman, J., 1993, Laminar analysis of the number of neurons, astrocytes, oligodendrocytes and microglia in the visual cortex (area 17) of 6- and 12-month-old rhesus monkeys fed a human infant soy-protein formula with or without taurine supplementation from birth, Dev. Neurosci., in press.

21. Pion, P.D., Kittleson, M.D., Rogers, Q.R, and Morris, J.G., 1987, Myocardial failure in cats associated with low plasma taurine: a reversible cardiomyopathy, *Science* 237:764-768.
22. Sanides, F., and Hoffmann, J., 1969, Cyto- and myeloarchitecture of the visual cortex of the cat and the surrounding integration cortices, *J. Hirnforsch.* 11:79-104.
23. Shaw, C., Needler, M.C., and Cynader, M., 1984, Ontogenesis of muscimol binding sites in cat visual cortex, *Brain Res. Bull.* 13:331-334.
24. Shaw, C., Needler, M.C., Wilkinson, M., Aoki, C., and Cynader, M., 1985, Modification of neurotransmitter receptor sensitivity in cat visual cortex during the critical period, *Dev. Brain Res.* 22:67-73.
25. Siassi, F., and Siassi, B., 1973, Differential effects of protein-calorie restriction and subsequent repletion on neuronal and nonneuronal components of cerebral cortex in newborn rats, *J. Nutr.* 103:1625-1633.
26. Sturman, J.A., 1993, Taurine in development, *Physiol. Rev.* 73:119-148.
27. Sturman, J.A., and Messing, J.M., 1991, Dietary taurine content and feline reproduction and outcome, *J. Nutr.* 121:1195-1203.
28. Sturman, J.A., Gargano, A.D., Messing, J.M., and Imaki, H., 1986, Feline maternal taurine deficiency:effect on mother and offspring, *J. Nutr.* 116:655-667.
29. Sturman, J.A., Moretz, R.C., French, J.H., and Wisniewski, H.M., 1985, Postnatal taurine deficiency in the kitten results in a persistence of the cerebellar external granule cell layer: correction by taurine feeding, *J. Neurosci. Res.* 13:521-528.
30. Sturman, J.A., Moretz, R.C., French, J.H., and Wisniewski, H.M., 1985, Taurine deficiency in the developing cat: persistence of the cerebellar external granule cell layer, *J. Neurosci. Res.* 13:405-416.
31. Tanaka, K., Freeman, R.D., and Ramoa, A.S., 1987, Dark-reared kittens: GABA sensitivity of cells in the visual cortex, *Exp. Brain Res.* 65:673-675.
32. Tusa, R.J., Palmer, L.A., and Rosenquist, A.C., 1978, The retinotopic organization of area 17 (striate cortex) in the cat, *J. Comp. Neurol.* 177:213-236.
33. Warnke, R., and Levy, R., 1980, Detection of T and B cell antigen hybridoma monoclonal antibodies: a biotin-avidin-horseradish peroxidase method, *J. Histochem. Cytochem.* 8:771-776.
34. Weibel, E.R., 1969, Stereological principles for morphometry in electron microscopic cytology, *in:* "International Reviews of Cytology. Biological Applications", Bourne, G.H., and Danielli, J.F. eds., Academic Press, New York, pp. 235-302.
35. Weibel, E.R., and Bolender, P., 1973, Stereological techniques for electron microscopic morphometry, *in:* "Principles and Techniques of Electron Microscopy", Hyatt, M.A., ed., Van Nostrand Reinhold, New York, pp. 237-296.
36. Xu, Y., Lu, P., Imaki, H., and Sturman, J.A., 1992, Feline maternal taurine deficiency: a quantitative morphometric and immunohistochemical study of newborn visual cortex, *Dendron* 1:129-145.
37. Xu, Y., Lu, P., Imaki, H., and Sturman, J.A., 1994, Feline maternal taurine deficiency: A quantitative morphometric and immunohistochemical study of 8-week-old kitten visual cortex, *Biomed. Lett.*, in press.

EFFECTS OF POSTNATAL TAURINE DEPRIVATION ON VISUAL CORTEX DEVELOPMENT IN RHESUS MONKEYS THROUGH ONE YEAR OF AGE

Martha Neuringer,[1,2] Thomas Palackal,[3]
John A. Sturman,[3] and Humi Imaki[3]

[1]Departments of Medicine & Ophthalmology
Oregon Health Sciences University
Portland, OR 97201
[2]Division of Neuroscience
Oregon Regional Primate Research Center
Beaverton, OR 97006
[3]Department of Developmental Biochemistry
N. Y. State Institute for Basic Research in Developmental Disabilities
Staten Island, NY 10314

INTRODUCTION

In rhesus monkeys and humans, the effects of dietary taurine deprivation are age-dependent. Infants require a dietary source of taurine to maintain normal plasma and tissue taurine concentrations whereas adults generally do not. In both human and monkey infants, taurine-free infant formulas produce a drop of approximately 50% in plasma taurine concentrations compared to either taurine-supplemented formulas or breast milk, which is rich in taurine. We have examined the effects of dietary taurine deprivation in rhesus monkeys through the first postnatal year in order to define the time course of these effects, their persistence, and the period of vulnerability, particularly with respect to the morphology and function of the retina and central visual system.

In rhesus infants fed a taurine-free human infant formula from birth, taurine concentrations were reduced ~30-50% in plasma and most tissues at 3 and 6 months of age, but by 12 months did not differ from those in taurine-supplemented monkeys (9, 17, 18). The retina was particularly resistant to taurine depletion, with concentrations reduced by only about 20% at 3 and 6 months. Nevertheless, 3-month-old taurine-deprived infants showed degenerative changes in the retina, particularly disorganization and disruption of cone photoreceptor outer segments in the fovea and perifovea (4). These effects became

Taurine in Health and Disease, Edited by R. Huxtable
and D.V. Michalk, Plenum Press, New York, 1994

progressively less severe with age but were still present in some taurine-deprived monkeys at 12 months (5), indicating a persistent effect of early taurine deprivation.

Investigation of effects on the visual system in these animals has proceeded with studies of the morphology of the visual cortex. The primary visual area of the cerebral cortex, also known as Brodmann's area 17 or striate cortex, receives information from the retina via the lateral geniculate nucleus. Area 17 contains a topographical map of the visual field and is involved in the analysis of such basic visual features as form, orientation, motion and depth. It consists of six principal layers, some of which, most notably layer IV, have prominent subdivisions. Each of these layers and sublayers has a distinct pattern of synaptic connections and specific functions in visual processing. Layers IV-A, IV-Ca and IV-Cb receive the major inputs from the lateral geniculate nucleus and relay information to layer IV-B, a dense fiber layer, and layers II and III. These layers in turn send outputs to higher visual cortical areas for further processing. Layers II and III also send efferents to layers V and VI, which in turn project to subcortical areas involved in eye movements, visual attention and spatial orientation.

In area 17 of 3-month-old taurine-deprived rhesus infants, we previously found poor differentiation of pyramidal neurons and astrocytes and changes in their distribution within the cortical laminae (8, 12). The numerical density of neurons decreased in layer IV-Ca and V-2 but increased in layer IV-B. As layer IV-Ca normally contains a considerably higher density of neurons than IV-B, the differentiation of these layers was somewhat blurred. Among glial cells, only astrocytes showed significant quantitative differences in the taurine-deprived infants. Astrocyte density was lower in layer IV-A and higher in IV-Ca in compared to taurine-supplemented infants. We have now completed quantitative morphological examination of the visual cortex for taurine-deprived and taurine-supplemented rhesus infants at 6 and 12 months of age (15).

METHODS

Rhesus monkeys were fed soy protein-based human infant formulas (Isomil®, Ross Laboratories) from the day of birth. For the taurine-deprived groups the formula contained no detectable taurine, whereas for the taurine-supplemented groups it contained 70 μmol/100 ml taurine, the approximate level in rhesus milk. Five monkeys received each diet until 3 months, 5 until 6 months, and 5 until 12 months of age. In addition, the reversibility of the effects of taurine deprivation was tested by feeding 5 monkeys the taurine-free formula from birth until 6 months of age and then transferring them to the supplemented formula until 12 months (reversal group, R). At the end of the assigned feeding period, the animals were killed for biochemical and morphological evaluation.

The primary visual cortex was examined by quantitative morphometric methods. Sections were taken from the dorsolateral surface of area 17 for staining by rapid Golgi, Bodian and Nissl methods. For stereological analysis, five 1 mm^2 blocks from each animal were postfixed in osmium tetroxide, and half micron sections were stained with toluidine blue. For each of the cortical layers and sublayers, its thickness and the numerical densities of neurons, oligodendrocytes, astrocytes and microglia were determined by the methods of O'Kusky and Colonnier (10, 11). These values were used to calculate the number of each cell type in each layer under 1 mm^2 of cortical surface.

RESULTS

At 6 and 12 months of age, no differences were found in the number or numerical density of neurons in any of the cortical layers. However, there were some significant

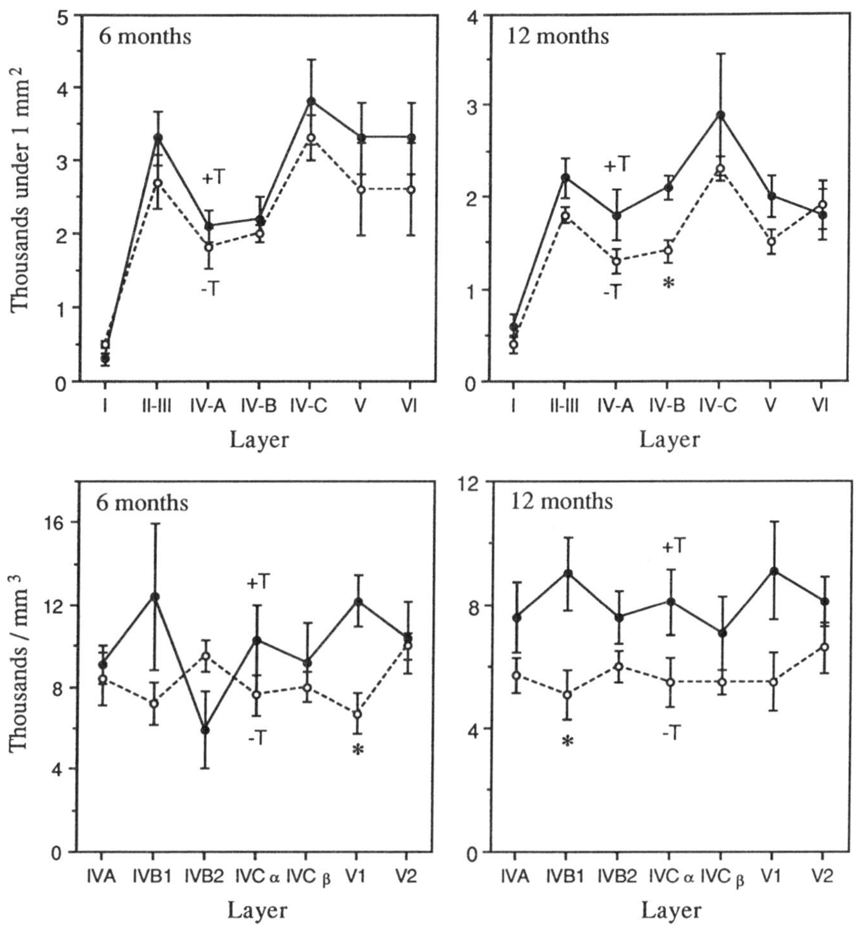

Figure 1. The number (upper graphs) and numerical density (lower graphs) of astrocytes in layers of the primary visual cortex of taurine-supplemented (+T, closed circles) and taurine-deprived (-T, open circles) rhesus monkeys at ages 6 and 12 months. For numerical density, only the subdivisions of layer IV and V are shown. Asterisks indicate significant differences between the groups in individual layers ($p<0.03$).

differences in quantitative measurements of the three glial cell types. Astrocytes were consistently lower, both in numerical density and in number, in the taurine-deprived than in the supplemented group at both 6 and 12 months in almost all layers (Figure 1), although the differences were significant only for numerical density in layer V-1 at 6 months and for both number in layer IV-B and density in layer IV-B1 at 12 months.

The overall thickness of the cortex did not differ between groups at either age. However, significant increases were found at 12 months in the thickness of layers I and VI in the taurine-deprived group compared to both the supplemented and reversal groups. The increased thickness of layer I in the deprived group corresponded to significant increases in the number of oligodendrocytes in this layer (Figure 2). Oligodendrocytes were the most numerous cell type in this layer in the deprived group, more numerous than neurons (1800 vs. 1600 per mm^2); in contrast, there were twice as many neurons as oligodendrocytes in this layer in the taurine-supplemented group (2200 vs. 1100 per mm^2).

In the reversal group, the total number of astrocytes was similar to that for the 12-month taurine-deprived group (Figure 3); values for the individual cortical layers were similar to those in the deprived group or intermediate between the deprived and supplemented groups. In addition, some differences emerged which were unique to this group. The overall number and density of neurons was 22-28% lower than in the other two groups (Figure 3), with significant density differences in layers IV-Ca and V-1 (Figure 4). This pattern resembled that found in the 3-month-old taurine-deprived group, which had lower neuronal densities in layers IV-Ca and V-2 than its age-matched supplemented group. Effects in the reversal group were most striking in layers II-III, where there were significantly lower numbers of neurons, astrocytes and microglia than in the supplemented group (Figure 5).

DISCUSSION

These data suggest that the primary visual cortex, like the retina, shows some persistent effects of postnatal taurine deprivation despite the normalization of tissue taurine levels by 12 months of age. It is not clear if these effects are specific to the primary visual cortex or represent a more general change in cortical development. Our preliminary results for frontal cortex indicate that the effects of taurine deprivation are less prominent and do not follow the same pattern seen in visual cortex. To the extent that the effects are specific to visual cortex, they could be due to diminished input from the retina, representing a secondary effect of the retinal morphological changes in these animals. Postnatal deprivation of light or pattern stimulation alters the pattern of synaptic connections and thus the physiological responsiveness of neurons in area 17, and these effects are largely irreversible if the lack of normal visual input is not corrected within an early critical period (1, 7, 19). However, visual deprivation does not produce the morphological effects seen here (3, 10, 11).

It is interesting that the most prominent effects were on astrocytes, given the normally high concentration of taurine in these glial cells and their postulated role in osmoregulation in nervous tissue (16).

The reasons for the findings in the reversal group are unclear. For those parameters, such as the number and density of astrocytes, which resembled the taurine-deprived group, the results suggest a critical period during the first 6 months in which these aspects ofcortical development may be persistently altered by taurine deprivation. The unique results

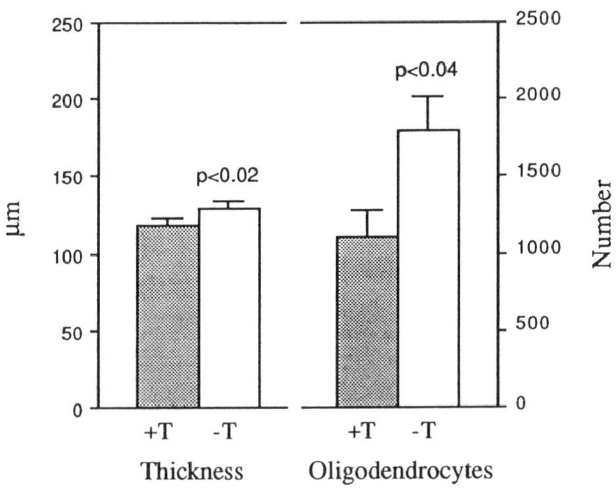

Figure 2. The thickness and number of oligodendrocytes in layer I of the primary visual cortex in taurine-supplemented (+T, shaded bars) and taurinedeprived (-T, open bars) rhesus monkeys at 12 months. Both values were significantly higher in the deprived group.

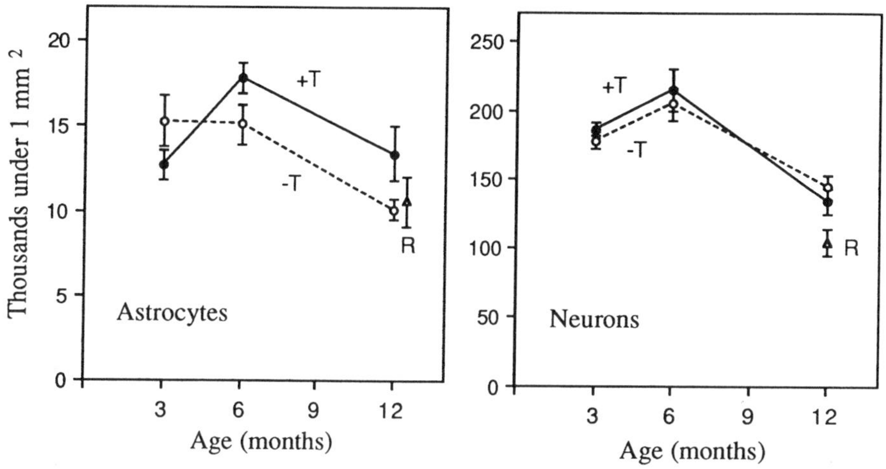

Figure 3. The total number of astrocytes and neurons in a 1 mm² column of visual cortex at 3, 6 and 12 months of age in taurine-supplemented (+T, closed circles) and taurine-deprived (-T, open circles) monkeys and in the reversal group (R, triangles) at 12 months.

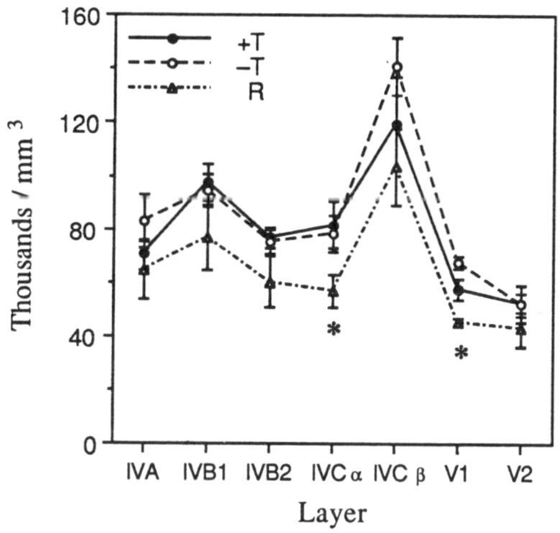

Figure 4. Neuronal densities in subdivisions of visual cortex layers IV and V in the taurine-supplemented (+T, closed circles), taurine-deprived (-T, open circles) and reversal (R, open triangles) groups at 12 months. Asterisks indicate layers for which the reversal group had significantly lower values ($p<0.05$) than either the supplemented or the deprived group.

for this group are more difficult to explain. It is noteworthy that in layers II-III, where these effects were most evident, the numbers of neurons, astrocytes and microglia all normally decrease between 6 and 12 months, and in all three cases this decrease was exaggerated in the reversal group. O'Kusky and Colonnier (10) noted that these layers undergo the greatest degree of postnatal change in several parameters, including thickness and the number and density of synapses, all of which show a peak at 6 months followed by a decline to adult values. These layers also appear to be sensitive to alterations in physiological responsiveness induced by visual deprivation over a longer postnatal period than other layers of visual cortex (6).

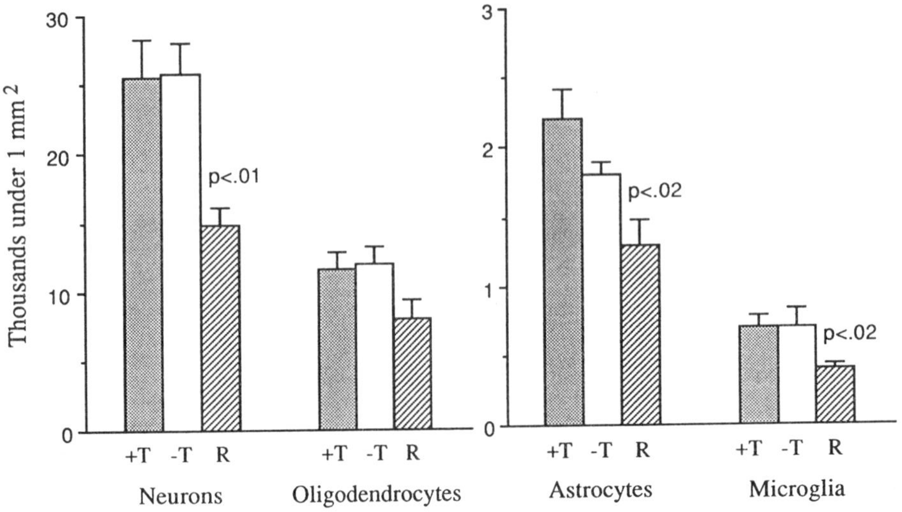

Figure 5. Numbers of neurons and glial cells in layers II-III of visual cortex in the taurine-supplemented (+T, shaded bars), taurine-deprived (-T, open bars) and reversal (hatched bars) groups at 12 months. P-values are given for significant differences between the reversal and supplemented groups.

The effects of early taurine deprivation described here are considerably milder and less extensive than those found in the kitten (13, 14, 20). This is due in part to the special sensitivity of cats to taurine deprivation, but it is also due to the different stages of development involved. In kittens, deprivation was imposed during both prenatal and postnatal development. Adult cats, unlike adult monkeys, are dependent on dietary sources of taurine, and therefore the pregnant female can be depleted of taurine during gestation. In addition, kittens are born at a much earlier stage of brain development than rhesus monkeys. Furthermore, studies of kittens have examined the visual cortex at birth and at 8 weeks but not at the much later time points which were assessed in these monkeys. Thus, the effects described here occurred despite initiation of taurine deprivation at a relatively late stage of brain development and despite several months of recovery from taurine depletion.

The rhesus infant's visual system develops at approximately four times the rate of the human infant (2), so that a 6-month period of susceptibility to dietary taurine deprivation is equivalent to two years in the life of a human infant. However, the human infant, like the kitten, is considerably less advanced than the rhesus monkey in retinal and brain development at birth, and so may be more vulnerable to effects of taurine deprivation in the early postnatal period. Premature human infants, with their even greater immaturity and lower brain taurine concentrations, are likely to be still more susceptible to any alterations in neural development induced by taurine depletion. Thus, the presence of even subtle effects of taurine deprivation in the developing rhesus monkey brain reinforces the need to provide dietary taurine in formulas and parenteral feeding preparations for human infants.

ACKNOWLEDGEMENTS

This research was supported by National Institutes of Health grants HD-18678 and RR-00163 and by grants from the Medical Research Foundation of Oregon and the National

Retinitis Pigmentosa Foundation. Experimental diets were provided by Ross Laboratories. This is publication 1901 of the Oregon Regional Primate Research Center.

REFERENCES

1. Blakemore, C. and Vital-Durand, F., 1984, Development of the monkey's geniculo-cortical system, *in:* "Development of Visual Pathways in Mammals", J. Stone, B. Dreher and D.H. Rapaport, eds., Alan R. Liss, New York, pp. 463-467.
2. Boothe, R.G., Dobson, V. and Teller, D.Y., 1985, Postnatal development of vision in human and non-human primates, *Ann. Rev. Neurosci.*,8: 495-545.
3. Boothe, R.G., Vassdal, E. and Schneck, M., 1986, Experience and development in the visual system: anatomical studies, *in:* "Developmental Neuropsychobiology", W.T. Greenough and J.M. Juraska, eds., Academic Press, New York, pp. 295-315.
4. Imaki, H., Moretz, R., Wisniewski, H., Neuringer, M. and Sturman, J., 1987, Retinal degeneration in 3-month-old rhesus monkey infants fed a taurine-free human infant formula, *J. Neurosci. Res.*,18: 602-614.
5. Imaki, H., Jacobson, S.G., Kemp, C.M., Knighton, R.W., Neuringer, M. and Sturman, J., 1993, Retinal morphology and visual pigment levels in 6- and 12-month-old rhesus monkeys fed a taurine-free human infant formula, *J. Neurosci. Res.*,36: 290-304.
6. LeVay, S., Hubel, D.H. and Wiesel, T.N., 1980, Development of ocular dominance columns in normal and visually deprived monkeys, *J. Comp. Neurol.*,191: 1-51.
7. Movshon, J.A. and Van Sluyters, R.C., 1981, Visual neural development, *Ann. Rev. Psychol.*,32: 477-522.
8. Neuringer, M., Palackal, T., Kujawa, M., Moretz, R.C. and Sturman, J.A., 1990, Visual cortex development in rhesus monkeys deprived of dietary taurine, *in:* "Taurine: Functional Neurochemistry, Physiology and Cardiology", H. Pasantes-Morales, D.L. Martin, W. Shain and R. Martin del Rio, eds., Wiley-Liss, New York, pp. 415-422.
9. Neuringer, M., Sturman, J.A., Imaki, H. and Palackal, T., 1992, Regression of taurine depletion in rhesus monkeys deprived of dietary taurine throughout the first year, *in:* "Taurine: Nutritional Value and Mechanisms of Action", J.B. Lombardini, S.W. Schaffer and J. Azuma, eds., Plenum Press, New York, pp. 207-214.
10. O'Kusky, J. and Colonnier, M., 1982a, Postnatal changes in the number of neurons and synapses in the visual cortex (area 17) of the macaque monkey:A stereological analysis of normal and monocu-larly deprived animals, *J Comp Neurol*,210: 291-306.
11. O'Kusky, J. and Colonnier, M., 1982b, Postnatal changes in the number of astrocytes, oligodendro-cytes, and microglia in the visual cortex (area 17) of the macaque monkey:A stereological analysis in normal and monocularly deprived animals, *J Comp Neurol*,210: 307-315.
12. Palackal, T., Kujawa, M., Moretz, R.C., Neuringer, M. and Sturman, J.A., 1991, Laminar analysis of the number of neurons, astrocytes, oligodendrocytes and microglia in the visual cortex (area 17) of 3-month-old rhesus monkeys fed a human infant soy-protein formula with or without taurine supplementation from birth, *Devel. Neurosci.*,13: 20-33.
13. Palackal, T., Moretz, R., Wisniewski, H. and Sturman, J.A., 1986, Abnormal visual cortex develop-ment in kittens associated with maternal dietary taurine deprivation, *J. Neurosci. Res.*,15: 223-239.
14. Palackal, T., Moretz, R.C., Wisniewski, H.M. and Sturman, J.A., 1988, Ultrastructural abnormal-ities in the visual cortex of kittens from taurine-deficient mothers, *Brain Dysfunction*,1: 71-89.
15. Palackal, T., Neuringer, M. and Sturman, J., 1993, Laminar analysis of the number of neurons, astrocytes, oligodendrocytes and microglia in the visual cortex (area 17) of 6- and 12-month-old rhesus monkeys fed a human infant soy-protein formula with or without taurine supplementation from birth, *Dev. Neurosci.*,15: 54-67.
16. Pasantes-Morales, H. and Schousboe, A., 1988, Volume regulation in astrocytes:a role for taurine as osmoeffector, *J. Neurosci. Res.*,20: 505-509.
17. Sturman, J.A., Messing, J.M., Rossi, S.S., Hofmann, A.F. and Neuringer, M., 1988, Tissue taurine content and conjugated bile acid composition of rhesus monkey infants fed a human infant soy-protein formula with or without taurine supplementation for three months, *Neurochem. Res.*,13: 311-316.
18. Sturman, J.A., Messing, J.M., Rossi, S.S., Hofmann, A.F. and Neuringer, M., 1991, Tissue taurine content, activity of taurine synthesis enzymes and conjugated bile acid composition of taurine-deprived and taurine-supplemented rhesus monkey infants at 6 and 12 months of age, *J. Nutr.*,121: 854-862.
19. Wiesel, T.N., 1982, Postnatal development of the visual cortex and the influence of environment, *Nature*,299: 583-591.

20. Xu, Y., Lu, P., Imaki, H. and Sturman, J.A., 1992, Feline maternal taurine deficiency: a quantitative morphometric and immunohistochemical study of newborn visual cortex, *Dendron*,1: 129-145.

THE ROLE OF TAURINE IN MAMMALIAN HEARING

W.E. Davies[1], N.J. Harding[1], I.S. Kay[2] and P.C. Hopkins[1]

[1] Department of Pharmacology, University of Birmingham, England
[2] Department of Biological Sciences, Manchester Metropolitan University, England

INTRODUCTION

Taurine was first reported to be present in the organ of Corti of the mouse cochlea by Davies and Owen in 1985 (6). Although it is the most abundant amino acid in this organ no categoric function has been attributed to it there and the concentration in mouse at least appears to remain at a constantly high level during development (6) unlike brain (1) where it decreases with development in most species until adulthood.

The evidence for its essentiality in hearing is rather conflicting. Davies et al. (5) showed that ferrets whose whole body taurine had been depleted by the administration of β-alanine, a competitive taurine transport antagonist, exhibited a marked decrease in their hearing threshold as measured by the brain-stem auditory evoked response (BAER). Vallecalle-Sandoval et al. (16) on the other hand were unable to show any deficits in taurine-deprived cats, a species that might be expected to be more susceptible to taurine deprivation than the ferret, although they found changes in some BAER parameters. The BAER peaks of the taurine deprived cats showed a significant decrease in their latency, in contrast to Tyson et al. (15) who showed a shortening of peak latencies in premature babies fed a taurine supplement. There could therefore, be a considerable disparity in the essentiality of taurine in the hearing of different species. Consistent with this view is the fact that guinea pigs treated with β-alanine, unlike ferrets, show no hearing deficits (Davies and Gorodkin, unpublished observations) (9).

With regard to the function of taurine in the cochlea, a variety of possibilities exist. It is released in vivo by sound stimulation (8) or by perfusing the cochlea with high potassium artificial perilymph (3,10). There is, however, some doubt as to whether this taurine is derived from neural elements or supporting elements as there are claims that the release is not calcium dependent (3,10) and Bobbin et al. (2) have shown that taurine is still released from cochleae where most of the organ of Corti has been destroyed by ethacrynic acid and kanamycin treatment.

Another possibility for the role of taurine in the cochlea is that of an osmotic regulator. It has long been accepted that the cochlea contains an active mechanical amplifier that increases the neural output of the auditory nerve in response to sound by several orders of

Taurine in Health and Disease, Edited by R. Huxtable
and D.V. Michalk, Plenum Press, New York, 1994

magnitude (12). Such an active process is almost certainly provided by the outer hair cells which outnumber the inner hair cells by about 3:1 in most species and whose direct output to the brain is minimal. Isolated outer hair cells have been shown to be able to shorten and lengthen in response to sound and to efferent stimulation (4,17). Very little is known about the molecular mechanism of this contractile process, its control or its precise contribution to the basilar membrane elasticity and the output of the inner hair cell. It is clear that any change in the turgor of these cells would have a profound effect on the elasticity and the dampening properties of the basilar membrane and hence the output of the inner hair cells to the auditory nerve.

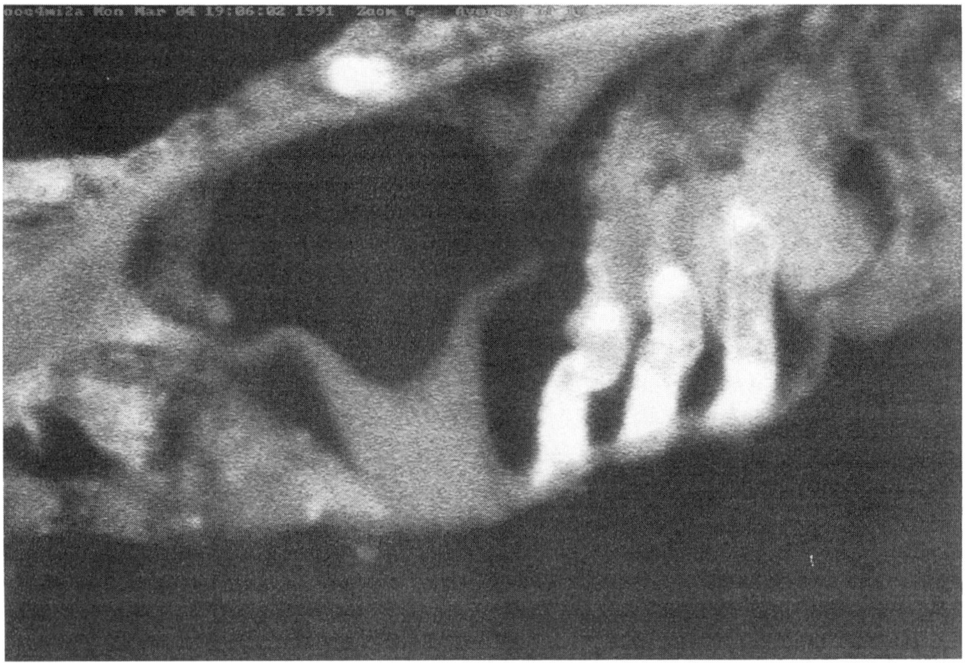

Figure 1. Transverse section through the organ of Corti (x1500) showing the localisation of taurine in the apical and basal regions of the outer hair cells. Non-specific staining can be seen in the vas spirale.

Yet another possibility for the role of taurine in the cochlea is in the control of ion channels. During sound stimulation of the cochlea, the major current flow is that of potassium ions from the scala media across the organ of Corti to scala tympani (7). Entry of the ions into the hair cells is through channels located within the hair bundles but their nature and their dependency on other ions is poorly understood. There is also a 100 fold calcium gradient across the basilar membrance (14), although this may be consequent to the potential gradient across the basilar membrane.

In an attempt to further our knowledge on the role of taurine in hearing, we embarked on two series of experiments, one consisting of the antibody localisation of taurine in the cochlea of the guinea pig and one involving the maintenance of cats on a taurine-free diet while monitoring their hearing thresholds.

METHODS

The histochemical localisation of taurine using a monoclonal antibody was basically as described by Harding and Davies (1993). Pigmented guinea pigs (albino animals often show abnormal auditory physiology and morphology) were anaesthetised and perfused upwards through the carotid arteries with 4% paraformaldehyde. After perfusion, the bullae containing the cochleae were removed, opened and the fixation completed by immersion in the perfusing fluid overnight at 4°C. The organs of Corti were then hand-dissected away from the modiolus (the central bony structure to which the organ is attached) and pieces 1 - 2 mm subjected to various antibody incubations. Typically, a piece of the basilar membrane, upon which the organ rests was washed three times for 2 min each with 2% Triton X-100 in 0.1 M sodium phosphate buffer pH 7.4, incubated at room temperature for 1 h with normal sheep serum (diluted x 50 with phosphate-buffered saline/Triton X-100) to remove non-specific staining and then incubated for 20 h at room temperature with a taurine/keyhold limpet haemocyanin antibody (11). After washing three times with phosphate-buffered saline, the whole structure was incubated for 1 h at room temperature in fluorescein linked sheep anti mouse antibody and washed three times with Triton X-100, phosphate-buffered saline. The complete structure was then dehydrated by successive immersion for a few minutes in 50%, 75% and 100% polyethylene glycol 400 (carbowax) in water. The organ of Corti was then embedded in carbowax at -25°C and sections at 25 μ cut in a cryostat. Control sections were obtained from pieces of basilar membrane treated in an identical manner with the omission of the taurine antibody incubation. The sections were viewed by laser scanning confocal microscopy (Biorad MRC 600 connected to an inverted Nikon Diaphot). Fluorescence was measured at k-515 nm in an optical section of full width half maximum of approximately 1 micron excited by an argon laser (k=488 nm).

TAURINE DEPRIVATION OF CATS

Three young male adult cats weighing between 3.7 and 4.3 kg were treated with a taurine-supplemented (500 mg/day) normal diet for 4 weeks. They were then weaned for 6 weeks on a diet containing 50% taurine free chow (Special Diets Services, Witham, Essex) before being maintained for 96 weeks on a taurine-free chow only. These animals were assessed weekly for whole blood taurine levels and hearing thresholds. Taurine was determined by the HPLC method of Hopkins et al., (1989) and the hearing thresholds ascertained by brain stem evoked potential measurement.

The brain stem audiometry was carried out on animals under general anaesthesia (Rompun, 0.76 mg/kg and Saffan, 1.25 mg/kg). Body temperature was maintained at 37°C by means of a thermostatically controlled heating blanket. The evoked responses were recorded with Ag/AgCl electrodes placed at the vertex and right mastoid referenced to a third electrode placed on the left mastoid. Acoustic stimuli were delivered free-field from a speaker placed 20 cm away from the animal in a sound proof chamber. Square wave clicks of 100 μs duration were employed at a repetition rate of 11 per s. The resulting potentials were amplified (Biodata PA400 at a gain of 2500) and filtered (0.008-10 kHz). They were sampled at a rate of 20 kHz for 10 ms following stimulation and an average of 1000 sweeps provided the final ABER. The resulting ABER was stored on a Tandon PCA20 micro-computer for subsequent off-line analyses.

One cat sacrificed after 65 weeks. After 96 weeks, the remaining cats were sacrificed by the administration of an overdose of Nembutal and immediate perfusion of the outer and inner ears with a fixative consisting of 2.5% glutaraldehyde in 0.05 BES buffer adjusted to 7.4 with NaOH. The cochlea were immediately removed from the temporal bone and stored at 4°C in the same fixative for at least one week or until processed for scanning electron microscopy essentially as described by Osborne et al., (1988) (13). The specimens were dehydrated in ethanol and acetone, dried by the critical point technique with liquid CO_2 and sputter-coated (Polaron) with platinum to a nominal depth of 25 nm. Specimens were examined at 40 kV in a JOEL 120 CX11 transmission electron microscope with a TEMS-CAN scanning attachment.

RESULTS

Figure 1 shows a transverse section through the organ of Corti at the level of the 2nd turn at a magnification of 1500. The only structures that appear positive are the three outer hair cells and the spiral artery (vas spirale). The latter staining is a non-specific staining, present also in sections incubated in the visualising antibody only, and caused by the affinity of sheep antibody for guinea pig immunoglobulins. The staining in the three outer hair cells was completely specific and was entirely absent from cells immersed only in the FITC-linked antibody.

The results on the taurine-deprived cats were equivical. The blood taurine levels reduced from a normal of about 639-719 μM of whole blood to a level of about 24-64 μM after 5 weeks of taurine-free diet after which the levels remained fairly constant. With regards to the hearing thresholds, no statistically significant changes were found in two of the three animals after 96 weeks. The third however, showed a gradual elevation of the threshold, reaching a deficit of 33 dB after 65 weeks of taurine deprivation, at which time it was sacrificed for the electron microscopic examination of the cochleae.

Figure 2 shows the scanning electron micrograph of the cochleae of the three animals along with the cochlea of a control animal of the same age. It can be seen that two of the animals appear to have perfectly normal cochleae whilst the third which had a 33 dB hearing deficit shows a huge drop-out of the outer hair cells. In line with what is seen in ototoxic or sound damaged cochleae, the inner hair cells appear to be unchanged.

DISCUSSION

The obvious accumulation of taurine in the outer hair cells would suggest that the role of taurine in the cochlea is pertinent to the motor activity of these cells. The relative absence of taurine from the inner hair cells, which have no known motor function and which are likely to have a similar array of ionic channel response to sound as the outer hair cells, might also suggest that its role is later in the transduction process than channel opening. A function as part of the unique contractile process exhibited by these cells or a role as cell volume controller are both possibilities. Our lack of knowledge of this particular contractile process with its very short contraction and relaxation time along with its two driving forces (sound stimulation and efferent pathway activation) makes it even more difficult to assess the importance of taurine in hearing.

With regard to the taurine-deprived cats, the numbers of animals involved and the different response by the different animals make it impossible to draw any conclusions on the essentiality of taurine for adequate hearing. An identical deprivation in the three animals produced a similar lowered blood taurine levels but vastly different hearing thresholds. In view of the findings of Vallecalle-Sandoval et al. (1993) who showed little change in the threshold of animals deprived of taurine for 16 months, it seems prudent to conclude that the response of the one animal in this study is an aberrent one. It is however, a possibility that a decreased taurine level might make an animal more vulnerable to some other auditory insult such as an infection.

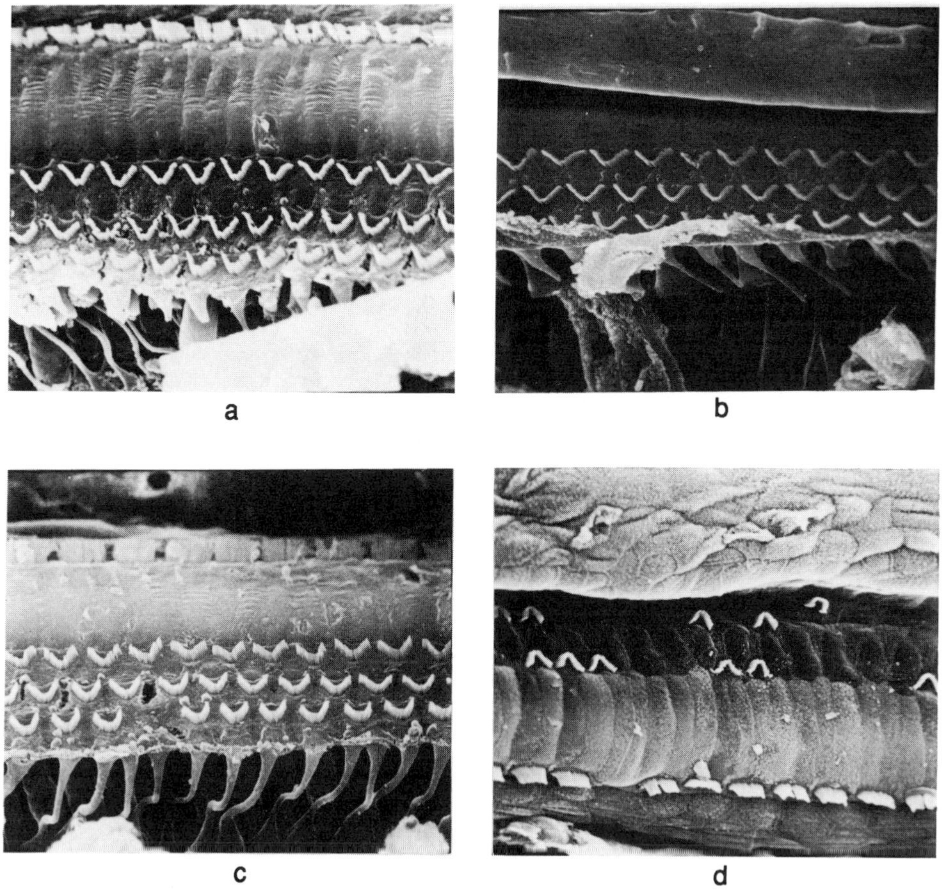

Figure 2. Scanning electron micrograph of the organ of Corti of four cats: **a:** control; **b; c:** two animals which showed no change in their hearing threshold after 96 weeks of taurine deprivation and **d:** the cat which showed a 33 dB hearing deficit after 65 weeks. While a complete array of outer hair cells can be seen in micrographs **a, b,** and **c**, massive hair cell loss can be seen in **d**.

ACKNOWLEDGEMENT

The expert help of S.D. Comis and M.P. Osborne with the electron microscopy is gratefully acknowledged.

REFERENCES

1. Agrawal, H.C., Davis, J.M., and Himwich, W.A. 1968, Developmental changes in mouse brain: weight, water content and free amino acids, *J.Neurochem.*, 15:97.
2. Bobbin, R.P., Ceasar, G., and Fallon, M. 1990, Potassium-induced release of GABA and other substances from the guinea pig cochlea, *Hearing Res.*, 46:83.
3. Bobbin, R.P., Ceasar, G., and Fallon, M. 1991, Changing cation levels (Mg^{2+}, Ca^{2+}, Na^{+}) alters the release of glutamate, GABA and other substances from the guinea pig cochlea, *Hearing Res.*, 54:135.
4. Brownell, W.E., Bader, C.R., Bertrand, D., and de Ribaupierre, Y. 1985, Evoked mechanical responses of isolated hair cells, *Science*, 227:194.

5. Davies, W.E., Kay, I.S., and Birnso, O.V. 1990, Taurine function in the auditory system, *in:* "Taurine: Functional Neurochemistry, Physiology, and Cardiology", Pasantes-Morales, H., Martin, D.L., Shain, W. and del Río, R.M., eds., Wiley-Liss, New York, pp. 397-405.
6. Davies, W.E. and Owen, C. 1985, The nature of neurotransmitters in the mammalian lower auditory system, *in:* "Auditory biochemistry", Drescher, D.G., ed., Thomas, Springfield,IL, pp. 244-257.
7. Davis, H. 1957, Biophysics and physiology of the inner ear, *Physiol.Rev.*, 37:1.
8. Drescher, D.G. and Drescher, M.J. 1985, HPLC analysis of presumptive neurotransmitters in perilymph, *in:* "Auditory Biochemistry", Drescher, D.G., ed., C.C. Thomas, Springfield, Illinois, pp. 50.
9. Horner, K.C., Huang, W., and Erre, J.P. 1993, The effect of a taurine modified diet on normal and hydropic ears of the guinea pig, *Hear.Res.*, 70:1-8.
10. Jenison, G.L., Bobbin, R.P., and Thalmann, R. 1985, Potassium-induced release of endogenous amino acids in the guinea pig cochlea, *J.Neurochem.*, 44:1845.
11. Magnusson, K.R., Madl, J.E., Clements, J.R., Wu, J.-Y., Larson, A.A., and Beitz, A.J. 1988, Colocalization of taurine- and cysteine sulfinic acid decarboxylase-like immunoreactivity in the cerebellum of the rat with monoclonal antibodies against taurine, *J.Neurosci.*, 8:4551-4564.
12. Neeley, S.T. and Kim, D.O. 1986, A model for active elements in cochlear biomechanics, *J.Acoust.Soc.Am.*, 79:1472.
13. Osborne, M.P., Comis, S.D., and Pickles, J.O. 1988, Further observations on the fine structure of tip links between stereocilia of the guinea pig cochlea, *Hearing Res.*, 35:99.
14. Russell, I.J. 1987, The physiology of the organ of Corti, *Brit.Med.Bull.*, 43:802.
15. Tyson, J.E., Lasky, R., Flood, D., Mize, C., Picone, T., and Paule, C.L. 1989, Randomized trial of taurine supplementation for infants $\leq$1,300-gram birth weight: Effect on auditory brainstem-evoked responses, *Pediatrics*, 83:406-415.
16. Vallecalle-Sandoval, M.-H., Heaney, G., Sersen, E., and Sturman, J.A. 1993, Comparison of the brainstem auditory evoked response (BAER) in taurine-deficient and taurine-supplemented cats, *Nutr.Res.*, 13:475-487.
17. Zenner, H.-P. 1986, Motile responses in outer hair cells, *Hearing Res.*, 22:83.

THE PLASMA CONCENTRATION AND RENAL HANDLING OF TAURINE IN HEALTHY CHILDREN AND IN PEDIATRIC PATIENTS WITH DISTURBED SULFUR METABOLISM

D. Rödenbeck, D.J. Byrd, and
J. Brodehl

Kinderklinik der Medizinischen
Hochschule Hannover
Konstanty-Gutschow-Straße 8
D-30623 Hannover, Germany

INTRODUCTION

In humans, taurine biosynthesis in liver is considered to be negligible. Therefore taurine could be conditionally essential for human infants and children (6). However, only a few studies have reported taurine values in healthy children as a function of age or in children suffering from diseases that might influence sulfur metabolism.

MATERIALS AND METHODS

The taurine concentration in blood and urine was investigated in 142 healthy children: 34 term newborns, 36 infants, 46 toddlers and 26 school children and adolescents. The probands were separated into 4 age groups: I) 0-8 weeks; II) 2-11 months; III) 12-60 months; and IV) 6-17 years. In addition, the study also included 12 children with infantile nephropathic cystinosis, 13 children with classic cystinuria, 6 children with homocystinuria (cystathionine ß-synthase deficiency) and 122 children with liver cirrhosis. All children had a normal protein intake. Morning fasting venous blood samples were centrifuged immediately.

Taurine excretion was measured in 24 h urine samples. Samples were deproteinized with sulphosalicylic acid (5%). Taurine concentrations were determined on LC 2000 and LC 6000 amino acid analysers (Biotronic, Munich, Germany) with ninhydrin detection and peak area quantification with the help of the integrator-calculator CR 2 A (Shimadzu, Kyoto, Japan) against an external standard. Values are expressed as means $\pm$ standard deviations. Data were analysed using the Mann-Whitney U-test with P set at < 0.05.

Taurine in Health and Disease, Edited by R. Huxtable and D.V. Michalk, Plenum Press, New York, 1994

RESULTS

Healthy Children

The values for taurine plasma levels adhered to a logarithmic normal distribution and were not influenced by age and sex. The cumulated mean value in all healthy probands was 79.4 ± 43.1 μMol/L.

There was a significant difference in urinary taurine excretion between the four age groups (Table 1). The mean urinary excretion rate in infants was 40% lower than that found in newborns. After that, the urinary taurine excretion increased slowly with age until a rate similiar to that in newborns was attained in school children (Group IV).

The age dependency of the urinary taurine excretion was reflected in the age dependency of the renal clearance of taurine (Table 1).

The values for the percent tubular reabsorption of taurine fluctuated widely (90.4 - 98.8%) but the means in the different groups did not differ significantly.

Table 1. Urinary taurine excretion and renal clearance rate in healthy children

Group	Age Range	N	Excretion (μMol/24 h/1.73 m^2)	N	Renal Clearance (ml/min/1.73 m^2)
I	0-8 week	20	436 ± 216	8	3.55 ± 1.08
II	2-11 month	17	173 ± 161[a]	6	2.56 ± 1.84
III	1-5 year	36	212 ± 36	10	3.32 ± 1.18
IV	6-17 year	32	481 ± 32[b]	7	5.42 ± 2.27[c]

Values are means ± SD for N children. Significant differences: [a]II versus I; II versus III ($p < .01$); [b]IV versus III ($p < .01$); [c]IV versus III ($p < .05$)

Nephropathic Cystinosis

In the 12 patients with nephropathic cystinosis both the plasma taurine mean value of 49.6 ± 15.8 μMol/L and the urinary excretion rates (Age group III: 149 ± 78 μMol/min/1-.73m^2; group IV: 257 ± 231 μMol/min/1.73m^2) were significantly lower than in the reference group. However, neither the mean renal clearance (Age group III: 2.41 ± 1.75 mL/min/1,73m^2; group IV: 5.21 ± 3.16 mL/min/1.73m2) nor the mean renal tubular reabsorption rates (Group III: 94.9 ± 4.7%; group IV: 89.9 ± 11.7%) were significantly different from those of age-matched controls.

Cystinuria

The mean concentration of taurine in plasma was significantly reduced to 48.4 ± 21 μMol/L. However, in these patients the urinary excretion rates, the renal clearance and the tubular reabsorption rate were in the normal range.

Homocystinuria

In these probands only a slight, insignificant decrease of taurine plasma levels was seen (54.8 ± 11.9 μMol/L). Renal taurine excretion and tubular reabsorption did not differ from controls.

Liver Cirrhosis

In the 122 patients with cirrhosis of the liver and raised plasma methionine levels the mean concentration of taurine in plasma was significantly reduced to 46.6 ± 24.4 μMol/L. The urinary excretion rates and the renal clearance rates varied over a wide range (coefficient of variation in the range of 90%) such that a statistical work-up was not possible. The mean tubular reabsorption rate was not different from normal.

DISCUSSION

Healthy Children

The mean plasma taurine level in children was found to vary between 57 and 88 μMol/L. No influence of age or sex was seen, except in premature babies or in newborns during the first hours of life. The mean value for taurine in plasma, 84 ± 20 μMol/L, measured in a larger group of 80 adults (15) corresponds well with our mean value, 79.4 ± 43.1 μMol/L, found in children.

Few data on urinary excretion of taurine in children have been published, and therefore no normal range has been established. O'Brien and Butterfield (11) described renal excretion ranging between 19 and 264 μMol/24 h/1.73m^2 in 3 children 5-6 weeks of age. Mårtensson et al. (10) reported a mean excretion rate of 725 ± 19 μmol/d/1.73m^2 in 10 adults, which is comparable with the mean value of our age group IV. The age dependency of taurine excretion was suggested first by Fowler et al.(5). Others (1,12) confirmed his findings but their values were not related to body surface area. The increased taurine excretion in our age group I may be a phenomenon of the immature renal tubule. A temporary inbalance between the glomerular and tubular system was supposed (2). Qualitative and age-related quantitative differences in protein intake, and therefore of the sulfur amino acids, may have influence on the taurine excretion.

Age dependency of renal taurine clearance has not been investigated in former studies. In an extensive review, Jacobsen and Smith (8) published a clearance rate between 0.2 and 20 ml/min to be normal and described a normal range of 0.1 - 3.8 ml/min/1.73m^2 for newborns aged 0 - 8 weeks. The reabsorption rate of 90.4 - 98.8% found in the present study is much lower than the rates for most of the carboxylic amino acids. The wide fluctuation of tubular reabsorption levels has also been seen in other studies (11,14) where ranges of 83.1 - 95.3%, respectively 95 - 99% were found. An age-dependency of tubular reabsorption of taurine was not seen.

Nephropathic Cystinosis

Cystinosis, due to defective carrier-mediated transport of cystine across the lysosomal membrane, is characterized by a massive accumulation of cystine in lysosomes. We found plasma taurine decreased in nephropathic cystinosis. In contradiction to previously published data, Crawhall et al. (4) reported plasma taurine to be elevated by 100% in cystinosis. Urinary taurine excretion in our patients was also measured to be sub-normal as reported by others (16). We found a renal clearance rate for taurine in the lower normal range [see also (9)] while a normal tubular reabsorption rate was found. These latter observations contrast sharply to the severe hyperaminoaciduria (Fanconi syndrome) seen in these patients, another indication that the renal handling of taurine is very different from that of the carboxylic amino acids. The reduced plasma concentration and the low normal urinary excretion in patients with nephropathic cystinosis may reflect a reduced taurine synthesis from immobilized cystine.

Cystinuria

While the plasma concentration of taurine in our cystinuric patients was significantly decreased (10), the renal handling of this amino acid was completely normal. The reason for the lowered plasma concentration is not obvious; even though patients with this renal tubular transport defect lose a massive amount of cystine via the urine the plasma concentration of the taurine precursor cystine is only minimally affected.

Homocystinuria

Only a few patients with this rare disease could be studied. It is surprising that the plasma concentration of taurine and the renal taurine excretion were more or less normal in patients in whome the metabolic pathway from methionine to taurine is blocked. The findings of others (7,13) has not been uniform.

Liver Cirrhosis

In our patients the mean plasma level of taurine was reduced by some 40%. This finding has been confirmed in our hospital in a different group of children with liver cirrhosis (3). Liver failure could cause a complex disturbance in sulfur and taurine metabolism, because the liver is the first organ that contacts alimentary taurine, and is also an organ that produces taurine, and has a daily loss of taurine by bile acid conjugation. The plasma concentration of the taurine precursor methionine in children with liver cirrhosis is negatively correlated to hepatic blood flow. Thus a reduced blood flow in the portal vein along less and damaged hepatocytes may reduce methionin uptake and metabolism. Increased formation of taurine-conjugated bile acids may decrease taurine reserves.

REFERENCES

1. Berger, H., Brenner, M. and Vetterli-Buchner, H., 1959, Aminoacidurie und Hyperaminoacidurie. S. Karger, Basel-New York.
2. Brodehl, J., 1969, Der renale Transport der Aminosäuren im Säuglings- und Kindesalter. Beihefte zum Archiv für Kinderheilkunde, 61. Heft; Enke, Stuttgart.
3. Byrd, D.J., Wiltfang, A., Rodeck, B., Latta, A., Burdelski, M. and Brodehl, J., 1993, The plasma amino acid profile and its relationships to standard quantities of liver function in infants and children with extra-hepatic biliary atresia and preterminal liver cirrhosis, *Eur. J. Clin. Chem. Clin Biochem.* 31: 197-204.
4. Crawhall, J. C., Lietman, P. S., Schneider, J. A. and Seegmiller, J. E., 1968, Cystinosis: Plasma cystine and cysteine concentrations and the effect of D-penicillamine and dietary treatment, *Am. J. Med.* 44: 330-339.
5. Fowler, D. I., Norton, P. M., Cheung, M. W. and Pralt, E. L., 1957, Observations on the urinary amino acid excretion in man: Influence of age and diet, *Arch. Bioch. Bioph.* 68: 452-466.
6. Gaull, G. E., 1986, Taurine as a conditionally essential nutrient in man, *J. Am. Coll. Nutr.* 5: 121-125.
7. Gerritsen, T. and Waisman, H. A., 1964, Homocystinuria, an error in the metabolism of methionine, *Pediatrics* 33: 413-420.
8. Jacobsen, J. G. and Smith, L. H., 1968, Biochemistry and physiology of taurine and taurine derivatives, *Physiol. Rev.* 48: 424-511.
9. Linneweh, F., Schaumlöffel, E., Graul, E. H., Hundeshagen, H., Kirsten, E., Kirsten, R. and Barthelmai, W., 1964, Über den Cystin-Metabolismus bei der Cystinose, *Klin. Wschr.* 42: 999-1005.
10. Mårtensson, J., Denneberg, T., Lindell, Å. and Textorius, O. 1990 Sulfur amino acid metabolism in cystinuria: A biochemical and clinical study of patients, *Kidney Int.* 37: 143-149.
11. O'Brien, D. and Butterfield, L. J., 1963, Further studies on renal tubular conservation of free amino acids in early infancy, *Arch. Dis. Childhood* 38: 437-442.

12. Parvy, P. R., Bardet, J. I., Rabler, D. M. and Kamoun, P. P., 1988, Age-related reference values for free amino acids in first morning urine specimens, *Clin. Chem.* 34:2092-2095.
13. Reddy, G. S. and Wilcken, G. E., 1982, Experimental homocystinemia in pigs: Comparison with studies in 16 homocystinuric patients, *Metabolism* 31:778-783.
14. Scriver, C. R. and Davies, E., 1965, Endogenous renal clearence rates of free amino acids in prepubertal children, *Pediatrics* 36:592-598.
15. Scriver, C. R., Gregory, D. M., Sovetts, D. and Tissenbaum, D., 1985, Normal plasma free amino acid values in adults: The influence of some common physiological variables, *Metabolism* 34:868-873.
16. Seegmiller, J. E. and Howell, R. R., 1961, Cystine metabolism in De Toni-Fanconi Syndrome with cystinosis, *Clin. Res.* 9:189.

BREAST MILK TAURINE AND ITS POSSIBLE INFLUENCE ON THE DEVELOPMENT OF BREAST MILK INDUCED JAUNDICE OF THE NEONATE - A HYPOTHESIS

M. Demirkol,[1] and H. Bohles[2]

[1]University Children's Hospital
Istanbul (Capa), Turkey
[2]Department of General Pediatrics
Johann-Wolfgang-Goethe University
Frankfurt/Main, Germany

INTRODUCTION

Breast milk-induced jaundice is a well-known entity in otherwise healthy breast-fed newborns (5). Unconjugated bilirubin characteristically increases after the fourth day of life and peaks around the second week. The pathogenesis of breast milk-induced jaundice is largely unknown although some hypotheses have become quite popular such as a cholestatic effect of breast milk pregnane-3,20- diol (2), or the glucuronyltransferase blocking effect of large amounts of long chain fatty acids in breast milk with high lipase activity (15).

Based on the observations of Yamada et al. (22) that breast-fed infants have higher serum bile acid levels than formula-fed infants, as well as higher taurine bile acid conjugate concentrations, we studied the influence of the breast milk taurine concentration on the development of breast milk-induced jaundice.

STUDY DESIGN AND METHODS

To the study were admitted 74 full-term infants (42 boys; 32 girls) born in the Department of Obstetrics, Istanbul Faculty of Medicine from July to September 1989. Exclusion criteria were a birth weight < 2500 g and any disease or malformation. All children were regularly and exclusively breast-fed. The infants were followed and their stool and blood samples obtained during regular home visits by the main investigator (M.D.). In case of a discernible development of jaundice, blood was obtained for bilirubin determination. Breast milk-induced jaundice was defined according to the criteria of Dahms et al. (4); namely, when there was exclusively indirect hyperbilirubinemia, no blood group incompatibility and when the hemoglobin concentration, reticulocyte count and erythrocyte morphology were normal. In cases of hyperbilirubinemia, serum concentrations were monitored on a daily basis. In infants considered to have jaundice, breast-feeding was interrupted for two days at the end of the first week of life in order to assure the diagnosis (11). During those two days the infants were formula-fed and breast milk was manually expressed. The serum bilirubin concentration was determined before and after this interruption. Infants that could not be assigned to the breast milk-induced jaundice group were followed as controls. Nine infants with ABO-incombatibility were excluded from the study. The study was approved by the ethical committee of the University of Frankfurt.

Stool Samples

After passage of meconium, stools were collected for three days (period 1). This was repeated at the end of the first, third and sixth week of life (periods 2 to 4). Thus 4 stool samples were obtained from each child. On the second day of each period, a small quantity of breast milk was expressed at 9 a.m. prior to feeding. All infants had the same kind of nappies (diapers). Stool samples were carefully obtained during each collection period. Before chemical analysis, stool samples were homogenized with distilled water (1:1; w:v) and stored at -20°C. Milk samples were deep frozen immediately after expression.

Chemical Analyses

Fecal Bile Acid Determination. Bile acids in homogenized stool samples (2 g) were extracted with methanol-HCl (21) and further purified using Sep-pak columns (18,19). Bile acids were separated using capillary gas-liquid chromatography (7,17). Lithocholic, desoxycholic, chenodesoxycholic and cholic acids served as standards. The total of all chromatographically separated bile acids was taken as 100%. Individual bile acid concentration was calculated as a percentage of the total bile acid concentration.

Glycine and Taurine Conjugation. Glycine and taurine conjugates were determined after enzymatic cleavage with cholylglycine hydrolase from *Clostridium perfringens* (EC. 3.5.; Sigma) followed by quantitative amino acid analysis (Biotronic LC 5000).

Taurine and Glycine in Breast Milk. Free taurine and glycine were determined in breast milk after protein precipitation with sulfosalicylic acid.

Statistical analysis of the data was performed with the nonparametric U-test according to Mann and Whitney.

RESULTS

After exclusion of 9 icteric infants with ABO incompatibility, a total of 65 neonates (36 boys; 29 girls) were investigated. Twelve infants (18.5 %) developed breast milk-induced jaundice. Serum bilirubin concentrations rose to 12.7 ± 3.4 mg/dl at the end of the first week of life and decreased after a two day interruption of breast feeding to 9.3 ± 2.6 mg/dl ($p < 0.001$). No infant had serum bilirubin concentrations requiring phototherapy and there were no abnormalities with respect to direct bilirubin (1.3 ± 1.0 mg/dl), hemoglobin (19.8 ± 1.5 g/dl), reticulocyte count (3.0 ± 2.2%), or erythrocyte morphology.

The concentrations of the individual bile acids in stools are shown in Table 1. When compared to controls only the chenodesoxycholic acid concentrations were significantly decreased ($p < 0.001$) during period 1. The mean lithocholic acid concentration was always higher in the stools of jaundiced infants. The ratio of glycine to taurine bile acid conjugates

Table 1. Fecal bile acids in breast mild-induced jaundice

	n	C (%)	CDC (%)	D (%)	L (%)	C/CDC	G/T
1.Period							
A	53	55.0 ± 17.5	32.3 ± 13.1	6.9 ± 5.6	6.4 ± 4.7	2.4 ± 2.0	0.9 ± 0.4
B	12	66.7 ± 143.1	15.3 ± 5.9[b]	5.6 ± 1.8	13.7 ± 7.4	4.7 ± 1.7[a]	0.8 ± 0.3
2.Period							
A	49	71.4 ± 12.2	25.9 ± 12.9	2.7 ± 5.8	0	3.9 ± 2.9	2.6 ± 1.4
B	12	69.2 ± 15.5	27.0 ± 14.3	2.4 ± 1.3	1.6 ± 0.7	3.8 ± 2.8	3.1 ± 1.7
3.Period							
A	37	74.1 ± 5.9	25.3 ± 5.9	0.7 ± 1.2	0	3.1 ± 1.0	2.9 ± 2.1
B	10	66.6 ± 16.4	30.0 ± 12.5	3.1 ± 2.1	0.9 ± 0.6	3.1 ± 2.5	2.3 ± 1.9
4.Period							
A	33	68.8 ± 18.1	31.2 ± 18.1	0	0	3.7 ± 3.5	2.9 ± 2.5
B	11	76.5 ± 10.6	22.5 ± 12.0	0.5 ± 0.2	0.5 ± 0.2	4.1 ± 2.7	1.1 ± 0.3

Cholic acid (C), chenodesoxycholic acid (CDC), desoxycholic acid (D) and lithocholic acid (L) as well as the ratios of C/CDC and glycine/taurine conjugation of bile acids in stools of patients without (A) and with (B) breast milk-induced jaundice. The results are presented as percent (%) of total fecal bile acids. [a] $p<0.5$, [b] $p<.001$ compared with corresponding nonjaundiced group.

showed no significant alteration throughout the study period; nevertheless, the ratio was highest in infants with breast milk jaundice at the end of the first week (period 2) (3.1 ± 1.7 versus 2.6 ± 1.4).

The concentrations of free taurine and glycine in breast milk are listed in Table 2. During period 2, breast milk from mothers of jaundiced infants showed significantly higher taurine concentrations (49.5 ± 21.6 μmol/dl) than in those without icterus (30.0 ± 6.3 μmol/dl)($p < 0.05$). In addition, during period 2 there was a significant linear correlation between the taurine concentration of breast milk and the glycine/taurine ratio of bile acid conjugates in stools of jaundiced infants ($r = 0.78$; $p < 0.01$) (Figure 1).

DISCUSSION

The 18.5% frequency of breast milk-induced jaundiced infants in our study population corresponds to the data given in the literature (1,10). Bile acid analysis in feces reflects the state of the enterohepatic circulation (12). The lower concentrations of chenodesoxycholic compared to cholic acid found during all sampling periods may be explained by the fact that this apolar bile acid is already passively absorbed in the upper jejunum. In contrast, the polar cholic acid is actively absorbed later in the ileum (12). The better resorption of chenodesoxycholic acid is followed by a higher serum concentration. Our results are, therefore, in accordance with those of Yamada et al. (22), who demonstrated chenodesoxycholic acid to be the dominant serum bile acid in infants. The secondary bile acids, namely desoxycholic and lithocholic acids, were found primarily in the stools of children who developed breast milk-induced jaundice. Two speculative theories can be proposed to explain the occurrence of these compounds: Either (a) desoxycholic and lithocholic acids are produced in the maternal gut, partially resorbed and passed on to the

Table 2. Breast mild glycine and taurine concentrations and their G/T ratios measured in mothers of infants without (A) and with (B) breast milk-induced jaundice

	n	Taurine (μmol/dl)	Glycine (μmol/dl)	G/T
1.Period				
A	53	34.5 ± 5.5	15.6 ± 9.5	0.5 ± 0.3
B	12	31.8 ± 6.9	29.4 ± 21.6	0.9 ± 0.7
2.Period				
A	54	30.0 ± 6.3	14.4 ± 11.3	0.5 ± 0.3
B	12	49.5 ± 21.6[a]	11.1 ± 6.3	0.2 ± 0.1
3.Period				
A	45	28.2 ± 9.0	17.7 ± 14.3	0.8 ± 0.3
B	12	33.9 ± 9.9	16.2 ± 12.9	0.5 ± 0.4
4.Period				
A	37	24.9 ± 13.5	9.9 ± 7.8	0.5 ± 0.7
B	11	33.9 ± 13.3	6.3 ± 5.1	0.2 ± 0.1

[a] $p < 0.05$ compared with corresponding nonjaundiced groups

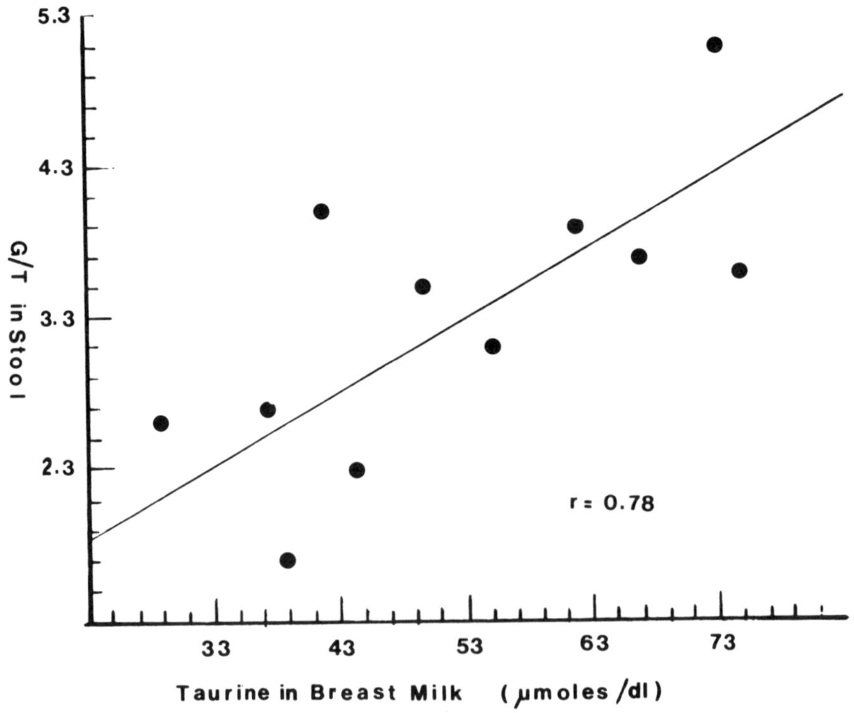

Figure 1. Correlation between the taurine concentration of breast mild and the glycine/taurine-ratio (G/T) of bile acid conjugates in stools of breast milk-jaundiced infants (4=0.78; $p<0.01$).

fetus, who then secretes them into the bile (8,13); or (b) lithocholic acid is a primary bile acid in fetal bile acid metabolism, which is initiated by mitochondrial degradation of the cholesterol side chain (3). Increased stool concentrations might therefore be indicative of increased fetal bile acid synthesis and a delayed switch-over to mature bile acid formation in jaundiced infants. The result would be persistence of only a minimal bile flow, as is normal during fetal life.

A major finding in our study was that the taurine concentration was highest in breast milk at the time when breast milk-induced jaundice developed. It is well-known that the hepatic taurine pool as well as the amount of taurine bile acid conjugates is very much influenced by the amount of taurine ingested (9). As taurine bile acid conjugates are more lipid-soluble than glycine conjugates, more unconjugated bilirubin, representing up to 10% of total bilirubin in the small intestine (20), may have been reabsorbed. Our data are therefore well in accordance with those of Gartner and Lee (6), who suggested that hyperbilirubinemia in breast-fed infants may result from exaggerated intestinal reabsorption of bilirubin. This theory is based on an observation made as early as 1937, when Ross et al. (16) demonstrated that jaundiced infants had lower amounts of bilirubin in their stools than anicteric infants. Poland and Odell (14), using agar feedings that kept bilirubin from being reabsorbed, were able to lower the serum bilirubin concentrations. With respect to this idea of an enhanced bilirubin resorption, we suggest the increased formation of taurine bile acid conjugates, resulting from a high taurine intake with breast milk, to be the pathogenetic basis of this hyperbilirubinemia.

ACKNOWLEDGMENTS

The study was supported by the Alexander von Humboldt Foundation.

REFERENCES

1. Adams, J.A., Hey, D.J., and Hall, R.T. 1985, Incidence of hyperbilirubinemia in breast vs. formula-fed infants, *Clin.Pediatr.*, 24:69-71.
2. Arias, I.M., Gartner, L.M., Seifert, B., and Furman, M. 1964, Prolonged neonatal unconjugated hyperbilirubinemia associated with breast feeding and a steroid, pregnane-3,20 β-diol in maternal milk that inhibits glucuronide formation in vitro, *J.Clin.Invest.*, 43:2037-2046.
3. Bock, H.E., Kaufmann, W., and Löhr, G.W. 1981, "Pathophysiologie", Georg Thieme Verlag, Stuttgart.
4. Dahms, B.B., Krauss, A.N., Gartner, L.M., Klain, D., Soodalter, J., and Aulk, P.A.M. 1973, Breastfeeding and serum bilirubin values during the first 4 days of life, *J.Pediatr.*, 83:1049-1054.
5. Gartner, L.M. and Arias, I.M. 1966, Studies of prolonged neonatal jaundice in the breast fed infant, *J.Pediatr.*, 68:54-55.
6. Gartner, L.M. and Lee, K.S. 1993, Jaundice in the breast-fed infant: new concepts of pathogenesis, *in:* "Human milk: Its Biological and Social Values", Freier, S. and Eidelman, A.I., eds., Excerpta Medica, Amsterdam, pp. 153.
7. Grundy, S.M., Ahrens, E.H., and Miettinen, T.A. 1965, Quantitative isolation and gas-liquid chromatographic analysis of total fecal bile acids, *J.Lipid Res.*, 6:397-409.
8. Hammons, J.L., Jordan, W.E., Stewart, R.L., Taulbee, J.D., and Berg, R.W. 1988, Age and effects on fecal bile acids in infants, *J.Pediatr.Gastroenterol.Nutr.*, 7:30-38.
9. Järvenpää, A.-L., Rassin, D.K., Kuitunen, P., Gaull, G.E., and Räihä, N.C.R. 1983, Feeding the low birth-weight infant. III. Diet influences bile acid metabolism, *Pediatrics*, 72:677-683681.
10. Kuhr, M. and Paneth, N. 1982, Feeding practices and early neonatal jaundice, *J.Pediatr.Gastroenterol.Nutr.*, 7:30-38.
11. Lascari, A.D. 1986, "Early" breast-feeding jaundice: clinical significance, *J.Pediatr.*, 108:156-158.
12. Lepage, G. and Roy, C.C. 1988, Fecal bile acids as a means of assessing the integrity of the enterohepatic circulation of bile acids and their microbial degradation, *J.Pediatr.Gastroenterol.Nutr.*, 7:6-19.

13. Lester, R., St.Pyrek, J., Littler, J.M., and Adcock, E.W. 1983, Diversity of bile acids in the fetus and newborn infant, *J.Pediatr.Gastroenterol.Nutr.*, 2:335-364.
14. Poland, R.L. and Odell, G.B. 1971, Physiologic jaundice; the enterohepatic circulation of bilirubin, *N.Engl.J.Med.*, 284:1-6.
15. Poland, R.L., Schultz, G.E., and Garg, G. 1980, High milk lipase activity associated with breast milk jaundiced infants, *Pediatr.Res.*, 14:1328-1331.
16. Ross, S.G., Waugh, T.R., and Malloy, H.T. 1937, The metabolism and excretion of bile pigment in icterus neonatorum, *J.Pediatr.*, 11:397-404.
17. Setchell, K.D.R. 1983, Liquid-solid extraction, lipophilic gel chromatography and capillary column gas chromatography in the analysis of bile acids form biological samples, *in:* "Bile Acids in Gastroenterology", Dowling, R.H., Hofmann, A.F. and Roda, E., eds., MTP Press, Boston, pp. 1-18.
18. Setchell, K.D.R. and Worthington, J. 1982, A rapid method for the quantitative extraction of bile acids and their conjugates from serum using commerically available reversed-phase octadecylsilane bonded silica cartridges, *Clin.Chim.Acta*, 125:135-144.
19. Shackelton, C.H.L. and Whitney, J.O. 1980, Use of Sep-pak cartridges for urinary steroid extraction: evaluation of the method for use prior to gas chromatographic analysis, *Clin.Chim.Acta*, 107:231-243.
20. Siegenthaler, W. 1986, "Klinische Pathophysiologie", Georg Thieme Verlag, Stuttgart.
21. van den Ende, A., Rädecker, C.E., Mairuhu, W.M., and van Zanten, A.P. 1982, Improved extraction procedures for determination of bile acids in faeces, *Clin.Chim.Acta*, 121:95-109.
22. Yamada, M., Tazawa, Y., and Nakagawa, M. 1985, Alterations of serum bile acid profile in breastfed infants with prolonged jaundice, *J.Pediatr.Gastroenterol.Nutr.*, 4:741-745.

TAURINE IN THE NEWBORN AND INFANT UNDERGOING SURGERY

C. Stratmann, H. J. Goebel, M. Abel,[1]
and J. Holzki

Department of Anesthesiology
Children's Hospital and
[1]University of Cologne
Cologne, Germany

INTRODUCTION

The importance of taurine for the newborn and infant is still under discussion. According to theory, and both experimental and clinical trials, it seems to be necessary for the regular development of the central nervous system of newborn and infants (1,2). Clinical data in pediatric patients undergoing surgery are missing.

PATIENTS AND METHODS

A study was performed in 45 infants up to 5 months of postnatal age who required total parenteral nutrition for at least one week after gastrointestinal surgery. The body weight on the day of surgery ranged from 1,100 to 4,130 g. The patients were randomized into two groups: the taurine group comprised 24 patients, the control group 21. Both groups received standardized parenteral nutrition with amino acids either from the taurine-free commercially available solution or from the taurine-containing test solution. The amino acid infusion was started at the first postoperative day with 2 g/kg body weight and increased to 2.5 g/kg body weight at the second postoperative day and up to 3 g/kg body weight from the third postoperative day on. Beside routine parameters in blood and urine, plasma amino acid concentrations were measured on postoperative days 1, 2, 4, 6, and 8. At postoperative day 14 only 12 patients who were still under total parenteral nutrition were investigated, 4 in the control group and 8 in the taurine group.

RESULTS

Routine parameters were within normal limits for both groups. Developmental differences were not observed. The nitrogen balance was positive from the first postoperative day. Plasma amino acid levels were different for some amino acids, reflecting the different composition of the applied amino acid solutions. Plasma taurine concentrations were statistically different in the two groups ($p < 0.01$; post-hoc-test). After an initial decrease in both groups, there was an increase in the taurine group after postoperative day 4, whereas in the control group the decrease continued (Fig. 1). Two weeks after surgery, when only 8 patients in the taurine group and 4 patients in the control group required total parenteral nutrition, this trend continued. In both groups the taurine levels did not reach the preoperative values. No side effects related to the infused taurine were observed.

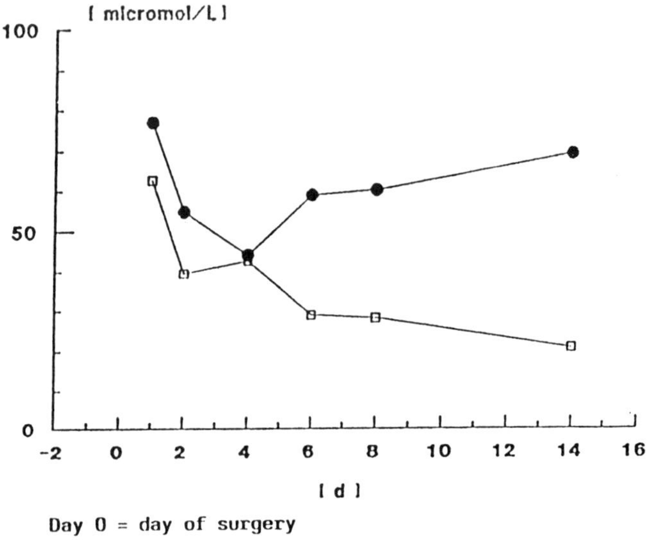

Figure 1. Mean values of the taurine plasma levels. In the control group, there is a continuous decrease of the values, whereas in the taurine group the values increase after postoperative day 4. Normal range according to (3): 19-265 μMol/L. ■ = taurine group. □ = control group.

CONCLUSION

In the newborn and infant undergoing surgery followed by total parenteral nutrition for at least one week, the external supply of taurine seems to be necessary for maintaining normal plasma levels. Discussing the importance of taurine intake in total parenteral nutrition in infants, we have to rely on plasma levels, as there is no defined taurine deficiency syndrome in man. But since a delay of cerebral development in infants due to taurine deficiency under long-term total parenteral nutrition has been observed by some authors (1), it seems to be useful to maintain normal taurine plasma levels to prevent possible damage.

REFERENCES

1. Gaull, G.E., 1989, Taurine in pediatric nutrition: review and update, *Pediatrics* 83:433-442.
2. Hayes, K.C., and Carey, R.E., 1975, Retinal degeneration associated with taurine deficiency in the cat, *Science* 188:949-951.
3. Pohlandt, F., 1977, Referenzwerte für Plasmaaminosäurekonzentrationen bei Neugeborenen. *Mschr. Kinderheilk.* 125:441-443.

TAURINE AND THE LUNG: PHARMACOLOGICAL INTERVENTION BY AEROSOL ROUTE

Jesús Covarrubias

Hospital Infantil
Mexico City, Mexico

INTRODUCTION

The last ten years have been productive in demonstrating the range of important roles of taurine in several species, including man. These findings constitute almost a subscience from which the practical fruits are yet to come.

The possible clinical actions of taurine, based on its described physiological actions, are many. Beautifully described by Huxtable (7) in a recent monograph, they are divided basically into nonmetabolic and metabolic ones (Table 1). In theory, it should be possible to modify pathological situations using taurine as a product that translates the physiological actions into therapeutical ones. Some of these possibilities have been tried with different degrees of success.

For practical purposes but based on solid scientific considerations we need to simplify the taurine therapeutic scenario by saying that there are only two main actions: membrane stabilization and calcium modulation, forgetting, just for practical purposes, any other considerations. For instance, is taurine an antioxidant compound? The answer is yes, but perhaps its effects are coming from the membrane stabilization and/or modulation of calcium homeostasis. Following this hypothesis, and ignoring the well-proved "chloramine story" we still can present taurine as an indirectly acting antioxidant.The same considerations can be made for radioprotection or metabolism and energy production. If we think for a moment in the long list of physiological processes that are calcium-mediated we can easily conceive the large number of therapeutic implications without the need of any other extra potential taurine actions.

WHAT HAS BEEN DONE UP TO NOW?

Leaving aside many interesting but incomplete clinical studies, the first real fact is that the cardiovascular area can be seen as a success because there is a good number of studies that prove the role of taurine, through its positive inotropic action, as an adjuvant therapy of congestive heart failure; and more important, the treatment of ischemic cardiac conditions in which the cardiac protection that taurine supplies through its calcium regulation and scavenger actions is outstanding (1,2,9). However, recent high quality publications have not been enough to convince doctors worldwide. What are the problems?

Table 1. Physiological actions of taurine

Non metabolic actions	Metabolic actions
Osmoregulation	Antioxidation
Calcium modulation	Radioprotection
	Metabolism and energy production

BARRIERS

Unfortunately, taurine is not patentable and lacks the attention of the pharmaceutical industry. Without this imput it is difficult to translate a project into a product.

Other reasons that have delayed development of taurine include the wrong doses, or wrong dosage forms, that have not taken into account the modern needs for good compliance. More than any other aspect, however, is the fact that taurine is more a drug that protects rather than modifies; more a prophylactic than a therapeutic. This is a difficult concept to be sold.

REASONS TO FOLLOW THE SEARCH

In Table 2, a list of good attributes for taurine is given, which we cannot ignore, specially when we know that "the real new and good drug" is the one that appears every ten years with a cost of 50 to 100 million dollars and which will be expensive and restricted in its use for the first number of years; and for the rest, the "new molecules" are a sort of "me-too-products" that eventually increase the quality of life of the producers.

In the process of selecting a field of research for taurine, we took into consideration the most clear action that should be a "need" for a target organ. The lungs were selected.

Table 2. Positive attributes of taurine

It is a naturally occurring substance
The most abundant nonmetabolized amino acid
A nontoxic substance even at high dosage
It can be administered by different routes
Enough well-known pharmacokinetics
Complete preclinical support
No formal contraindications
Very stable product
Good organoleptic characteristics
Inexpensive

OXIDANTS AND LUNG DISEASE

All tissues are vulnerable to oxidative damage, but due to the site of location, anatomy and function, the epithelial surface of the lungs is one of the most exposed tissues. The continuous contact with oxidizing atmospheric agents and xenobiotics creates a constant battle against the natural antioxidants, enzymes, and vitamins at the level of the lung epithelium. The oxidative damage creates an inflammatory process that is the basis of a variety of acute and chronic pulmonary diseases.

Some examples of pathological conditions in which the epithelial damage is mediated by oxidizing agents are: asthma, chronic bronchitis, pulmonary idiopathic fibrosis, the oxidative damage caused by cigarette smoke, and the pulmonary antioxidant deficiency in AIDS patients. In all of these cases, the ready access of the respiratory epithelial surface to therapeutic intervention (via aerosol) permits the direct administration of antioxidants correcting an oxidant-antioxidant imbalance directly at the site of the disease.

Because of the special properties of taurine, and due to the failure of vitamin E, catalase, superoxide dismustase, and other antioxidants, we decided to study this amino acid.

PRECLINICAL SUPPORT

Excellent animal data were generated by Ronald E. Gordon and co-workers from Mount Sinai Hospital in New York (5,6). They demonstrated, with bleomycin and paraquat -two model systems of oxidant lung injury - that taurine antagonizes damage. Also they gave evidence in other papers of how it prevents also acute lung injury induced by NO_2, and finally, the damage of ozone as well. At the same time, Melanie A. Banks and colleagues described the taurine uptake by alveolar macrophages and lung type II cells (3). S.N.Giri and Q. Wong (4) concluded that taurine alone or together with niacin ameliorates completely or partially the chemically-induced lung fibrosis. All these data, together with the evidence of Michio Kurachi and colleagues (8) that taurine administration suppresses the bronchial response to PAF, provides a basis for going to the human model.

Even if taurine is a naturally occurring substance, hospital ethics committees require a complete preclinical file, so the following data were prepared (Table 3).

Table 3. Taurine, preclinical studies.

Perinatal toxicity and teratogenesis in the rabbit
Subacute toxicity by intravenous route in the dog
Chronic toxicity by the oral route in the dog
Acute toxicity in the rat and mouse
Effect on fertility and reproductive cycle in the rat

Because our intention was to give taurine inhaled, other pieces of information were required: first, to test taurine toxicity in the rat, male and female, giving 300 and 1000mg/kg/day by aerosol route during 26 consecutive weeks. All the clinical, laboratory, and pathologic examination indicated a lack of toxicity of the compound. The second was to perform the pharmacodynamics of taurine by the aerosol route to determine its possible role in the field of asthma.

The actions of taurine on both bronchial anaphylaxis in sensitized guinea pigs and in vitro on bronchial tissue were studied.

The first study indicated that taurine as well as sodium chromoglycate is efficacious in inhibiting the acute bronchial reactivity in guinea pigs sensitized with treatments that induced IgE antibody production. The second study confirmed that the tracheal contraction is a c-dependent phenomenon and that taurine antagonizes it.

CLINICAL STUDIES

We prepared an sterile acqueous isotomic solution for aerosol containing 250 mg of taurine or placebo per vial of 6 ml to be administered with De Vilbiss equipment.

In the first study we tested, in 10 adult males with well-controlled chronic asthma, the ability of taurine to prevent the bronchoconstriction induced by the inhalation of ultrasonic nebulized distilled water in comparison against placebo and disodium chromoglycate.

RESULTS

There was no significant difference among the mean baseline value of FEV1 on the three study days. After placebo inhalation, distilled water bronchial challenge caused an average 31.4% decrease in FEV1 ($p < 0.04$); with mean baseline value of FEV1 2.50 ± 0.8, and after challenge 1.74 ± 0.22. After 250 mg inhaled taurine, distilled water bronchial challenge induced an average 5% decrease in FEV1 (not significant); with mean baseline value of FEV1 2.39 ± 0.8 and after challenge 2.27 ± 0.92. After 20 mg of chromoglycate inhalation, the same challenge induced an average 1% decrease in FEV1 (not significant); with mean baseline value of FEV1 2.46 ± 0.80 and after challenge 2.44 ± 0.82.

COMMENTS

Treatment with inhaled taurine significantly prevents bronchoconstriction induced by ultrasonic nebulized distilled water similar to protective effect of chromoglycate. This preliminary success created the condition to further explore taurine behavior using methacoline as a challenger. Two double blind-placebo controlled studies using 250 or 500 mg in a single dose a day during 2 months in the treatment of children or adult population affected by moderate chronic asthma were completed recently with good results.

Perhaps the main scope of this presentation is to create interest in other clinicians to generate other data that could confirm the activity and safety of this impressive amino acid.

REFERENCES

1. Azuma, J., Sawamura, A., Awata, N., Hasagawa, H., Ogura, K., Harada, H., Ohta, H., Yamauchi, K., Kishimoto, S., Yamagami, T., Ueda, E., and Ishiyama, T. 1983, Double-blind randomized crossover trial of taurine in congestive heart failure, *Curr.Ther.Res.*, 34:543-557.
2. Azuma, J., Takihara, K., Awata, N., Sawamura, A., Ohta, H., Hamaguchi, T., Tanaka, Y., Fukuda, K., and Kishimoto, S. 1985, Taurine and failing heart: experimental and clinical aspects, *in:* "Taurine: Biological Actions and Clinical Perspectives", Oja, S.S., Ahtee, L., Kontro, P. and Paasonen, M.K., eds., Alan R. Liss, New York, pp. 195-213.

3. Banks, M.A., Martin, W.G., Pailes, W.H., and Castranova, V. 1989, Taurine uptake by isolated alveolar macrophages and type II cells, *J.Appl.Physiol.*, 66:1079-1086.
4. Giri, S.N. and Wang, Q. 1992, Taurine and niacin offer a novel therapeutic modality in prevention of chemically-induced pulmonary fibrosis in hamsters, *Adv.Exp.Med.Biol.*, 315:329-340.
5. Gordon, R.E. and Heller, R.F. 1992, Section IV: Taurine and its action as an anti-oxidant in lung. Taurine protection of lungs in hamster models of oxidant injury: A morphologic time study of paraquat and bleomycin treatment, *Adv.Exp.Med.Biol.*, 315:319-328.
6. Gordon, R.E., Heller, R.F., and Del Valle, J.R. 1989, Membranes perturbations and mediation of gap junction formation in response to taurine treatment in normal injured alveolar epithelia, *Exp.Lung Res.*, 15:895-908.
7. Huxtable, R.J. 1992, Physiological actions of taurine, *Physiol.Rev.*, 72:101-163.
8. Kurachi, M., Hongoh, K., Watanabe, A., and Aihara, H. 1987, Suppression of bronchial response to platelet activating factor following taurine administration, *in:* "The biology of taurine: methods and mechanisms", Huxtable, R.J., Franconi, F. and Giotti, A., eds., Plenum Press, New York, pp. 189-198.
9. Milei, J., Ferreira, R., Llesuy, S., Forcada, P., Covarrubias, J., and Boveris, A. 1992, Reduction of reperfusion injury with preoperative rapid intravenous infusion of taurine during myocardial revascularization, *Am.Heart J.*, 123:339-345.

TAURINE LEVELS IN PLASMA AND PLATELETS IN INSULIN-DEPENDENT AND NON-INSULIN-DEPENDENT DIABETES MELLITUS: CORRELATION WITH PLATELET AGGREGATION

Flavia Franconi,[1] Federico Bennardini,[1]
Antonella Mattana,[1] Mauro Miceli,[2] Mila Ciuti,[2]
Maurizio Milan,[2] Alfredo Gironi,[2] Giancarlo Bartomomei,[2]
Roberto Anichini,[2] and Giuseppe Seghieri[2]

[1]Institute of Biochemistry, Facoltà di Farmacia
University of Sassari, Sassari, Italy
[2]Diabetes Unit, Spedali Riuniti, Pistoia, Italy

INTRODUCTION

In human volunteers it has been shown that taurine supplementation decreases collagen-induced thromboxane release and platelet aggregation (6). In addition, we found that, while in vitro taurine did not modify platelet aggregation per se, it potentiated the effect of anti-aggregating agents such as aspirin, papaverine, or sodium nitroprusside (13). These observations seem to indicate a possible role for taurine in the modulation of platelet function.

In diabetes mellitus there is an increase in platelet aggregation (2,5,11,15) and therefore it was interesting to evaluate the effect of taurine on platelet-rich plasma obtained from diabetic patients, as well as the taurine levels both in plasma and platelets obtained from healthy controls and diabetic patients. Moreover in insulin-dependent diabetes mellitus the effect of taurine supplementation was studied.

MATERIALS AND METHODS

The outpatients attending the Clinic of the Diabetes Unit of the Hospital of Pistoia entered the study over a 3 month period. Patients were excluded if they had an urinary albumin excretion rate > 20 μg/min, and/or glycated hemoglobin levels of above 8%, measured by HPLC. Patients were classified as being affected with insulin-dependent diabetes mellitus and non-insulin-dependent diabetes mellitus, according to the National

Diabetes Data Group Criteria (16). The healthy group was made up of hospital staff or blood donors. The healthy group was divided according to age into a "young" and an "older" subset to match it with insulin-dependent diabetes mellitus and non-insulin-dependent diabetes groups. The diet of the groups under study was assessed by means of self-administered standard questionnaires. Quantitative and qualitative differences in diet between controls and diabetic subjects were not significant (Tables 1 and 2). The control and diabetic groups differed in plasma glucose and glycated hemoglobin levels (Tables 1 and 2). No significant differences were observed in plasma total cholesterol, HDL-cholesterol, LDL-cholesterol, apoproteins A_1 and B, triglycerides, plasma and urine uric acid or in plasma creatinine between control and diabetic groups.

Table 1. Main characteristics of controls and patients with insulin-dependent diabetes mellitus

	Insulin-dependent patients	Control Subjects
Number	39	34
Sex (M/F)	17/22	16/18
Age (years)	32 ± 1.8	38 ± 2.6
Duration of diabetes (years)	9.1 ± 0.2	----
Body mass index (Kg/m^2)	24.1 ± 0.4	23.8 ± 0.4
Plasma glucose (mmol/l)	8.1 ± 0.1[a]	4.9 ± 0.01
Glycated hemoglobin (%)	6.9 ± 0.04[a]	4.4 ± 0.02
Albumin excretion rate (μg/min)	15.2 ± 0.6	13.2 ± 0.02
Caloric intake (Kj/kg/day	120.6 ± 3.9	136.8 ± 6.5
Protein caloric intake (Kj/kg/day)	12.8 ± 0.5	14.5 ± 0.6

Values are mean ± SEM; [a] $p < 0.001$ compared to control group

The above parameters were measured following standard techniques. Taurine was assayed in both plasma and platelets treated according to Hayes et al. (6) in duplicate by a HPLC method, using homoserine as internal standard after derivation with O-phthalaldehyde (10).

Platelet aggregation was determined in platelet-rich plasma prepared as previously described (13) and measured according to Born and Cross (1) using an aggregometer (Daichii model PA-3320, Kyoto, Japan).

RESULTS AND DISCUSSION

As shown in Table 3 the two control groups, which differed for age had similar taurine levels, both in plasma and in platelets, while insulin-dependent diabetic patients had reduced taurine levels both in plasma and in platelets. In non-insulin-dependent patients, taurine was, on the contrary, reduced only in platelets, while plasma concentrations were similar to control values.

Taurine added to platelet-rich plasma obtained from healthy patients does not modify the dose-response curve of arachidonic acid, and also the addition of taurine to platelet-rich plasma obtained from patients with non-insulin-dependent diabetes mellitus does not alter the aggregation induced by arachidonic acid. The whole dose-response curve is, on the contrary, shifted to the right when taurine is pre incubated with platelet-rich plasma of insulin-dependent diabetes mellitus patients (Fig. 1).

After 3 months of taurine supplementation (1.5 g/day per os), taurine levels in insulin-dependent diabetes mellitus patients were significantly increased in both plasma and in platelets (Table 4), while platelet aggregation induced by arachidonic acid was decreased (Table 5).

Table 2. Main characteristics of health controls and patients with non-insulin-dependent diabetes mellitus

	Non-insulin-dependent patients	Control Subjects
No.	13	29
Sex (M/F)	7/6	13/16
Age (years)	55 ± 2.5	50.6 ± 1.1
Duration of diabetes (years)	10 ± 2.2	----
Therapy (diet alone/oral/insulin)	4/4/5	----
Body mass index (KG/M^2)	28 ± 1.1	26 ± 0.56
Plasma glucose (mmol/l)	9.8 ± 0.3[a]	5.1 ± 0.2
Glycated hemoglobin (%)	7.8 ± 0.5[a]	5.4 ± 0.1
Caloric intake (Kj/kg/day)	121 ± 1.1	132 ± 1.1
Protein caloric intake (Kj/kg/day)	13 ± 0.1	14 ± 0.1

Values are mean ± SEM; [a] $p < 0.01$ compared to control group

Table 3. Plasma and platelet taurine levels in controls and in diabetic patients

	Control	Insulin-dependent diabetes mellitus patients	Non-insulin-dependent diabetes mellitus patients
Plasma (μmol/liter)	(31) 93.3 ± 6.3	(38) 65.6 ± 3.1*	----
	(29) 103.9 ± 7.2	----	(13) 139.6 ± 22.1
Platelets (μmol/mg protein)	(17) 0.99 ± 0.16	(17) 0.64 ± 0.007*	----
	(29) 0.95 ± 0.1	----	(13) 0.50 ± 0.1*

Values are mean ± SEM; number of experiments in brackets; * $p < 0.01$ vs controls

In insulin-dependent diabetes, taurine is depleted both in plasma and in platelets. The reason why there is a reduction in plasma and in platelet taurine is still a matter of debate: the diet factor cannot be considered responsible since both diabetic and control subjects were given the same protein intake; the intestinal absorption of taurine does not seem to be modified because taurine supplementation increased taurine levels; the absence of clinical nephropathy seems to exclude the hypothesis that taurine depletion may be related to impairment of renal function; an alteration of biosynthesis seems unlikely, because it is of minor importance in the maintenance of taurine levels in humans (9); a higher consumption of taurine may imply an increased anti-hypoxic effect (4,12); and taurine may well behave as a scavenger of hypochlorous acid and reduce lipid peroxidation (8/20). Recent studies show evidence of the possible role of oxygen radicals in the pathogenesis of diabetic complications (19).

Table 4. Effects of taurine supplementation on plasma and platelet taurine levels in insulin-dependent patients

	Base-line	After supplementation with taurine (1.5 g daily)
Plasma (μmol/liter)	67.9 ± 4.2	126.8 ± 12.3*
Platelets (μmol/mg protein)	0.64 ± 0.07	0.99 ± 0.14*

Values are mean ± SEM of 17 experiments, each performed in duplicate; * p <0.01 vs base-line value

Table 5. ED_{50} (mmol) of arachidonic acid in platelet-rich plasma obtained from healthy controls and insulin-dependent patients measured on maximal aggregation

Control subjects	Base-line	After supplementation of taurine (1.5 g daily)
(28) 0.77 ± 0.02	(17) 0.44 ± 0.07*	(17) 0.72 ± 0.04**

Values are mean ± SEM; * p < 0.001 vs controls and ** p < 0.001 vs base-line

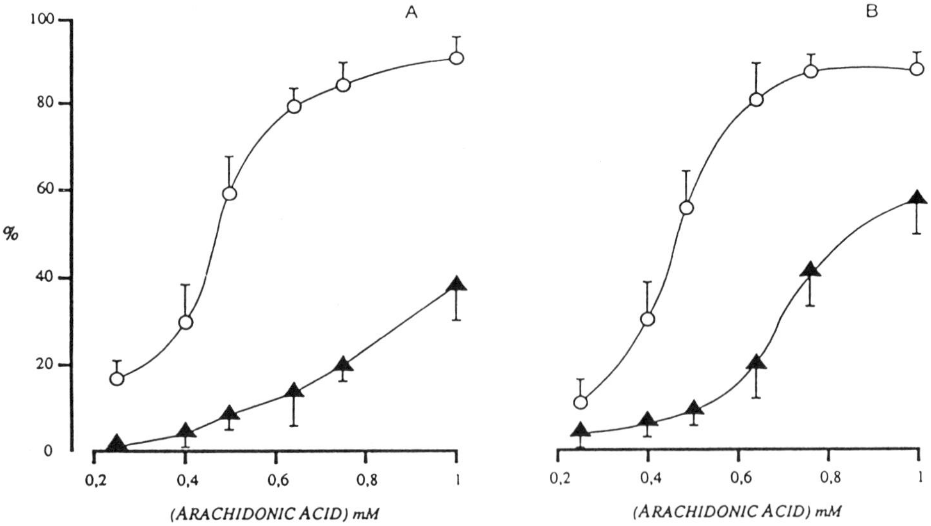

Figure 1. Dose-effect curves of arachidonic acid on platelet-rich plasma from diabetic patients after preincubation with (filled triangles) or without (open circles) taurine (10 mM). Values are means ± SEM for 6 experiments. Analysis of variance (ANOVA) revealed a significance of 0.09. **A:** % of maximal aggregation; **B:** Rate

Taurine supplementation increased plasma and platelet taurine levels and decreased platelet aggregation when the aggregating agent was arachidonic acid. Aggregation induced by arachidonic acid depends on thromboxane formation and generation of peroxides (7). Since taurine supplementation in healthy humans decreases platelet aggregation and thromboxane formation (6) it is likely that this can happen in insulin-dependent diabetic patients as well. The in vitro experiments suggest that plasma taurine is more important in the regulation of platelet function than platelet taurine content itself because only in platelet-rich plasma obtained from insulin-dependent patients does the addition of taurine reduce the aggregation.

Many differences exist between insulin-dependent and non-insulin-dependent diabetes, including protein metabolism (3). The reduction in plasma taurine in insulin-dependent diabetes, as well as the different effects of taurine pre incubated in vitro can be considered among these differences. The reason why in non-insulin-dependent patients taurine is reduced only in platelets remains unknown. It is possible that diabetes could bring about a change in the distribution of taurine, as shown in studies carried out in diabetic rats, where no variation in plasma levels were detected (17). In conclusion our results suggest that taurine supplementation can decrease platelet aggregability in patients affected with insulin-dependent diabetes mellitus.

REFERENCES

1. Born, G.U.R., and Cross, M.J., 1963, The aggregation of blood platelets, *J. Physiol.* 168:178.
2. Davi, G., Catalano, I., Averna, M., Notarbartolo, A., Strano, A., Ciabattoni, G., and Patrono, C., 1990, Thromboxane biosynthesis and platelet in type II diabetes mellitus, *N. Engl. J. Med.* 322:1769.
3. De Fronzo, R.A., and Ferrannini, E., 1992, Insulin actions in vivo: protein metabolism, *in:* "International Textbook of Diabetes Mellitus", Alberti K.G.M.M., De Fronzo, R.A., Keen, H., and Zimmet, P., eds., John Wiley, New York, pp. 467.
4. Franconi, F., Stendardi, I., Failli, P., Matucci, R., Baccaro, C., Montorsi, L., Bandinelli, R., and Giotti, A., 1985, The protective effects of taurine on hypoxia performed in the absence of glucose and on reoxygenation in the presence of glucose in guinea-pig heart, *Biochem. Pharmacol.* 34:2611.
5. Halushka, P.V., Rogers, R.C., Loadholt, C.B., and Colwell, J.A., 1981, Increased platelet thromboxane synthesis in diabetes mellitus, *J. Lab. Clin. Med.* 97:87.
6. Hayes, K.C., Pronezuk, A., Addesa, A.E., and Stephan, Z.F., 1989, Taurine modulates platelet aggregation in cats and humans, *Am. J. Clin. Nutr.* 49:1211.
7. Hember, M.E., Cook, H.W., and Lands, W.E.M., 1979, Prostaglandin biosynthesis can be triggered by lipid peroxides *Arch. Biochem. Biophys.* 193:340.
8. Huxtable, R.J., 1992, Physiological actions of taurine, *Physiol. Rev.* 72:101.
9. Jacobsen, J.B., and Smith, L.H., 1963, Comparison of decarboxylation of cysteine sulfinic acid-1-C_{14} and cysteic acid-1-C_{14} by human and rat liver and brain, *Nature* 200:575.
10. Jones, B.N., and Gilligan, J.P., 1983, O-phthalaldehyde pre-column derivation and reversed-phase high-performance liquid chromatography of polypeptide hydrolysates and physiological fluids, *J. Chromatogr.* 226:471.
11. Kaneti-Roa, A., Goldberg, R.E., and Walsh, P.N., 1984, Platelet coagulant activities in diabetes mellitus: evidence for platelet coagulant hyperactivity and platelet volume, *J. Lab. Clin. Med.* 103:82.
12. Malcangio, M., Bartolini, A., Bennardini, F., Malmberg-Aiello, P., Franconi, F., and Giotti, A., 1991, Effect of I.C.V. taurine on the impairment of learning, convulsion and death caused by hypoxia, *Psychopharmacology* 98:316.
13. Miceli, M., Bennardini, F., Mattana, A., Covarubias, J., Seghieri, G., and Franconi, F., 1992, Taurine modifies platelet sensitivity to aspirin, indomethacin, sodium nitroprussiate and papaverine, *Platelets* 3:265.
14. Milei, J., Ferreira, R., Llesuy, S., Forcada, P., Covarubias, J., and Boveris, A., 1992, Reduction of reperfusion injury with preoperative rapid intravenous infusion of taurine during myocardial revascularization, *Am. Heart J.* 123:339.
15. Mustard, J.F., and Packham, M.A., 1984, Platelets and diabetes mellitus, *N. Engl. J. Med.* 311:665.
16. National Diabetes Data Group, 1979, Classification and diagnosis of diabetes mellitus and other categories of glucose intolerance, *Diabetes* 28:1039.

17. Reibel, D.K., Shaffer, J.E., Kocsis, J.J., and Neely, J.R., 1979, Change in taurine content in the heart and other organs of diabetic rats, *J. Mol. Cell. Cardiol.* 11:827.
18. Waitzman, M.B., 1979, Proposed metabolic disfunctions in diabetic microthromboses and microangiopathy, *Metabolism* 28 Suppl.1:401.
19. Wolff, S.P., and Dean, J.R., 1991, Glucose autoxidation and protein modification. The potential role of "autoxidative glycosylation" in diabetes, *Biochem. J.* 245:243.
20. Wright, C.E., Tallan, H.H., Lin, Y.Y., and Gaull, H., 1986, Taurine: biological update, *Ann. Rev. Biochem.* 55:427.

LONG-TERM EFFECT OF TAURINE IN CONGESTIVE HEART FAILURE: PRELIMINARY REPORT

Junichi Azuma and the Heart Failure
Research with Taurine Group

The Third Department of Internal Medicine
Osaka University Medical School
Suita, Osaka 565, Japan

INTRODUCTION

Taurine is found in high concentrations in the heart, where levels are approximately 100 times that found in plasma. In most species, the intracellular taurine pool of the heart is very large, although significant species differences do exist. Taurine levels in hearts are normally fairly stable; however, elevated levels have been observed in patients suffering from congestive heart failure (11) while ischemic heart disease is associated with a decline in taurine content (7).

MODULATION OF CALCIUM TRANSPORT IN HEART

Recent interest in the role of taurine in the heart stems from reports that taurine is beneficial in treating patients with chronic congestive heart failure (2,3), and that cats fed a taurine-deficient diet develop a cardiomyopathy which is reversible upon dietary supplementation with taurine (15). Although both humans and cats have a low capacity to synthesize taurine, cats are particularly susceptible to taurine deficiency because of their requirement for taurine in the conjugation of bile acids. Interestingly, rats have high amounts of total body taurine. Hence, it may be providence that cats chase rats.

A large volume of literature is now available indicating that taurine modulates myocardial contraction; a biphasic contractile response as a function of the extracellular calcium. Hearts made hypodynamic by reductions in medium calcium content exhibit improved contractile performance upon exposure to pharmacological dose of taurine, while the same concentrations of taurine reduce contractile performance in hearts perfused with elevated levels of medium calcium. Since the ability of taurine to alter calcium movement has been proposed, the cardioprotective effect against several types of calcium overload-induced heart failure has been explored, including the cardiomyopathic hamster (1), cardiotoxicity induced by isoproterenol and doxorubicin (14,10), and calcium paradox- or oxygen paradox-induced myocardial injury (19,13).

Using fura-2 as a probe to measure intracellular free calcium content, we have demonstrated that addition of taurine to normal medium increased by 30% the calcium transient of normal myocytes, while reducing beating rate and having no effect on calcium content during diastole. By comparison, raising medium calcium concentration resulted in elevations not only in the calcium transient but also in free calcium during diastole (18). These differences in the cellular response to taurine and elevated medium calcium concentration indicate that the two agents increase intracellular calcium level by a different mechanism. The positive inotropic action of taurine is not mediated by an increase in cyclic AMP or inhibition of Na^+K^+-ATPase activity, indicating that the mechanism of action of taurine is different from that of well known positive inotropic agents, such as catecholamines, phosphodiesterase inhibitors and digitalis glycosides.

A likely explanation for these observations is that taurine can raise or lower intracellular calcium content, depending upon the status of the heart. The possible mechanisms underlying physiological and pharmacological actions of taurine on myocardial calcium transport are discussed by Schaffer et al (16).

CLINICAL USE OF TAURINE IN HEART FAILURE

The major goal in the treatment of patients with heart failure is to improve the quality and quantity of life. Presently, the general treatment of overt chronic congestive heart failure involves the elimination of excess salt and water (by diuretics and restriction of sodium intake), the reduction in workload (by vasodilators and restriction of physical activity), the reduction of the neurohormonal activation (by angiotensin converting enzyme inhibitors and possibly ß-adrenergic receptor antagonists) and the direct improvement of pumping performance (digitalis glycosides and other inotropic agents). Although several newer inotropic agents have been introduced, they generally have not been shown to improve symptoms consistently. In fact, most of them are associated with increased mortality, presumably because of induced arrhythmia and enhanced energy expenditure linked to inotropic stimulation (12). Based on these findings, there is a perception that long-term administration of positive inotropic agents that act by increasing either intracellular calcium, or cyclic AMP levels by stimulating the ß-adrenergic receptor or inhibiting phosphodiesterase, or both, may ultimately be deleterious (6).

We have performed several clinical studies examining the utility of taurine therapy in the management of congestive heart failure, and taurine has been found to be effective in treating congestive heart failure (2,3,4,5).

LONG-TERM EFFECT OF TAURINE IN MILD CONGESTIVE HEART FAILURE

Preliminary Report

One of the major goals of therapy in congestive heart failure is to relieve the symptoms during daily activities for a long period. We have started long-term (more than one year), randomized but open clinical trials to determine whether oral taurine added to conventional therapy can further improve clinical signs and quality of life measures in patients with chronic symptomatic, but mild, congestive heart failure. This is not a double-blind study, and has not been completed; therefore, we should be aware that this report is preliminary and that the results may include some bias.

METHODS

Patient Eligibility

Patients with chronic heart failure were enrolled at 24 study centers. All the patients had subjective evidence of reduced exercise capacity, as demonstrated by exertional symptoms (New York Heart Association functional class II or III) that had been stable despite treatment with digoxin, diuretics or an angiotensin-converting enzyme inhibitor.

Study Protocol

After the initial assessments of patients who fulfilled the criteria for enrollment, they were randomly assigned either to receive additional 3 g of taurine per day or none for one year of the study period. Any changes in the background therapy for heart failure were allowable if there was evidence of worsening heart failure severe enough to require a therapeutic intervention, or of improving symptoms enough to either discontinue or reduce a dose of background medications. Information about outcomes was obtained at three, six and twelve months from all patients.

Data were compared by the chi-square statistic, Fisher's exact test or Wilcoxon rank-sum test. All the analyses were two-sided, and differences with P values less than 0.05 were considered statistically significant.

RESULTS

After 103 patients had been enrolled, 54 patients were randomized to the taurine group. Clinical signs and symptoms of the patients were rather mild (mainly in New York Heart Association class II) and fairly stable before the entry into the study. By April, 1993, 25 patients receiving taurine and 23 patients without taurine completed the study. The clinical characteristics of these patients of two study groups were similar (Table 1). Four patients failed to reach the one-year end point. Of them, after the assessment of 6 months, one patient receiving taurine and another without taurine withdrew because of the patient's preference. One patient receiving taurine died of hepatoma. Data from these three patients were not included for the analysis. Another patient who was not receiving taurine died from worsening heart failure within 3 months.

During the study, both the patients and physicians were asked about any changes in the patients' overall condition. After 3 months of the study, more patients in the taurine than in the non-taurine group were considered by the investigators to have symptomatic improvement (60% versus 20%). When the patients assessed their own overall progress after one year of the study, seventeen patients receiving taurine reported that they felt better; that is, moderately better (30%) and slightly better (44%), as compared with those without taurine reported only slightly better (24%) ($p<0.01$) (Table 2).

After one year of the study, breathlessness on exertion became better in 39% ($p<0.01$) of the patients receiving taurine, but it improved in only 15% of the patients without taurine and 10% of those worsened (Figure l-a). Lack of energy, and swelling in ankles, legs etc. improved less frequently and deteriorated more frequently in the patients without taurine (improved rate of 38 and 10%, and deteriorated rate of 19 and 10%, respectively) than in those with taurine; none worsened in the taurine group, and 48 and 35% of those improved ($p<0.01$ and $p<0.05$, respectively) (Figure 1b,c). Deterioration in New York Heart Association functional class was reported in two patients, 10% of those without taurine, but none receiving taurine; however, improvement in the functional class

Table 1. Baseline characteristics of the study patients

Characteristics	Non-Taurine (N=21)	Taurine (N=23)
Age (yr)	71.1 ± 1.4	71.8 ± 1.4
Sex (M/F)	8/13	13/10
NYHA class		
II	16	20
III	5	3
Cause of heart failure		
Ischemic	24	13
Valvular	7	3
Cardiomyopathy	0	2
Hypertensive	7	9
Others	1	1
Concomitant medication		
Digitalis glycosides	14 (67)	14 (61)
Diuretics	15 (71)	19 (83)
Vasodilators	13 (62)	16 (70)
ACE inhibitors	1 (5)	3 (13)

Percentages receiving a given medication are indicated in parentheses

was reported 22% of those receiving taurine, but only 10% of those without taurine, although the difference was not statistically significant ($p < 0.1$).

When the patients' clinical course was assessed by investigators, the rate of overall improvement was 74% of those receiving taurine but only 24% of those without taurine. One patient receiving taurine required a positive inotropic agent, a vasodilator and an angiotensin-converting enzyme inhibitor for the treatment of worsening congestive heart failure after 3 months of taurine treatment, and finally completed the one year of study with taurine administration when he felt better; among those without taurine, three patients worsened by the investigators' assessment ($p < 0.01$)(Table 3).

DISCUSSION

Is a depression of cardiac contractility the primary defect in heart failure? If so, drug-induced increases in cardiac contractility should be beneficial for patients with heart failure. However, it is increasingly evident that oral inotropic agents that increase the intracellular concentration of cyclic AMP by stimulating the ß-adrenergic receptor or inhibiting phosphodiesterase are not necessarily effective in the long-term management of heart failure, and none have produced consistent improvement in symptoms or exercise tolerance, and many have been shown to cause an increase in mortality. The increase in intracellular calcium produced by these positive inotropic agents is believed to compromise the viability of myocardial cells.

Table 2. Patients' assessment of their overall improvement after one year of the study

Rate of improvement	Non-Taurine (N=21)	Taurine (N=23)
Much better	0	0
Moderately better	0	7 (30)
Slightly better	5 (24)	10 (44)
Unchanged	14 (67)	5 (22)
Slightly worse	1 (5)	0
Moderately worse	0	1 (4)
Much worse or death	1 (5)	0

U-test: <0.01 The number of patients is shown, with the percentages given in parentheses. Three patients withdraw before the final assessments, but their own assessments were moderately better at 6 months

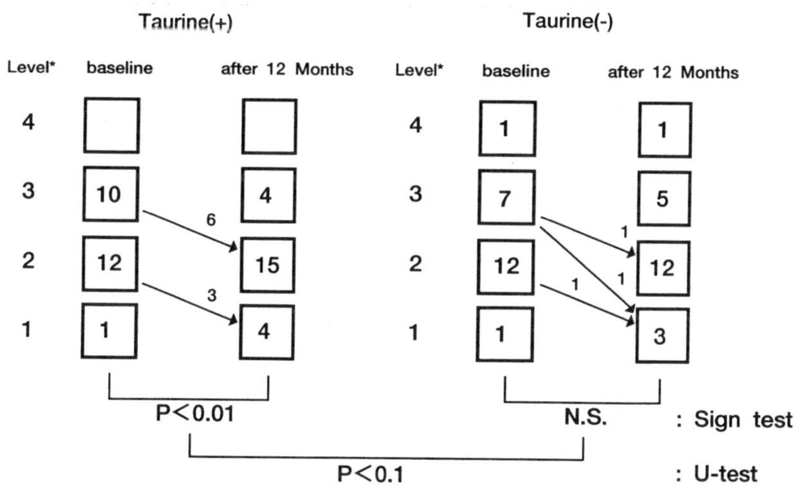

Figure. 1a. Breathlessness on exertion. The difficulty of climbing one flight of stairs was assessed. Level 1: Without breathlessness; Level 2: Slowly without stopping to rest; Level 3: Stopping to rest many times; Level 4: Very difficult.

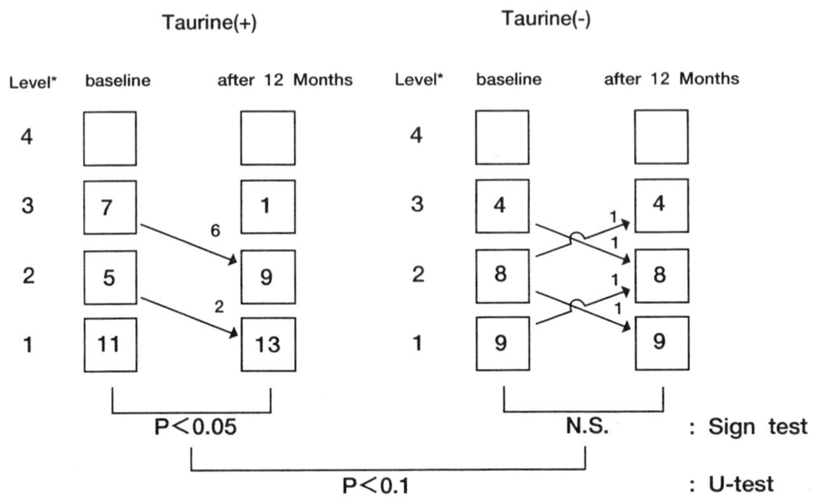

Figure 1b. Lack of energy. The general level of tiredness, fatigue or low energy was assessed. Level 1: No lack of energy; Level 2: Mild fatigue after walking, doing light work or any unusual activity; Level 3: Severe fatigue after walking, doing light work, or during daily activities; Level 4: Fatigue even at rest.

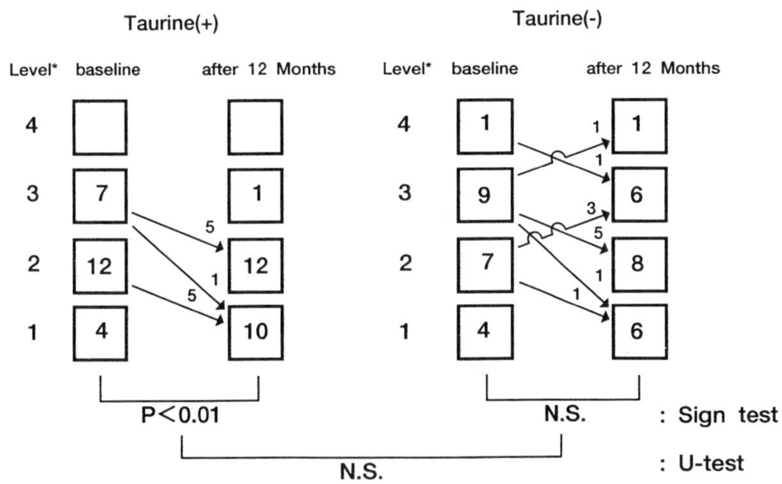

Figure 1c. Effect of taurine on edema. The presence of swelling in the lower extremities was assessed. Level 1: None; Level 2: Sometimes in the evening; Level 3: Usually in the evening; Level 4: All day long.

Table 3. Doctors' assessment of patients' overall progress after one year of the study

Rate of progress	Non-Taurine (N=21)	Taurine (N=23)
Much better	0	1 (4)
Moderately better	0	2 (9)
Slightly better	5 (24)	14 (61)
Unchanged	13 (62)	5 (22)
Slightly worse	2 (10)	0
Moderately worse	0	1 (4)
Much worse or death	1 (5)	0

U-test: <0.01 The number of patients is reported, with the percentage being given in parentheses.

Some recently developed agents, such as pimobendan, have been shown to exert their positive inotropic action by enhancing myofilament sensitivity to calcium. Interestingly, Galler et al. (9) and Steele et al. (17) found in skinned heart muscle fiber that part of the positive inotropic effect of taurine can be traced to a dose-dependent enhancement of myofilament sensitivity to calcium ion. Galler et al. proposed that taurine weakens charge-dependent protein-protein interactions because of its dipolar character. Thus, troponin-I and actin binding, which inhibits the force-generating myosin and actin interaction, could be weakened by taurine. This could lead to the calcium-sensitizing effect of taurine. It is hoped that this important property of taurine may obviate the harmful effects associated with an inotropic agent.

We examined whether taurine or guanidinoethane sulfonate (a taurine transport inhibitor) could affect the status of congestive heart failure produced by artificially induced aortic regurgitation in rabbits (20). In this study we focused our attention on the ability of taurine to prevent rapid progression of heart failure and hence to prolong life expectancy. Administration of taurine prevented a reduction in cardiac function when examined 8 weeks after induction of regurgitation. It also delayed the onset of symptoms associated with heart failure. Taurine content of the left ventricle of the heart failure rabbits was significantly increased, with further increase in myocardial taurine levels produced by taurine administration. The cumulative mortality rate at 8 weeks in the taurine-treated heart failure group was 10% compared with 53% in the untreated heart failure group and 75% in the guanidinoethane sulfonate-treated heart failure group. An important question is whether the increase in myocardial taurine content in the failing heart is a causal factor of heart failure, or whether this is secondary tp factors related to the disease process, such as the compensatory mechanism.

Recently, the first evidence was reported that therapy with a positive inotropic agent, vesnarinone, can favorably alter the natural history of chronic heart failure (8). This agent inhibits cyclic AMP phosphodiesterase, but, at the same time, it prolongs the duration of the action potential by exerting an effect on ion channels. Six months of therapy with 60 mg of vesnarinone per day resulted in lower morbidity and mortality and improved the quality of life of patients with congestive heart failure. However, a higher dose (120 mg) increased mortality. The improvement in quality of life by the lower dose was not associated with a change in either the ejection fraction or the New York Heart Association class. Vesnarinone increases mortality when used in high doses that increases cardiac contractility, but reduces mortality when used in low doses that are hemodynamically inconsequential. However, the mechanism by which vesnarinone reduced the combined risk of worsening heart failure and mortality remains unclear.

Neurohormonal activation is a well-documented accompaniment of congestive heart failure, which is one of the principal pathophysiological mechanisms of heart failure development and can adversely influence survival. This mechanism has applied to explain the efficacy of angiotensin-converting enzyme inhibitors. The clinical efficacy of digitalis could be related primarily to its modulating action on neurohormonal systems, including sympathetic-parasympathetic autonomic balance and baroreceptor-reflex function, rather than its effect on cardiac contractility.

Impaired diastolic function, possibly caused by abnormal calcium handling, is one of the potential causes of heart failure and also appears to trigger hypertrophy and the progression of left ventricular dysfunction. Taurine not only improves the heart by altering intracellular calcium movement, but lacks many of the adverse side effects of newer inotropic agents, such as increased heart rate and induction of arrhythmia. It also has an ability to either increase calcium availability for contraction or to protect against calcium overload injury. These characteristics of taurine are potentially important therapeutic considerations in the long-term management of patients with congestive heart failure.

The beneficial effect of taurine against disease progression in heart failure may be related to its modulating action on calcium transport systems in the heart. Other possible mechanism by which taurine benefits heart failure may include the reduction of the neurohormonal activation, the modulation on sympathetic-parasympathetic autonomic balance, or both. There is now an urgent need to resolve the mechanism of action of taurine against the progression of heart failure which may provide clues in understanding the fundamental mechanisms responsible for the progression of heart failure. Long-term, randomized, placebo-controlled, double-blind clinical trials, in which the combined end-point of major cardiovascular morbidity and mortality is defined, are necessary to determine definitively the effect of taurine before this agent will find a general approval as a medication for chronic congestive heart failure.

ACKNOWLEDGEMENTS

We thank Dr. Y. Tachi and Ms. E. Tamura for their laborious assistance.

REFERENCES

1. Azari, J., Brumbaugh, P., and Huxtable, R., 1980, Prophylaxis by taurine in the hearts of cardiomyopathic hamsters, *J. Mol. Cell Cardiol.* 12:1353-1366.
2. Azuma,J., Hasegawa, H., Sawamura, A., Awata, N., Harada, H., Ogura, K., and Kishimoto S., 1982, Taurine for treatment of congestive heart failure, *Int. J. Cardiol.* 2:303-304.
3. Azuma, J., Sawamura,A., and Awata, N., 1992, Usefulness of taurine in chronic congestive heart failure and its prospective application, *Jap. Circ. J.* 56:95-99.
4. Azuma, J., Sawamura, A., Awata, N., Ohta, H., Hamaguchi, T., Harada, H., Takihara, K., Hasegawa, H., Yamagami, T., Ishiyama, T., Iwata, H., and Kishimoto, S., 1985, Therapeutic effect of taurine in congestive heart failure: A double-blind cross-over trial, *Clin. Cardiol.* 8:276-282.
5. Azuma, J., Katsume, H., Kagoshima, T., Furukawa, K., Awata, N., Ishiyama, T., Yamagami, T., Ishikawa, H., Iwata, H., Kishimoto, S., and Yamamura, Y., 1989, Clinical evaluation of taurine in congestive heart failure: A double-blind comparative study using CoQ_{10} as a control drug, *in:* "Taurine and the Heart", Iwata, H., Lombardini, J.B. and Segawa, T., eds., Kluwer Academic Publishers, Boston, Dordrecht and London, pp. 75-97.
6. Colucci, W.S., Wright, R.F., and Braunwald, E., 1986, New positive inotropic agents in the treatment of congestive heart failure: Mechanism of action and recent clinical developments, *N. Engl. J. Med.*,314:349-358.

7. Crass III, M.F., and Lombardini, J.B., 1977, Loss of cardiac muscle taurine after acute left ventricular ischemia, *Life Sci.* 21:951-958.
8. Feldman, A.M., and the Vesnarinone Study Group, 1993, Effects of vesnarinone on morbidity and mortality in patients with heart failure, *N. Engl. J. Med.* 329:149-55.
9. Galler, S., Huttzler, C., and Haller, T., 1990, Effects of taurine on Ca^{2+}-dependent force development of skinned muscle fibre preparations, *J. Exptl. Biol.* 152:255-264.
10. Hamaguchi, T., Azuma, J., Awata, N., Ohta, H., Takihara, K., Harada, H., Kishimoto, S., and Sperelakis, N., 1988, Reduction of doxorubicin-induced cardiotoxicity in mice by taurine, *Res. Commun. Chem. Pathol. Pharmacol.* 59:21-30.
11. Huxtable, R., and Bressler, R., 1974, ELevation of taurine in human congestive heart failure, *Life Sci.* 14:1353-1359.
12. Katz, A.M., 1989, The myocardium in congestive heart failure, *Am. J. Cardiol.* 63:12-15.
13. Kramer, J.H., Chovan, J.P., and Schaffer, S.W., 1981, The effect of taurine on calcium paradox and ischemic heart failure, *Am. J. Physiol.* 240:H238-246.
14. Ohta, H., Azuma, J., Awata, N., Hamaguchi, T., Tanaka, Y., Sawamura, A., Kishimoto, S., and Sperelakis, N., 1988, Mechanism of the protective action of taurine against isoprenaline induced myocardial damage, *Cardiovasc. Res.* 22:407-413.
15. Pion, P.D., Kittleson, M.D., Rogers, Q.R., and Morris, J.G., 1987, Myocardial failure in cats associated with low plasma taurine: A reversible cardiomyopathy, *Science* 237:764-768.
16. Schaffer, S.W., Ballard, C., and Azuma, J., 1993, Mechanisms underlying physiological and Pharmacological actions of taurine on myocardial calcium transport. *This book.*
17. Steele, D.S., Smith, G2.L., and Miller, D.J., 1990, The effects of taurine on Ca^{2+} uptake by the sarcoplasmic reticulum and Ca^{2+} sensitivity of chemically skinned rat heart, *J. Physiol.* 422:499-511.
18. Takahashi, K., Harada, H., Schaffer, S.W., and Azuma, J., 1992, Effect of taurine on intracellular calcium dynamics of cultured myocardial cells during the calcium paradox, *in:* "Taurine: Nutritional Value and Mechanisms of Action", Lombardini, J.B., Schaffer, S.W., and Azuma, J., eds., Plenum Press, New York, pp.121-128.
19. Takihara, K., Azuma, J., Awata, N., Ohta, H., Sawamura, A., Kishimoto, S., and Sperelakis, N., 1985, Taurine's possible protective role in age-dependent response to calcium paradox, *Life Sci.* 37:1705-1710.
20. Takihara, K., Azuma, J., Awata, N., Ohta, H., Hamaguchi, T., Sawamura, A., Tanaka, Y., Kishimoto, S., and Sperelakis, N., 1986, Beneficial effect of taurine in rabbits with chronic congestive heart failure, *Am. Heart J.* 112:1278-1284.

NEUROACTIVE AMINO ACIDS IN SYNAPTOSOMES FROM FOCAL AND NONFOCAL TEMPORAL LOBE TISSUE OF PATIENTS WITH INTRACTABLE COMPLEX PARTIAL SEIZURES

P.-L. Lleu, D. Labiner, M. Weinand, and R. J. Huxtable

College of Medicine
University of Arizona Health Sciences Center
Tucson, Arizona 85724

INTRODUCTION

The putative involvement of disturbances in amino acid patterns in the pathogenesis of epilepsy has been vigorously debated (13,34,36,47,50). Amino acid levels in whole tissue removed from epileptic foci at surgery or autopsy have been compared with levels in surrounding nonfocal tissue or with tissue removed at autopsy or surgery from nonepileptic subjects. The advantages and disadvantages of each procedure have been argued at length (13,34,36,47,50). No clear pattern of changes has emerged, however (2,5,7,10,12,13,15,1-8,19,21,23,26,30,32,35,36,38,40,48,50). The literature on changes in amino acids in epilepsy has been reviewed (9,16).

As epilepsy involves a disturbance in neuronal excitability, it would appear that better insight as to the involvement of amino acids in the pathogenesis of the condition might be derived from a study of synaptosomal rather than whole tissue concentrations of amino acids. To the best of our knowledge, no such study has been performed on humans. We now report such a study on surgically removed human brain tissue.

METHODS

Patients

Brain samples were obtained from twelve adult patients undergoing temporal lobectomy for medically intractable complex partial epilepsy. Some characteristics of the patients are shown on Table 1. All patients had a standard preoperative evaluation and corticography to determine the site of their epileptic focus. In all patients the medial temporal stuctures were epileptogenic with the lateral temporal cortex being nonepileptogenic. Nonfocal (lateral temporal cortex) tissue was removed en bloc while focal tissue was removed by suctioning, utilizing a Cavitron ultrasonic aspirator. Approximately half of each sample was sent for standard pathological analysis while the remaining nonfocal and focal tissue was retained for neurochemical analysis.

A 27 year-old woman with probable Sturge-Weber syndrome underwent resection of an occipital lesion. The occipital cortex overlying the calcific lesion was removed half en bloc and half by suctioning.

All patients had barbiturate anesthesia and were operated on by the same surgeon. This study was approved by the University Human Subjects Committee.

Table 1. Characteristics of patients undergoing temporal lobectomy for intractable complex partial seizures

Patient	Gender	Age	Side of operation
M.L.	Female	29	Right
B.T.	Female	20	Left
C.J.	Female	19	Left
G.C.	Female	33	Left
C.C.	Female	39	Left
L.L.	Female	43	Right
Mean age		**31 ± 9**	
G.R.	Male	39	Left
V.R.	Male	50	Left
H.W.	Male	50	Left
L.J.	Male	38	Left
W.D.	Male	33	Right
T.M.	Male	38	Right
Mean age		**41 ± 6**	
Mean age of whole population		**36 ± 9**	

Rats

Six male Sprague-Dawley rats (380-500 grams) obtained from the Division of Animal Resources at the University of Arizona, were anesthetized with pentobarbital (50 mg/kg). On each animal, one side of the cerebral cortex was removed en bloc and the other side removed by suction with a Cavitron ultrasonic aspirator. The procedure used on a given side was alternated from one rat to the next. Rats were then killed by decapitation, following a protocol approved by the Institutional Animal Care and Use Committee.

Synaptosomal Preparation

Insufficient tissue was obtained from human surgery for isolation of the P_2B fraction. The P_2 fraction was, therefore, isolated. Immediately after removal of the tissue, it was placed in ice-cold 0.9% saline until synaptosomes were prepared. Synaptosomes (P_2 fraction) were prepared by the method of Gray and Whittaker (11). Tissue was homogenized in 0.32 M sucrose. The homogenate was centrifuged for 10 min at 1000 x g. The supernatant was centrifuged at 18,000 x g for 20 min. P_2 pellets were then resuspended in 0.32 M sucrose and protein concentration determined (31).

Determination of Amino Acids

Aliquots of P_2 fraction were and deproteinized with 3.5% sulfosalicylic acid. Deproteinized samples were filtered and 20 μl aliquots were mixed with 20 μl *o*-phthaldehyde solution (prepared by mixing 9 ml of solution of 343 mg of sodium tetraborate decahydrate with a solution of 54 mg *o*-phthaldehyde in 1 ml HPLC-grade methanol and 50 μl mercaptoethanol). One minute after the addition of *o*-phthaldehyde, 20 μl sample was injected onto an Altex ultrasphere ODS C18 reverse phase column (4.6 mm x 25 cm) equipped with a Beckman model 157 fluorescence detector. The column was developed with a mixture of two solvents. Solvent A consisted of 80% 40 mM sodium acetate (adjusted to pH 6.8 with sulfuric acid) and 20% methanol plus 1% tetrahydorfuran. Solvent B consisted of 20% 10 mM sodium acetate (adjusted to pH 6.8 with sulfuric acid) and 80% methanol. A mixture of 80% solvent A and 20% solvent B was run for 10 min, then a linear gradient from 20 to 50% B was run over the next 12 min, followed by a gradient from 50 to 100% B over the next 2 min. Aspartate, glutamate, serine, glutamine, glycine, threonine, taurine, alanine and GABA were eluted at 3.1, 3.8, 8.4, 9.1, 14.8, 15.4, 21.3, 23.8 and 24.7 min, respectively.

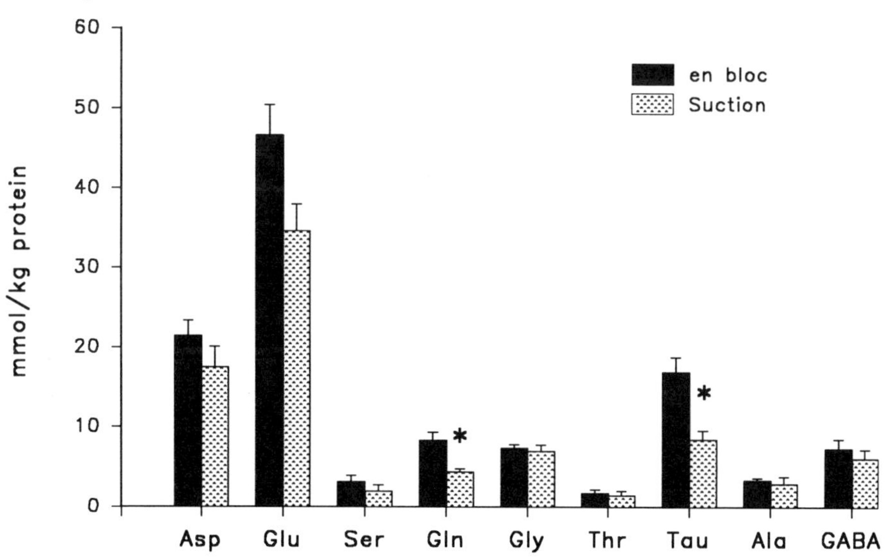

Figure 1. Free amino acid concentrations in synaptosomes (P_2 fraction) from temporal lobe of patients with intractable complex partial seizures: focal versus nonfocal tissue. Data are shown as means $\pm$ SD for 12 patients. $^*p < .05$ compared to nonfocal tissue.

Statistics

Amino acid concentrations from focal (suctioned) and nonfocal (en bloc) tissue were compared using the Student's paired t-test.

RESULTS

Temporal lobe tissue was removed from 12 patients undergoing temporal lobectomy to treat medically intractable complex partial epilepsy. The characteristics of the patients are given on Table 1. The group contains equal numbers of males and females, with the average age of the males being higher. Synaptosomes (P_2 fraction) were separately prepared from both focal and nonfocal temporal lobe tissue taken from each patient. Compared to nonfocal tissue, synaptosomes from focal tissue had significantly lower concentrations of six of the free amino acids determined (Fig. 1). Glutamine (16% of nonfocal) showed the biggest decrease, followed by taurine (27%) and alanine (40%). Four of these six amino acids are neurochemically significant: glutamate and GABA are neurotransmitters, taurine is a "neuromodulator" (17), with a number of transmitter-like actions, and glutamine serves as a glial buffer or reservoir for neuronal glutamate content (42-44). The lowering in synaptosomal glutamine, therefore, suggests that glial glutamine is also extremely low.

The combined concentrations of the neuroinhibitory amino acids, GABA plus taurine, were decreased by 59% in synaptosomes from focal as compared to nonfocal tissue. The neuroexcitatory amino acids, aspartate plus glutamate, showed a 52% fall in total concentration. The net change in the ratio of excitatory to inhibitory amino acids in the two preparations was statistically significant: 4.40 ± 1.23 in nonfocal synaptosomes versus 6.02 ± 1.94 in focal synaptosomes ($p < 0.05$).

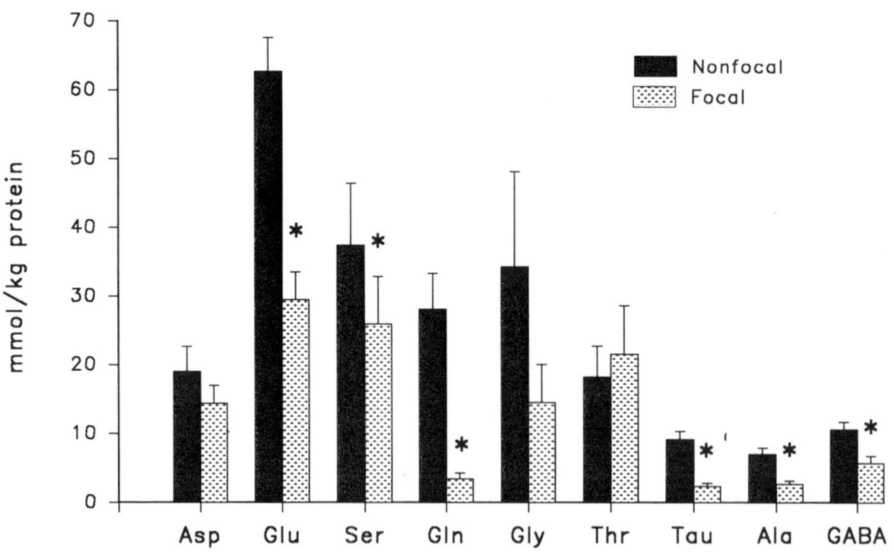

Figure 2. Free amino acid concentrations in synaptosomes (P_2 fraction) prepared from rat cerebral cortex. Tissue removed by suction is compared to tissue removed en bloc. Data are means $\pm$ SD for 6 preparations per point. $^{*}p < .05$ compared to tissue removed en bloc.

A major caveat in interpreting these findings, however, is the marked difference in the way the tissue was surgically removed. Nonfocal tissue was removed en bloc, while focal tissue was removed by suction. To examine the effect of removal method on synaptosomal amino acid content, therefore, we removed cortical tissue from the brains of separate groups of rats by the en bloc and suction methods, isolated P_2 fractions, and compared the resulting amino acid levels (Fig. 2). Again, there was a general tendency for amino acid levels to be lower in synaptosomes from tissue removed by suction. However, the difference was statistically significant only for taurine (reduced by 50%) and glutamine (reduced by 47%).

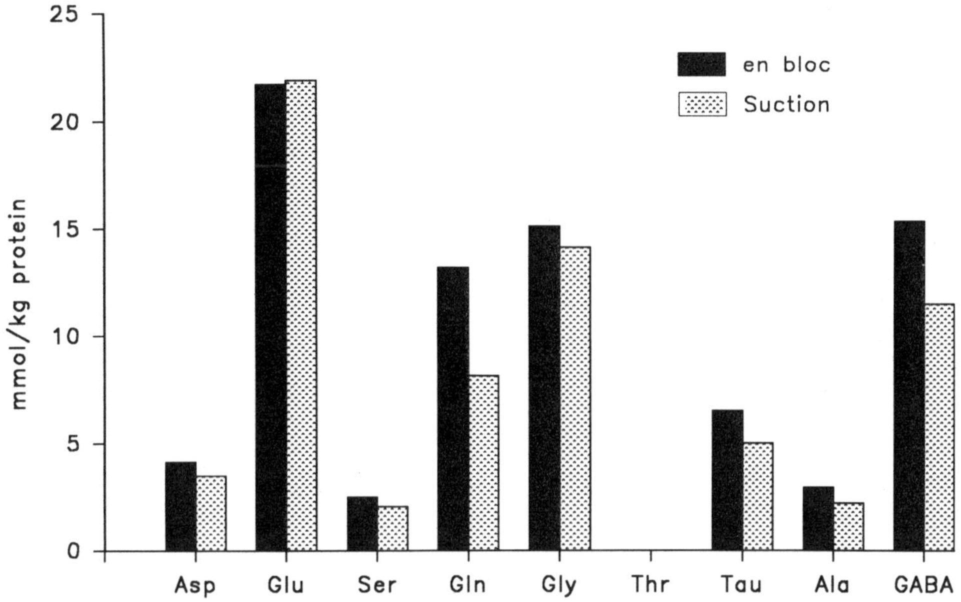

Figure 3. Free amino acid concentrations in synaptosomes (P_2 fraction) prepared from occipital cortex of a patient with Sturge-Weber syndrome. Tissue removed by suction is compared to tissue removed en bloc.

It is difficult to find human controls for comparison of removal methods. However, we did compare removal methodologies in a single patient with Sturge-Weber syndrome undergoing resection of an epileptic focus in the occipital region (Fig. 3). The seizures experienced by the patient were probably due to calcifications secondary to the syndrome. In this patient, the general pattern of amino acids in the occipital cortex was similar to that found in the temporal lobe (Fig. 1). The absolute concentrations of amino acids were lower, except in the case of GABA. We again found that amino acid concentrations in synaptosomes prepared from suctioned tissue tended to be lower (Fig. 3). Compared to the six amino acids showing lower concentrations in focal temporal lobe tissue, the Sturge-Weber patient showed no marked changes in glutamate or taurine levels in synaptosomes from suctioned tissue, while the decrements in the other four amino acids were less. The biggest decrement was in glutamine, the suctioned tissue having only 63% the level of that found in tissue removed en bloc. For the other amino acids, the average concentration in the suctioned tissue was 84% of that in the tissue removed en bloc.

Glutamine and glutamate share a close metabolic relationship. Although glutamine is stored largely in the glia, one of its functions is to "buffer" glutamate concentrations in the neuron. It is transported from the glia to the neurons and hydrolyzed to glutamate (3,14,39). A disturbance in the regulation of glutamate, therefore, could show itself as an alteration in the glutamine/glutamate ratio. This ratio is significantly depressed in synaptosomes from focal as compared to nonfocal tissue (Table 2), but is not significantly depressed in synaptosomes from rat cortex obtained by suction as compared to cortex obtained en bloc.

Table 2. Synaptosomal (P_2) glutamine:glutamate ratios

Species	Pathology	Area of Brain	Glutamine/glutamate ratio	
			En bloc	Suction
Human	Complex partial seizures	Temporal lobe	0.45 ± 0.23	0.10 ± 0.08*
Rat		Cerebral cortex	0.178 ± 0.035	0.128 ± 0.022
Human	Sturge-Weber syndrome	Occipital cortex	0.608	0.372

*$p < 0.05$ compared to en bloc tissue

DISCUSSION

Taurine and glutamine are the two amino acids showing the greatest depression in concentration in synaptosomes prepared from focal tissue. Both these amino acids are largely glial in origin (1,41,46), and both are involved in glutamate regulation.

Glutamine is a storage form of glutamate, and serves to buffer neuronal glutamate levels (46). Glutamate released during neuronal depolarization is partially taken up by glia and converted to glutamine. The relationship between the two amino acids, therefore, is analogous to that existing between ATP and phosphorylcreatine. Excessive loss of glutamate should result in glutamine levels falling before a fall in glutamate levels.

Taurine and glutamate levels are, in general, highly correlated in the brain. In rat cortex and inferior colliculus, the concentrations of the two amino acids show correlation coefficients of 0.74 and 0.58, respectively (19). Rassin (37) found a correlation coefficient of 0.99 in the occipital lobe of rhesus monkies. Van Gelder (45) reported similarly high correlations for a number of species. The basis of this correlation is unclear. However, glutamate and other excitotoxins stimulate the release of taurine (24,27-29). An increase in the extracellular levels of glutamate, therefore, could lead to a depletion of cellular taurine.

This is the first report on amino acid levels in synaptosomes from human focal tissue. However, there is a body of confusing literature on whole tissue amino acid levels in brains of epileptics. Perry and Hansen (35), for example, found increases in glutamate, glycine and GABA levels but no change in glutamine, aspartate or taurine levels in human focal tissue. Their control was tissue from individuals with brain tumors. Sherwin, et al. (38) found increases in aspartate and glutamate levels in spiking cortex compared to nonspiking cortex. They found no differences in GABA, taurine or glycine levels.

Unlike the confusion with other amino acids, altered levels of glutamate in epilepsy have frequently been found in brain (35,38) and cerebrospinal fluid (48). Other studies have documented increased release of glutamate from epileptic foci of animals (8,25,26). Huxtable et al. (20) found plasma glutamate to be markedly elevated in several kinds of epilepsy in humans, compared to nonepileptic controls. The levels of other amino acids varied depending on the type of epilepsy. Van Gelder et al. (49) saw significant increases in plasma glutamate levels in individuals with absence (3/s spike-wave) epilepsy as well as in their first degree relatives. In these studies, the differences in amino acid levels could not be attributed to diet, age or medications. Janjua et al. (22) have likewise shown an increase in plasma glutamate levels in individuals with either primary generalized or partial epilepsy as well as their first degree relatives, when compared to nonepileptic controls. Our current study, however, is the first to demonstrate alterations in the subcellular levels of amino acids in human epileptic tissue.

These earlier findings plus the ones we report here are consistent in pointing to an imbalance in glutamate regulation in epilepsy. Extracellular glutamate is excitatory, so a failure to maintain a sufficiently high glutamate concentration ratio (intracellular to extracellular) across the synaptosomal membrane predisposes to epilepsy. Glutamine levels could also be lowered as a result of depression of glutamine synthetase activity. The activity of this enzyme is slightly depressed in the brains of genetically seizure-susceptible rats compared to seizure resistant rats (6). Glutamate is also the precursor of GABA, the major inhibitory amino acid in the brain. A deficiency in cell glutamate, therefore, is consistent with a decrease in GABA levels, as we observe (Fig. 1). GABA levels may also be affected by taurine. In one report, taurine, added in vitro to brain homogenates, stimulated glutamic acid decarboxylase activity in preparations from genetically seizure-susceptible rats, but was without effect in preparations from seizure-resistant rats (4). This suggests the possibility that lowered taurine levels in epilepsy might correlate with lowered GABA levels.

Many studies on amino acids in epilepsy have used tissue from a mixed population. In some cases, autopsy tissue has been employed, although amino acid levels can change rapidly in the brain following death (33). Earlier studies have all examined whole tissue amino acid levels. We have analyzed surgically-removed tissue from a highly selected group of epileptic patients, all with severe epilepsy and all refractory to medication. Amino acid analyses have been performed on a neurally important fraction, the synaptosomal fraction, rather than on whole tissue. Our results support the existence of a disturbance in glutamate regulation as a contribution to the pathogenesis of epilepsy. The earlier findings on elevated plasma levels of glutamate suggest a general defect in regulation that becomes clinically relevant only in the brain.

CONCLUSIONS

We have examined brain tissue from a highly selected group of epileptic patients, all with severe temporal lobe seizures that were refractory to medication. After allowances have been made for the unavoidable methodological complications, our data suggest that there is an imbalance in glutamate regulation in the focal tissue of these severely epileptic patients. We hypothesize that this imbalance is caused by an abnormally high neuronal efflux of glutamate, and that this efflux contributes to the hyperexcitability of the tissue. The lowered levels of glutamine, taurine and GABA are a consequence of the loss of glutamate.

REFERENCES

1. Almarghini, K., Remy, A., and Tappaz, M. 1991, Immunocytochemistry of the taurine biosynthesis enzyme, cysteine sulfinate decarboxylase, in the cerebellum: Evidence for a glial localization, *Neuroscience*, 43:111-119.
2. Avoli, M. 1991, Excitatory amino acid receptors in the human epileptogenic neocortex, *Epilepsy Res.* 10:33-40.
3. Berl, S. and Clarke, D.D. 1978, Metabolic compartmentation of the glutamate-glutamine system: glial contribution, *in:* " Amino Acids as Chemical Transmitters", Fonnum, F. ed., Plenum Press, New York, pp. 691-708.
4. Bonhaus, D.W., Laird, H., Mimaki, T., Yamamura, H.I., and Huxtable, R.J. 1983, Possible bases for the anticonvulsant action of taurine, *in:* " Sulfur Amino Acids: Biochemical and Clinical Aspects", Kuriyama, K., Huxtable, R.J. and Iwata, H. eds., A.R.Liss, New York, pp. 195-209.
5. Bonhaus, D.W., Lippincott, S.E., and Huxtable, R.J. 1984, Subcellular distribution of neuroactive amino acids in brains of genetically epileptic rats, *Epilepsia*, 25:564-568.
6. Carl, G.F., Thompson, L.A., Williams, J.T., Wallace, V.C., and Gallagher, B.B. 1992, Comparison of glutamine synthetases from brains of genetically epilepsy prone and genetically epilepsy resistant rats, *Neurochem.Res.* 17:1015-1019.
7. Craig, C.R. and Hartmann, E.R. 1973, Concentration of amino acids in the brain of cobalt epileptic rat, *Epilepsia*, 14:409-414.

8. Dodd, P.R., Bradford, H.F., Abdul-Ghani, A.S., Cox, D.W.G., and Continho-Netto, J. 1980, Release of amino acids from chronic epileptic and sub-epileptic foci in vivo, *Brain Res.* 193:505-517.
9. Emson, P.C. 1978, Biochemical and metabolic changes in epilepsy, *in:* " Taurine and Neurological Disorders", Barbeau, A. and Huxtable, R.J. eds., Raven Press, New York, pp. 319-338.
10. Fumero, S. and Mondino, A. 1976, Longitudinal changes of brain amino acid content occurring before, during and after epileptic activity, *Brain Res.* 24:1-9.
11. Gray, E.G. and Whittaker, V.P. 1962, The isolation of nerve endings from brain: an electron microscopic study of cell fragments derived by homogenization and centrifugation, *J.Anat.* 96:79-88.
12. Hamberger, A., Haglid, K., Nyström, B., and Silfvenius, H. 1993, Co-variation of free amino acids in human epileptogenic cortex, *Neurochem.Res.* 18:519-525.
13. Hansen, S., Perry, T.L., Wada, J.A., and Sokol, M. 1973, Brain amino acids in baboons with light-induced epilepsy, *Brain Res.* 50:480-483.
14. Hertz, L., Kvamme, E., McGeer, E.G., and Schousboe, A. 1984, " Glutamine, Glutamate, and GABA in the Central Nervous System", Alan R. Liss, Inc. New York.
15. Hiramatsu, M., Edamatsu, R., Suzuki, S., Shimada, M., and Mori, A. 1990, Regional excitatory and inhibitory amino acid levels in epileptic El mouse brain, *Neurochem.Res.* 15:821-825.
16. Huxtable, R.J. 1981, Insights on function: metabolism and pharmacology of taurine in the brain, *in:* " The Role of Peptides and Amino Acids as Neurotransmitters", Lombardini, J.B. and Kenny, A. eds., A.R.Liss, New York, pp. 53-97.
17. Huxtable, R.J. 1989, Taurine in the central nervous system and the mammalian actions of taurine, *Prog.Neurobiol.* 32:471-533.
18. Huxtable, R.J., Bonhaus, D., and Lehmann, A. 1986, Excitatory and inhibitory amino acids in the genetically seizure-susceptible rat, *in:* " Neurotransmitters in Epilepsy", Nistico, G., Morselli, P., Lloyd, K.G., Fariello, R.G. and Engel, J. eds., Raven Press, New York, pp. 369-385.
19. Huxtable, R.J., Laird, H., Bonhaus, D., and Thies, A.C. 1982, Correlations between amino acid concentrations in brains of seizure-susceptible and seizure-resistant rats, *Neurochem.International*, 4:73-78.
20. Huxtable, R.J., Laird, H., Lippincott, S.E., and Walson, P. 1983, Epilepsy and the concentrations of plasma amino acids in humans, *Neurochem.International*, 5:125-135.
21. Huxtable, R.J. and Laird, H.E. 1978, Are amino acid patterns necessarily abnormal in epileptic brains? Studies on the genetically seizure-susceptible rat, *Neuroscience Letters*, 10:341-345.
22. Janjua, N.A., Itano, T., Kugoh, T., Hosokawa, K., Nakano, M., Matsui, H., and Hatase, O. 1992, Familial increase in plasma glutamic acid in epilepsy, *Epilepsy Res.* 11:37-44.
23. King, L.J., Carl, J.L., and Lao, L. 1974, Brain amino acids during convulsions, *J.Neurochem.* 22:307-309.
24. Kontro, P. and Oja, S.S. 1987, Taurine and GABA release from mouse cerebral cortex slices: effects of structural analogues and drugs, *Neurochem.Res.* 12:475-482.
25. Koyama, I. 1972, Amino acids in the cobalt induced epileptogenic and nonepileptogenic cat's cortex, *Can.J.Physiol.Pharmacol.* 50:740-752.
26. Koyama, I. and Jasper, H. 1977, Amino acid content of chronic undercut cortex of the cat in relation to electrical afterdischarge: comparison with cobalt epileptogenic lesions, *Can.J.Physiol.Pharmacol.* 55:523-536.
27. Lehmann, A., Hagberg, H., Huxtable, R.J., and Sandberg, M. 1987, Reduction of brain taurine: effects on neurotoxic and metabolic actions of kainate, *Neurochem.International*, 10:265-274.
28. Lehmann, A., Hagberg, H., Nystrom, B., Sandberg, M., and Hamberger, A. 1985, In vivo regulation of extracellular taurine and other neuroactive amino acids in the rabbit hippocampus, *in:* " Taurine: Biological Actions and Clinical Perspectives", Oja, S.S., Ahtee, L., Kontro, P. and Paasonen, M.K. eds., A. R. Liss, New York, pp. 289-311.
29. Lehmann, A., Isacsson, H., and Hamberger, A. 1983, Effects of in vivo administration of kainic acid on the extracellular amino acid pool in the rabbit hippocampus, *J.Neurochem.* 40:1314-1320.
30. Lehmann, A., Sandberg, M., and Huxtable, R.J. 1986, *In vivo* release of neuroactive amines and amino acids from the hippocampus of seizure-resistant and seizure-susceptible rats, *Neurochem.International*, 8:513-520.
31. Lowry, O.J., Rosebrough, N.J., Farr, A.L., and Randall, R.J. 1951, Protein measurement with folin phenol reagent, *J.Biol.Chem.* 193:265-275.
32. Oja, S.S. and Kontro, P. 1983, Free amino acids in epilepsy: possible role of taurine, *Acta Neurol.Scand.* 67, Suppl.93:5-20.
33. Perry, T.L., Berry, K., Hansen, S., Diamond, S., and Mok, C. 1971, Regional distribution of amino acids in human brain obtained at autopsy, *J.Neurochem.* 18:513-519.

34. Perry, T.L. and Hansen, S. 1976, Taurine content of epileptogenic foci in human brain, *in:* " Taurine", Huxtable, R. and Barbeau, A. eds., Raven Press, New York, pp. 275-281.
35. Perry, T.L. and Hansen, S. 1981, Amino acid abnormalities in epileptogenic foci, *Neurology,* 31:872-876.
36. Perry, T.L., Hansen, S., Kennedy, J., Wada, J., and Thompson, G.B. 1975, Amino acids in human epileptogenic foci, *Arch.Neurol.* 32:752-754.
37. Rassin, D.K. 1982, Taurine, cysteine sulfinic acid decarboxylase and glutamic acid in brain, *in:* " Taurine in Nutrition and Neurology", Huxtable, R.J. and Pasantes-Morales, H. eds., Plenum Press, New York, pp. 257-268.
38. Sherwin, A., Robitaille, Y., Quesney, F., Olivier, A., Villemure, J., Leblanc, R., Feindel, W., Andermann, E., Gotman, J., Andermann, F., Ethier, R., and Kish, S. 1988, Excitatory amino acids are elevated in human epileptic cerebral cortex, *Neurology,* 38:920-923.
39. Tapia, R. and Gonzalez, R.M. 1978, Glutamine and glutamate as precursors of the releasable pool of GABA in brain cortex slices, *Neuroscience Letters,* 10:165-169.
40. Tapia, R., Pasantes, H., Perez De La Mora, M., Ortega, B., and Massieu, G. 1967, Free amino acids and glutamate decarboxylase activity in brain of mice during drug induced convulsions, *Biochem.Pharmacol.* 16:1211-1227.
41. Tappaz, M., Almarghini, K., and Do, K. 1994, Cysteine sulfinate decarboxylase in brain: Identification, characterization and immunocytochemical localization in astrocytes, *in:* " Taurine in Health and Disease", Huxtable, R.J. and Michalk, D. eds., Plenum Press, New York, pp. This volum
42. Van Gelder, N.M. 1978, Glutamic acid and epilepsy: the action of taurine, *in:* " Taurine and Neurological Disorders", Barbeau, A. and Huxtable, R. eds., Raven Press, New York, pp. 387-402.
43. Van Gelder, N.M. 1981, The role of taurine and glutamic acid in the epileptic process: a genetic predisposition, *Rev.Pure Appl.Pharmacol.Sci.* 2:293-316.
44. Van Gelder, N.M. 1981, Glutamic acid in nervous tissue and changes of the taurine content: its implication in the treatment of epilepsy, *in:* " Amino Acid Neurotransmitter", DeFeudis, F.V. and Mandel, P. eds., Raven Press, New York, pp. 115-125.
45. Van Gelder, N.M. 1982, Changed taurine-glutamic acid content and altered nervous tissue cytoarchitecture, *in:* " Taurine in Nutrition and Neurology", Huxtable, R.J. and Pasantes-Morales, H. eds., Plenum Press, New York, pp. 239-256.
46. Van Gelder, N.M. 1983, Metabolic interactions between neurons and astroglia: glutamine synthetase, carbonic anhydrase, and water balance, *in:* " Basic mechanisms of neuronal hyperexcitability", Anonymous Alan R. Liss, Inc., New York, pp. 5-29.
47. Van Gelder, N.M. and Courtois, A. 1972, Close correlation between changing content of specific amino acids in epileptogenic cortex of cats, and severity of epilepsy, *Brain Res.* 43:477-484.
48. Van Gelder, N.M., Edmonds, H.L., Jr., Hegreberg, G.A., Chatburn, C.C., Clemmons, R.M., and Sylvester, D.M. 1980, Amino acid changes in a genetic strain of epileptic beagle dogs, *J.Neurochem.* 35:1087-1091.
49. Van Gelder, N.M., Janjua, N.A., Metrakos, K., MacGibbon, B., and Metrakos, J.D. 1980, Plasma amino acids in 3/sec spikewave epilepsy, *Neurochem.Res.* 5:659-671.
50. Van Gelder, N.M., Sherwin, A.L., and Rasmussen, T. 1972, Amino acid content of epileptogenic human brain: focal versus surrounding regions, *Brain Res.* 40:385-393.

PARTICIPANTS

Azuma, J., Third Department of, Internal Medicine, Osaka University Medical School, Osaka 553, Japan

Böhles, Dr. Hans, Zentrum der Kinderheilkunde, Theodor Stern Kai 7, 6000 Frankfurt/Main 70, Germany

Budreau, Andrea, Department of Pediatrics, University of Tennessee, Memphis, TN 38103, USA

Cantafora, A., Laboratory of Metabolism and, Pathological Biochemistry, Istituto Superiore di Sanità, 00161 Roma, Italy

Cavallini, Prof. Doriano, Dipartimento di Scienze Biochimiche, Universita di Rome "La Sapienza", I-00185 Roma, Italy

Chesney, Russell W., Department of Pediatrics, University of Tennessee, Memphis, TN 38103, USA

Conte Camerino, Diana, Dipartimento di Farmacobiologico, Facoltà Farmacia, Universität Bari, Bari 70125, Italy

Covarrubias, Prof. Jesus, Children Hospital of Mexico, Blvd. Adolto López Mateos 2742, Col. Altavista, C.P. 01060, México, D.F.

Davies, W. Ewart, Department of Pharmacology, University of Birmingham, The Medical School, Vincent Drive, Birmingham B15 2TJ, England

Duprè, Prof. Silvestro, Dipartimento di Scienze Biochimiche, Universita' Degli Studi di Roma, "La Spezia", Piazzale Aldo Moro, 5, I-00185 Roma, Italy

Fellman, Jack H., Professor of Biochimistry, University of Oregon, Health Science Center, Portland, Oregon 97201, USA

Franconi, Prof. Flavia, Istituto di Chimica Biologica, Universita'di Sassari, Via Muroni, 23/a, Sassari, Italia

Grimble, Robert F., Institute of Human Nutrition, University of Southampton, Southampton S09 3 TU, U.K.

Hansson, Elisabeth, Institute of Neurobiology, University of Göteborg, S-400 33 Göteborg, Sweden

Huxtable, Ryan J., Department of Pharmacology, College of Medicine, University of Arizona, Tucson, AZ 85624, USA,

Jones, Deborah, Department of Pediatric, University of Tennessee, 848 Adams Avenue, Memphis, TN 38103, USA

Kuriyama, Kinya, Department of Pharmacology, Kyoto Prefectural, University of Medicine, Kyoto 602, Japan

Lake, Dr. Norma, Department of Physiology, McGill University Montreal, 3655 Drummond Street, Montreal,Quebec, Canada H3G IY6

Lehmann, Anders, Associate Director, Gastrointestinal Pharmacology, Astra Hässle AB, S-431 83 Mölndal, Sweden

Leibach, Prof. Frederick H., Medical College of Georgia, School of Medicine, Department of Biochemistry, Augusta, Georgia 30912-2100, U.S.A.

Lombardini, John B., Department of Pharmacology and, Ophthalmology and Visual Sciences, Texas Tech University, Health Sciences Center, Lubbock, TX 79430, USA

Magnusson, Dr. Kathy, Department of Anatomy and, Neurobiology, Colorado State University, Fort Collins, Colorado 80523, USA

Martin, David L., Laboratory of Neurotoxicology, and Nervous System Disorders, Wadsworth Center for Laboratories, and Research, Albany, NY 12201-0509, USA

Meldrum, M.J., Department of Pharmacodynamics, College of Pharmacy, University of Florida, JHMHC Box 100487, Gainsville, Florida 32610

Michalk, Herrn Dr. Dietrich, Geschäftsführender Direktorund Universitätprofessor der Klinik, Poliklinik für Kinderheilkunde, Universität Köln, D-5000 Köln 41, Germany

Moran, Dr. Julio, Departamento de Neurociencias, Centro de Investigaciones en Fisiologia Celular, UNAM, AP 70-600, Mexico 20, D.F., Mexico

Morris, James G., Department of Physiological Sciences, School of Veterinary Medicine, University of California, Davis, CA 95616, USA

Nakagawa, Masao, Second Department of Medicine, Kyoto Prefectural, University of Medicine, Kyoto 602, Japan

Nakanishi, Takeshi, Fifth Department of Internal Medicine, Hyogo College of Medicine, Nishinomiya 663, Japan

Neuringer, Martha, Oregon Regional Primate, Research Center, Beaverton, OR 97006, USA

Dr. Mizuho Ogasawara, Taisho Pharmaceutical Co., Research Center, 403 Yoshino-cho 1-chome, Ohmiyashi, Saitama 330, JAPAN

Oja, S.S., Department of Biomedical Sciences, Tampere Brain Research Center, University of Tampere, SF-33101 Tampere, Finland

Pasantes-Morales, Herminia, Institute of Cellular Physiology, National University of Mexico, Mexico City, Mexico

Rebel, G., Centre de Neurochimie, Université L. Pasteur, 67084 Strasbourg, France

Rodenbeck, Dr. D., Kinderklinik der Medizinischen, Hochschule Hannover, Konstanty-Gutschow-Straße 8, D-30623 Hannover, Germany

Saransaari, Pirjo, Tampere Brain Research Center, Department of Biomedical Sciences, University of Tampere, Box 607, SF-33101 Tampere, Finland

Satoh MD, PhD, Hiroyasu, Department of Pharmacology, Nara Medical University, Kashihara, Nara 634, JAPAN

Schaffer, S.W., Department of Pharmacology, University of South Alabama, School of Medicine, Mobile, AL 36688, USA

Schuller-Levis, Georgia, Institute for Basic Research, in Developmental Disabilities, Immunology Laboratory, 1050 Forrest Hill Road, Staten Island, N.Y. 10314-6399, U.S.A.

Sgaragli, G.P., Istituto di Scienze Famacologiche, Via E.S. Piccolomini 170, I-53100 Siena, Italy

Stipanuk, Martha H., Division of Nutritional Sciences, Cornell University, Ithaca, NY 14853, USA
Stratmann, Dr. Cornelia, Krankenanstalten der Stadt Koln, Kinderkrankenhaus, Amsterdamer Straße 59, 5000 Koln 60, Germany
Sturman, John A., Department of Developmental Biochemistry, 1050 Forrest Hill Road, Staten Island, NY 10314-6399, USA
Tappaz, M., Neurochimie Fonctionelle, INSERM U.171, Centre Hospitalier Lyon-Sud, 69310 Pierre Bénite, France
Trachtman, Howard, Schneider Children's Hospital, Division of Pediatric Nephrology, 271-16 76th Avenue, New Hyde Park, NY 11042, USA
Trenkner, Ekkhart, Department of Pathology, Columbia University College of, Physicians and Surgeons, New York, NY 10032, USA
Wu, Jang-Yen, Department of Physiology, and Cell Biology, University of Kansas, Lawrence, KS 66045, USA

INDEX